AF352296

WASTE MANAGEMENT SERIES 5

OLIVE PROCESSING WASTE MANAGEMENT
Literature Review and Patent Survey
Second Edition

Waste Management Series

WASTE MANAGEMENT SERIES 5

OLIVE PROCESSING WASTE MANAGEMENT
Literature Review and Patent Survey
Second Edition

Michael Niaounakis
European Patent Office, The Netherlands

Constantinos P. Halvadakis
Department of Environment, University of the Aegean, Greece

ELSEVIER

Amsterdam – Boston – Heidelberg – London – New York – Oxford
Paris – San Diego – San Francisco – Singapore – Sydney – Tokyo

ELSEVIER B.V.
Radarweg 29
P.O. Box 211, 1000 AE
Amsterdam, The Netherlands

ELSEVIER Inc.
525 B Street, Suite 1900
San Diego, CA 92101-4495
USA

ELSEVIER Ltd
The Boulevard, Langford Lane
Kidlington, Oxford OX5 1GB
UK

ELSEVIER Ltd
84 Theobalds Road
London WC1X 8RR
UK

Second edition 2006

British Library Cataloguing in Publication Data
A catalogue record is available from the British Library.

ISBN-10: 0-08-044851-8
ISBN-13: 978-0-08-044851-0
ISSN Series: 1478-7482

♾ The paper used in this publication meets the requirements of ANSI/NISO Z39.48-1992 (Permanence of Paper).
Printed in Italy.

M. Niaounakis and C. P. Halvadakis
Olive Processing Waste Management: Literature Review and Patent Survey.

First published under the title: "Olive-Mill Waste Management: Literature Review and Patent Survey", February 2004 by Typothito Publications, Athens, Greece.

Statement

Contents

Foreword to the First Edition

Olive-milling like every human activity and industrial process results in a low-entropy desired product and a high-entropy unwanted by-product or waste termed olive-mill waste. The production of olive oil, viewed in a holistic perspective, begins with the picking of olives and ends after their processing in olive-mills. Olive-mill technology at present generates a variety of waste in both energy and mass forms. In addition to solid waste generated in the olive groves by annual pruning of olive trees, a considerable amount of solid waste is generated during milling in the form of leaves and small twigs brought to the mill with the olives and in the form of crushed olive stones and sizable remnants of olive pulp (flesh) following olive oil extraction. Leaves and twigs can be used as animal feed (mainly for goats) or in the production of compost after mixing with other appropriate materials. Liquid waste is known as olive-mill wastewater (OMWW), since during olive milling and olive oil extraction substantial amounts of added water as well as olive juice (or olive vegetation water) combine with small amounts of unrecoverable oil and fine olive pulp particles to constitute this type of waste. Gaseous waste consists of fumes produced during malaxation of crushed olives and exhaust gases from burners providing thermal energy to the mill. Finally, energy waste consists of thermal energy losses and acoustic energy (noise) of utilized machinery.

From an environmental point of view, OMWW is the most critical waste emitted by olive-mills in terms of both quantity and quality. There is archaeological evidence that this effluent has been damaging delicate shoreline environments for thousands of years around the Mediterranean. Pollution from olive oil production is often a problem in poor communities in Southern Europe and North Africa where sophisticated solutions to the problem are too expensive. The problems created in managing this waste have been extensively investigated during the last 50 years without finding a solution, which is technically feasible, economically viable, and socially acceptable. The prevalent waste management strategy up to date has been traditional wastewater treatment processes aimed at reducing pollution loads to legally accepted levels for disposal into environmental media (mainly land and water bodies). Recently, Spain has adapted a manufacturing process for olive oil production, which minimizes the utilization of water and, therefore, of generated wastewater, the so-called two-phase olive-mill extraction technique. The pollution load is, however, the same since it originates from the olives and not from the water utilized during olive processing. In view of the above, it is apparent that a new strategy for olive waste management must be adopted. Up to now the emphasis has been on detoxifying OMWW prior to disposal. However, the present trend is towards further utilization of OMWW by recovering useful by-products.

OMWW contains most of the water-soluble chemical species of the olive fruit. Critical chemical species like water-soluble phenols and polyphenols appear to be an obstacle during treatment — being recalcitrant — but can be industrially beneficial if isolated. If one considers that the pollution load is merely the remnants of olives (a natural product), it is preferable to adopt a waste utilization management strategy. It is not surprising that during recent years a number of patents have appeared following such strategy. OMWW management is presently approached by both in-house process modification combined with waste minimization, and end-of-pipe waste utilization.

The focus of the present study is to evaluate the existing technologies and to develop environmental criteria for disposing and/or reusing olive-mill wastes in general, and wastewater in particular. The prior art is critically reviewed by both discussing the extensive literature coverage — more than 1000 references are cited and commented upon, including journals, patents, conference proceedings, dissertations, theses, technical notes, reports of projects — and by recording the traditional techniques still being used by smaller olive-mills, which have been passed down from generation to generation. It is a fact that most of the literature or know-how comes from countries around the Mediterranean. A substantial part of the literature collection consists of patents. Despite their technological importance, until now patents appear to have been cited rarely in scientific journals or books.

More emphasis has been given to OMWW and to the new by-product, known as "alperujo" in Spain, generated by the two-phase extraction process (2POMW). OMWW represents the still unsolved problem of the olive-mill industry, both for its extent and significance. 2POMW represents a new type of problem due to its consistency (thick sludge that contains pieces of stone and pulp of the olive fruit as well as vegetation water) and its steadily increasing production, especially in Spain. The rest of the olive wastes, such as olive cake, leaves, and twigs, do not represent a serious environmental problem and have only been commented upon briefly. The various olive oil extraction systems have been described shortly together with the effects each one of them has on the environment.

The wastewater (brines) arising from the table olive industry has also not been reviewed. It was considered that brines constitute a different type of wastewater.

It is not the intention of the present study to propose any solution. Instead, it defines the problems faced by the olive-mill industry, makes proposals for discharge/reuse of olive-mill wastes on the basis of each treatment technique and shows the current trends in the olive-mill waste management.

Some of the conclusions of the present study are:

- Most of the technologies reviewed in this study have been tested on a small scale only. In depth assessment of these results and subsequent full-scale applications has yet to be carried out.
- The olive-mill waste management can be viewed as: (i) extracting valuable materials (e.g. irrigation water, compost, fodder, fuel, antioxidants etc.); and (ii) in lowering pollution load for final disposal to natural receiving bodies (surface water, land, and sea). The double nature of olive-mill waste (as a pollutant of streams or a resource to be recycled) causes antagonism between agriculture and environmental groups, because of their different point of view on this topic.
- Most of the treatment processes are focused on both bioremediation, as a means of reducing the polluting effect of OMWW and transformation into valuable products, together with modification of the technology used in oil extraction. The presence of large amounts of phenolic compounds constitutes one of the major obstacles in the detoxification of OMWW. These recalcitrant compounds decelerate the process, hinder removal of part of COD, and detract from its economic viability. Nowadays, the trend is towards turning this problem to a benefit by extracting these compounds. Recent studies have shown that the abundant phenolic antioxidant fractions of olive oil have a potent inhibitory ability on reactive oxygen species. There is an increasing body of evidence indicating the involvement of oxygen-derived free radicals in several pathologic processes, such as cancer and atherosclerosis. OMWW has a powerful antioxidant activity, and thus might be a cheap source of natural antioxidants. Up to now the antioxidant compounds of OMWW have not been effectively exploited, due to the impracticality of extracting usable amounts of antioxidant compounds using conventional technology.
- The problem of olive-mill waste is further aggravated by the lack of a common policy among the olive oil producing countries. Every country has its own legislation/regulations that often vary greatly among them with a consequent non-uniform application of generally accepted guidelines. For this reason, there is a need for a unified strategy behavior among the EU member states.

To the best of our knowledge this is the first extensive and all-encompassing review to appear on the subject of olive-mill wastewater. A few earlier reviews can be found in the literature, but these are mostly partial in scope and outdated. This is quite surprising given the environmental impact of this waste. It is hoped that this review will increase public awareness and will further provide a valuable information resource for olive oil producers, researchers, and policy makers dealing with the problem of olive-mill wastes.

The present publication has been financed by EU Regional Directorate-General (ERDF Innovative Actions 2000–2006, Programme 2001 GR 16 0 PP 209). The project title was "North Aegean Innovative Actions and Support (NAIAS)", while this work was a deliverable of Action 7.6 "Innovative Olive-Mill Waste Management Systems".

The authors would like to acknowledge the help of various individuals who contributed in certain aspects of the present work. Specifically, thanks are due to D. Schaelicke and M. Karatzas for aiding in management issues and compiling supplement information; C. Tzoutzoumitros and M. Hadjimanolakis for perfecting certain figures and chemical formulae; W. Bolger for helpful comments regarding the text; and to the editorial board of Dardanos Publications for their efforts into materializing the book form of this work.

Dr. M. Niaounakis
Professor C. P. Halvadakis
Mytilene, January 2004

Foreword to the Second Edition

In the first edition titled "Olive-Mill Waste Management" emphasis was given to olive-mill waste. In the second edition, the original title has been modified to encompass all types of by-products generated during olive tree cultivation and olive fruit processing. In the case of olive tree cultivation, information is presented referring to pruning and harvest residues but does not include wasted fertilizers, herbicides, and insecticides which constitute a subject matter of their own. In the case of olive processing, information is presented referring to olive-milling wastewaters, solid, gaseous and energy wastes, and table olive processing wastewaters. In addition, information has been included concerning the management of used olive oil from cooking or other activities. Literature references and patents published or located since the first edition have been examined and incorporated where appropriate.

Two entirely new parts have been added. Part IV presents information on the characterization, environmental effects, treatment processes, and uses of waste generated during table olive processing. Part V gives an economical and legislative overview concerning olive-mill waste.

Finally, corrections and suggestions from colleagues and other interested researchers have been taken care of in this edition.

We acknowledge the valuable contribution of several people who helped us during the formation of the second edition. Many thanks go to E. Karatzas, Research Associate of the Waste Management Laboratory — University of the Aegean, for his valuable help with various aspects of this edition. To a lesser extent, we would like to also thank other members of the above mentioned laboratory who contributed i.e. G. Giouzepas, M. Hadjimanolakis, and D. Balabanis.

We are also grateful to our colleagues for providing valuable comments and recommendations during the preparation of the manuscript, namely: L. Di Giovacchino, L. Gilles, A. Bourgonje, N. Azbar, T. Colliner, E. Z. Panagou, A. Giannes, and M. Lorenz.

Our thanks are also due to W. Bolger for contributing to this edition. Many thanks also go to Ms Jasmin Bakker of Elsevier Amsterdam; the professional work and cooperation of Cepha Imaging Pvt. Ltd., Bangalore, India is greatly appreciated.

Dr. M. Niaounakis
Professor C. P. Halvadakis
The Hague, May 2005

Part I
Background Information

Chapter 1

Introduction

Olive and Oil Production Statistics

The olive tree is member of the family *Oleaceae*, which comprises 30 species such as jasmine, ash, lilac, and privet. The only edible species is *Olea europeaea* L, which is cultivated for its plump, fleshy, and oil-containing fruits. There are more than 850 million productive olive trees worldwide, which occupy a surface of about 8,514,300 ha (FAOSTAT, 2004)[1]. There are 1000 inventoried varieties; 139 of which are included in the World Catalogue of Olive Varieties published by IOOC. These 139 varieties coming from 23 countries account for 85% of the olives grown worldwide. Olive cultivation is widespread throughout the Mediterranean region and is important for the rural economy, local heritage, and environment — see Fig. 1.1. The countries around the Mediterranean basin and in the Middle-East provide 98% of the total surface area for olive tree culture and total productive trees, and 99% of the total olive production, with Spain being first as regards total culture surface (2,400,000 ha) and number of productive trees (180,000,000), followed by Italy (1,140,685 ha), and Greece (765,000 ha). Olives are also cultivated in California USA, Australia, Iran, Argentina, and Peru. The world production of olives for the year 2004 was 15,340,488 metric tons — see Table 1.1. The average world production of olive oil for the harvesting years 1999/2000–2002/2003 was 2,564,800 metric tons (IOOC, 2004)[2]. The European Union (EU) is the largest olive oil producer with 80.2% of the total — see Fig. 1.2. The average production of olive oil in EU for the same period was 2,056,200 tons with Spain accounting for 978,800 (47.6%), Italy 633,700 (30.8%), and Greece 405,600 tons (19.7%) of the EU total — see Fig. 1.3. Apart from EU, other significant olive oil producers are Syria (4.9%),

---

[1]http://apps.fao.org; last accessed March 2005.

[2]http://www.internationaloliveoil.org; last accessed March 2005.

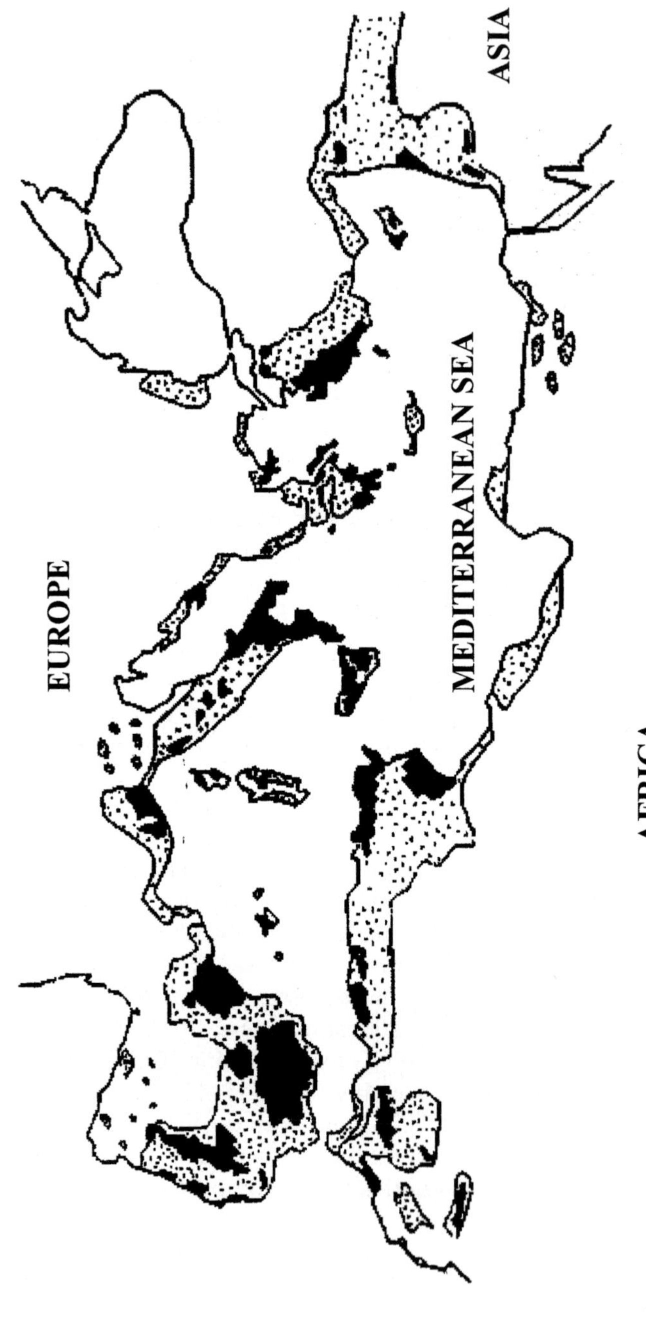

Fig. 1.1. Olive-oil-making regions in the Mediterranean basin. Shaded areas: high production regions. Dotted areas: low production regions (adapted from González-López J. et al., 1994; source IOOC, 1991).

Table 1.1. World olive production, yield and area harvested for the year 2004. FAOSTAT data, 2004; last accessed, March 2005

Countries	Production (metric tons)	Area harvested (ha)	Yield (hg/ha)*
Albania	30,000	28,500	10,526
Algeria	170,000	200,000	8500
Argentina	95,000	33,000	28,788
Australia	1800	1000	18,000
Azerbaijan, Republic of	600	1500	4000
Brazil	5	10	5000
Chile	18,000	7000	25,714
China	2500	300	83,333
Croatia	33,000	15,000	22,000
Cyprus	27,500	8600	31,977
Egypt	320,000	50,000	64,000
El Salvador	3500	5000	7000
France	24,231	17,352	13,964
Gaza Strip (Palestine)	0	0	0
Greece	2,300,000	765,000	30,065
Iran, Islamic Rep of	43,000	15,000	28,667
Israel	25,000	14,000	17,857
Italy	3,149,830	1,140,685	27,613
Jordan	85,000	65,000	13,077
Kuwait	11	–	–
Lebanon	180,000	58,000	31,034
Libyan Arab Jamahiriya	148,000	100,000	14,800
Macedonia, The Fmr Yug Rp	15,500	6200	25,000
Malta	1	3	3333
Mexico	14,200	4900	28,980
Morocco	470,000	500,000	9400
Palestine, Occupied Tr.	125,000	90,000	13,889
Peru	38,100	7900	48,228
Portugal	270,000	360,000	7500
Serbia and Montenegro	1000	1500	6667
Slovenia	800	800	10,000
Spain	4,556,000	2,400,000	18,983
Syrian Arab Republic	950,000	500,000	19,000
Tunisia	350,000	1,500,000	2333
Turkey	1,800,000	597,000	30,151
United States of America	77,110	14,500	53,179
Uruguay	3300	1500	22,000
Uzbekistan	100	100	10,000
West Bank	0	0	0
World	15,340,488	8,514,300	18,017

*Hectogram per hectare.

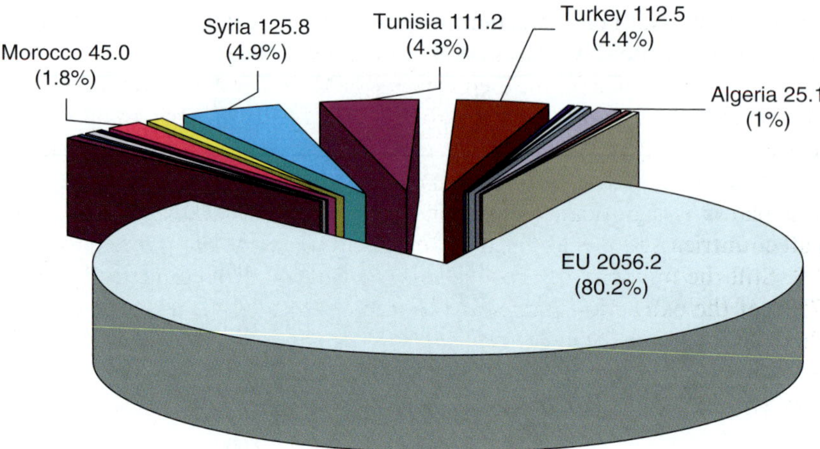

Fig. 1.2. Average world production of olive oil (1000 tons) for the harvesting years 1999/ 2000–2002/2003. IOOC data, December 2004; last accessed, March 2005.

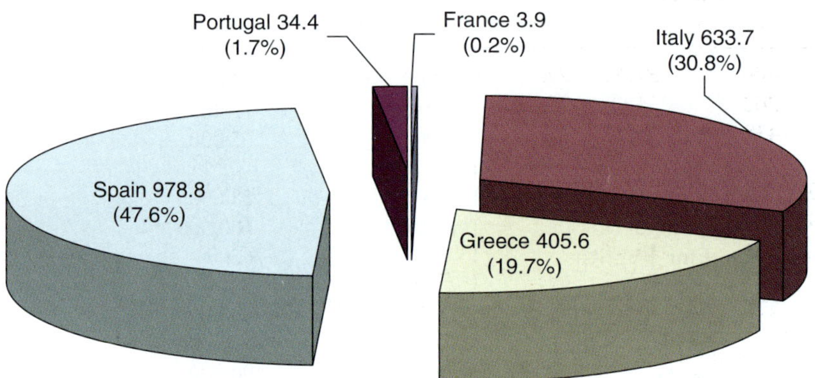

Fig. 1.3. Average production of olive oil (1000 tons) in EU for the harvesting years 1999/ 2000–2002/2003. IOOC data, December 2004; last accessed, March 2005.

Turkey (4.4%), and Tunisia (4.3%) and to a lesser extent, Morocco (1.8%) and Algeria (1%). The corresponding production of table olives in EU was 651,400 tons with Spain being again the first followed by Italy and Greece – see Chapter 11: "Table Olives".

The olive sector in EU involves about 2.5 million producers — roughly one-third of all EU farmers — with 1,160,000 in Italy, 840,000 in Greece, and 380,000 in Spain, and is characterized by intense fragmentation. Olive production offers the advantage of providing seasonal employment in winter, complementary with other agricultural activities, and provides significant off-farm employment in the associated milling and processing industry. The olive oil sector consists of growers, cooperatives, mills, refiners, blenders, seed-oil extraction plants, and companies involved in various aspects of marketing. There are about 12,000 olive-mills in EU. The majority of olive-mills is small enterprises (SMEs), in many cases family owned and with less than 10 workers — see Table 1.2. In Spain, where production is more concentrated — mainly in Andalusia — in geographical terms, the mills are fewer in number but have a greater throughput (cooperatives). The number of seed-oil extraction plants is significantly smaller than the corresponding number of olive-mills in all countries and it is about the same in Spain, Italy, and Greece (40–50) – see Table 1.3. Still the majority of those plants are SMEs in all countries but Portugal, where 75% of the extraction plants are characterized as large and one of them falls under the IPPC directive[3]. On the refining side, however, the number of installations remains limited and stable because of the size and complexity of the plant and machinery required (working paper of the Directorate-General for Agriculture)[4].

Table 1.2. Structure of olive-mills by annual throughout (tons); source: working paper of the Directorate-General for Agriculture; figures communicated by the Member States of EU

	Spain 1999–2000		Italy 1998–1999		Greece 1999–2000		Portugal 1998–1999	
	Number	%	Number	%	Number	%	Number	%
0–100	**640**	**37.32**	**993**	**16.34**	**871**	**39.02**	**857**	**92.20**
0–20	202	11.78	201	3.31	140	6.27		
20–100	438	25.54	792	13.03	731	32.75		
>100	**772**	**45.01**	**4450**	**73.24**	**1344**	**60.22**	**70**	**7.50**
100–500	585	34.11			1236	55.38	67	7.20
500–1000	187	10.90			108	4.84	3	0.30
>1000	**231**	**13.47**			**17**	**0.76**	**2**	**0.20**
Various*	**72**	**4.20**	**633**	**10.42**				
Total**	**1715**		**6076**		**2232**		**929**	

*Indeterminate structure. France has more than 140 approved mills.
**Total is the sum of the numbers in bold.

Table 1.3. Olive processing plants but mills in EU (1998–1999); source working paper of the Directorate-General for Agriculture

	Spain	Italy	Greece	Portugal
Refineries	29	13	27	8
Seed-oil extraction plants	53	45	42	13
Bottling/canning plants	440	300	90	49
Table olive packing stations	404	53	256	30

---

[3]Council Directive 96/61/EC on Integrated Pollution Prevention and Control (IPPC).
[4]http://europa.eu.int/comm/agriculture/markets/olive/reports/rep_en.pdf.

Olive processing (olive oil and table olives) is one of the fastest growing agro-food sectors in EU with an average annual growth rate higher than 4% (IOOC, 2004). The production is likely to continue increasing because of the substantial increment recorded in olive tree cultivation.

Olive Growing and Environmental Effects

Intense fragmentation is a feature of olive cultivation. Many small holdings, often farmed on a part-time basis, constitute a non-inconsiderable part of the EU olive growing area. The area covered by olive groves or plantations in the EU is approximately 5,163,000 ha, roughly 4% of the utilizable agricultural area of which 48% are in Spain, 22.5% in Italy, and 20% in Greece (European Commission, Directorate-General for Agriculture, 2000). The cultivated area has been more than doubled since 1980.

Olive trees range from ancient large-canopied trees to modern dwarf varieties planted in dense lines. Tree density and planting patterns depend partly on local conditions and tradition but water availability is also a determining factor. Most frequently, 70–150 olive trees are planted per hectare. According to variety and climatic conditions, an olive tree yields from 15 to 40 kg of olives per year.

Three broad types of olive farming can be distinguished:

- Low-input traditional groves and scattered trees, often with ancient olive trees and typically planted on terraces which are managed with few or no chemical inputs, but with a high labor input.
- Intensified traditional plantations, which to some extent follow traditional patterns but are under more intensive management making systematic use of fertilizers and pesticides and with more intensive weed control and soil management. There is a tendency to intensify further by means of irrigation, increased tree density, and mechanical harvesting.
- Intensive, modern plantations of smaller tree varieties planted at high densities and managed under an intensive and highly mechanized system, usually with irrigation.

As a result of their particular plantation characteristics and farming practices, the low-input traditional plantations have potentially the highest natural value (biodiversity and landscape value) and most positive effects (such as water management in upland areas), as well as the least negative effects on the environment. These plantations are also the least viable in economic terms and hence most vulnerable to abandonment.

The intensified traditional and modern intensive systems are inherently of least natural value and have potentially, and in practice, the greatest negative environmental impacts, particularly in the form of soil erosion, run-off to water bodies, degradation of habitats and landscapes, and exploitation of scarce water resources.

Soil erosion is probably the most serious environmental problem associated with olive farming (as distinct from olive processing). Inappropriate weed-control and soil-management practices, combined with the inherently high risk of erosion in many olive farming areas, is leading to desertification on a wide scale in some of the main producing regions, as well as considerable run-off of soils and agro-chemicals into water bodies (Beaufoy G., 2000)[5].

The mix of ancient and modern helps explain the differing farm sizes, ownership characteristics and processing structures that exist within the EU. Likewise, large differences in production systems occur within each producing region. The average holding size is as low as 1 ha in Italy, though olive holdings in Spain are larger (6 ha on average).

Wide fluctuations in production are a feature of olive growing. They are linked to the uncertainties of the climate (viz. drought in Spain in 1995–1996 and frost in Greece in 2001–2002) and alternate bearing, a characteristic of olive trees whereby, olive bumper crops tend to be followed by lower production the following year (working paper of the Directorate-General for Agriculture)[6]. Therefore, more olive oil and waste are generated every other year.

The Problem of Olive Processing Wastes

Both, olive tree culture and olive processing industry produce large amounts of by-products. It has been estimated that pruning alone produces 25 kg of by-products (twigs and leaves) per tree annually. It must also be considered that leaves represent 5% of the weight of olives in oil extraction. The manufacturing process of the olive oil usually yields next to olive oil (20%), a semi-solid waste (30%), and aqueous liquor (50%).

The crude olive cake is composed of a mixture of olive pulp and olive stones. The olive cake is collected in central seed-oil extraction plants (about one for every 65 olive-mills) where the residual oil (pomace- or seed-oil) is extracted with hexane after being dried in rotary driers using hot air of 60°C. Through this process there is an additional annual seed-oil production of about 170,000 tons and a production of stones of 1,600,000 tons per year which are usually used as solid fuel (Vlyssides A.G. et al., 1998). However, since the introduction of the two-phase extraction system, the market of olive cake has been declining and at the same time, both the limited storage life and the high transportation costs of this waste are raising the problem of olive cake disposal.

The aqueous liquor comes from the vegetation water and the soft tissues of the olive fruits. The mixture of this water-based by-product with the water used in the different stages of oil production makes up the so-called "olive-mill waste water"

---

[5]http://europa.eu.int/comm/agriculture/envir/index_en.htm#publications.

[6]See note 4.

(OMWW). Furthermore, olive washing water, waters from filtering disks, and from washing of equipment and rooms are to be included into this wastewater.

The quantity of OMWW produced in the process ranges from 0.55 to 2 l/kg of olives, depending on the oil extraction process. Essentially, the OMWW composition is water (80–83%), organic compounds (15–18%), and inorganic compounds (mainly, potassium salts and phosphates) 2%, and it varies broadly depending on many parameters such as olive variety, harvesting time, climatic conditions, oil extraction process, etc. (Fiestas Ros de Ursinos J.A. and Borja-Padilla R., 1990). Thanks to the presence of large amounts of proteins, polysaccharides, mineral salts, and other useful substances for agriculture, such as humic acids, OMWW has a high fertilizing power. Therefore, OMWW might be used as natural, low-cost fertilizer available in large amounts. Unfortunately, besides these useful substances for agriculture, OMWW also contains phytotoxic and biotoxic substances, which prevent it from being disposed of. The phytotoxic and antibacterial effects of OMWW have been attributed to its phenolic content — see Chapter 2: "Characterization of olive processing waste", section: "Antimicrobial activity of OMWW". In fact, the presence of such substances causes OMWW to be non-biodegradable, and consequently, unsuitable for further use as fertilizer, or as irrigation water. Besides, OMWW is an acidic — pH 4.5–5 — dark colored liquid smelling strongly of oils.

The maximum biochemical oxygen demand (BOD_5) and chemical oxygen demand (COD) concentration of OMWW can reach values up to 100 and 220 g/l, respectively (Balice V. et al., 1990). The large volumes of OMWW, which are produced every year, aggravate these characteristics. The polluting load of the olive-mills is considerable (2800–3600 tons BOD_5 per day assuming a milling season lasting 100 days), and it has been reported to be 5–10 (Boari G. et al., 1984) or even 25–80 (Schmidt A. et al., 2000) times larger than that of domestic sewage. The OMWW production in Spain in the early nineties ($2–3 \times 10^6$ m^3/year) was equivalent with the pollution of $10–16 \times 10^6$ inhabitants in the short milling period (November–March).

Estimations of the total amount of OMWW produced annually range from 7 to over 30 million m^3 — see Table 1.4. This large divergence of results can be partly explained by the fact that the production of olives varies from one year to another due to weather conditions and plagues that can affect the olive trees. But the main reason is that the provided data are only rough estimations. There is a lack of clear and comprehensive information concerning the quantity of the wastes actually produced by the olive-mill industry and where they are produced. Taking into account that in Spain is produced more than one-third of the world's olive oil production by the two-phase extraction system where no process water is used, it can be reasonably assumed that the total OMWW production is 10–12 million m^3/year.

Should OMWW be directly discharged into fresh water or into the sea, it would destroy the self-purifying capabilities of these environments, and seriously alter their biological balance for a long time in future. Furthermore, the poor biodegradability of OMWW inhibits a possible spreading thereof onto the fields, as a customary practice, since the non-biodegradable organic compounds contained in

Table 1.4. Estimates of waste generated from olive oil processing

	OMWW (m^3/year)	Olive cake (m^3/year)	References
Spain	$2–3\times10^6$		Cabrera F. et al., 1996
	2.1×10^6		Paredes C. et al., 1999b
	2.8×10^6	1.6×10^6	European Commission-DG for Environment, 2001
Italy	$1.5–2\times10^6$		EP520239 (1992)
	1.5×10^6		EU project: AIR3-CT94-1987 "BIOWARE"
	1×10^6		ES2051238 (1994)
	800,000		Visioli F. et al., 1995a
	2.4×10^6	1.6×10^6	European Commission-DG for Environment, 2001
	1.6×10^6		Di Giovacchino L., personal communication 2004
Greece		250,000	Papaioannou D., 1988
	200,000–250,000		EU project: AIR3-CT94-1987 "BIOWARE"
	1.5×10^6		Iconomou D. et al., 2000
	1.4×10^6	800,000	European Commission-DG for Environment, 2001
	785,000	275,000	Papafotiou M. et al., 2005
Tunisia	700,000		BADIS, 1994
	550,000	300,000	Mekki H. et al., 2003
Portugal	60,000–350,000		EU project: AIR3-CT94-1987 "BIOWARE"
	200,000	100,000	European Commission-DG for Environment, 2001
Mediterranean	$10–12\times10^6$		Cabrera F. et al., 1996
Total	30×10^6		Fiestas Ros de Ursinos J.A. 1981b; Fiestas Ros de Ursinos J.A. and Borja-Padilla R., 1992; Sayadi S. and Ellouz R., 1995
	$10–12\times10^6$		Cabrera F. et al., 1996

it would reach the water bed and pollute it. There is archaeological evidence that this effluent has been damaging delicate shoreline environments for thousands of years around the Mediterranean (Hadjisavvas S., 1992). The Roman author Varro[7] (I, 55) had observed that where the *amurca* — the watery residue obtained when the oil is drained from olive fruits — flowed from the olive presses onto the fields the ground became barren. These features were decisive in determining the decision, which now forbids direct dumping of OMWW into superficial waters as used to happen traditionally in the past.

Rozzi A. and Malpei F. (1996) described the peculiar problems related to the treatment and disposal of OMWW and showed that abatement of pollution owing to OMWW is a complex problem, which has different solutions according to local factors such as the oil extraction process, the possibility to store the waste and the ratio between the pollution load caused by the olive-mills and that caused by the local population. The peculiarities, which make the treatment of OMWW particularly difficult, can be summarized as follows:

- high organic load (OMWW is among the "strongest" industrial effluents, with COD up to 220 g/l);
- seasonal nature of the production, which requires storage of OMWW (often impossible in small mills);
- high regional scattering of olive-mills;
- small size of the majority of olive-mills, where costs due to OMWW are not properly integrated in the business management;
- presence of organic compounds which are difficult to degrade by microorganisms (long chain fatty acids and phenolic compounds of the C_7 and C_9 phenylpropanoic family);
- high percentage of dissolved mineral salts and of solids in suspension (Beccari M. et al., 1996).

For all the above-mentioned reasons, flexible and efficient treatment plants are needed; they should ensure not only a significant reduction of BOD_5 and COD values, but also the possibility of selectively recovering some valuable compounds. A regionally integrated solution strategy is also needed and a few treatment schemes have been presented. The recovery and exploitation of the by-products can be

---

[7]Varro Marcus Terentius (c.116-27 B.C.), Roman scholar and author. Varro was a soldier but suffered some military misfortunes. He was a true scholar interested in literature and antiquities and was commissioned by Caesar to supervise the collection and arrangement of a great library of Greek and Latin literature destined for public use. He excelled in learning and assembled an enormous collection of writings. He was a prolific author, with almost 700 works on a wide variety of topics (literature, history, philosophy, biography, military history, music medicine, rhetoric, grammar, antiquities, and technical treatises). Only nine of his works remain, three on the Latin language and three on agriculture. *Res Rusticae* was begun in his 80th year and is addressed to his wife Fundanias, who had just purchased a farm. His three books cover agriculture and livestock including game birds and bees. Each book is cast in the form of a dialogue. His work was a source for Virgil and Pliny as well as later agricultural writers (Palladius and Vegetius).

economically justifiable while a realistic consideration of the local realities must be kept in mind (Mendia L. et al., 1986).

The efforts to find a solution to the OMWW problem are more than 50 years old (Fiestas Ros de Ursinos J.A. and Borja-Padilla R., 1992). There are many different types of processes that have been tested, which can be classified into three separate general categories:

1. decontamination processes
   a. physical processes
   b. thermal processes
   c. physico-chemical processes
   d. biological processes
   e. combination of processes
2. recycling and recovery of valuable components
3. production system modification.

None of the decontamination techniques on an individual basis allow the problem of disposal of OMWW to be solved to a complete and exhaustive extent, effectively, and in an ecologically satisfactory way. At the present state of the OMWW treatment technology, industry has found no interest in supporting on a wide scale any traditional process (physical, chemical, thermal, or biological). This is because of the high investment and operational costs, the shortness of the production period (3–5 months), and the small size of most of the olive-mills (Garrido Fernández A., 1975; Arpino A. and Carola C., 1978; Boari G. et al., 1984).

In case of a total recovery of the organic mass, the problems of the disposal remain unsolved which, if oriented to a reuse of said residues as fertilizer or for irrigation purposes, shows the same difficulties as shown by the original OMWW, in that the biotoxic and phytotoxic principles were only concentrated not eliminated. Several techniques also exist, which make it possible for some potentially valuable organic compounds contained in OMWW to be extracted. These techniques use specific solvents and ultrafiltration/reverse osmosis techniques, which in turn, require that complex chemical facilities are available.

Furthermore, the double nature of OMWW (as a pollutant of streams or a resource to be recycled) causes antagonism between agricultural and environmental groups, because of their different point of view on this topic.

The manufacture of olive oil has undergone evolutionary changes. The traditional discontinuous pressing process is being replaced initially by the continuous centrifuge using a three-phase system and later by a two-phase system. In all cases, the olives are firstly washed and then crushed and ground. The three-phase extraction method was developed in the 1970s in order to reduce labor costs and increase processing capacity and yield. While classical methods can process around 8–10 tons of olives per day, the three-phase continuous systems can process 30–32 tons per day with a fraction of the labor requirement. However, this technology also uses 50% more water than the simple pressing method (average 80–100 l of water per 100 kg of fruit processed) and generates twice more OMWW per unit mass of fruit processed (1.3–2 l/kg compared to 0.5–1 l/kg in the classical method). As a consequence, more recently, the two-phase process, which uses much less water than the three-phase process, was developed. The two-phase process uses no process water, and delivers oil as the liquid phase and a very wet, olive cake (2POMW) as the solid phase using a more effective centrifugation technology. However, this process has also inherent environmental problems associated with it, in that although it produces no wastewater as such, it combines the wastewater that is generated with the solid waste to produce a single effluent stream of semi-solid nature (~30% by mass). This doubles the amount of "solid" waste (*alperujo or alpeorujo*) requiring disposal, and it cannot be composted or burned without some form of (expensive) pretreatment. It is also not economically profitable to produce more oil by solvent extraction — see Chapter 4: "The effect of olive-mill technology". More than 4 million tons of 2POMW are annually generated in Spain (Junta de Andalucía, 2002; Alburquerque J.A. et al., 2004).

The use of the two-phase system was implemented for the first time in Spain during the 1991–1992 harvesting season. As its introduction coincided with a dry season, the fact of eliminating the addition of water for the oil separation was very well received at that time by the Spanish olive oil sector (Alvarado C.A., 1998). Although in Spain a vast majority (about 90%) of the olive-mills has presently adopted fully operational two-phase decanters, which can be retro-fitted to existing decanters at a relatively affordable cost, the current penetration rates of two-phase systems in other countries, with the exception of Croatia, are negligible ($<< 5\%$).

Current Practices for Olive Processing Waste Management

Spain

About 75–80% of the average annual production of olive oil in Spain comes from the Region of Andalusia, where are located most of the 1700 olive-mills that operate in Spain.

Until the year 1980, the majority of olive-mills were traditional press systems and evaporation ponds were used for the liquid effluent. In the early 1980s, the three-phase extraction system started to dominate. In 1982, in Spain a law forbade river disposal of OMWW and subsidized construction of storage ponds to promote evaporation during the summer period. Around 100 evaporation ponds were constructed, which improved the water quality, but raised annoyances in ambient air quality because of odor problems. In 1992, the two-phase extraction system was introduced in the region of Andalusia. Nowadays, almost all olive-mills in Spain use two-phase centrifugal decanters. There is still some liquid effluent from the process, but existing evaporation ponds are more than adequate to handle it. Since olive-mills have already started to use water recycling, it is expected that eventually most of the evaporation ponds can be closed down. However, the semi-solid residue

(2POMW) has reached an amount of more than 4 million tons/year and a lot of effort has been put on finding a solution for its management (Alburquerque J.A. et al., 2004). One of the options seems to be extraction after drying and the use of the final — extracted — solid residue as solid cake fuel. Today about 800,000 tons of this waste are exported (Sousa M., 2003).

Italy

In Italy, 5000–6000 olive-mills are operating with most common extraction technology still based on simple pressure (source: "La filiera olio di oliva", ISMEA-Istituto di servizi per il Mercato Agricolo e Alimentare, Rome 2003).

Italy is the only olive oil producing country with a special legislation for the disposal and/or recycling of olive processing wastes. Land spreading of wastes arising from olive processing is specifically regulated under the Law no. 574 of 11/11/1996 on OMWW and olive cake — see Table 8.5. However, the prescriptions of the law have been criticized because they make the inspections quite difficult as the regional and provincial authorities, from which the inspection depend, do not know the exact dates and places of the spreading (Burali A. and Boeri G.C., 2003).

A typical disposal scheme applied in Italy for the treatment of olive-mill wastes is outlined in Fig. 1.4.

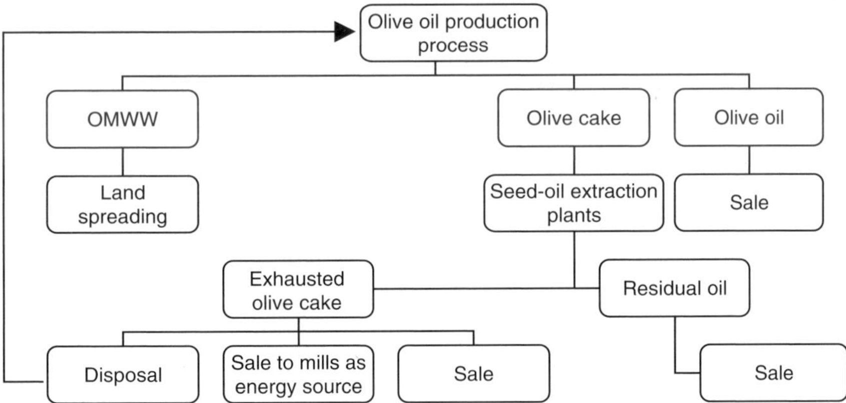

Fig. 1.4. Typical disposal scheme applied in Italy for the treatment of olive-mill wastes.

Greece

In Greece there are 2786 olive-mills, 70% of which are of a three-phase centrifugal type and the rest of classical type or combinations thereof (Georgacakis D. et al., 2002). In addition, there are 40–45 (active only 32) seed-oil extraction plants, more than 200 enterprises of standardization-packaging plants and around 25 refineries.

There are only a very small number of olive-mills that uses two-phase centrifugal decanters. Some olive oil producers tried this technology, but they had to abandon it because there was no viable alternative for the management of 2POMW, while the existing extraction plants cannot handle it and do not accept it (Vlyssides A.G., 2003).

In Greece there is no specific regulation regarding the discharge of OMWW. The olive oil producing prefectures have their own environmental requirements and, on the gained local experience and the results of sponsored research projects, they encourage different waste management approaches. Nowadays, the issuing of an olive-mill operation permit is subject to measures taken to treat the olive-mill waste. More specifically, the Prefecture of Lesvos has stipulated that OMWW must be pretreated with lime before disposal in the natural recipients. However, this solution was not enforced and the olive-mills were granted a two-year extension of the validity of their operation permits. The Prefecture of Chios decided to construct open ponds, large enough to accommodate the entire quantity of wastewater produced in one olive cultivation season. Twelve of the fourteen olive-mills on the island dispose of their wastewater in such mud ponds. The Prefecture of Samos has granted all its olive-mills a two-year extension with regard to issuing an operation permit. Meanwhile, a wastewater management technique is due for evaluation for real-scale application by an olive-mill on Samos. This method — proposed by Professor Georgacakis D. of the Agricultural University of Athens — initially includes pretreatment/fractionization of OMWW by natural sedimentation. Separate management of the individual fractions then takes place (Georgacakis D. and Christopoulou N., 2002). A general conclusion drawn from research to date is that there is no single technical solution that can ensure a satisfactory level of treatment efficiency whose application cost will be within the economic means of each individual olive-mill owner. This conclusion accounts especially in the case of Greece, given its geographical distribution and the size of its olive-mill plants. In other parts of the country, evaporation ponds (lagoons) are commonly used for the treatment and disposal of OMWW, optionally after neutralization with lime. In practice, all the generated OMWW results in creeks (58%), or in sea and rivers (11.5%), or in soil (19.5%).

Turkey

In Turkey too, there is no specific regulation regarding the discharge of OMWW. The Turkish water pollution control regulation oversees protection of the water resources against pollution and sets discharge standards both for protection of the receiving media and for effluents of olive-mills. The biggest and main obstruction for the safe disposal of OMWW is that olive-mills are small and scattered in a large geographical area.

In regard to solid olive-mill waste, the Ministry of Environment in Turkey has permitted the combustion of dried solid cake only in olive-mill beginning in 2003, with the condition that the gas emission limits are met (Azbar N. et al., 2004).

Tunisia

In Tunisia, a common way of dealing with OMWW is to convey it from the mills to a central point and discharge it into a purpose-built lagoon. Here, the volume reduces by evaporation, providing that the lagoon base has been sealed (thereby preventing possible groundwater contamination); this can be a very reasonable way of containing the problem. Recently, in the Sfax area of Tunisia, a new facility has been built to receive OMWW. Four lagoons have been constructed with a combined surface area of 50 ha and a total storage capacity of 40,000 m^3. A charge of around 7 Tunisian Dinars per ton of OMWW is levied for reception at these lagoons (Skerratt G. and Ammar E., 1999).

Portugal

In Portugal there are around 1000 olive-mills most of which use the traditional discontinuous pressing process, although over the last few years several units have introduced continuous solid–liquid centrifugation systems.

The olive oil sector has been subject to a specific intervention that started in 1997 and was completed in 1999 with the signing of an agreement. Both the Ministry of Environment and the Ministry of Agriculture were involved, while the agreement was technically supported by a University that did exhaustive characterization of the sector, studied technical solutions for OMWW and performed cost–profit analysis for their implementation. The olive-mills are subjected to monitoring under the agreement and the new legislation that has been produced (regulation for the use of OMWW in irrigation, interpretation for excluding the olive cake from classicization as "waste" and selection of representative sample for air emission characterization) (Figueira F., 2003).

The use of OMWW for irrigation is also subject to restrictions similar to those applicable in Italy. Namely, the limits for the spreading of OMWW on soil for agricultural use are 50 m^3/ha · y from a traditional press system and 80 m^3/ha · y from the three-phase centrifugation system. Furthermore, it is forbidden to spread within 300 m from a drinking water source; within 200 m from a habitation center; over territories where in the same moment some crops are being grown; over soils where there may be any kind of contact with groundwater, or where the groundwater flow is within 10 m from the surface. It is also forbidden to discharge in surface waters and in the sea.

France

The annual olive-oil production comes from four regions: Provence-Alpes-Côte d'Azur (61%), le Languedoc-Roussillon (17%), Rhône Alpes (12%), and Corse (10%). In France there are more than 25,000 olive farms and 152 mills and cooperatives (source: Afidol, May 2001).

Land spreading is the disposal practice most commonly used in France (Le Verge S. and Bories A., 2004). The creation of evaporation ponds has been encouraged as an alternative disposal treatment. The construction costs of an evaporation pond are subsidized up to 30% by the Water Agency and supplementary by regional and departmental authorities (Ferrieres B., 2004). The norms of construction of evaporation ponds are regulated by a ministerial decree (Arrêté 26/02/2002) concerning the pollution control of farming effluents (JO 21/03/2002).

Cyprus

There are 35 olive-mills in Cyprus today, with an average capacity of 1000 tons of olives per year, producing around 7500 tons/year of olive oil. Due to the small size of olive-mills in Cyprus, it is rather unreasonable to assume that each mill will have its own liquid waste treatment facility.

Existing permitting system provides for liquid and solid waste conditions. Since facilities are SMEs they do not have to comply with Emission Limit Values (EVLs) relevant to treatment of wastes. The permit conditions are based on techniques/practices rather than treatment technologies of the wastes. The most useful practice is the storage of OMWW in artificial ponds and remaining there for evaporation (evaporation rate is about 550 mm per year). Most of the plants are situated in the peripheries of villages. No discharge in the sea or in the surface waters and rivers is allowed. It is estimated that 95% of stones are used for heating, 85% of OMWW are stored in ponds and/or discharged to soil and, approximately 10% are discharged in central industrial treatment facilities, especially constructed and operated for SMEs (Hadjipanayiotou C., 2003).

Croatia

There are about 4 million olive trees, covering 16,000 ha — 94% private farmsteads, 0.5% of total planted agricultural land — and 41,000 olive growers. During recent years, the annual production of olive oil ranges from 2000 to approximately 5000 tons/year. There are 86 olive-mills most of which use two-phase systems. There is no seed-oil extraction plant and 2POMW is usually applied to the soil as conditioner/compost (Miocic S. and Milic I., 2003).

Malta

There are presently five olive-mills in operation: two of these mills have a productive capacity of 0.5 and 0.4 tons/ha, respectively; the largest mill has a productive capacity of 3.5 tons/ha; and the two smaller ones have a productive capacity of 0.15 tons/ha each. The viable amount of olive oil that can be produced locally in the existing mills is estimated at 1052 tons.

It is a typical practice in Malta to recycle the generated olive waste. The crude olive cake is left to dry and then it is mixed with natural manure for composting and used as fertilizer in the farmer's fields. Some of it is left in a cylindrical form and wrapped in newspapers to absorb water and then dried. This is in turn used as combustion fuel, instead of wood logs. OMWW produced from the traditional plant is sprayed back in the orchard.

The operation of an olive-mill requires: (i) planning permit — for the building up of the pressing room; (ii) Public Sewer Discharge Permit — this binds them with the following effluent limit values as established in LN 139 of 2002 and the Sewerage Master Plan for Malta and Gozo - November 1992. In the case of non-compliance, a fine of 240 € "for every day the default continues after the expiration of the said time". In any case, no mill raised any complaints from the public. In short, the olive oil sector in Malta, being a cottage industry, generates relatively small amounts of waste, which is reused within the same industry (Vasallo C., 2003).

Terminology

The terms used for olive-mill wastes are neither standardized nor country specific. This causes some confusion in the publications which makes it sometimes difficult to identify clearly the particular by-products concerned. The Spanish term for OMWW is *alpechín*; the name *alpechín*, comes from the Latin *faecinus*, and alludes to the latter characteristic. Other Spanish terms such as *murga*, *morga* or *amorca* as well the French term *margine* come from the Latin *amurca*, which means stinking juice. The Italians refer to OMWW as *acqua di vegetazione*, while the covered basin in an olive-mill (generally underground), where OMWW is collected and stored, is called *inferno* or hell. The Turks refer to it as *kara su* or black water, due to its appearance; the Arabs call it *zubar*, and the Greeks call it *liozumia* or olive juice (in Crete they call it *katsigaros*). The most common terminology used in the Mediterranean area is shown in Table 1.5. The description of each term is given in the "Glossary".

Reviews in the Prior Art

Despite the serious environmental problems caused by OMWW, there are only few reviews — mostly partial and/or outdated — on this subject in literature. In the earliest of them, preliminary investigations are discussed aimed at identifying the most effective purification methods and profitable recovery of residues (Carola C. et al., 1975). A following review considers the characteristics of OMWW; the problems of its disposal through the public waterway system due to its high pollution; possible systems for reducing the pollution load, e.g. by collection in evaporation ponds, by discharge on the land with subsequent percolation and evaporation or irrigation and fertilization; production of fertilizers; and purification methods, e.g. aerobic or anaerobic biological purification, chemical purification by CaO, Al, or Fe salts, purification and use as a substrate for growing feed yeast (Fiestas Ros de Ursinos J.A., 1977). A later study provides a review on the pollution

Table 1.5. Terminology used for the olive-mill wastes

	Pressing (traditional or classical system)	Three-phase centrifugation	Two-phase centrifugation
Olive-mill wastewater (OMWW)	acqua di vegetazione/Italian água ruça/ Portugese alpechín/Spanish amorca/Spanish amorgi (αμόργη)/ancient Greek amurca/Latin mourga (μούργα)/Greek kara su/Turkish katsigaros(κατσίγαρος)/Greek liozumia (λιοζούμια)/Greek margine/French morga/Spanish mrar/Maltese murga/Spanish olive lees/English veget abilna voda/Croatian vegetable water/English	acqua di vegetazione/Italian água ruça de 3 fases/ Portugese alpechín/Spanish amorca/Spanish kara su/Turkish katsigaros(κατσίγαρος)/Greek liozumia(λιοζούμια)/Greek Margine/French morga/Spanish mrar/Maltese murga/Spanish veget abilna voda/Croatian zubar/Arabic	água residual de 2 fases/Portugese* alpechín-2/Spanish* margine-2/French* jamila-2/Italian* mrar/Maltese*
Olive-mill (semi-) solid waste i) crude or raw	bagaço húmido/Portugese crude olive cake /English eleopirina (ελαιοπυρήνα)/ Greek grignons/French olive husks/English jefet/Arabic jift/Arabic marc/French maxx/Maltese merc/Spanish	bagaço húmido de 3 fases/Portugese crude olive cake/English eleopirina (ελαιοπυρήνα)/Greek grignons/French husks/English jefet/Arabic jift/Arabic marc/French maxx/Maltese merc/Spanish	"two-phase olive-mill waste (2POMW)" alpeorujo/Spanish alperujo/Spanish bagaço húmido de 2 fases/ Portugese foot cake/English komina/Croatian maxx/Maltese oil-foot/English orujo de dos fases/Spanish orujo humedo/Spanish

	orujo/Spanish pomace/English sansa (vergine)/Italian stemphyla (στέμφυλα)/ancient Greek trester/English	orujo de tres fases/Spanish pomace/English sansa (vegine)/Italian	sansa (vegine)/Italian
ii) exhausted or deoiled	bagaço seco/Portugese deoiled olive cake/English exhausted olive cake/English sansa esausta/Italian	bagaço seco de 3 fases/ Portugese deoiled olive cake/English exhausted live cake/English orujillo/Spanish sansa esausta/Italian	bagaço seco de 2 fases/Portugese exhausted 2POMW orujillo/Spanish sansa esausta/Italian
iii) destoned	bagaço sem caroço/Portugese destoned olive cake/English pasta/Maltese pirinoxilo (πυρηνόξυλο)/Greek sansa denocciolata/Italian	bagaço sem caroço/Portugese destoned olive cake/English pasta/Maltese pirinoxilo (πυρηνόξυλο)/(Greek) sansa denocciolata/Italian	
iv) stone(s)	(caroço de) bagaço extractado/ Portugese ghadma/Maltese koštice/Croatian nocciolino/Italian orujo eco y egrasado/Spanish pirina (πυρήνα)/Greek/Turkish pit(s)/English stone(s)/English	(caroço de) bagaço extractado/ Portugese ghadma/Maltese koštice/Croatian nocciolino/Italian orujo eco y degrasado/Spanish pirina (πυρήνα)/Greek/Turkish pit(s)/English stone(s)/English	
Residual oil	aceite de orujo/Spanish óleo de bagaço/Portugese olio di sansa/Italian pomace oil/English seed-oil/English zejt ta'l-ghadma/Maltese	aceite de orujo/Spanish óleo de bagaço/Portugese olio di sansa/Italian pirineleo (πυρηνέλαιο)/Greek pomace oil/English seed-oil/English zejt ta'l-ghadma/Maltese	aceite de orujo/Spanish óleo de bagaço/Portugese olio di sansa/Italian pomace oil/English seed-oil/English zejt ta'l-ghadma/Maltese

*Liquid fractions from secondary 2POMW (*alperujo*) treatments (second decanting, repaso, etc.).

produced by OMWW and proposes various solutions: evaporation basins; spreading on soil for evaporation and percolation, or for irrigation and fertilization; incineration; use as animal feed; fermentation substrate; and chemical or biological purification with possible re-use in olive-mills (Janer del Valle M.L., 1980). A good review of the treatment options for OMWW can be found in the Proceedings of the International Symposium: "Olive by-products valorization" held in Seville in 1986 (FAO, 1986). Another study reviews the methods described in literature emphasizing the most important features and constraints of each of these processes (Hamdi M., 1993a, 48 references). In a later, paper, the main processes utilized in the treatment and disposal of OMWW are reviewed and the most interesting ones are described and commented upon (Rozzi A. and Malpei F., 1996). A lately published critical review examines the available treatment and disposal alternatives of olive-mill wastes, with emphasis on the present-day techniques of waste management. Waste characteristics, treatment options with regard to the economic feasibility, and challenges of existing waste disposal practices in olive growing countries are also discussed. Most interesting, this article provides a comparative economical analysis of the various treatment schemes (Azbar N. et al. 2004) — see Chapter 12: "Economic evaluation".

Chapter 2

Characterization of Olive Processing Waste

The Olive Fruit

The olive fruit is technically a drupe and it consists essentially of three parts: epicarp, mesocarp, and endocarp — see Fig. 2.1. The epicarp (skin or peel or epidermis) is covered with wax and remains green throughout the growth phase, and then it may turn purple and brown or black when ripe according to variety. The mesocarp (pulp or flesh) has low sugar content (3–7.5%) and high oil content (15–30%) that varies according to the variety and ripeness of the fruit. The endocarp (stone or pit) is hard and made of fibrous lignin. Its ovoid shape and the extent to which it is furrowed are varietal characteristics. The endocarp encloses the olive kernel (seed) that accounts for approximately 3% of fruit weight and contains 2–4% of total fruit oil. The chemical compositions of the different parts of the olive fruit are given in Table 2.1. The olive fruit weighs from 2–12 g. The average composition of an olive is water (50.0%), oil (22.0%), sugars (19.1%), cellulose (5.8%), proteins (1.6%), and ash (1.5%) (IOOC, 2002).

The olive fruit contains also high concentrations of hydrophilic and lipophilic phenolic compounds in the range 1–3% of the fresh pulp weight (Garrido Fernández A. et al., 1997). The main lipophilic phenols are cresols. The main hydrophilic phenols are phenolic acids, phenolic alcohols, flavonoids, and secoiridoids. The phenolic compounds classified as secoiridoids are characterized by the presence of either elenolic acid, or elenolic acid derivative in their molecular structure. Oleuropein, demethyloleuropein, ligstroside, and nuzhenide are the most abundant secoiridoids glucoside in olive fruit (Gariboldi P. et al., 1986; Garrido Fernández A. et al., 1997; Servili M. et al., 2004). Oleuropein is responsible for the bitter taste of the olive fruit and is concentrated in the mesocarp.

There are hundreds of cultivars for olives: for oil, for table, and for both uses. They have different shapes and sizes, various ratios between stone and pulp, and average oil content. Yield depends on many factors that influence the biennial

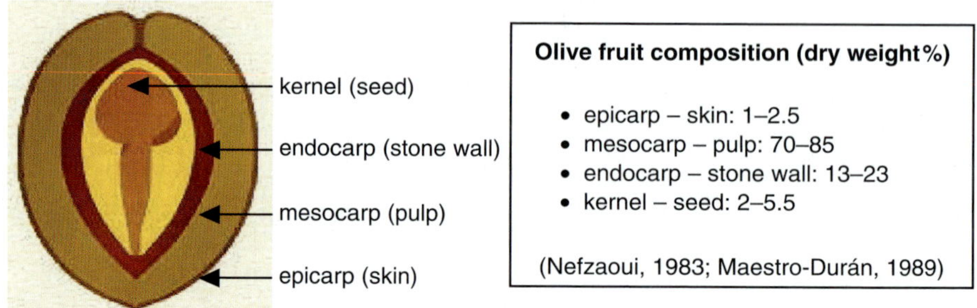

Fig. 2.1. Cross section of the olive fruit (Maymone B. et al., 1961a,b).

Table 2.1. A representative chemical composition of olive fruit (%) (EU project: FAIR-CT96-1420 "IMPROLIVE")

Components	Olive pulp	Stone	Seed
Water	50–60	9.3	30
Oil	15–30	0.7	27.3
N-contg. compounds	2–3	3.4	10.2
Sugar	3–7.5	41	26.6
Cellulose	3–6	38	1.9
Minerals	1–2	4.1	1.5
Polyphenols	2.25–3	0.1	0.5–1
Other compounds	–	3.4	2.4

rhythm of production, including climate and cultivation practices. The majority of the oil is contained in the pulp and for this reason olive fruits with a high ratio pulp/stone are preferred in olive oil production.

Olive By-Products

Both olive tree culture and olive processing-related industries produce large amounts of by-products. The main olive by-products are depicted schematically in Fig. 2.2.

Pruning and Harvest Residues

Olive trees are usually subjected to severe pruning every second year and light pruning in the alternate year. It has been estimated that pruning alone produces

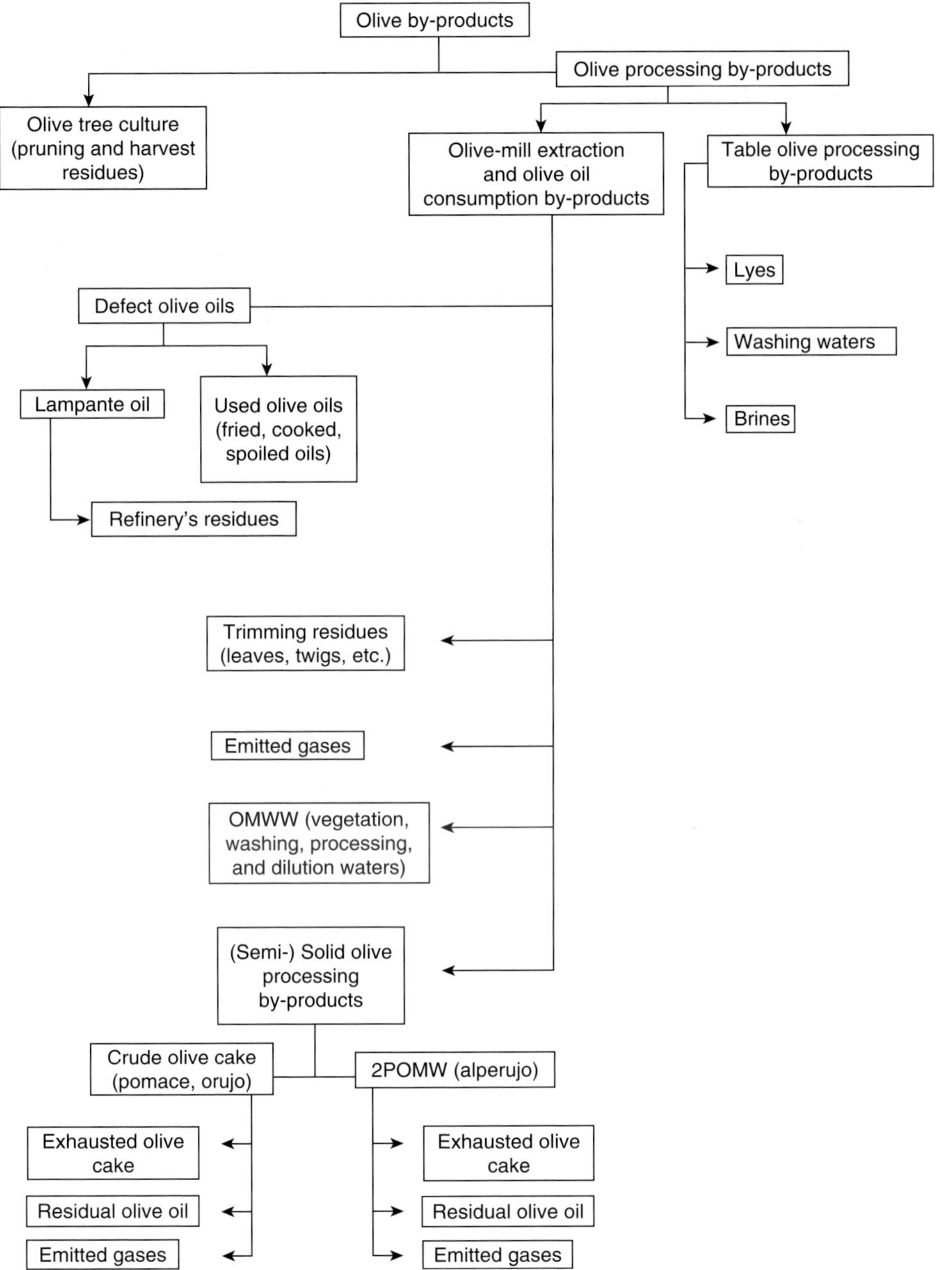

Fig. 2.2. Main olive by-products.

25 kg of by-products (twigs and leaves) per tree annually. After separation of the large branches, the leaves and twigs (less than 3 cm in diameter) can be distributed to ruminants (Sansoucy R. et al., 1985).

There are differences between leaves collected at the olive-mill, which have a small proportion of wood, and branches on which the proportion of wood can be considerable. On branches with less than 4 cm in diameter the proportion of leaves is about 50% (Civantos L., 1981a,b).

The chemical composition of olive leaves, obtained both from pruning or olives cleaning from different mills and dried using different procedures, is shown in Table 2.2 (Molina Alcaide E. et al., 2003). The organic matter (OM) content is variable (76.4–92.7 g/100 g DM); the crude protein (CP) content is low (6.31–10.9 g/100 g DM) being the amino acid N proportion relatively important (89.9 g/100 g TN) in the only analyzed sample. The proportion of the N attached to the cell walls is high although variable (49.2 and 35.4 g/100 g TN, respectively, for leaves from olives cleaning and pruning). The crude fat (CF) is variable in leaves from the olives cleaning (2.28–9.57 g/100 g DM).

The chemical composition of leaves and twigs varies according to many factors, such as olive variety, climate conditions, tree age, wood proportion, etc. In general, these different by-products have relatively homogeneous and well-defined characteristics — see Table 2.3:

- the dry matter (DM) in green leaves comes to about 50%, that in dry leaves to about 90%;
- their total crude protein (CP) content is low i.e. 7 to 8% in dry or ensiled leaves, slightly higher in green leaves;
- their ether extract (EE) content of about 6% is higher than that of traditional fodders;
- their crude fat (CF) content is variable and relatively small;
- their cell wall constituents increase considerably with wood percentage, especially lignocellulose contents; lignin level seems to be stable at 18 to 19%.

The crude protein content seems to be less in branches than in green leaves and is comparable to that of dry leaves. Naturally the crude fiber content is markedly higher than that of leaves.

Information on the content of phenolic compounds in olive leaves is scarce (Le Tutour B. et al., 1992; Delgado-Pertiñez M., 1994, 1997). Even so, its phenolic compound content appears to be very variable with reported values in the range of 1.4–6.4 g/100 g DM (Molina Alcaide E. and Nefzaoui A., 1996; Pinelli P. et al., 2000) and 0.14–4.3 g/100 g DM (Molina Alcaide E. et al., 2003) — see Table 2.2. The following polyphenols have been detected in olive leaf tissue: hydroxytyrosol, tyrosol, tocopherol (Lucas R. et al., 2002), hydroxytyrosol glucoside, elenolic acid derivative, caffeic acid, oleuropein (Gariboldi P. et al., 1986; Capasso R. et al., 1996), verbascoside, and the flavonoids: rutin, luteolin-7-glucoside, luteolin-4'-glucoside, apigenin-7-rutinoside, and apigenin-7-glucoside — see also Chapter 10: "Uses", section: "Recovery of organic compounds".

Table 2.2. Chemical composition (g/100 g DM) of different OLs[a]

	DM (g/100g fresh matter)	OM	CP	CF	ADFN (g/100 g TN)	NDF	ADF	ADL	TEP	TCT	FCT	PBCT	FBCT
OL-A[b]	88.7	76.4	6.31	9.57		36.2	31.7	19.9	0.141				
OL-B[b]	60.7	83.8	7.00	3.21	53.3	41.3	33.3	19.0	2.53	0.669	0.201	0.340	0.128
OL-C[b]	50.4	92.7	10.9	2.28		34.9	29.0	18.4	4.29				
OL-D[b]	49.0	84.1	7.69	6.37		39.8	34.2	21.1	1.70				
OL-G[b]	67.6	85.4	7.00	5.29		38.9	32.7	19.5	2.38				
OL-I[b]	60.6	87.8	8.00	2.53		38.6	33.0	17.9	1.99				
OL-L[c]	94.0	89.0	7.88	8.03	45.0	39.9	22.5	14.1		0.575	0.175	0.360	0.030
OL-P[c]	91.1	89.8	9.00		35.4		31.2			1.11	0.353	0.635	0.125

[a]OL: olive leaves; A, B, C, D, G, I, and L: indicate different mills; P: pruning; DM: dry matter; OM: organic matter; CP: crude protein; CF: crude fat; ADFN: acid detergent insoluble nitrogen; NDF: neutral detergent fiber; ADF: acid detergent fiber; ADL: acid detergent lignin; TEP: total extractable polyphenols; TCT: total condensed tannins; FCT: free condensed tannins; PBCT: protein-bound condensed tannins; FBCT: fiber-bound condensed tannins (Molina Alcaide E. et al., 2003).
[b]Dried at 60 °C.
[c]Air-dried.

Table 2.3. Indicative chemical composition of olive tree leaves and branches (Sansoucy R. et al., 1985; adapted from Alibes X. and Berge Ph., 1983; many sources cited)

By-product	DM	OM	CP	CF	EE	NDF	ADF	ADL
Green branch	68	90	7.7	24.5	11.2	–	–	–
Dry branch	87–92	91.5	7–9	23–29	6	–	–	–
Green leaves	50–58	95	11–13	15–18	7	47	28	18
Air-dried leaves	95	95	7–11	13–23	5	40–45	28–35	18
Leaves with 8.8% wood	87	92	7.7	19	–	48	34	19
Leaves with 11.4% wood	93	92	8.7	19	–	–	–	–
Leaves with 15% wood	74	95	6.7	30	–	56	44	19
Leaves with 22.6% wood	93	92	7.8	21	–	51	35	18
Leaves ensiled with 8.8% wood	46	91	7.7	–	–	–	32.5	19

Types of Olive-Mill Extraction By-Products

The olive oil industry produces large amounts of by-products. It is estimated that for every 100 kg of treated olives 35 kg of solid waste (olive cake) and from 55 to 200 l liquid waste are produced depending on the oil extraction process. Leaves represent 5% of the weight of olives in oil extraction.

Olive-Mill WasteWater (OMWW)

The OMWW composition is not constant — both qualitatively and quantitatively — and it varies according to:

i.   composition of the vegetation water;
ii. olive oil extraction process;
iii. storage time.

i. The composition of the vegetation water varies according to:
- olive variety,
- maturity of the olives,
- olive's water content,
- cultivation soil,
- harvesting time,
- presence of pesticides and fertilizers,
- climatic conditions.

ii. Water use in the mills varies widely, both because of equipment requirements (the centrifugal mill needs substantially greater quantities of water) and local operational conditions and practices. The water used in the different stages of oil production plus olive washing water reduce the concentration of the various components already present in the vegetation water.

iii. Storage gives rise to substantial changes in composition caused by the aerobic and anaerobic fermentation of several organic compounds with consequent emission of volatiles, increase of the acidity, precipitation of suspended solids, etc.

Olive oil manufacturers operate in annual seasons of less than one hundred days, process widely variable batches of olives daily, use rather different technological systems and operational habits and produce very different amounts of OMWW. The type and origin of OMWW samples frequently are not reported and it cannot be ascertained whether the characteristics reflected total plant waste or a partial waste stream. The inherent difficulty, coupled with the almost complete absence of data pertaining to corresponding waste volumes, significantly limit the usefulness of the reported data (Tsonis S.P. et al., 1987). It is only possible to obtain an idea of the range of values for each parameter, many of which differ by more than an order of magnitude. This variation is seen both between samples from a similar source and between samples from different countries — see Table 2.4 — and processes — see Tables 2.5 and 2.6. As it can be seen in Table 2.4, the Spanish OMWW has a COD, a BOD_5, and dry matter content all of which are approximately half that of the Italian OMWW (Knupp G. et al., 1996). The substantial range in the reported values can be only partially justified by differences in the organic (oil and sugar) content of the olives due to variety and degree of ripening and their nutrient (TKN and Total-P) content as affected by the availability of soil nutrients.

The OMWW is characterized by the following special features and components (Vásquez-Roncero A. et al., 1974b; López C.J., 1993):

- intensive violet-dark brown up to black color,
- strong specific olive oil smell,
- high degree of organic pollution (COD values up to 220 g/l),
- pH between 3 and 6 (slightly acid),
- high electrical conductivity,
- high content of polyphenols (0.5–24 g/l),
- high content of solid matter.

Table 2.4. Spanish OMWW sample vs Italian OMWW
(Knupp G. et al., 1996)

Parameter	Spanish OMWW[a]	Italian OMWW[b]
COD (g/l)	49.0	80.4
BOD_5 (g/l)	4.2	11.5
Dry matter (g/l)	35.1	73.0
pH (24°C)	4.9	5.2

[a]Origin: Area of Granada.
[b]Origin: Borgomaro, Liguria.

Table 2.5. Average results of the characteristic parameters carried out
on fresh OMWW samples obtained from olive-mills processing olives by
pressure and 3-phase centrifugation systems (Di Giovacchino L. et al., 1988)

Parameter	Pressure system	3-Phase centrifugation system
pH	5.27	5.23
Dry matter (g/l)	129.7	61.1
Specific weight	1.049	1.020
Oil (g/l)	2.26	5.78
Reducing sugars (g/l)	35.8	15.9
Total phenols (g/l)	6.2	2.7
o-Diphenols (g/l)	4.8	2.0
Hydroxytyrosol (mg/l)	353	127
Precipitate with alcohol (g/l)	30.4	24.6
Ash (g/l)	20	6.4
COD (g O_2/l)	146.0	85.7
Organic nitrogen (mg/l)	544	404
Total phosphorous (mg/l)	485	185
Sodium (mg/l)	110	36
Potassium (mg/l)	2470	950
Calcium (mg/l)	162	69
Magnesium (mg/l)	194	90
Iron (mg/l)	32.9	14.0
Copper (mg/l)	3.12	1.59
Zinc (mg/l)	3.57	2.06
Manganese (mg/l)	5.32	1.55
Nickel (mg/l)	0.78	0.57
Cobalt (mg/l)	0.43	0.18
Lead (mg/l)	1.05	0.42

Organic Compounds OMWW contains various amounts of sugars depending
on the variety of olives, the climatic conditions during growth and the extraction
methods used. Sugar levels are generally of the order 1.6–4 % (w/v), but can be
higher in rare cases (Fiestas Ros de Ursinos J.A., 1961a, 1967; Fernández-Bolaños J.
et al., 1983). The sugars constitute up to 60% of the dry substance and comprise, in
decreasing amount, fructose, mannose, glucose, saccharose, and traces of, sucrose,
and some pentose.

Both the stone and pulp of olives are rich in phenolic compounds. These
compounds, once released or formed during processing of olives, are distributed
between the water and oil phases. Another part of the phenols is trapped in the olive
cake. The distribution of the released amount of the phenols between water and oil
is dependent on their solubilities in these two phases. The olive phenols are amphiphilic in nature and are more soluble in the water than in the oil phase. Due to their

Table 2.6. Characteristics of OMWW[1] as reported in the literature (Table compiled by Tsonis S.P. et al., 1987)

| Type of mill | BOD$_5$[2] | COD[2] | TKN | Total-P | Solids | | | Oil and grease | pH | References |
| | | | | | TS | VTS | SS | | | |
	g/l	g/l	gN/l	gP/l	g/l	g/l	g/l	g/l		
Press	41–62	152–162			86–126	40–77			4.6–4.9	Telmini M. et al., 1976
	9.2–9.6	98–119	0.56–0.66	0.01–0.23	96–108	64–82		0.2–0.3	4.9–5.2	Raimundo M.C., Oliveira J.S.,1976
	29	93	0.009	trace	60	52	0.6		4.9	Bradley R.M. and Baruchello L., 1980
	12–41	50–68	0.10–0.14	0.001	66–72	25–26		5–14	6.4–7.0	Curi K. et al., 1980
	90–100	120–130	0.8–3.2	1.1	120	105	1.0	0.3–10	4.5–5.0	Fiestas Ros de Ursinos J.A. et al., 1981
	42	103	0.18	0.08	64	46	36	8.9	4.5	Tsonis and Grigoropoulos S.G., 1983
	1.5–1.8	6.4–90			28–31	4–20	2.2–7.7	1.2–4.0	5.0–5.4	Muezzinoglu A. and Oslu, 1986
	38–46	62–74	0.12–0.14	1.2–1.4	95–123	42–49	14–24	4.7–6.5	4.2–4.6	Curi K. et al., 1982
	1.5–100	6.4–162	0.009–3.2	trace-1.4	28–126	4–82	0.6–24	0.2–10	4.2–7.0	Range
Centrifuge	13–14	39–78			19–37	12–20			5.2–5.3	Telmini M. et al., 1976
	19	60	0.02	trace	32		0.4		5.0	Bradley R.M. and Baruchello L., 1980
	23–44	25–48	0.27–0.64	0.02–0.26	14–45	12–41	2.7–17	5–23	4.7–5.2	Fiestas Ros de Ursinos J.A. et al., 1981
	24	45	0.19	0.16	35	24	22	12	4.8	Tsonis S.P. and Grigoropoulos S.G., 1983
	13–44	25–78	0.02–0.64	trace-0.26	14–37	12–41	0.4–22	5–23	4.7–5.3	Range

[1]The type of OMWW, whether total or a specific stream, is not always mentioned.
[2]BOD$_5$ and COD values of 50 and 100 Kg/ton olives, independent of mill type have also been reported (Rozzi A. and Di Pinto A.C., 1986).

low partition coefficients (K_p)[8], only a fraction of the phenols enters the oil phase (EU project: FAIR-CT97-3039). In general, the concentration of the phenols in the olive oil ranges from 50 to 100 μg/g of oil depending on the olive variety. This amount corresponds to 1–2% of the total phenolic content of the olive fruit, while the rest is lost in OMWW (~53%) and the olive cake (45%) depending on the extraction system — see also Chapter 4: "The effect of olive-mill technology", section: "Olive oil production systems" (Rodis P.S. et al., 2002).

Phenolic compounds are present in OMWW at concentrations in the range from 0.5 to 24 g/l, and are strictly dependent on the processing system used for olive oil production (Ragazzi E. and Veronese G., 1967a,b; Sorlini et al., 1986; Borja-Padilla R. et al., 1990a,b). Phenolic compounds generically include a great many organic substances that have the common characteristic of possessing an aromatic ring with one or more substitute hydroxyl group and a functional chain. The prevalent classes of hydrophilic phenols identified and quantified in OMWW include phenolic alcohols, phenolic acids, phenyl alcohols, secoiridoids, flavonoids, and lignans. So far, more than 30 phenolic compounds have been identified in OMWW — see Table 2.7. The presence of these recalcitrant organic compounds constitutes one of the major obstacles in the detoxification of OMWW.

A group of phenolic compounds found in OMWW are derived from cinnamic acid: the parent unsubstituted cinnamic acid, *o*- and *p*-coumaric acid (4-hydroxycinnamic acid), caffeic acid (3,4-dihydroxycinnamic acid), and ferulic acid (4-hydroxy-3-methoxycinnamic acid). Another group of phenolic compounds found in OMWW are derived from benzoic acid: the parent unsubstituted benzoic acid, protocatechuic acid, and β-3,4-dihydroxyphenyl ethanol derivatives, such as tyrosol and hydroxytyrosol (Pompei C. and Codovilli F., 1974). Other phenols found in OMWW include catechol, 4-methylcatechol, *p*-cresol, and resorcinol (Capasso R. et al., 1992a, b; Vinciguerra V. et al., 1993). Some of these polyphenols, particularly hydroxytyrosol and catechol are responsible for several biological effects, including antibiosis (Paredes M.J. et al., 1986; Rodríguez M.M. et al., 1988), ovipositional deterrence (Girolami V. et al., 1981) and phytotoxicity (Capasso R. et al., 1992) (see Fig. 2.3).

Of all the polyphenols considered, hydroxytyrosol is worth noting as the main natural polyphenolic compound in OMWW (Ragazzi E. and Veronese G., 1967a,b Vásquez-Roncero A. et al., 1974b; Capasso R. et al., 1992a). Possibly it arises from the hydrolysis of oleuropein by an esterase during the milling process (Amiot M.J. et al., 1989). Hydroxytyrosol is characterized by major bio-antioxidant activity. Since it is commercially unavailable, a method for its chromatographic purification has been developed to produce it from OMWW (Ragazzi E. and Veronese G., 1967a–c, 1982; Capasso R. et al., 1992a,b) — see Chapter 10: "Uses", section: "Recovery of organic compounds".

---

[8]The partition coefficient is defined as: $K_p = C_{oil}/C_{water}$, where C_{oil} and C_{water} are the equilibrium concentrations of the phenolic compound in the oil and water phase, respectively.

Table 2.7. Simple and complex phenols identified in OMWW

Phenolic compounds	Synonyms	References
Caffeic acid	3-(3,4-Dihydroxyphenyl)-2-propenoic acid; 3,4-dihydroxycinnamic acid	Vásquez-Roncero A. et al., 1974b; Balice and Cera O., 1984; Cichelli A. and Solinas M., 1984; Visioli F. et al., 1995a, 1999; Aramendia M.A. et al., 1996; Servili M. et al., 1999b; Sayadi S. et al., 2000; Lesage-Meessen L. et al., 2001; Della Greca M. et al., 2001, 2004
Catechol	Benzocatechol; *o*-dihydroxybenzene; *o*-benzenediol; 1,2-benzenediol; catechin; 1,2-dihydroxybenzene; *o*-hydroxyphenol; 2-hydroxyphenol; oxyphenic acid; phthalhydroquinone	Vásquez-Roncero A. et al., 1974b; Servili M. and Montedoro G.F., 1989; Capasso R. et al., 1992a, b; Vinciguerra V. et al., 1993; Capasso R., 1999; Fiorentino F. et al., 2003; Della Greca M. et al., 2001, 2004
Cinnamic acid	3-Phenyl-2-propenoic acid; β-phenylacrylic acid	Balice V. and Cera O., 1984; Chichelli A. and Solinas M., 1984; Servili M. and Montedoro G.F., 1989; Ramos-Cormenzana A. et al., 1996
o-Coumaric acid	*o*-Hydroxycinnamic acid	Servili M. and Montedoro G.F., 1989; Aramendia M.A. et al., 1996
p-Coumaric acid	3-(4-Hydroxyphenyl)-2-propenoic acid; p-hydroxycinnamic acid; β-(4-hydroxyphenyl)-acrylic acid	Vásquez-Roncero A. et al., 1974b; Balice V. and Cera O., 1984; Cichelli A. and Solinas M., 1984; Servili M. and Montedoro G.F., 1989; Ramos-Cormenzana A. et al., 1996; Aramendia M.A. et al., 1996; Sayadi S. et al., 2000; Lesage-Meessen L. et al., 2001; Bianco et al., 2003; Della Greca M. et al., 2001, 2004
Cresol	Methylphenol; cresylic acid; cresylol; tricresol; hydroxytoluene	Capasso R. et al., 1992a, b; Vinciguerra V. et al., 1993

(*continued*)

Table 2.7. Simple and complex phenols identified in OMWW — Cont'd

Phenolic compounds	Synonyms	References
Demethyloleuropein		Visioli F. et al., 1999; Servili M. et al., 1999b
3,4-Dihydroxyphenyl-acetic acid		Aramendia M.A. et al., 1996; Knupp G. et al., 1986; Della Greca M. et al., 2001, 2004
2-(3,4-Dihydroxyphenyl)-1,2-ethandiol	3,4-Dihydroxyphenyl glycol	Della Greca M. et al., 2001, 2004; Knupp G. et al., 1996
3,4-Dihydroxyphenyl ethanol-elenolic acid dialdehyde	3,4-DHPEA-EDA	Visioli F. et al., 1999; Lo Scalzo and Scarpati, 1993; Servili M. et al., 1999b; Della Greca M. et al., 2004
2-(3,4-Dihydroxyphenyl) ethanol 3β-D-glucopyranoside		Bianco A. et al., 1998; Della Greca M. et al., 2004
2-(3,4-Dihydroxyphenyl) ethanol 4β-D-glucopyranoside		Bianco A. et al., 1998; Della Greca M. et al., 2004
1-*O*-(2-(3,4-Dihydroxy)-phenylethyl)-(3,4-dihydroxy)phenyl-1,2-ethandiol		Della Greca M. et al., 2000, 2004
D(+)-Erythro-1-(4-hydroxy-3-methoxy)-phenyl-1,2,3-propantriol		Della Greca M. et al., 2004
Ferulic acid	3-(4-Hydroxy-3-methoxyphenyl)-2-propenoic acid; 4-hydroxy-3-methoxycinnamic acid; 3-methoxy-4-hydroxycinnamic acid; caffeic acid 3-methyl ether	Vásquez-Roncero A. et al., 1974b; Martínez-Nieto L. et al., 1993; Rodriquez M.M. et al., 1988; Servili M. and Montedoro G.F., 1989; Pérez J. et al., 1992; Sayadi S. et al., 2000; Della Greca M. et al., 2001, 2004
Gallic acid	3,4,5-Trihydroxybenzoic acid	Aramendia M.A. et al., 1996

Homovanillic alcohol	Homovanillyl achohol 2-(4-hydroxy-3-methoxy) phenylethanol; 4-hydroxy-3-methoxyphenethyl alcohol; 4-(2-hydroxyethyl)guaiacol; 2-(4-hydroxy-3-methoxyphenyl)ethanol; 4-(2-hydroxyethyl)-2-methoxyphenol	Della Greca M. et al., 2001, 2004
4-Hydroxybenzoic acid	*p*-Hydroxybenzoic acid	Balice V. and Cera O., 1984; Cichelli A. and Solinas M., 1984; Servili M. and Montedoro G.F. 1989; Aramendia M.A. et al., 1996; Visioli F. et al., 1999; Servili M. et al., 1999b; Della Greca, 2004
4-Hydroxyphenylacetic acid	*p*-Hydroxyphenylacetic acid	Knupp G. et al., 1996; Aramendia M.A. et al., 1996; Della Greca M. et al., 2001, 2004
1-*O*-(2-(4-Hydroxy) phenylethyl)-(3,4-dihydroxy)phenyl-1, 2-ethandiol		Della Greca M. et al., 2000
p-Hydroxyphenethyl-β-D-glucopyranoside		Limiroli R. et al., 1996; Della Greca M. et al., 2004
Hydroxytyrosol	2-(3,4-Dihydroxyphenyl)ethanol; 4-(2-hydroxyethyl)benzene-1,2-diol; 3,4-DHPEA	Vásquez-Roncero A. et al., 1974b; Pompei C. and Codovilli F., 1974; Capasso R. et al., 1994a; Knupp G. et al., 1996; Aramendia M.A. et al., 1996; Visioli F. et al., 1999; Servili M. et al., 1999b; Ceccon L. et al., 2001; Della Greca M. et al., 2001, 2004; Allouche N. et al., 2004
4-Methylcatechol	3,4-Dihydroxytoluene; homocatechol	Capasso R. et al., 1992a, b; Vinciguerra V. et al., 1993
Oleuropein		Vásquez-Roncero A. et al., 1974b; Servili M. et al., 1999b; Mulinacci N. et al., 2001
Protocatechuic acid	3,4-Dihydroxybenzoic acid	Balice V. and Cera O., 1984; Cichelli A. and Solinas M., 1984; Knupp G. et al., 1996; Aramendia M.A. et al., 1996; Della Greca M. et al., 2001, 2004

(*continued*)

Table 2.7. Simple and complex phenols identified in OMWW — Cont'd

Phenolic compounds	Synonyms	References
Resorcinol	1,3-Benzenediol; *m*-dihydroxybenzene; resorcin	Capasso R. et al., 1992a, b; Vinciguerra V. et al., 1993; Della Greca M. et al., 2004
Sinapic acid	sinapinic acid; 3,5-dimethoxy-4-hydroxycinnamic acid; 4-hydroxy-3,5-dimethoxycinnamic acid	Della Greca M. et al., 2001, 2004
Syringic acid	4-Hydroxy-3,5-dimethoxybenzoic acid; 3,5-dimethoxy-4-hydroxybenzoic acid; suringic acid	Balice V. and Cera O., 1984; Cichelli A. and Solinas M., 1984; Servili M. and Montedoro G.F., 1989; Ramos-Cormenzana A. et al., 1996; Aramendia M.A. et al., 1996; Della Greca M. et al., 2001, 2004
3,4,5-Trimethoxybenzoic acid		Balice V. and Cera O., 1984; Cichelli A. and Solinas M., 1984
Tyrosol	*p*-Hydroxy phenyl ethanol; 4-hydroxy phenylethanol; 1-hydroxy-2-(4-hydroxyphenyl) ethane; *p*-HPEA	Pompei C. and Codovilli F., 1974; Servili M. and Montedoro G.F., 1989; Capasso R. et al., 1992a, b, 1994a; Visioli F. et al., 1995a, 1999; Ramos-Cormenzana A. et al., 1996; Knupp G. et al., 1996; Aramendia M.A. et al., 1996; Servili M. et al., 1999b; Ceccon L. et al., 2001; Della Greca M. et al., 2001, 2004; Allouche N. et al., 2004
Vanillic acid	4-Hydroxy-3-methoxybenzoic acid; 3-methoxy-4-hydroxybenzoic acid; protocatechuic acid 3-methyl ester; 4-hydroxy-meta-anisic acid; para-vanillic acid	Balice V. and Cera O., 1984; Cichelli A. and Solinas M., 1984; Knupp G. et al., 1996; Ramos-Cormenzana A. et al., 1996; Aramendia M.A. et al., 1996; Visioli F. et al., 1999; Servili M. et al., 1999b; Lesage-Meessen L. et al., 2001; Bianco A. et al., 2003; Della Greca M. et al., 2001, 2004
Veratric acid	3,4-Dimethoxybenzoic acid; 3,4-dimethylprotocatechuic acid	Balice V. and Cera O., 1984; Cichelli A. and Solinas M., 1984
Verbascoside		Visioli F. et al., 1995a; Servili M. et al., 1999b; Mulinacci N. et al., 2001; Romero Barranco C. et al., 2002

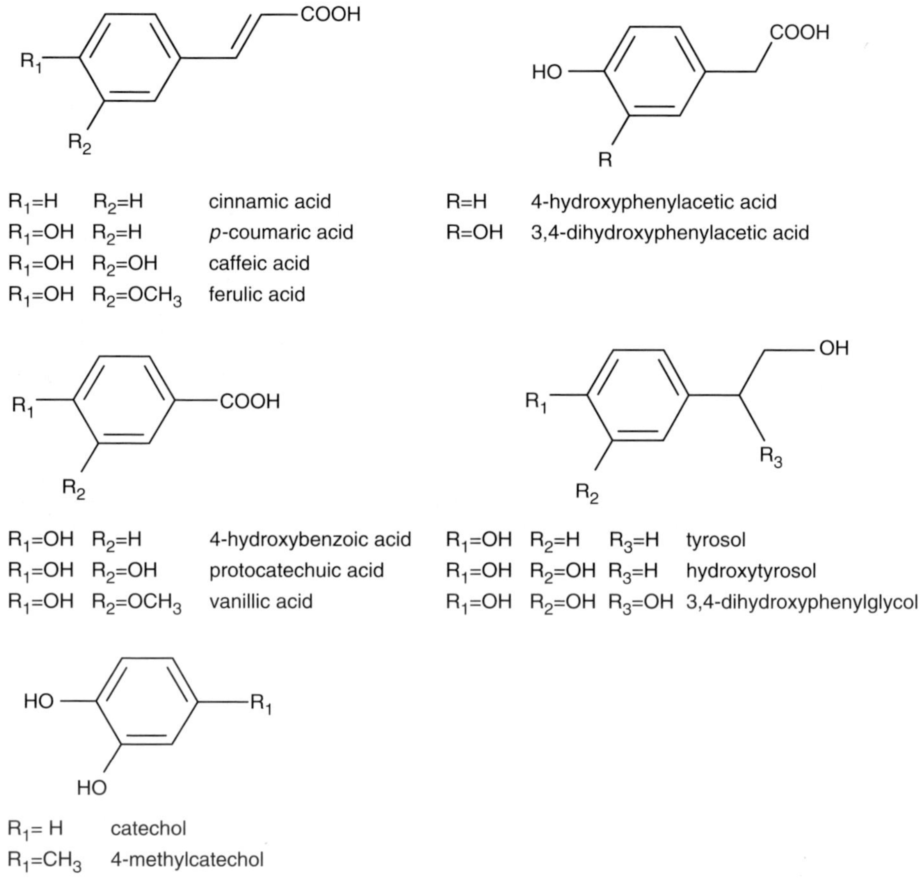
COOH
R1
R2
R1=H R2=H cinnamic acid
R1=OH R2=H p-coumaric acid
R1=OH R2=OH caffeic acid
R1=OH R2=OCH3 ferulic acid
COOH
HO
R
R=H 4-hydroxyphenylacetic acid
R=OH 3,4-dihydroxyphenylacetic acid
R1
COOH
R2
R1=OH R2=H 4-hydroxybenzoic acid
R1=OH R2=OH protocatechuic acid
R1=OH R2=OCH3 vanillic acid
OH
R1
R3
R2
R1=OH R2=H R3=H tyrosol
R1=OH R2=OH R3=H hydroxytyrosol
R1=OH R2=OH R3=OH 3,4-dihydroxyphenylglycol
HO
R1
HO
R1= H catechol
R1=CH3 4-methylcatechol

Fig. 2.3. Phenolic compounds of OMWW described in the literature.

Tyrosol could also be of particular interest, considering that it can be obtained in crystals from OMWW by a simple chromatographic procedure (Capasso R. et al., 1992a). Tyrosol and its derivatives, on the other hand, are substantially resistant to air/oxygen, bacterial and enzymatic degradation, and are of a highly polluting nature.

From the quantitative and qualitative point of view, remarkable differences are noted among the results reported by different authors when studying the phenolic acid composition of OMWW. Thus, whereas some authors emphasize caffeic acid as being especially abundant (Vásquez-Roncero A. et al., 1974b), other authors do not detect large amounts of this compound (Balice V. and Cera O., 1984; Rodríguez M.M. et al., 1988). Veratric acid despite been described in the literature as a typical OMWW compound (Balice V. and Cera O., 1984) it could not be identified by other researchers (Knupp G. et al., 1996). While 4-hydroxybenzoic acid, 4-hydroxyphenylacetic acid, and protocatechuic acid were clearly identified in the extracts of Italian OMWW, they could only be found in trace amounts in Spanish OMWW (Knupp G. et al., 1996). Therefore, because OMWW has a highly variable composition, all the qualitative and quantitative evaluations of its various compounds should be treated with caution.

OMWW can contain, under certain conditions, small amount of oleuropein, demethyloleuropein, and verbascoside (Servili M. et al., 1999b). The chemical structures of demethyloleuropein (1) and verbascoside (2) are given in Fig. 2.4. Oleuropein and demethyloleuropein are the predominant secoiridoids of olive fruit, which also contains verbascoside (a derivative of caffeic acid) (Ryan D. et al., 1998, 1999; Servili M. et al., 1999a). In unripe green olives, oleuropein is present as the major *o*-diphenolic compound, while in ripe olives, demethyloleuropein predominates (Lo Scalzo R. and Scarpati M.L., 1993). OMWW was found to contain high concentrations of secoiridoid derivatives, such as 3,4-DHPEA-EDA — the dialdehydic form of elenolic acid (EDA) linked to 3,4-dihydroxyphenylethanol (3,4-DHPEA or hydroxytyrosol) (Lo Scalzo R. and Scarpati M.L., 1993; Servili M. et al., 1999b). In fact, the secoiridoid aglycones, such as 3,4-DHPEA-EDA, are generated by hydrolysis of oleuropein and demethyloleuropein during the olive oil mechanical process (Amiot M.J. et al., 1986, 1989). Oleuropein and its derivatives are readily degraded into breakdown products (e.g. upon exposure to air/oxygen, certain enzymes, or bacteria) that are substantially non-polluting and non-toxic. It is estimated that during the processing for the extraction of olive oil, almost 80% of all

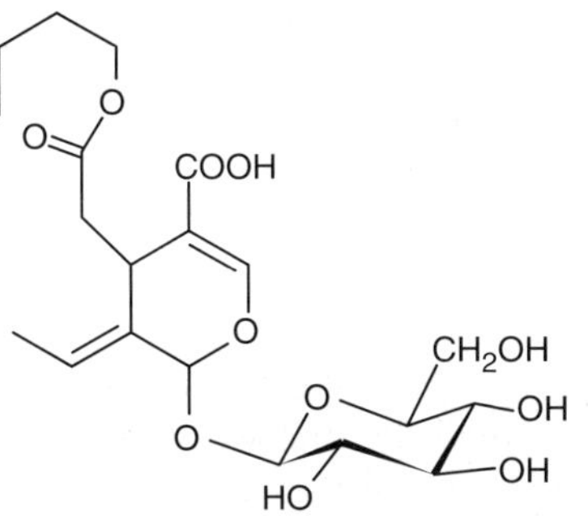

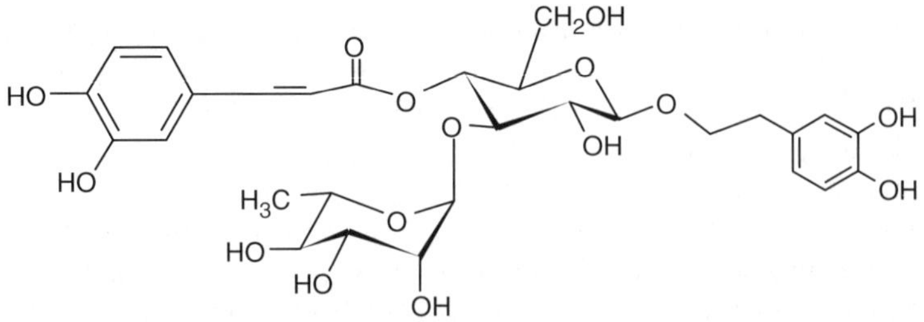

Fig. 2.4. Chemical structure of demethyloleuropein (1) and verbascoside (2).

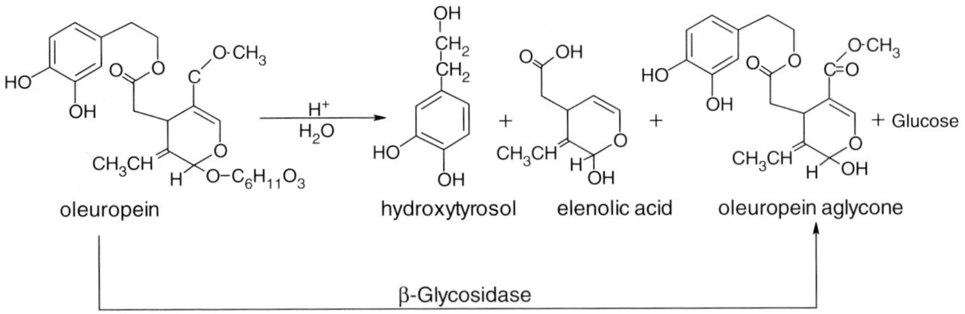

Fig. 2.5. The structure of oleuropein and its hydrolysis products.

oleuropein is degraded upon crushing the olives (Vásquez-Roncero A. et al., 1976). The structure of oleuropein and its hydrolysis products are given in Fig. 2.5.

Various low-molecular weight phenols were isolated and identified by ultrafiltration, nanofiltration, and reverse osmosis fractionation of OMWW from a Ligurian mill (Della Greca M. et al., 2001). Along with catechol, tyrosol, and hydroxytyrosol, the following aromatic compounds from the reverse osmosis fraction were also confirmed as constituents of OMWW: 4-hydroxybenzoic acid, protocatechuic acid, vanillic acid, 4-hydroxy-3,5-dimethoxybenzoic acid, 4-hydroxyphenylacetic acid, 3,4-dihydroxyphenylacetic acid, 2-(4-hydroxy-3-methoxy)phenylethanol, 2-(3,4-dihydroxyphenyl)-1,2-ethandiol, *p*-coumaric acid, caffeic acid, ferulic acid, sinapic acid, 1-*O*-(2-(3,4-dihydroxy)phenylethyl)-(3,4-dihydroxy)phenyl-1,2-ethandiol, 1-*O*-(2-(4-hydroxy)phenylethyl)-(3,4-dihydroxy)phenyl-1,2-ethandiol, D(+)-erythro-1-(4-hydroxy-3-methoxy)-phenyl-1,2,3-propantriol, *p*-hydroxyphenethyl- β-D-glucopyranoside, 2-(3,4-dihydroxyphenyl)ethanol 3β-D-glucopyranoside, and 2-(3,4-dihydroxyphenyl)ethanol 4β-D-glucopyranoside. A later study performed on the nanofiltered fraction of the same effluent isolated and characterized further low molecular weight phenols from OMWW. A new lignan: 1-hydroxy-2-(4-hydroxy-3-methoxyphenyl)-6-(3-acetyl-4-hydroxy-5-methoxyphenyl)-3,7-dioxabicyclol [3.3.0] octane, the secoiridoid: 2H-pyran-4-acetic acid, 3-hydroxymethyl-2,3-dihydro-5-(methoxycarbonyl)-2-methyl-, methyl ester, the phenylglycoside: 4-(β-D-xylopyrano-syl-(1→6))-β-D-glucopyranosyl-1,4-dihydroxy-2-methoxybenzene and the lactone: 3-[1-(hydroxymethyl)-1-propenyl]-δ-glutarolactone were isolated and identified for the first time in OMWW. The structures were established on the basis of spectroscopic data including two-dimensional NMR, as components of OMWW (Della Greca M. et al., 2004).

OMWW contains also relatively high concentrations of flavonoids. The flavonoids are polyphenolic compounds possessing 15 carbon atoms; two benzene rings joined by a linear three-carbon chain — see Fig. 2.6.

The main flavonoids detected in OMWW are anthocyanins (cyanidin-3-glucoside, cyanidin-3-rutinoside), flavones (luteolin, luteolin-7-glucoside, apigenin, apigenin-7-glucoside), and flavonols (quercetin, rutin) (Durán Barrantes M.M., 1990; Servili M. et al., 1999b) — see Table 2.8.

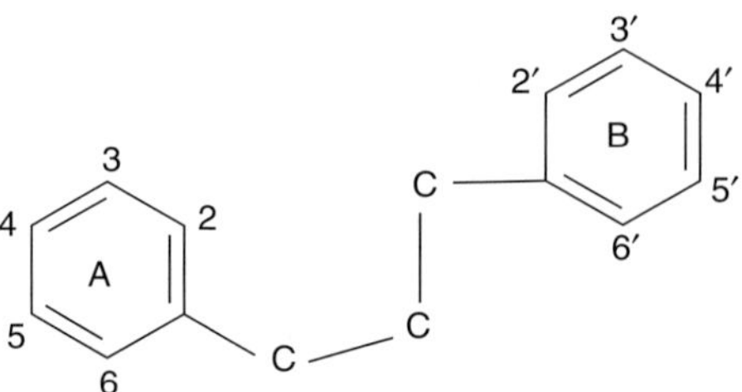

Fig. 2.6. Generic structure of flavonoids. The skeleton can be represented as the $C_6 - C_3 - C_6$ system.

The phenolic compounds contained in OMWW were classified by Hamdi roughly in two groups (Hamdi M., 1992). The phenolic compounds of the first group contain simple phenolic compounds, not autooxidated tannins (of low molecular weight), and flavonoids. The main phenolic acids are syringic acid, 4-hydroxyphenylacetic acid, vanillic acid, veratric acid, caffeic acid, protocatechuic acid, *p*-coumaric acid, and cinnamic acid (Balice V. and Cera O., 1984; Cichelli A. and Solinas M., 1984). The polyphenols of the second group, which contain darkly colored polymers, result from the polymerization and autooxidation of phenolic compounds of the first group. The color of OMWW depends on the ratio between the two groups of polyphenols. It was observed that OMWW becomes blacker when it has been stored for some time. This change in color might be a result of the oxidation and subsequent polymerization of tannins giving darkly colored polyphenols (Hamdi M., 1992). Ragazzi E. et al. (1967) found that the major component in the colored fraction is a molecule of polymeric nature derived from several low molecular phenolic compounds, while Pérez J. et al. (1987) showed that it is chemically related to lignin and humic acids. This lignin-like macromolecule is probably produced by enzyme-initiated dehydrogenative polymerization of phenols detected in the wastes, as has been suggested by Saiz-Jiménez C. et al. (1986). Recently, a dark polymeric pigment, named polymerin, was isolated from OMWW and characterized. The pigment can be considered as a polyphenolglycosilated polymer whose aromatic core seems to be very similar to a vegetable catecholmelanine pigment (Capasso R. and Martino de, 1998; Capasso R. et al., 2000). Polymerin comprised carbohydrate (52.40%), melanin (26.14%), protein (10.40%), and minerals (11.06%), mainly K^+ but also Na^+, Ca^{2+}, Mg^{2+}, Zn^{2+}, Fe^{3+} and Cu^{2+}, bound via carboxylate anions and other nucleophilic functional groups present. Distribution of polymerin's relative molecular weight was shown by gel filtration to be between approximately 500 and 2 kDa. Results indicated that one fraction consists of a protein, melanin, and polysaccharide aggregate, bound together in a supramolecular structure via covalent and hydrogen bonding, while another comprised free polysaccharide only (Capasso R. et al., 2002a). Polymerins could be used in agriculture as bioamendments, macro- and micro-element biointegrators and, due to their similarity with humic acids, as a biofilter for toxic metal removal — see Chapter 3: "Environmental effects", section: "Leaching" and Chapter 10: "Uses", section: "Use as fertilizer/soil conditioner".

Table 2.8. The main flavonoids detected in OMWW

Anthocyanins		Durán Barrantes M.M. et al., 1990
Apigenin	2-(*p*-Hydroxyphenyl)-5,7-dihydroxychromone; 5,7-dihydroxy-2-(4-hydroxyphenyl)-4H-1-benzopyran-4-one; 5,7-dihydroxy-2-(4-hydroxyphenyl)-chromen-4-on; 4H-1-benzopyran-4-one, 5,7-dihydroxy-2-(4-hydroxyphenyl)-(9CI); 4′,5,7-trihydroxyflavone; apigenol; chamomile; spigenin; apigeninidin; versulin	Durán Barrantes M.M. et al., 1990
Apigenin-7-glucoside		Durán Barrantes M.M. et al., 1990
Cyanidin		Durán Barrantes M.M. et al., 1990
Cyanidin-3-glucoside	Chrysanthemin; asterin; kuromanin	Durán Barrantes M.M. et al., 1990
Cyanidin-3-rutinoside		Durán Barrantes M.M. et al., 1990
Flavones		Durán Barrantes M.M. et al., 1990
Luteolin	Digitoflavone; cyanidenon 1470	Vásquez-Roncero A. et al., 1974b; Durán Barrantes M.M. et al., 1990; Visioli F. et al., 1999
Luteolin-7-glucoside	Cynaroside	Durán Barrantes M.M. et al., 1990; Visioli F. et al., 1999; Servili M. et al., 1999b
Quercetin	Meletin; sophoterin; cyanidenolon 1522	Vásquez-Roncero A. et al., 1974b; Durán Barrantes M.M. et al., 1990; Visioli F. et al., 1999
Rutin	Birutan; rutoside; 3,3′,4′,5,7-pentahydroxy-flavone; phytomelin; eldrin; quercetin-3-rutinoside; ilixathin; melin; myrticolorin; sophorin; osyritrin	Servili M. et al., 1999b; Romero Barranco C. et al., 2002

A more recent study demonstrated, by using gel filtration, the presence of three phenolic fractions in OMWW, namely, a high molecular mass fraction ($M > 250$ kDa), a medium molecular mass fraction ($M = 13$ kDa) and monomeric phenols (Allouche N. et al., 2004).

OMWW was found to contain exploitable quantities of oleanolic acid (3-β-hydroxy-28-carboxyoleanene) and maslinic acid (2-α,3-β-dihydroxy-28-carboxyoleanene) — see Chapter 10: "Uses", section: "Recovery of organic compounds". Both triterpenic acids possess various biological effects; namely, the oleanolic acid has a carcinogenic promoter–inhibitory effect and an effect of promoting wound–healing; the maslinic acid (also known as cratzegolic acid) has an anti-inflammatory and antihistamic effect.

Pesticide residues used for the control of *Bactrocera oleae* (Gmelin) (Insecta: Diptera: Tephritidae) were detected in OMWW. A study has been carried out on the presence of pesticide residues (deltamethrin, fenthion, formothion, and their metabolites) on olives, oil, and OMWW, after adulticide and larvicide treatments for the control of *Bactrocera oleae* (formerly *Dacus oleae*) (Leandri A. et al., 1993). Deltamethrin was found to have a faster degradation rate than fenthion, leaving no traces on olives 18 days after treatment. Fenthion, on the contrary, was still present on the drupes at harvest 28 days after the treatment. Formothion was not found in definable quantities 3 days after treatment, while its metabolites, dimethoate, and omethoate were found on olives at harvest in levels of 4–23 ppb.

On the basis of the most recent literature in the field is reasonable to conclude that a detailed knowledge of all compounds to be found in OMWW, as well further studies on their phytotoxicity, is an important step for future developments in solving environmental problems relating to the disposal of OMWW.

Inorganic Compounds Paredes C. et al. (1999b) characterized OMWW and its sludge for agricultural purposes. Ten samples of OMWW taken from different mills in southern Spain and other ten of OMWW sludge from evaporation ponds were analyzed. The aim was to study the composition of these wastes and to find relationships, which would make it possible to use easily determinable parameters to ascertain their composition. Compared with other organic wastes, these materials had high potassium concentration, similar organic matter content, and notable levels of nitrogen, phosphorus, calcium, magnesium, and iron. The highest potassium concentrations were observed in OMWW, while the sludges showed higher levels of the other nutrients, especially iron. The dry matter of OMWW was significantly correlated with most of the parameters studied but, in the sludges, the only correlation was between the ash content and the total organic carbon and total nitrogen concentrations.

Arienzo M. and Capasso R. (2000) analyzed the content, composition, and physicochemical status of metal cations and inorganic anions in raw OMWW processed by pressure as well as centrifuge. Table 2.9 shows the concentration values of the metal cations (C_{c1} and C_{c2}) and inorganic anions (C_{a1} and C_{a2}) in OMWW obtained from both processing systems. In the case of OMWW obtained by pressure,

Table 2.9. Concentrations of cations (C_{c_1} and C_{c_2}) and anions (C_{a_1} and C_{a_2}) in OMWW obtained by pressure (OMWWP) and centrifuge (OMWWC), respectively (Arienzo M. and Capasso R., 2000)

Cations (g/l)			Anions (g/l)		
Cation	OMWWP (C_{c_1})	OMWWC (C_{c_2})	Anion	OMWWP (C_{a_1})	OMWWC (C_{a_2})
K^+	17.10 ± 0.31	9.80 ± 0.12	Cl^-	1.63 ± 0.06	1.3 ± 0.02
Mg^{2+}	2.72 ± 0.13	1.65 ± 0.11	$H_2PO_4^-$	1.07 ± 0.06	0.85 ± 0.04
Ca^{2+}	2.24 ± 0.14	1.35 ± 0.010	F^-	0.57 ± 0.01	0.530 ± 0.01
Na^+	0.40 ± 0.17	0.162 ± 0.08	SO_4^-	0.53 ± 0.05	0.420 ± 0.02
Fe^{2+}	0.129 ± 0.05	0.033 ± 0.01	NO_3^-	0.023 ± 0.01	0.0109 ± 0.008
Zn^{2+}	0.0630 ± 0.001	0.0301 ± 0.001			
Mn^{2+}	0.0147 ± 0.001	0.0091 ± 0.001			
Cu^{2+}	0.0086 ± 0.001	0.0098 ± 0.001			

K was the predominant metal (17.10 g/l) followed in decreasing order by Mg (2.72 g/l), Ca (2.24 g/l), Na (0.40 g/l), Fe (0.129 g/l), Zn (0.063 g/l), Mn (0.0147 g/l), and Cu (0.0086 g/l). Lower concentration levels of cations were detected in OMWW samples obtained by centrifuge due to the dilution of the water during the centrifugal processing of the olive oil. With regard to the anions, the prevailing anion proved to be Cl^- followed by the biacid phosphate H_2PO_4, which was in this form as a consequence of the acid $pH = 5.1$. In OMWW samples obtained by pressure, the anions F^- and SO_4^- presented very similar concentrations, whereas in the OMWW samples obtained by centrifuge the concentrations of the same anions were slightly different. With respect to the other anions, NO_3^- ions were present at very low concentrations in both kinds of wastewater. Most of the metal cations found to be bound to the organic polymeric fraction. The organic polymeric fraction is composed of polysaccharides, polymeric polyphenols, and proteins to which K and Na are essentially bound by single electrostatic bonding, whereas all other ions are more strongly bound, even under chelated form by means of anionic functional groups and/or having Lewis base properties natively bound to said polymer. This trend was consistent with the chemical nature of the metals analyzed. In fact, except for K and Na, all of the remaining cations are bivalent and possess strong chelating properties. Na is more reactive toward the negative sites of the polymer than K. The relative abundance of charged sites of the organic polymeric fraction could explain the consistent binding of K to the polymer. The relative molecular weight was substantially estimated in the range between 1000 and 30,000 for $\sim$75% and in the range from 30,000 to 100,000 for $\sim$25%. The free residual cations proved to be neutralized by the inorganic counter anions.

The copper content in OMWW from a mill in Alcala la Real, Spain was determined to be 0.36 mg Cu/l (WO9211206, 1992). In the reductive environment of OMWW copper is present in monovalent form as crystals of copper oxide

Table 2.10. Characteristic parameters of OMWW (Fiestas
Ros de Ursinos J.A., 1986a; Martínez J. et al., 1986)

pH	4.5–6
EC_{25} (dS/m)	8–22
BOD_5 (g/l)	35–100
COD (g/l)	40–195
Lipids (g/l)	0.3–23
Organic matter (g/l)	40–165
Mineral matter (g/l)	5–14
Polyphenols (mg/l)	3,000–24,000
N (g/l)	5–15
P (g/l)	0.3–1.1
K (g/l)	2.7–7.2
Ca (g/l)	0.12–0.75
Mg (g/l)	0.10–0.40
Na (g/l)	0.04–0.90
Solids (%)	5.5–17.6

(cuprite Cu_2O). Cuprite is insoluble in water and has a red color. The red color, which may be observed on the surface of OMWW lagoons, may be caused by cuprite. In addition to the fact that copper is used against fungus attacks, copper is also toxic towards algae and other lower vegetation, but in concentrations of one-tenth of what has been found in OMWW. Fertilizer being used on trees is the real cause of copper presence in OMWW and its content is enriched in the bottom sludge of a lagoon. If this sludge is used as fertilizer it may poison the soil. The cause is claimed to be the contents of polyphenols in OMWW. But, polyphenols are produced in nature and is a natural conserving agent with a temporary toxicity, whereas the toxic action of copper is permanent.

The general characteristics of OMWW are shown in Table 2.10 -cf. Tables 2.5, 2.6, 2.9. The essential properties of OMWW depend on the process and the quantity of the added water. They range generally in the following limits:

- pH = 4–6;
- BOD_5 = 35–110 g/l;
- COD = 40–220 g/l;
- TOC = 25–45 g/l;
- toxic compounds are phenols, tannins, and dyes;
- phenolic compounds are present in OMWW at concentrations in the range from 0.5 to 24 g/l;
- the phenols comprise at least 30 compounds;
- the sugars constitute up to 60% of the dry substance and comprise, in decreasing amount, fructose, mannose, glucose, and saccharose;
- potassium is the predominant inorganic material (~ 4 g/l) and is a very important nutritional compound;

- one ton of processed olives produced a polluting load equivalent to that of 50–100 inhabitants; the average BOD_5 concentration of undiluted OMWW is 120–150 kg/m^3 and the dilution of OMWW with processing waters does not affect substantially the polluting load.

Microbial Content of OMWW OMWW contains a variable and high number of bacteria and fungi. Among the bacterial strains identified are several species of *Acinetobacter*, *Pseudomonas*, and *Enterobacter*. However, much of the microbial activity is represented by 71 strains, showing different metabolic patterns. The pathogenic *Klebsiella pneumoniae* ss *pneumoniae* has also been isolated from untreated and treated OMWW (EU project: AIR3-CT94-1987).

Seven aerobic bacterial strains were isolated from an Italian OMWW (Di Gioia D. et al., 2002). The results of the 16S rDNA restriction analysis demonstrated that these strains are distributed among four different groups. One strain of each group was taxonomically characterized by sequencing the amplified 16S rDNA and the four strains were assigned to the genera *Comamonas*, *Pseudomonas*, *Ralstonia* and *Sphingomonas*. The isolated bacterial strains exihibited a biodegradation potential towards the monocyclic aromatic compounds of OMWW. Millan B. et al. (2000) studied the microbial composition of OMWW from four disposal ponds. Among the fungal members, 12 different genera (*Acremonium*, *Alternaria*, *Aspergillus*, *Chalara*, *Fusarium*, *Lecytophora*, *Paecilomyces*, *Penicillium*, *Phoma*, *Phycomyces*, *Rhinocladiella*, and *Scopulariopsis*) were found. Members of five genera (*Chalara*, *Fusarium*, *Paecilomyces*, *Penicillium*, and *Scopulariopsis*) were widely distributed, and were able to grow efficiently in undiluted OMWW as a sole source of nutrients. Strains of *Fusarium*, *Paecilomyces*, *Penicillium*, and *Scopulariopsis* showed a marked capacity for OMWW detoxification, depleting its antibacterial activity almost completely.

Knupp G. et al. (1996) evaluated the problems of identifying phenolic compounds during the microbial degradation of OMWW. The purification of OMWW was carried out by biodegrading phenolic compounds and the metabolites were investigated during fermentation prior to its safe disposal. In addition to the well-known compounds, 3,4-dihydroxyphenyl glycol was also identified in untreated Spanish and Italian OMWW samples using a gas chromatography-mass spectrometry (GC-MS) method. The qualitative composition of the Italian and Spanish samples differs. First results of degradation tests of reference substances showed that *Arthrobacter* was capable of completely transforming added tyrosol to the corresponding 4-hydroxyphenylacetic acid while no traces of tyrosol could be identified after 139 h of fermentation. In contrast, only traces of phenylacetic acid were produced by *Bacillus pumilus* after 139 h of fermentation of tyrosol.

Antimicrobial Activity of OMWW The antimicrobial effect of OMWW has been described by many authors (Caro N.de and Ligori C.N., 1959; Ramos-Cormenzana A. et al., 1996) and this activity has been related to its phenolic content, which constitutes from 0.5 to 24 g/l, and is strictly dependent on the processing system used for olive oil production (Ragazzi E. and Veronese G., 1967a,b; Ragazzi E. et al., 1967a,b; Sorlini C. et al., 1986; Borja-Padilla R. et al., 1990b).

Several phenolic compounds were detected in n-propanol extracts of OMWW that had bactericidal effects on *Bacillus megaterium* ATCC 33085, inhibiting sporulation and germination at 5.6 mmol/l total phenolics (expressed as syringic acid). The biological effect was increased in the presence of high glucose and NaCl concentrations and after β-glucosidase hydrolysis (Rodríguez M.M. et al., 1988). Diluted OMWW solutions have been shown to decrease counts in sporulated soil bacteria. However, contradictory results were also obtained concerning the nature of the substances responsible for the antibacterial activity of OMWW, the inhibitory effect of the phenolic acids found in OMWW and of the fatty acids present in olive oil (González-López J. et al., 1994). The antibacterial activity of the phenolic acids (tested separately or in mixtures at the started concentrations) when they were tested against *B. megaterium* and against a collection of bacteria isolated from unpolluted soil and OMWW polluted soil, did not coincide with the inhibitory effect of the OMWW. On the other hand, although antibacterial activity has not been detected in olive oil, its fatty acids (linoleic, oleic, linolenic, lauric, and myristic acid) are capable of inhibiting the growth of *B. megaterium*.

OMWW constituents were more effective on bacteria than yeast (Moreno E. et al., 1987, 1990), whereas the antibacterial activity was higher on Gram-positive than on Gram-negative bacteria. Ethyl acetate and *n*-propanol extracts were the most active against *B. megaterium*. Propanol was the solvent of choice to extract the antibacterial phenols from OMWW (Rodríguez M.M. et al., 1988), although the propanolic extract was less active than OMWW itself (Moreno E. et al., 1987, 1990).

OMWW is highly toxic to both phytopathogenic *Pseudomonas syringae* pv. *savastanoi* (Gram-negative) and *Corynebacterium michiganense* (Gram-positive) and shows bactericidal activity in its original concentration (in raw form) (Capasso R. et al., 1995). Among the main polyphenols, present in the wastewater, 4-methylcatechol proved to be the most toxic to *P. syringae* pv. *savastanoi* at 10^{-4} mol/l, and also demonstrate bactericidal activity, while on *C. michiganense* it is only slightly active; catechol and hydroxytyrosol are less active on *P. syringae* pv. *savastanoi*, but inactive on *C. michiganense*; tyrosol and its synthetic isomers 1,2- and 1,3-tyrosol are completely inactive on both bacteria. Among the derivatives of polyphenols considered, acetylcatechol and guaiacol (*o*-methoxyphenol) are selectively toxic for *P. syringae* pv. *savastanoi*, while *o*-quinone is strongly toxic for both bacteria. The minor carboxylic polyphenols of OMWW at 10^{-4} mol/l are all inactive on the bacteria. In addition, OMWW, catechol, 4-methylcatechol, and the less abundant carboxylic polyphenols proved to be toxic on Hep2 human cells. Capasso R. et al. (1995) investigated further the possibility of using the active polyphenols in agriculture in an integrated pest management program for the protection of the olive plant — see Chapter 10: "Uses", section: "Use as herbicide/pesticide".

OMWW and its phenolic extracts showed deterrent action at high concentrations to oviposition by *Dacus oleae* (Bactrocera oleae) (Gmelin) females (Capasso R. et al., 1994b). Catechol was found to be the most potent repulsive phenol, whereas tyrosol and hydroxytyrosol were inactive — Chapter 10: "Uses", section: "Use as herbicide/ pesticide".

Olive Cake

The chemical composition of olive cake (*pomace*) produced by olive-mills varies within very large limits according to type, condition, and origin of olives as well as to olive oil extraction process — see Table 2.11.

Crude olive cake (*orujo*) obtained by pressure contains crushed stones, skin, pulp, water (~25%), and a remaining quantity of oil (4.5–9%). Crude fat (CF) and neutral detergent fiber (NDF) are the most variable components. Lignin content is particularly high. Crude protein content (CP) is generally low, and a substantial part is linked to cell wall components. Aminoacid composition is similar to that of barley grain with a deficit in glutamic acid, proline, and lysine (Nefzaoui A. et al., 1985).

The exhausted olive cake (*orujillo*) is a dry material (8–10% moisture) composed of ground olive stones and pulp. The exhausted olive cake has a high lignin, cellulose, and hemicellulose content.

Steam explosion has been used to recover the main components of olive cake. The various water-soluble non-carbohydrate compounds generated during steam explosion, such as sugar degradation compounds (furfural and hydroxymethylfurfural), lignin degradation compounds (vanillic acid, syringic acid, vanillin, and syringaldehyde), and the simple phenolic compounds characteristic of olive fruit (tyrosol and hydroxytyrosol), were identified. The amount of hydroxytyrosol solubilized was higher than that of the other compounds, and increased with increasing steaming temperature and time (Fernández-Bolaños J. et al., 1998).

Table 2.11. Characteristics of olive cakes (Vlyssides A.G. et al., 1998)

Parameter	Pressure system	3-Phase system	2-Phase system
Moisture %	27.21 ± 1.048	50.23 ± 1.935	56.80 ± 2.188
Fats and oils %	8.72 ± 3.254	3.89 ± 1.449	4.65 ± 1.736
Proteins %	4.77 ± 0.024	3.43 ± 0.017	2.87 ± 0.014
Total sugars %	1.38 ± 0.016	0.99 ± 0.012	0.83 ± 0.010
Cellulose %	24.14 ± 0.283	17.37 ± 0.203	14.54 ± 0.170
Hemicellulose %	11.00 ± 0.608	7.92 ± 0.438	6.63 ± 0.366
Ash %	2.36 ± 0.145	1.70 ± 0.105	1.42 ± 0.088
Lignin %	14.18 ± 0.291	0.21 ± 0.209	8.54 ± 0.175
Kjeldahl Nitrogen %	0.71 ± 0.010	0.51 ± 0.007	0.43 ± 0.006
Phosphorous as P_2O_5 %	0.07 ± 0.005	0.05 ± 0.004	0.04 ± 0.003
Phenolic compounds %	1.146 ± 0.06	0.326 ± 0.035	2.43 ± 0.15
Potassium as K_2O %	0.54 ± 0.045	0.39 ± 0.033	0.32 ± 0.027
Calcium as CaO %	0.61 ± 0.059	0.44 ± 0.043	0.37 ± 0.036
Total Carbon %	42.90 ± 3.424	29.03 ± 2.317	25.37 ± 2.025
C/N ratio	60.79 ± 5.352	57.17 ± 5.033	59.68 ± 5.254

Felizon B. et al. (2000) applied also steam-explosion under different steam conditions, followed by fractionation to separate the main components of olive cake. In the water-soluble fraction, the main compounds were carbohydrates. Glucose represented a significant part of the total monosaccharide content, especially under conditions of mild severity, followed by arabinose, but the solubilization of sugars occurred predominantly in the oligomeric fraction. Mannitol was also found in significant amounts (1.5%), similar to that in the initial material. In the ethyl acetate extract, low molecular weight phenols were identified, the most abundant being hydroxytyrosol, which is present in the olive pulp. Hydroxytyrosol is abundant and has great antioxidant activity, reaching $149\,mg/100\,g$ of dry olive cake. The procedure used in this study obtained all the hydroxytyrosol residual present in the by-product. The constitutive polymers were quantified in the insoluble fraction, and the sugar composition showed that cellulose was associated with a high proportion of xylans and other polysaccharides rich in arabinose and galactose. This cellulose was nearly amorphous, as it was highly susceptible to hydrolytic enzymes. The extractables in dilute alkali (not true lignins) increased as steaming became more severe; the residual "lignin" in this fraction decreased. Enzymatic hydrolysis of the insoluble fraction using a cellulolytic complex was also studied. The slight increase in the extent of saccharification was not proportional to the high alkaline delignification. However, when the residues were efficiently delignified with chlorite treatment, the susceptibility to enzymatic hydrolysis greatly increased.

Two-Phase Olive-Mill Waste (2POMW)

The characteristics of 2POMW are obviously very different from those of olive cake coming from press systems and three-phase centrifuges. 2POMW is a thick sludge that contains pieces of stone and pulp of the olive fruit as well as vegetation water. It has a moisture content in the range of 55–70%, while traditional olive cake has a moisture content around 20–25% in press systems and 40–45% in three-phase centrifuges (Alba-Mendoza J. et al., 1990; Alburquerque J.A. et al., 2004). It contains also some residual olive oil (2–3%) and 2% ash with 30% of potassium content. The greater moisture, together with the sugars and fine solids that in the three-phase system were contained in the vegetation water, give 2POMW a doughy consistency and make transport, storage, and handling difficult — it cannot be piled and must be kept in large ponds.

2POMW is rich in K, which is a common characteristics in olive-mill wastes. However, 2POMW is poor in P, Ca, and Mg, compared to municipal solids wastes and sewage sludges, although similar to other vegetable wastes and manures in this respect — see Table 2.12. Furthermore, 2POMW contains an intermediate level of nitrogen, between those of OMWW and OMWW sludge, most of which is organic. The main micro-nutrient is Fe while Cu, Mn, and Zn levels are lower than that of Fe. Except in the case of K, both the macro- and micro-nutrient content is lower in 2POMW than in most manures and other organic soil amendments. The main organic constituents of 2POMW are lignin, hemicellulose,

Table 2.12. Main characteristics of 2POMW samples (dry weight)
(Alburquerque J.A. et al., 2004)

Parameters	Mean	Range	CV (%)
Moisture (% fresh weight)	64.0	55.6–74.5	7.6
pH[a]	5.32	4.86–6.45	6.6
EC[a] (dS/m)	3.42	0.88–4.76	33.9
Ash (g/kg)	67.4	24.0–151.1	42.5
TOC (g/kg)	519.8	495.0–539.2	2.8
C/N (ratio)	47.8	28.2–72.9	22.1
TN (g/kg)	11.4	7.0–18.4	24.5
P (g/kg)	1.2	0.7–2.2	29.7
K (g/kg)	19.8	7.7–29.7	34.2
Ca (g/kg)	4.5	1.7–9.2	57.3
Mg (g/kg)	1.7	0.7–3.8	58.7
Na (g/kg)	0.8	0.5–1.6	36.6
Fe (g/kg)	614	78–1462	74.9
Cu (g/kg)	17	12–29	28.8
Mn (g/kg)	16	5–39	70.2
Zn (g/kg)	21	10–37	36.3

CV: Coefficient of variation.
[a]Water extract 1:10.

Table 2.13. Main components of the organic fraction of 2POMW samples (dry weight)
(Alburquerque J.A. et al., 2004)

Parameters	Mean	Range	CV (%)
Total organic matter (g/kg)	932.6	848.9–976.0	3.1
Lignin (g/kg)	426.3 [45.8%]	323.0–556.5	16.0
Hemicellulose (g/kg)	350.8 [37.7%]	273.0–415.8	12.7
Cellulose (g/kg)	193.6 [20.8%]	140.2–249.0	14.8
Fats (g/kg)	121.0 [13.0%]	77.5–194.6	28.9
Protein (g/kg)	71.5 [7.7%]	43.8–115.0	24.5
Water-soluble carbohydrates (g/kg)	95.8 [10.1%]	12.9–164.0	50.0
Water-soluble phenols (g/kg)	14.2 [1.5%]	6.2–23.9	41.0

and cellulose — see Table 2.13. Other important organic components are fats, hydrosoluble carbohydrates, and proteins. The high lignin content of 2POMW and the degree of binding of this component to other organic constituents in lignocellulosic materials may hinder the ability of microorganisms and their enzymes to degrade 2POMW, if used as a composting substrate (Alburquerque J.A. et al., 2004).

The most abundant phenolic compounds in 2POMW are tyrosol and hydroxytyrosol (Fernández-Bolaños J. et al., 2002) together with *p*-coumaric, caffeic

Table 2.14. Characteristics of 2POMW (Giannoutsou E. et al., 1997a)

	February 1997–June 1998			June 1998–December 1998	
	2POMW	Stone-free 2POMW	Oil reduced stone-free 2POMW	2POMW	Dried 2POMW (400°C)
pH	5.30	4.87	5.00	5.80	5.80
Ash (%w/w)	7.10	7.65	9.12	ND	ND
Lipids (%w/w)	4.34	7.18	6.38	7.46	12.48
Proteins (%w/w)	13.56	9.44	8.65	14.80	15.96
Sugars (%w/w)	2.31	1.48	1.21	1.30	1.87
Tannins (%w/w)	2.70	2.18	2.61	1.25	1.33
Nitrogen (%w/w)	2.48	2.10	1.96	3.16	3.08
LHV (kcal/kg)	27.61	15.04	22.45	ND	ND

ND: Not determined; LHV: Low heating value.

(Lesage-Meessen L. et al., 2001), and vanillic acid in less quantity. These compounds together with the lipid fraction has been related with the phytotoxic and antimicrobial effects currently attributed to olive-mill wastes.

Several 2POMW samples coming from different treatment processes (fresh, stone-free, deoiled, and dried 2POMW) were examined (Giannoutsou E. et al., 1997). The analysis concerned total sugars, total nitrogen, true protein, total lipids, moisture, total tannins, caloric content, and pH. The results showed that 2POMW has a high content of total ash and lipids, which is understandable taking into consideration that a small quantity of the oil remains in the waste, while tannins, sugars, and total nitrogen occur in lower levels — see Table 2.14.

Olives processed by double extraction of paste and fresh 2POMW, using two two-phase centrifugal decanters not requiring addition of water, gave an average yield of oil of 87% (83.3% from the paste and 3.6% from 2POMW). The quality of the olive oil extracted from the paste was superior to that from 2POMW, although the latter had a higher total phenol content as a result of the heating process (at 60°C) during malaxation. The content of C_5 volatile compounds in the headspace was higher in oils obtained by centrifugation of 2POMW, but the C_6 volatile compound content was lower. In some samples, the percentage of triterpene dialcohols (erythrodiol and uvaol) and contents of waxes, and aliphatic alcohols in oils obtained from 2POMW were above EU limits (Di Giovacchino L. et al., 2002).

Methanol extracts of 2POMW and olive pulp were analyzed by reverse phase HPLC and the eluted fractions were characterized by electrospray ionization mass spectrometry. This technique allowed the identification of some common phenolic compounds, namely, verbascoside, rutin, caffeoyl-quinic acid, luteolin-4-glucoside, and 11-methyl-oleoside. Hydroxytyrosol-1′-β-glucoside, luteolin-7-rutinoside, and oleoside were also detected. Moreover, this technique enabled the identification, for the first time in *Olea europaea* tissues, of two oleoside derivatives, 6′-β-glucopyranosyl-oleoside and 6′-β-rhamnopyranosyl-oleoside, and of 10-hydroxy-oleuropein. Also, an oleuropein glucoside that had previously been identified in olive leaves was now detected in olive fruit, both in olive pulp and 2POMW. With the exception of oleoside and oleuropein, the majority of phenolic compounds were found to occur in equivalent amounts in olive pulp and 2POMW. Oleoside was the main phenolic compound in olive pulp (31.6 mg/g) but was reduced to 3.6 mg/g in 2POMW, and oleuropein (2.7 mg/g in the pulp) almost disappeared (<0.1 mg/g in 2POMW). Both these phenolic compounds were degraded during the olive oil extraction process.

2POMW contains also exploitable amounts of oleanolic acid and maslinic acid — see section: "Olive-mill wastewater (OMWW)" as well, Chapter 10: "Uses", section: "Recovery of organic compounds".

The olive pulp cell walls contain about one third of arabinose-rich pectic polysaccharides. The L-arabinose-rich polysaccharides are named "arabinans". These polymers have a main structure of α-(1→5)-linked L-arabinofuranose units, substituted at either O-2 or O-3 or at both of these positions. The arabinans from 2POMW have been isolated and characterized by Cardoso S.M. et al. (2002, 2003). An arabinan (97% of Ara and 3% of hexuronic acid) was isolated from the alcohol-insoluble residue of 2POMW by treatment with a hot dilute acid (0.02 M HNO_3, at 80°), followed by graded precipitation with ethanol. It was separated from acidic pectic polysaccharides by anion-exchange chromatography and by size-exclusion chromatography while its molecular weight was estimated as 8.4 kDa. By methylation analysis and NMR spectroscopy, it was possible to propose the structure of Fig. 2.7.

Microbial Content of 2POMW   Eleven different isolates of bacteria were identified in 2POMW (EU project: FAIR-CT96–1420 "IMPROLIVE"). *Bacillus pumilus* was the most commonly found strain. The nine different isolates of yeasts were classified in four genera and eight species. The *Candida* genus was found to be the most predominant. *Saccharomyces cerevisiae* appeared in a low frequency, while *Candida valida* showed the highest frequency. The frequency of appearance of filamentous fungi was the lowest. The isolates of this group were found to belong in three different genera: *Rhizopus, Penicillium* and *Synchephalastrum,* and *Paecilomyces.* A great percentage of the isolated anaerobic bacteria seemed to be in close relation to *Lactobacillus acidophilus* and *Bifidobacterium* spp.

Experimental Techniques

A large number of chemical analyses are needed for the characterization of olive processing wastes; such analyses include the determination of COD, BOD_5, total solids, total phenols, individual phenols, total sugars, reducing sugars, tannins and

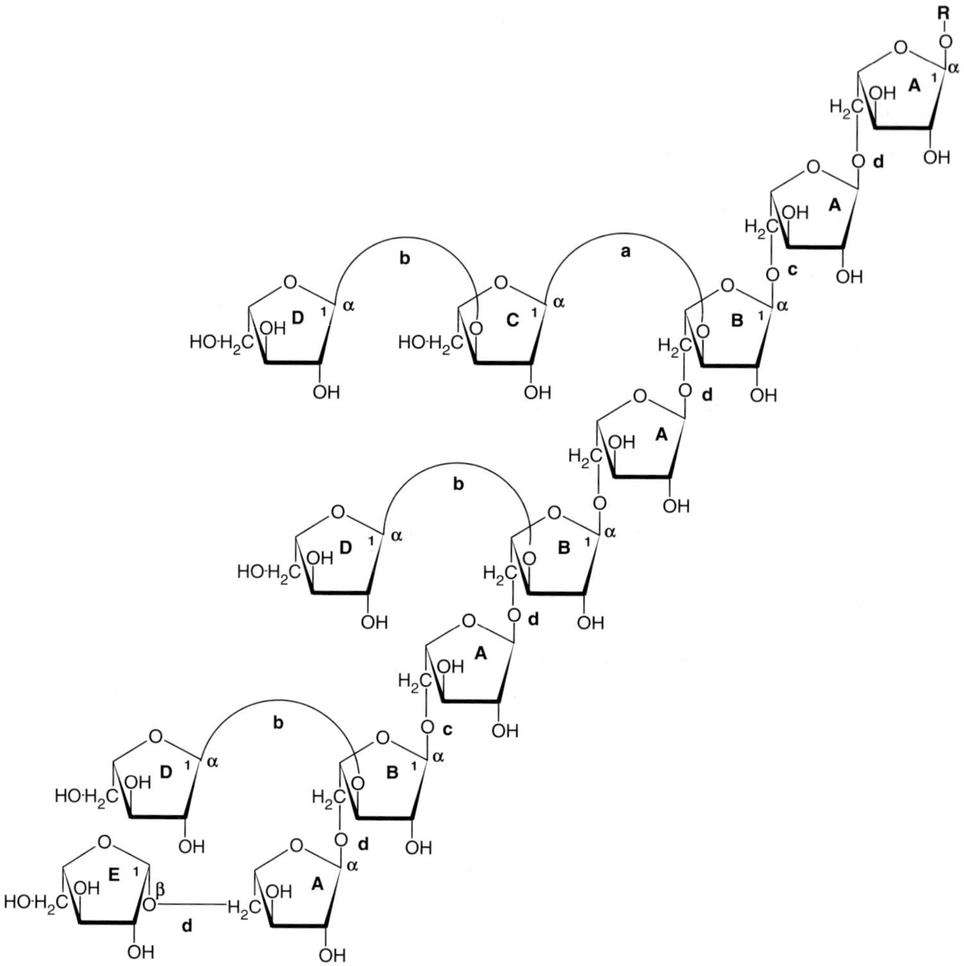

Fig. 2.7. Tentative structure of the olive arabinan (Cardoso S.M. et al., 2003). The linkage composition was established as 5:4:3:1 for (1→5)-Ara*f*, T-Ara*f*, (1→3,5)-Ara*f* and (1→3)-Ara*f*, respectively. ^{13}C NMR spectroscopy confirmed this linkage composition and allowed to assign the α anomeric configuration for the arabinofuranosyl residues, except for some terminally linked ones, that were seen to occur as T-β-Ara*f*. By 2D NMR spectroscopy (^{1}H and ^{13}C), it was possible to conclude that the T-β-Ara*f* was (1→5)-linked to a (1→5)-Ara*f* residue. Also, in the arabinan (1→5)-Ara*f* backbone, the branched (1→3,5)-Ara*f* residues were always adjacent to linear (1→5)-Ara*f* residues. According to the estimated molecular weight (8.4 kDa), it is possible to assume that it contains four arabinan structures of 13 Ara residues, occurring as side chains of a degraded pectic polysaccharide backbone.

lignins, total fats, individual fats, individual fatty acids, total organic carbon, total phosphorous, total nitrogen, metals, and ash. These analyses are multistep procedures, which are considered to be tedious and time consuming. In addition, most of them require the use of toxic solvents, which lead to the production of laboratory wastes, the disposal of which is problematic.

Chemical Oxygen Demand (COD)

The chemical oxygen demand (COD) can be determined according to the Soxhlet method 5520-D (*Standard Methods for the Examination of Water and Wastewater*, American Public Health Association, 1989).

Biochemical Oxygen Demand (BOD)

The biochemical oxygen demand (BOD) in a 5-days test period (5-d BOD or BOD_5) can be determined according to the method 5210-B (*Standard Methods for the Examination of Water and Wastewater*, American Public Health Association, 1989).

Total Organic Nitrogen

For the determination of total organic nitrogen, the semimicro Kjedahl method 4500-Norg C is used (*Standard Methods for the Examination of Water and Wastewater*, American Public Health Association, 1989).

Total Organic Carbon (TOC)

The total organic carbon (TOC) can be determined by photochemical oxidation of the organic compounds according to the method 5310-B (*Standard Methods for the Examination of Water and Wastewater*, American Public Health Association, 1989).

Lipid Characterization

For the determination of oil and grease, the open reflux-titrimetric method 5200-B can be used (*Standard Methods for the Examination of Water and Wastewater*, American Public Health Association, 1989). Alternatively, the lipid content of OMWW can be determined according to a Spanish Standard (1984). Crude lipids are extracted from OMWW with diethyl ether, the solvent is eliminated by evaporation and the residue is purified by re-extraction with hexane. The pure lipid is weighed and the lipid concentration in the sample is calculated.

Total Carbohydrate Characterization

The method of Dubois (Dubois M. et al., 1956) can be used to determine the total carbohydrate content.

Phenol Characterization

One of the most important analyte is the phenolic compound for which several methods have been described in the literature. Most of the analytical methods are laborious, providing different answers with different methods. Another problem is the unavailability of standards for a large number of phenolic compounds. A basic requirement is the development of a suitable analytical method for the rapid, sensitive, and unequivocal identification of the phenolic compounds. The characterization and quantification of phenols recovered from olive-mill wastes has been extensively reviewed by Obied H.K. et al. (2005).

Sample Preparation

Isolation of phenolic compounds from the sample matrix is generally a pre-requisite to any comprehensible analysis scheme. The ultimate goal is the preparation of a sample extract uniformly enriched in all components of interest and free from interfering matrix components. For the preparation of samples for the determination of the phenolic compounds, reference is being made to the work of Antolovich M. et al. (2000). Concerning the extraction process, two main techniques are reported in literature: (i) liquid-liquid extraction (LLE) and (ii) solid-phase extraction (SPE).

Liquid-Liquid Extraction (LLE)   A large number of solvents has been used for the extraction of phenols from olive processing wastes, including water (Servili M. 1999b), methanol (Gamel T.H. and Kiritsakis A., 1999; Cardoso S.M. et al., 2005), ethanol, ethyl acetate and, less commonly *n*-butanol, propanol, and *tert*-butyl methyl ether. Mixtures of methanol or ethanol with different levels of water are often used to extract phenols from olive cake (Romero Barranco C. et al., 2002). Methanol is able to disrupt cell walls and inhibit enzyme action, and its mixture with water provides a very good solvent for most phenolic compounds. A simple strategy to suppress enzyme action is to add methanol first followed by water (Bianco A. et al., 2003). A comprehensive study targeting hydroxytyrosol demonstrated that the extraction power of several solvents decreased in the order ethyl acetate > methyl isobutyl ketone > methyl ethyl ketone > diethyl ether (Allouche N. et al., 2004). It was further shown that best recovery of monomeric phenols occurred for OMWW with pH = 2. *o*-Diphenols were more extractable than monohydroxylated phenols due to more favorable coefficients.

During the LLE process, a solvent evaporation step is introduced in order to minimize the volume of the resulting organic extract. This step is prone to loss and/ or degradation of the more volatile analytes, introducing error into the measurements (Psillakis E. and Kalogerakis N., 2001).

Solid-Phase Extraction (SPE)   SPE has been used to separate phenolic compounds from OMWW (Servili M. et al., 1999b; Mulinacci N. et al., 2001) and olive cake (Servili M. et al., 1999b; Romero Barranco C. et al., 2002). C_{18} cartridges have been commonly used (Servili M. et al., 1999b; Romero Barranco C. et al., 2002). A sequence of organic solvents (*n*-hexane, ethyl acetate, ethyl ether, and acidic methanol (formic acid, pH 2.2) was employed for the selective recovery of phenols from freeze-dried OMWW (Servili M. et al., 1999b; Mulinacci N. et al., 2001). Freeze-dried OMWW were rehydrated with water containing 20 mg/l DIECA (diethyldithiocarbamate), to inhibit polyphenoloxidase and lipoxygenase activities; 2 ml was loaded on a 5/20 ml high-load C_{18} cartridge. In the case of olive cake, SPE gave higher recoveries of phenols than LLE (Servili M. et al., 1999b). With SPE, higher recoveries were achieved by elution with methanol than by elution with diethyl ether or ethyl acetate. For OMWW, highest recoveries of phenols were achieved by SPE with diethyl ether. The dissolution of quinones and melanoidins in methanol resulted in high background noise in HPLC chromatograms (Servili M. et al., 1999b).

A novel analytical method based on headspace Solid-Phase Microextraction (SPME) and Gas Chromatography-Mass Spectroscopy (GC-MS) has been used for the identification of the main volatile and semi-volatile organic compounds in OMWW samples (Psillakis E. and Kalogerakis N., 2001). SPME was based on extraction using a thin polymeric-coated fused-silica fiber, fitted in a special syringe-type holder for protection and sampling. Optimization of the SPME method was achieved by controlling several parameters, such as the SPME fiber and extraction temperature. Overall, the proposed technique proved to be an extremely fast, solvent-less, and simple detection method for the analysis of complex environmental samples such as OMWW.

Analytical Methods

Colorimetric Methods Traditional methods for the determination of total phenols have relied on direct measurement of absorption of radiation in the ultraviolet or, more commonly, colorimetric methods. The Folin-Ciocalteu method modified in various ways is the most common colorimetric method (Folin O. and Ciocalteau V., 1927; Swain T. and Hillis W.E., 1959; Singleton V.L. and Rossi J.A. Jr., 1965; Slinkard K. and Singleton V.L. 1977; Box J.D., 1983). It depends on measuring the absorbance of the blue reduction product of a phosphotungstic–phosphomolybdic complex in alkaline solution at 760 mm. The results are expressed as gallic acid, caffeic acid, tannic acid, or tyrosol equivalents. Any substance that is able to reduce the phosphotungstic–phosphomolybdic complex will interfere, including ascorbic acid, tocopherols, carotenes, reducing sugars, and phenolic amino acids (i.e. phenylalanine and tyrosine). The colorimetric methods are very simple and require few reagents, but they are limited by the low specificity of the reagent towards the phenols and usually tend to overestimate the amount of phenols present. In addition, there is generally no correlation between data for total phenols and those obtained by chromatographic techniques (Antolovich M. et al., 2000).

Chromatographic Methods   The main chromatographic methods used for the determination of the various phenolic compounds in olive processing wastes are: high performance liquid chromatography (HPLC) (Martínez-Nieto L. et al., 1992; Ceccon L. et al., 2001), reversed phase high performance liquid chromatography (RPLC) (Ryan D. et al., 1999; Vial J. et al., 2001), thin layer chromatography (TLC) (Vásquez-Roncero A. et al., 1974b; Capasso R. et al., 1992a), and Gas chromatography (GC) (Balice V. and Cera O., 1984; Hamdi M. and García J.L., 1991; Hamdi M., 1992, 1993b).

High performance liquid chromatography (HPLC):   HPLC methods for determining total phenols have been based on the summation of individual peak responses using calibration curves, available standards, and different wavelengths. One or more standard compounds have been employed as a reference and, in cases where commercial standards were not available, molecular weight correction factors were applied. The use of syringic acid as an internal standard and quantification by response factors (area of reference compound/area of syringic acid) has been tried. The complexity of the HPLC chromatogram and the number of overlapping peaks are the major limiting factors for the reliability of this technique for the quantification of total phenols (Obied H.K. et al., 2005).

A method was developed for the determination of simple phenolic compounds in OMWW by liquid chromatography (Ceccon L. et al., 2001). The sample under examination was acidified to pH = 2 to precipitate proteins, acetone was added to eliminate the colloidal fraction, and hexane was used for extraction to eliminate lipidic substances. The solution obtained was filtered and injected into the liquid chromatography system; the wavelength selected for the spectrophotometric detection was specific for phenolic compounds, so that carbohydrates, organic acids, and short-chain free fatty acids did not interfere. Recoveries of nine phenolic compounds spiked to a real sample were 90–100% for concentrations ranging from 20 to 2000 mg/l for each analyte.

HPLC avoids the need for derivatization and has been the standard for analysis of polyphenols from both qualitative and quantitative points of view, but it is time-consuming because of the length of the chromatographic run. In addition, as not all of the peaks revealed in the HPLC chromatograms have been identified, it is impossible to quantify the single phenols due to the absence of suitable standard compounds. This technique has been mainly associated with spectroscopic methods. The typical conditions for HPLC are reversed phase liquid chromatography (RPLC) using an octadecyl silica column with a suitable guard column and a binary pumping system, linear gradient elution and photodiode array detector (PDA) (Mulinacci N. et al., 2001). The mobile phase typically contains various combinations of water, methanol, or acetonitrile in different proportions and adjusted to an acidic pH by the addition of acetic acid, formic acid, or phosphoric acid. Different column brands and chromatographic systems (the pump and the detector) show different resolution abilities (Romero Barranco C. et al., 2002). RPLC coupled with mass detection (MS) is one of the most adapted techniques in the field of environmental analysis.

However, with very polar compounds, like phenolic acids present in OMWW, the low retention on classical RPLC columns, even when a high percentage of water is used in thermobile phase, often produces inadequate resolution. Another limitation encountered with OMWW samples is matrix effects, which is a consequence of high organic carbon content of such wastewater (> 100 mg/l). To overcome this problem, high dilution ratios or sample handling is required. An elegant way to solve the problem of poor retention and the need for a clean-up can be the use of porous graphitic carbon columns (PGC) instead of alkyl bonded silica taking advantage of its different retention behavior (Vial J. et al., 2001). PGC, involving the use of a tetrahydrofuran gradient as mobile phase, was coupled with mass detection (MS) for the analysis of six polar phenolic compounds of OMWW by liquid chromatography. The proposed PGC-LC-MS method was selective and linear for the six phenolic compounds analyzed with limits of quantification lower than 5 ppm in all cases. The precision was satisfactory (pooled RSD around 6%).

Thin layer chromatography (TLC):   TLC is a simple and versatile technique that can be used for the identification, separation, and isolation of polyphenols on both analytical and semi-preparative scales. The chromatogram contains the actual compounds, not their response. This permits subsequent elution and identification of each spot.   A TLC method has been developed to detect the major polyphenols in OMWW (Capasso R. et al., 1992a). The method involves reversed phase TLC (C_{18}-TLC) and silica gel high performance TLC (Si-HPTLC) on the organic extracts of OMWW, using catechol, 4-methylcatechol, tyrosol, hydroxytyrosol, and oleuropein as standards. Spots were visualized under UV light at 254 nm, by spraying with 10% sulfuric acid in methanol followed by phosphomolybdic acid (3% in methanol) and heating, and by spraying with aqueous ferric chloride and heating. The use of both reagents combines the high sensitivity of the first reagent and the specificity of the second reagent. By using Si-HPTLC analysis, it is possible to detect only tyrosol and hydroxytyrosol, whilst with C_{18}-TLC analysis all of the main polyphenols present, except oleuropein can be detected. Confirmation of the identities of the polyphenolic compounds was obtained by C_{18}-TLC and Si-HPTLC analysis of the acetylated organic extracts of OMWW together with the more stable acetyl derivatives of the phenols as standards. For the quantification of polyphenols could be used the Folin-Ciocalteau method after TLC separation (Ragazzi E. and Veronese G., 1973).

According to Obied H.K. et al. (2005) TLC can also be used for the screening of biological activity in a technique known as bioautography. It is widely applied for screening antibacterial, antifungal, and radical scavenging activities. For antibacterial and antifungal activities agar diffusion, direct application, and agar-overlay are the methods for application. This method has not been applied to olive processing wastes.

Gas chromatography (GC):   The phenolic compounds of OMWW are polar, and of limited volatility, so derivatization is often mandatory in GC (López Aparicio et al. 1977; Balice V. and Cera O., 1984). A very complex chromatogram resulted when GC-flame ionization detection (FID) was applied to an extract of OMWW after derivatization with bis(trimethylsilyl)trifluoroacetamide (Ceccon L. et al., 2001). FID was extensively employed in early work, but carbohydrate interference was a critical problem (Ceccon L. et al., 2001), and most GC work is now done using mass spectral detectors or tandem mass spectrometry (GC-MS). Knupp G. et al. (1996) used GC-MS for the identification of phenolic compounds during the microbial degradation of untreated Spanish and Italian OMWW samples. GC-MS was also applied for the identification of the sugar part of a new glycosidal polyphenol in OMWW (Della Greca M. et al., 2004) after methylation, hydrolysis, reduction, and acetylation. For large phenolic molecules, derivatization may increase the molecular mass of the analytes beyond the analyzing capacity of the mass detector. Thermal degradation, failure of derivatization of high molecular weight polyphenols and unsuitability for preparative scale analysis are other drawbacks. Hence, GC is not a popular technique for routine use in polyphenol analysis. It is more suitable for profile generation or structure elucidation, where its excellent resolving power is required (Obied H.K. et al., 2005).

Spectroscopic Methods Spectroscopic techniques include ultraviolet radiation (UV), nuclear magnetic resonance (NMR), and MS (mass spectroscopy).

Electron ionization (EI):   Electron ionization (EI) and fast atom bombardment-mass spectroscopy (FAB-MS) are the spectroscopic methods currently used for analyzing the main polyphenols naturally occurring in OMWW. EI-MS and FAB-MS are very suitable for analyzing catechol, 4-methylcatechol, tyrosol, and hydroxytyrosol, the main polyphenols in OMWW (Capasso R., 1999). The EI-MS method has also proved to be suitable for analyzing their acetyl derivatives, but does not allow the molecular ion of diacetyltyrosol and its isomers to be detected. FAB-MS is the only adequate ionization method for detecting the molecular ion of diacetyltyrosol and is more suitable than EI-MS for analyzing triacetylhydroxy-tyrosol. The mechanisms of fragmentation of the acetyl derivatives of tyrosol and hydroxytyrosol are determined using the EI and FAB ionization methods. In addition, FAB-MS, which was performed in both positive and negative ion modes, was shown to be the only adequate ionization method for analyzing oleuropein, a phenol glucoside which occurs naturally in olive leaves. The positive ion FAB-MS was shown to be much more suitable than EI-MS for analyzing the aglycone obtained following the hydrolysis of oleuropein by β-glucosidase. These results confirmed the versatility of FAB-MS for analyzing low molecular weight compounds, for which EI-MS proved to be an unsatisfactory method (Capasso R., 1999). The development of soft ionization techniques, such as atmospheric pressure ionization technique (API), for the investigation of polar, non-volatile, and thermoplabile molecules has facilitated the analysis of phenolic compounds by LC-MS (Vial J. et al., 2001). Aramendia M.A. et al. (1996) applied negative ion LC-APCI-MS to qualitatively and quantitatively analyze 15 phenolic compounds found in OMWW. Analytes were separated in a C_{18} phase by gradient elution with methanol-water containing formic acid. Mass spectral conditions were optimized by direct infusion of standards in the flow injection mode into the APCI mode source.

APCI still has the major drawback for polar thermolabile phenols that volatilization of the sample must occur before ionization. Electrospray ionization (ESI) overcomes lack of analyte volatility by direct formation or emission of ions from the surface of a condensed phase and sample ions are collected from the condensed phase inside the ion source and transferred to the mass analyzer. Hence, ESI eliminates the need for neutral molecule volatilization prior to ionization (Antolovich M. et al., 2000). Bianco A. et al. (2003) showed the high selectivity of HPLC-ESI-MS/MS in the analysis of OMWW. The study was also restricted to the negative mode. Despite lower intensity peaks in negative ion mode than in positive mode, negative ion mode was also chosen in the following study because clearer spectra were obtained. Methanol extracts of 2POMW and olive pulp were analyzed by reversed phase HPLC and the eluted fractions were characterized by ESI (Cardoso S.M. et al., 2005). The study demonstrates the utility of ESI spectra, particularly in the MS^n mode (Obied H.K. et al., 2005).

Limiroli R. et al. (1996) identified both free and glucosidal phenols from the vegetation water of olive fruit by H^1 NMR. Della Greca M. (2004) has identified four new compounds in OMWW using LC-MS and off-line NMR including two-dimensional NMR.

Capillary Zone Electrophoresis (CZE)   CZE is characterized by high separation efficiency, small sample and electrolyte consumption and rapid analysis, as the separation requires only several minutes. The last characteristic is the main advantage versus chromatographic methods, which makes CZE of great utility in routine analysis, control, and monitoring of processes in a number of industrial fields. CZE depends on the relative migration of ions under an electric field. Several phenolic compounds found in OMWW were quantitatively and qualita-tively analyzed on an uncoated fused-silica capillary electrophoresis column (67 cm × 75 µm i.d.) using 30 mM aqueous ammonium acetate buffer/methanol (90:10) and negative ESI-MS detection and compared with CZE-UV (Lafont F. et al., 1999). The total run time was 30 min. Quantitative analysis using *p*-chlorophenol as internal standard was carried out by single ion monitoring. Limits of detection ranged from 1 pg for 4-hydroxybenzaldehyde and protocatechuic acid to 386 pg for vanillic acid. The drawback of CZE is the use of buffer of high pH, which may be a problem for compounds unstable under these conditions as anthocyanins.

A CZE-DTA system has been used for the ultrasound-assisted extraction of 20 phenolic compounds from 2POMW (Priego-Capote F., 2004). Multivariate methodology was used to carry out a detailed optimization study of both the separation-determination and extraction steps in terms of resolution-analysis time and extraction efficiency, respectively. Consequently, the proposed method was able to extract the target analytes in 13 min; then, after dilution and centrifugation, the extract was injected into the CZE-DTA system for individual separation determination in 11 min. No clean-up of the extract was required. This method is less time consuming, more selective, and provides a larger information level than the Folin-Ciocalteau spectrophotometric method.

A technique similar to CZE is micellar electrokinetic capillary chromatography (MECC). The main similarity between CZE and MECC is the instrumentation; MECC is a hybrid between HPLC and CZE where both neutral and ionic species can be separated by the difference in the distribution between the moving buffer and the capillary coating (electroosmotic flow). MECC has been used for the separation of 10 phenolic acids (Pomponio R. et al., 2002), but it has not been applied to olive processing waste (Obied H.K. et al., 2005).

Determination of Partition Coefficient (K_p)

OMWW contains a number of phenols in quantities determined largely by their partition coefficients. The partition coefficient is defined as: $K_p = C_{oil}/C_{water}$, where C_{oil} and C_{water} are the equilibrium concentrations of a phenolic compound in the oil and water phase, respectively. The K_p of the phenolic compound between the oil and water phases can be determined experimentally according to the methodology described by Archer et al. (1994)[9].

Prediction of partitioning coefficients (K_p) between phases, though, may be feasible by using a general group contribution method for prediction of activity coefficients in a liquid-phase, such as the UNIFAC method. This method has enabled the prediction of vapor/liquid or liquid/liquid equilibrium, or the solubility of several substances, in aqueous or non-aqueous phases. The group contribution method is based on the concept of the solution of groups instead of molecules. Each molecule is considered as a mixture of simple groups (-CH_2-, -COOH, -OH, etc.), whose thermodynamic property parameters are known in the literature, and the various properties are found by the summation of the contributions of the various groups. Thus, the group contribution method has the advantage of predicting various thermodynamic properties through estimation of the effects of the various groups. The UNIFAC method was based on the universal quasi chemical activity coefficient (UNIQUAC) method, which is another method derived from an extension of Guggenheims's quasi-chemical theory of liquid mixture (reported by Rodis P.S. et al., 2002).

Analysis of Metal Cations and Inorganic Anions

The cations are quantitatively determined by atomic-absorption spectroscopy (AAS), whereas the anions are detected by ion chromatography (Arienzo M. and Capasso R., 2000).

---

[9]Archer M. H., Dillon V. M., Cambell-Platt G., and Owens J. D. (1994) The partitioning of diacetyl between food oils and water. *Food Chem.*, **50**, 407–409.

Attempts to separate soluble anions from OMWW by ion-exchange or to remove the oil fraction by solid-phase or solvent extractions were not completely satisfactory and erratic results were observed. Buldini P.L. et al. (2000) presented a simple and accurate procedure for the determination of inorganic anions in OMWW using on-line microdialysis of OMWW directly followed by the ion chromatography analysis of soluble chloride, nitrate, phosphate, and sulfate with conductimetric detection. OMWW is first of all sonicated at room temperature to make it homogeneous, then diluted and microdialized. Most of the organic load of the effluents is removed in a few minutes without using reagents, while soluble anion quantitation remains unaffected. The clear solution is analyzed for the inorganic anions content by direct injection on to an ion chromatograph equipped with a conductivity detector. In the absence of standards, the separation efficiency of microdialysis has been investigated by spiking wastewater samples as well as standard oil emulsions with varying amounts of inorganic anions and subjecting them to microdialysis for different periods of time prior to performing instrumental analysis. Excellent spike recoveries and low relative standard deviations are obtained for all the anions if a 10 min microdialysis time is overcome. Chloride, nitrite, nitrate, phosphate, and sulfate are not affected by the microdialysis procedure and their recovery is between 96 and 104% in wastewater as well as in standard oil emulsion. The dialysis membrane has been replaced after more than 100 analyses. The UV photolysis pretreatment of the same sample demonstrates the different information that can be obtained by the two sample pretreatment procedures.

Antioxidant Activity

Antioxidant activity has been assessed in many ways. In general, the antioxidant effectiveness is measured by monitoring the inhibition of a suitable substrate. After the substrate is oxidized under standard conditions, the extent of oxidation is measured by chemical, instrumental, or sensory methods. Hence, the essential features on any test are a suitable substrate, an oxidation initiator, and an appropriate measure of the end product. Antolovich M. et al. (2002) reviewed the major methodologies for the determination of antioxidant activity used by the food industry, with the diphenylpicrylhydrazyl (DPPH) radical assay being one of the more utilized due to its relative simplicity; it is, however, a lengthy procedure.

The limitation of many newer methods is the frequent lack of an actual substrate in the procedure. The combination of all approaches with the many test methods available explains the large variety of ways in which results of antioxidant testing are reported. The measurement of antioxidant activities, especially of antioxidants that are mixtures, multifunctional, or are acting in complex multiphase systems, cannot be evaluated satisfactorily by a simple antioxidant test without due regard to the many variables influencing the results. Several test procedures may be required to evaluate such antioxidant activities. A general method of reporting antioxidant activity independent of the test procedure has been proposed by Antolovich M. et al. (2002).

The antioxidant and anti-inflammatory activity properties of OMWW have been measured by Visioli F. et al. (1999). OMWW obtained by employing a benchtop mill were fractionated by liquid–solid extraction and further processed to yield three extracts. Extract 1 was obtained by fractionation of lyophilized OMWW on a chromatographic column filled with DUOLITE® XAD 1180 resin particles and elution with ethanol. Extract 2 was obtained by ethyl acetate extraction of hexane-washed OMWW. Extract 3 was obtained following a fractionation of extract 2 on a Sephadex LH-20 column. Multiple antioxidant assays (LDL oxidation, DPPH radical scavenging activity, superoxide anion scavenging, and protection of catalase against hypochlorous acid) and an anti-inflammatory activity assay (leukotriene B^4 production by human neutrophils) were performed. Extract 1 contained a complex mixture of phenolics including many polymers responsible for a high background absorption at 254 nm and exhibited low antioxidant activity and no anti-inflammatory activity. Extract 2 contained mainly low and medium molecular weight phenolics with elenolic acid as the principal constituent and showed good antioxidant and excellent anti-inflammatory activities. Extract 3 comprised hydroxytyrosol, tyrosol, and the unidentified derivative of the former and exhibited the most potent antioxidant activity and reasonable anti-inflammatory activity. The authors suggested that the extracts acted mainly as metal chelators and also had a potent free radical scavenging activity — see also Chapter 10: "Uses", section "Antioxidants".

Amro B. et al. (2002) investigated the antioxidative activity of different butanol extract fractions of olive cake. The residue left after evaporation of the ethanolic extract was dissolved in water and sequentially extracted with hexane, chloroform, and butanol. The butanol extract was fractionated in a silica gel column and nine fractions were collected. The fractions were examined using various measures of antioxidant activity [iron(III) reduction; inhibition of oxidation in refined soyabean oil; DPPH radical scavenging] and, consistent with previous studies, the antioxidant activity varied according to the test method. The first four fractions showed marked antioxidative activity in comparison with BHT(butylhydroxy toluene). Fractions tested also showed good hydrogen donating abilities, indicating that they had effective activities as radical scavengers.

Chemiluminescence is an alternative detection technique used for the determination of antioxidant activity, having the advantages of low detection limits, wide linear dynamic ranges, and speed of response. Luminol and lucigenin have been widely used for the determination of reactive oxygen species in a variety of biological systems and have been used indirectly to evaluate antioxidant activities. The chemiluminescene reactions provide a more rapid approach for measuring antioxidant activities when compared with standard methods (Atanassova D. et al., 2005a).

Atanassova D. et al. (2005a) described a rapid, simple and sensitive procedure for estimating the total phenolic/antioxidant levels of OMWW and 2POMW samples, using Co(II)ethylenediaminetetracetic acid (EDTA)-induced luminol chemiluminescence. A fair linear relationship was observed between the total phenolic content (measured by the classic Folin-Ciocalteu test and expressed as caffeic acid) and the antioxidant activity (measured by the luminol Co(II)/EDTA-enhanced chemiluminescence technique) for both samples.

Using thermogravimetric analysis (TGA), it is possible to estimate oil resistance to oxidation, by measuring weight gain percent due to reaction of a sample with oxygen during oxidation, and initial and final oxidation temperature.

Identification of Bacteria

Isolated bacteria can be identified using (Jones C.E. et al., 2000; EU project: FAIR-CT96–1420 "IMPROLIVE"):

- Standard microbiological tests;
- Biochemical growth differences (API);
- Polar lipid composition;
- Fatty acid composition;
- Molecular biological analyses;
- PCR-based 16S rRNA sequence analysis;
- Restriction fragment length polymorphism (RFLP);
- Single-stranded conformational polymorphism (SSCP).

Animal Feed Analysis

Two main types of laboratory analysis of nutritive value of feeds are used:

- Chemical evaluation;
- Weende system;
- van Soest system;
- Near infrared reflectance (NIR);
- *In vitro* digestion.

Weende System

After water is eliminated, feed is divided into five chemically defined components:

1. Crude fiber (CF), which approximates structural carbohydrate content.
2. Crude protein (CP) ($=N \times 6.25$), which approximates true protein content.
3. Ash, which approximates mineral content.
4. Ether extract (EE), which approximates lipid content.
5. Nitrogen-free extract (NEE), which approximates non-structural carbohydrate content. This is estimated by difference between total dry matter and the sum of the other four chemical components.

Detergent System (van Soest)

Extraction with neutral detergent recovers major plant cell wall components (cellulose, hemicellulose, lignin) and removes all other organic constituents.

Extraction of residue with strong acid detergent recovers cellulose; lignin and lignin-N-complexes and removes hemicellulose and fiber-bound protein.

Chapter 3

Environmental Effects

Effects on Soil

The environmental effects of olive-mill liquid wastes on soil are known since antiquity. The Roman author Varro (I, 55) had observed that where the amurca — the watery residue obtained when the oil is drained from olive fruits — flowed from the olive presses onto the fields, the ground became barren. Theophrastus[10] (IV, 16) (see Fig. 3.1) wrote that pouring olive oil over the roots could kill trees, young trees being more susceptible to this treatment than mature ones — see also Chapter 10: "Uses", section: "Use as herbicide/pesticide".

The uncontrolled disposal of OMWW on the land has the drawback of dispersing in the environment substances that are foul smelling and possibly pathogenic. In fact, higher application rates result in anomalous fermentations of the dispersed organic substances, which lead to changing the environmental conditions for microorganisms, the soil–air and the air–water balance and, therefore, to reduction of the soil fertility. However, if one could optimize the use of these wastes, they could be proved beneficial, as soil amendments, to the physical, chemical, and biological properties of the soil — see Chapter 8: "Biological processes", section: "Irrigation of agricultural land/Land spreading".

---

[10]Theophrastus (c.372–c.287 B.C.), Greek philosopher born in the island of Lesvos; Aristotle's successor as head of the Peripatetics. The school flourished under his leadership. He wrote on many subjects, but many of his treatises are lost. He did much to popularize science. His works on plants are perhaps the most important of his technical writings. His *History of Plants* and *Enquiry into Plants* presented the first thorough treatment of the science of botany and remained the definitive works on the subject through the Middle Ages. Also extant are portions of his *History of Physics*; nine scientific treatises including *On Stones, On Fire,* and *On Winds.*

Fig. 3.1. Theophrastus.

Effects on Soil Physical Properties

Porosity

The porosity corresponds to the volume of the soil occupied by water and air. Through the pores the soil exchanges water and air with the environment. These exchanges are indispensable for the development of the fauna and the microflora of the soil as well for the respiration of the roots.

Cox L. et al. (1997) studied the effect of OMWW on soil porosity in clay soil columns. Soil columns were hand packed with the unamended clay soil and with the same soil, which had been treated for three years with two different doses of OMWW (low dose: $300\,ml/m^2$ a year and high dose: $600\,ml/m^2$ a year). OMWW amendment resulted in an increase in the organic carbon content of the soils and a reduction in soil porosity, the later confirmed by mercury intrusion porosimetry (MIP) and scanning electron microscopy (SEM) studies. MIP and SEM data showed that the reduction in porosity is basically due to a reduction in larger pores (radius > 1 μm) and an important increase in finer pores (radius < 0.1 μm) (Cox L. et al., 1996). Similar results were reported by Zenjari B. and Nejmeddine A. (2001). These changes in porosity are attributed to the combined effect of the suspended and soluble organic matter and salts in OMWW and the solubilization–insolubilization of the soil carbonate minerals promoted by OMWW.

In the open field the application of OMWW on soil initially brings about a reduction in the microporosity (pores < 50 μm) in the surface layers of the soil. At the end of the winter with the resumption of the microbial activity the microporosity increases significantly compared to a non-treated soil (Pagliai M., 1996). The temporary decrease of the microporosity is not harmful neither for the microorganisms nor for the roots because of the reduction of their metabolic activity during winter. The macroporosity (pores > 50 μm) increases proportionally with the quantity of applied OMWW. However, excessive doses (more than 200 m³/ha) can cause structural damage accompanied by a decrease of the porosity, particularly in clay soils (Pagliai M. et al., 1993); in France and Italy the use of such quantities is forbidden (Italian law 574/1996; Le Verge S., 2004).

Aggregation

The aggregates of the soil have the tendency to disintegrate under the impact of the rain droplets forming a crust on the surface that obstructs the oxygenation of the soil and causes erosion. The application of OMWW contributes to the stabilization of the soil's aggregates, thanks to the binding action of certain organic components, in particular polysaccharides. The stabilizing effect remains for several months till the degradation of the organic compounds (Pagliai M. 1996; Le Verge S., 2004). It appears, therefore, that the application of OMWW could increase the stability of aggregates, prevent erosion phenomena, and the formation of surface crusts due to rain action, improve oxygenation of the surface profile of the soil in which root growth and microbial activity occur (cultivation layer), and contribute to a better hydraulic retention of the land due to its increased microporosity (Mellouli H.J. 1996; Colucci R. et al., 2002; Le Verge S., 2004). A laboratory study has shown that a surface layer of a sandy soil incorporated with OMWW is more effective in reducing evaporation losses (∼30%) than a surface on which OMWW is applied as a mulch (25 g/m²), while the application of a straw mulch (450 g/m²) was effective only during the initial stage of the evaporation (Mellouli H.J., 1998; Mellouli H.J. et al., 2000).

Olive cake contains 94% organic matter and, therefore, can be highly beneficial to agricultural soil. However, said waste contains oil that may increase soil hydrophobicity and decrease water retention and infiltration rate. Abu-Zreig M. and Al-Widyan M. (2002) investigated the impact of olive cake on water retention, saturated and unsaturated hydraulic conductivity, and capillary rise of three soils: loam, clay loam, and dune sand and under laboratory conditions. Application of the waste resulted in an increase in water retention and saturated hydraulic conductivity, but caused a decrease in capillary rise and unsaturated hydraulic conductivity for all soils tested. The increase in water retention has been observed at all levels of pressure potential and was significantly different at 3 bars or higher. The highest increase in saturated hydraulic conductivity occurred at 4% application rate at which about 300%, 200%, and only 12% increase was observed for loam, clay loam, and dune sand, respectively. Application of olive cake caused a significant decrease in the capillary rise ranging from 11.5% for dunes to 70% for clay loam soil.

Effects on Soil Chemical Properties

There are several studies on the chemical characteristics of OMWW (Della Monica M. et al., 1979; Potenz D. et al., 1985b; Senette C. et al., 1991; Marsilio V. et al., 1989; Levi-Minzi R. et al., 1992; Saviozzi A. et al., 1993; Proietti P. et al., 1995), and its humification index (Alianiello F., 1997; Alianiello F. et al., 1998).

Acidity

The application of OMWW at a moderate dose does not affect the acidity of the soil. Levi-Minzi R. et al. (1992) studied the evolution of acidity of an alkaline soil treated with various amounts of OMWW (80, 160, and 320 m^3/ha) for a period of 135 days. Because of its acidic character (pH = 5) OMWW had a temporary acidifying action shortly after their application; during the next fifteen days, the treated soil recovered its original acidity. Similar evolution patterns of the acidity are found in several other studies on various types of alkaline soil (Della Monica M. et al., 1978; Potenz D. et al., 1985b; Morisot A. et al., 1986; Monpezat G. de et al., 1999). This slight acidification is considered to be beneficial for the alkaline soils because it renders phosphorus and other elements more assimilable by the olive trees (Le Verge S., 2004).

The application of OMWW on acidic soils can cause acidification of the ground (Le Verge S., 2004). A study carried by Marsilio V. et al. (1989) showed that a dose exceeding 160 m^3/ha causes only a minimum acidification of the soil (0.03 units of pH) during the first 100 days; a distinct increase in the pH of the treated soil was observed after this period. As a measure of precaution Monpezat G. de et al. (1999) recommends the neutralization of OMWW with lime before its application on acidic soils.

Salinity

OMWW contains many acids, minerals, and organics that could destroy the cation exchange capacity of the soil. Higher levels of soil salinity due to potassium and sodium replacement of soil cations were detected in an alkaline soil after pollution with OMWW. The pH was practically unchanged and soil C/N ratio was increased (Paredes M.J. et al., 1986).

Sierra J. et al. (2001) studied the characterization and evolution of a soil affected by OMWW on a location used for 10 years as an uncontrolled OMWW disposal site. The study area included several evaporation ponds built on land without an impervious layer. The soil is formed by sedimentary materials (calcareous crusts and conglomerates). Once the disposal site was closed, the sediment remaining on the soil surface was removed. The use of a calcareous soil as a medium for OMWW disposal allowed the neutralization of the waste pH when passing through soil. The acidity of OMWW was compensated by soil carbonate alkalinity. The carbonates at the same time became bicarbonates and moved and accumulated in deeper horizons. An increase in salinity and in soluble phenolic compound contents was detected. The enrichment diminished in deeper layers, due to OMWW soil retention. Changes in electrical conductivity and phenolic compound content were observed down to 110–125 cm, where the OMWW flux was restrained by the sedimentary rock, which is more compact. Once the sediment remaining on the surface was removed, the salinity decreased quickly by rainfall leaching and biological activity, in time led to an effective decrease in electrical conductivity and phenolic compounds, although residual levels can be important even two years later. This similar evolution of conductivity and phenolic compounds is in accordance with the results obtained by Levi-Minzi R. et al. (1992), in an experiment undertaken for agricultural soils treated with OMWW and by Sierra J. et al. (2000), with leaching columns under laboratory conditions.

The application of OMWW at a moderate rate does not affect the salinity of the soil (Le Verge S., 2004). The application of an excessive dose ($320\,m^3/ha$) on a clay soil caused only a temporary increase of the salinity (Levi-Minzi R. et al., 1992). An experiment with an application dose of $200\,m^3/ha$ showed that the salinity increased slightly after 2.5 months (0.36‰ compared to 0.24‰ of a control soil) (Morisot A. et al., 1986).

Inorganic Chemical Compounds

A series of incubation experiments were performed in order to study the effects of OMWW in a calcareous soil (Pérez D.J. and Gallardo-Lara F., 1987, 1989; Gallardo-Lara F. et al., 2000). The first incubation experiment studied the effects of OMWW on nitrogen transformation in a calcareous soil (Pérez D.J. and Gallardo-Lara F., 1987). The application of this wastewater was shown to decrease NO_3^- formation in comparison with control assays during approximately the first half of the experimental period (6 weeks). Results were similar although were marked when OMWW plus ammoniacal nitrogen was applied as opposed to ammoniacal nitrogen alone. The incorporation of OMWW during the initial phases of study also reduced soil $N-NH_4^+$ levels both when residue only treatments were compared with controls and when OMWW plus ammoniacal nitrogen treatments were compared with ammoniacal nitrogen only. The second incubation experiment studied the effects of OMWW on sulfur transformation in a calcareous soil (Pérez D.J. and Gallardo-Lara F., 1989). In addition to raw OMWW, other preparations were tested including OMWW devoid of organic matter and deionized OMWW. The addition of OMWW to soil inhibits the formation of $S-SO_4^{2-}$ when OMWW plus elemental sulfur is compared to a treatment consisting of elemental sulfur applied alone. No such effect, however, was seen when the treatment with OMWW only is compared with control soils. Of the three types of OMWW tested, the least effective inhibitor of $S-SO_4^{2-}$ formation was OMWW in which all organic matter has been eliminated, while the deionized effluent yielded lowest levels of $S-SO_4^{2-}$. The exclusive application of OMWW on calcareous soils may raise $S-SO_4^{2-}$ levels in the middle run; however, when a sulfur deficient soil is fertilized with elemental sulfur, concurrent application of OMWW is unadvisable, given that it may interfere with soil $S-SO_4^{2-}$ formation. A pot experiment using calcareous soil was performed in a growth chamber to examine the effects of OMWW on the availability and post harvest soil extractability of K, Mg, and Mn (Gallardo-Lara F. et al., 2000). The experiment included 6 treatments: two rates of OMWW, two mineral fertilizer treatments containing K (which supplied K in amounts equivalent to the K supplied by the OMWW treatments), a K-free mineral fertilizer treatment, and a control. The pots were sown with rye-grass as the test plant, harvesting 3 times at intervals of one month. OMWW has demonstrated a considerable capacity for supplying K that can be assimilated by the plant, tending in fact to surpass the mineral potassium fertilizer tested. The application of OMWW tends to reduce the concentration of Mg in the plant, similarly to the effect of adding mineral potassium fertilizer. An enhancement of Mn availability takes place in the soil amended with OMWW, which on occasion has produced Mn concentrations in plant that could be considered phytotoxic or at least excessive. After harvesting the amount of exchangeable K in soil with added industrial wastewater was increased. However, these increases are lower than those in soil treated with mineral potassium fertilizer. The levels of exchangeable, carbonate-bound, organic-bound, and residual Mg in soil were higher in treatments incorporating OMWW than in those with added mineral K, with the opposite tendency occurring in the amount of Fe–Mn oxides-bound Mg in soil. Treatments based on OMWW, especially in high doses, increased the amount of exchangeable and carbonate-bound Mn in soil, in comparison with treatments adding mineral fertilizers with or without K. In contrast, the addition of industrial wastewater caused a drop in the amount of Fe–Mn oxides-bound and organic-bound Mn in soil.

Organic Chemical Compounds

OMWW contains on average about 6% of organic matter and 0.4% of mineral salts suspended or dissolved in an aqueous medium. The organic matter of OMWW contains compounds that are easily biodegradable by the microorganisms of the soil. The degradation of the organic matter produces volatiles substances that are foul smelling and possibly pathogenic. Mineralization of the organic matter produce higher contents of NO_3^--N in soil and increased NO_3^--N uptake by plants. OMWW contains also phenols that are assumed to be responsible for phytotoxicity and their bioconversion is very important for humic acid biosynthesis. OMWW has a high and unbalanced ratio of C/N and is often necessary to add other materials to optimize the C/N ratio (e.g. ~35) in order to cause more rapid microbial degradation in the soil and, minimize competition with agricultural crops for the nitrogen contained in the solution circulating in the soil (Paredes M.J. et al., 1986) — see also Chapter 2 "Characterization of olive processing waste", section: "Antimicrobial activity of OMWW" and Chapter 10: "Uses", section: "Use as fertilizer/soil conditioner".

Riffaldi R. et al. (1993) evaluated the changes in organic and inorganic compounds of soil amended with two doses of sludge obtained from OMWW during a 40-day incubation period. Differences between the amounts of organic components of the amended soil and those of the control, although related to doses and sampling time, disappeared at the end of the experimental period. On the contrary, the inorganic anion content was still different for the various processes, which suggest, especially for NO_3^- and SO_4^{2-}, a transient inhibition in the soil–sludge system.

Zenjari B. and Nejmeddine A. (2001) reported the effect of successive OMWW treatments on the chemical properties of clay soil profiles. The study showed that the clay soil has a very effective absorption/adsorption capacity. Over 99% of nutrients and 99% of phenols were removed after the first infiltration with OMWW. On the contrary, after the second infiltration the soil capacity to absorb/adsorb the anions was exhausted, while the phenol concentration was increased in the leachates which can present a risk of contamination of the groundwater.

The application of 2POMW to the soil is considered to have similar effects, although the available literature is still limited. The main organic constituents of 2POMW are lignin, hemicellulose, and cellulose — see Table 2.13. The high lignin content of 2POMW and the degree of binding of this component to other organic constituents in lignocellulosic materials may hinder the ability of microorganisms and their enzymes to degrade 2POMW, if used as a composting substrate (Alburquerque J.A. et al., 2004).

The use, therefore, of OMWW and/or 2POMW as a soil amendment requires knowledge of the effects that its application may produce on the status of the mineral nutrients in the plant-soil system — see Chapter 8: "Biological processes", section: "Irrigation of agricultural land/Land spreading" and Chapter 10: "Uses", section: "Use as fertilizer/soil conditioner".

Leaching

Although some research has been done to the effects of the addition of OMWW on soil characteristics, such as soil hydraulic properties or soil composition, information on the effect of these amendments on other compounds that are retained by the soil, such as pesticides or heavy metals, is scarce.

The discharge of OMWW in soils causes the release of heavy metals retained by them. This effect was simulated by leaching homogeneous soil columns with OMWW after passing solutions of Cu or Zn through the columns. Previous addition of a compost made from olive-mill sludge and plant refuse to the soil caused a significant reduction of the release of retained metal by OMWW. Previous addition of concentrated sugarbeet vinasse caused much less significant effects (Madrid L. and Díaz-Barrientos E., 1998b).

The effect of OMWW on the solubilization of some heavy metals present in a river's sediment was studied by equilibrating the sediment with solutions of various concentrations of the residue at various pH values (Bejarano and Madrid, 1992a,b). It was shown that at a given pH OMWW caused a nearly linear increase in dissolved lead (Pb) from the sediment as the OMWW concentration increased, and the lower the pH, the higher were the amounts released. Iron (Fe) and copper (Cu) were mobilized by OMWW at the higher pH values tested, but in more acid conditions the solubility of these two metals seems to be lower than in the absence of OMWW. For high OMWW concentrations, the concentrations of Fe and Cu tend to be a pH-independent value, which can correspond to an equilibrium distribution of metal-organic matter complexes between the two phases. OMWW does not show any mobilizing effect on manganese (Mn) or zinc (Zn) from the sediment, and in the case of Mn the sediment even removes part of the metal originally present in OMWW solutions. A later study by the same authors examined also the effect of OMWW on the solubilization of more heavy metals (Ni, Cd, Zn, Cu, Mn, Pb, and Fe) present in a sediment from Agrio river (Seville, Spain) at different pH values (Bejarano M. and Madrid L., 1996a–d). Metal solubilized by OMWW in solutions was compared with data from different fractions of metal speciation of the sediment. The data shows that the dominant effect is pH for all metals with the exception of Fe and Mn. Within a given pH, it is shown that the presence of OMWW causes mobilization of most metals studied at pH 5 except Cd and Zn and this effect is progressively less marked as pH decreases, so that at pH 4 mobilization is detected for Ni, Cu, Mn, and Pb, and at pH 3 is only noticeable for Ni and Mn. The joint effect of pH and of the presence of OMWW is the release of amounts of metals which are comparable to those metal fractions attributed to exchangeable and bound to carbonates.

The discharge of OMWW can affect sorption, degradation, and movement of pesticides in soil. Cox L. et al. (1996, 1997) studied the effect of OMWW on soil porosity and on leaching of the herbicides clopyralid (3,6-dichloropicolinic acid) and metamitron (4-amino-3-methyl-6-phenyl-1,2,4-triazin-5(4H)-one) in clay soil columns. Organic amendments used to enrich soils of low organic matter content can affect sorption and movement of pesticides in soils. Clopyralid moved more rapidly than metamitron in the unamended soil due to greater sorption and degradation of metamitron. Total amounts of clopyralid leached from the OMWW amended soils were significantly reduced (75 and 25% for the lower and higher dose, respectively) when compared with the unamended soil (100%), whereas metamitron did not leach at all from the amended soils. Sorption and degradation studies with soil slurry suggested this reduction may be mainly due to an increase in sorption and dehydration processes in amended soils, as a consequence of the increase in the organic carbon content. However, the decrease in mobility produced by OMWW amendment is greater than suggested from the sorption and degradation increases. The reduction in large size conducting pores and the increase in the small non-conducting pores, induced by OMWW amendment, produce an increase in the residence time of the herbicides in the immobile water phase, enhancing diffusion, sorption, and degradation processes, thereby retarding mobility. The retarding effect was more pronounced for metamitron than for clopyralid due to the higher sorptivity and degradability of the former herbicide. These results suggest the possible use of OMWW or similar wastewater amendment in reducing contamination of groundwater by pesticide drainage.

Albarrán A. et al. (2004) investigated the effects of the addition of exhausted 2POMW on the sorption, degradation, and leaching of the herbicide simazine [2-chloro-4,6-bis(ethylamino)-1,3,5-triazine] in a sandy loam soil. Simazine is a non-selective herbicide commonly used in olive-growing areas of Mediterranean regions at application rates close to 2 kg/ha. The soil was amended in the laboratory with exhausted 2POMW at two different rates (5 and 10% w/w). The results were compared with those of a previous study, where crude 2POMW was applied to the same soil (Albarrán et al., 2003). The addition of exhausted 2POMW increased the extent and strength of sorption of simazine, reduced herbicide biodegradation, and retarded the vertical movement of the herbicide through the soil and reduced the amount of herbicide available for leaching compared to the untreated soil. Therefore, amendment with exhausted 2POMW may be useful to prolong the residence time of the herbicide in the topsoil and to reduce the risk of groundwater contamination as a result of simazine leaching losses. Interestingly, the results were quantitatively different from those obtained for the crude 2POMW, illustrating the importance of the specific characteristics of the organic amendment in determining its effect on pesticide behavior.

Effects on Soil Biological Properties

Microbial Behavior

In nature OMWW is metabolized by microorganisms, insects, larvae, and earthworms present in the soil, to give a mixture of complex aromatic molecules known as humic or fulvic compounds or, more generally, as humic acids or humic extracts — see Fig. 3.2.

There are several studies on the effects of OMWW on the microflora of the soil (Paredes M.J. et al., 1986; Moreno E. et al., 1987, 1990; Lombardo N. et al., 1988; Flouri et al., 1990; Marsilio V. et al., 1989; Briccoli-Bati C. et al., 1990; Picci G. and Pera A., 1993) and the invertebrate community (Senette C. et al., 1991; Cicolani B. et al., 1992).

Marsilio V. et al. (1989) showed the beneficial influence a controlled disposal of OMWW can have on the populations of microorganisms; in a soil treated with 160 m^3/ha of OMWW, the number of microorganisms per gram of earth is multiplied 2.5 times after 15 days and 2.3 times after 100 days with reference to an untreated soil. This increase of the microflora and/or microfauna is accompanied by an accentuation of the respiration activity by more than 100%. The application of OMWW has a positive effect on the populations of mushrooms, actinobacteria, N$_2$-fixing bacteria, and cellulololytic bacteria. A negative effect has been recorded on the nitrite and nitrate bacteria after 15 days of the application. However, the

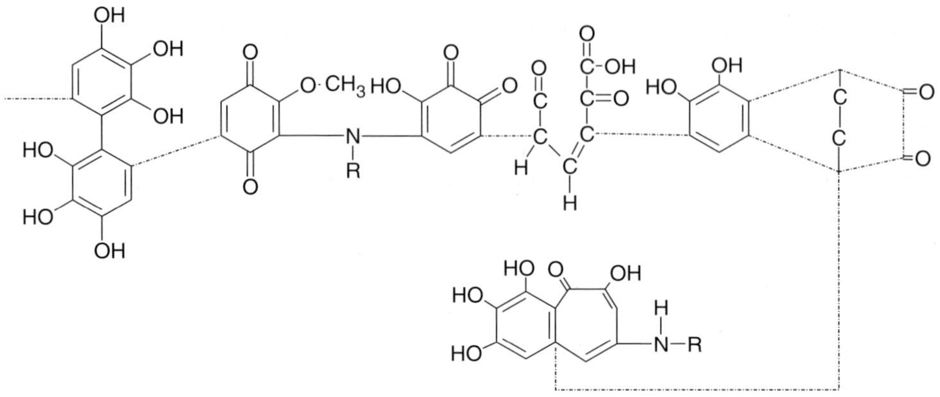

Fig. 3.2. Chemical formula of a humic acid.

population of the nitrate bacteria after 100 days was larger than that of the untreated soil. The increased amount of bacterial biomass may be attributed to the fermentation of the dispersed organic substances and the improved aeration of the soil.

A greater development of free-living N-fixers in soils treated with raw OMWW has been often recorded (Paredes M.J. et al., 1987; Flouri F. et al., 1990; Balis C., 1994). An increase of nitrogen fixation in soils treated with bioremediated in liquid culture (Balis C., 1994) or with composted OMWW has also been made evident (Tomati U. et al., 1995). Moreover, pure cultures of some strains of free-living N-fixers have been successfully cultivated in an OMWW medium (Tomati U. et al., 1995b; Balis C., 1994).

However, the uncontrolled disposal of OMWW can disturb the ecological balance of the soil (Moreno E. et al., 1987, 1990; Paredes M.J. et al., 1986). In fact, the higher ratios of disposed OMWW result in anomalous fermentations of the dispersed organic substances. Paredes M.J. et al. (1986) observed an increase in total microbial counts after soil pollution with OMWW. Pollution provoked an increase in coryneform bacteria and decrease in *Bacillus*. It was that the organisms responsible for the degradation experiments were among those whose number was increased by pollution.

Effects on Plants/Crops

There are numerous studies on the agronomic effects of spreading fresh, stored, or treated OMWW on soil cultivated with cereals or other annual crops (Albi Romero M.A. et al., 1960; Morisot A., 1979; Potenz D. et al., 1980; Morisot A. and Tournier J.P., 1986; Di Giovacchino L. et al., 1990, 2001, 2002; Bonari E. and Ceccarini L., 1991, 1993; Galoppini C. et al., 1994; García-Ortíz R. et al., 1993; Bonari E. et al., 1993; Caporali F. et al., 1996), with grapevine (Catalano L. and Felice M. de, 1989; Di Giovacchino L. et al., 1996, 2001, 2002) and with olive trees (Theophrastus c.372–c.287 B.C.; Catalano L. et al., 1985; Proietti P. et al., 1988; Marsilio V. et al., 1989;

Briccoli-Bati C. and Lombardo N. 1990; Briccoli-Bati C. et al., 1990) and on the photosynthesis of plants (Palliotti A. and Proietti P., 1992).

OMWW inhibits the germination of various seeds and early plant growth of several vegetable species (Wang T.S.C. et al., 1967; Pérez D.J. et al., 1986; Capasso R. et al., 1992b, 1995; Simone de et al., 1994, 1998; Ciafardini G. et al., 1998; Della Greca M. et al., 2001; Alliotta G. et al., 2002). It has also been reported that the direct application of raw OMWW on plants causes leaf and fruit abscission (Fiume F. and Vita G., 1977; Bartolini S. et al., 1994). The phytotoxicity of OMWW has been attributed by several authors to its phenolic content and some organic acids such as acetic acid and formic acid, which are often produced along with other microbial metabolites during storage. Many of the phenolic compounds present in OMWW have a considerable phytotoxic effect (Wang T.S.C. et al., 1967; Capasso R. et al., 1992b, 1995; Della Greca M. et al., 2001). The information available on the capacity of the aromatic fraction of this type of wastewater in seed germination is scarce.

Herbaceous plants (maize, tomatoes, and rye-grass) are usually used for the study of the effects of OMWW on plants because of their high sensitivity to toxicity and their short biological cycle. In fact, if no negative effects are detected on the herbaceous plants, one can reasonably assume that the application of OMWW is not going to disturb the development of the olive trees.

Pérez D.J. et al. (1986) studied the effects of OMWW on seed germination and early plant growth of different vegetable species. Three types of OMWW at different concentrations were tested: raw OMWW, OMWW with organic matter removed, and deionized OMWW. Results generally indicated an inhibitory effect on seed germination and early plant growth by all treatments containing any kind of OMWW. Of the three types of effluent, raw OMWW had the greatest depressive effect, followed by deionized OMWW and finally effluent with organic matter removed. Barley showed the least sensitivity to phytotoxic effects while tomato was the plant most affected.

OMWW exhibited phytotoxicity for seed germination on radish and wheat (Alliotta G. et al., 2002) and on tomato and vegetable marrow (Capasso R. et al., 1992b, 1995; Komilis D.P. et al., 2005). Phytotoxicity studies of the four main phenolic compounds — catechol, 4-methylcatechol, tyrosol, and hydroxytyrosol — isolated from OMWW on tomato (*Lycopersicon esculentum*) and vegetable marrow (*Cucurbita pepo*) plants showed that the phenolic compounds were selectively toxic, except for 4-methylcatechol and its acetate. OMWW remained phytotoxic even after total extraction of the polyphenols, suggesting that other chemical products contribute to the overall phytotoxicity (Capasso R. et al., 1992b). This is of high importance because OMWW can come into contact with the crop because of possible flooding during the winter.

Morisot A. and Tournier J.P. (1986) carried out nitrogen mineralization trials on experimental crops of rye-grass grown in pots in the greenhouse. The input of the equivalent of 40 mm (40 $1/m^2$) of OMWW on an established rye-grass crop resulted in 45% decrease in yield. When the rye-grass was sown immediately after the input of waste (amounts equivalent to 40 and 80 mm) the dry matter yield equaled one-third of the reference yield. These negative effects could be explained by the very high salinity of this effluent, its acidity, the presence of polyphenols or other toxic biodegradable substances and a lack of nitrate–nitrogen. Rye-grass sown 45 days after the input of waste showed positive effects (significant for 5 out of 10 crops). Soil analyses revealed an increase in the amount of exchangeable potassium. Nitrogen mineralization trials revealed a slight decrease of soil nitrate–nitrogen, limited to 0.2 mg N/g waste. Neither ammonification, nor nitrification inhibition could be ascribed to OMWW, within the conditions and amounts studied. Therefore, the disappearance of nitrate–nitrogen could be attributed to denitrification or reorganization.

While the herbaceous plants are sensitive to the phytotoxic effects of OMWW no such effects were observed on olive trees under normal conditions of OMWW use. In fact, no inhibitory effects were observed on the development of the olive trees during the hibernal dormancy period because of the inactivity of their roots. The use of doses of OMWW up to $200\,m^3/ha$ did not cause any negative reaction on the adult olive orchards (Morisot A., 1979; Catalano L. et al., 1985; Tamburino V. et al., 1999). On the contrary, the application of OMWW after the resumption of sap rising is to be avoided (Le Verge S., 2004). A high mortality among young olive trees in pots was found after their irrigation in March with $800\,cm^3$ of OMWW, while no toxicity was detected after applying it in the months of November and December (Briccoli-Bati C. and Lombardo N., 1990).

It has been reported that the use of doses of OMWW of up to $150\,m^3/ha$ on olive trees of 10 years old could be tolerated. The vegetative growth of these trees has been vigorous and their nutritional state improved. An increased production of buds has also been observed at doses of 20, 40, and $80\,m^3/ha$ during a study of six months on olive trees (Marsilio V. et al., 1989); on the other hand, a dose of $160\,m^3/ha$ caused a slight decrease in budding.

The land spreading of 2POMW on soil cultivated with crops is considered to be beneficial, although the available literature is limited. The uncontrollable application of 2POMW to the soil has been shown to have a detrimental effect on the soil structural stability (Tejada M. and González-López J., 1997). It may also negatively affect seed germination, plant growth, and microbial activity.

Tejada M. et al. (2003) studied the effects of foliar fertilization with 2POMW at different doses on the productivity and quality of maize crops (Zea mays, L. cv. Tundra) located in Lora del Rio, near Sevilla (Andalusia, Spain). Foliar fertilizer was applied four times during the season and three different concentrations were tested (15, 30, and $50\,cm^3/100\,l$). Foliar fertilization increased leaf soluble carbohydrate contents, chlorophyll A and B and carotenoids, and increased the leaf concentrations of N, K, Fe, Mn, and Zn. Yield was a 19% increase in grain protein content. In a complementary study 2POMW was applied at 0, 10, 20, 30, and 40 ton/ha rates, respectively, on a maize crop for 2 years (Tejada M. and González-López J., 2004). The results indicated that 2POMW has a high potential as soil amendment due to its organic matter and nutrient content. The application of 2POMW to the soil caused an increase in soil chemical, physical, and biological properties.

Mineralization of organic matter produced higher contents of NO_3^--N in soil and increased NO_3^--N uptake by plants. Yield parameters of the second experimental season were better than those of the first experimental season due to the residual effect of the organic matter after application in the first season. In fact, application of the 2POMW gave a significant grain gross protein content of about 18 and 20% for each experimental season, a significant grain soluble carbohydrate content of about 25% for both experimental seasons, a significant number of grains per corncob of about 17 and 21% for each experimental season, and a significant maize yield of about 16 and 18% for each experimental season over the control.

Although the direct application of olive-mill wastes (OMWW, 2POMW) is an inexpensive way for disposal and recovery of their mineral and organic contents as fertilizers, their uncontrolled disposal on the soil can be a source of pollution and unfavorable environment impact. This would be associated with the acidic pH, inhibition of seed germination and plant growth, antimicrobial properties, and frequent unbalanced C/N ratio (Alburquerque J.A. et al., 2004).

Effects on Water

Formerly, OMWW was usually discharged into nearby rivers and streams with a considerable impact on the receiving waters. As a result many rivers in Spain (Guadalquivir river), Italy (Vomano, Saline, and Foro rivers in Abruzzo), and Morocco (Sebu and Fez rivers) have become anoxic (Di Giovacchino L. et al., 1976; Cabrera F. et al., 1984; Zenjari B. and Nejmeddine A., 2001). As early as 1982, in Spain a law forbade river disposal of OMWW. Later, other Mediterranean countries adopted similar legislation. Despite the existing laws and regulations there is still uncontrolled disposal of OMWW directly into natural waters, or into the sea, or even in the sewerage system.

The main effects of OMWW on natural water bodies are related to their concentration, composition, and to their seasonal production. The most visible effect of OMWW pollution is the discoloring of natural waters. This change in color is attributed to the oxidation and subsequent polymerization of tannins giving darkly colored polyphenols, which are difficult to remove from the effluent (Hamdi M., 1992).

OMWW has a considerable content of reduced sugar. Should this be discharged directly into natural waters, the result would be an increase in the number of microorganisms that would use this as a substrate. The effect of this is also the consumption of oxygen dissolved in the water, and thus, they would reduce the share available for other living organisms. This may cause an imbalance of the whole ecosystem.

Another similar process can result from the high phosphorous content. Phosphorous encourages and accelerates the growth of algae and increases the chances to eutrophication, destroying the whole ecological balance in natural waters.

In contrast to nitrogen and carbon compounds, which escape after degradation as carbon dioxide and atmospheric nitrogen, phosphorous cannot be degraded but only deposited. This means that phosphorous is taken up only to a small extent via the food chain, plant – invertebrates – fish – prehensible birds.

The presence of such a large quantity of nutrients in OMWW provides a perfect medium for pathogens to multiply and infect waters, which have severe consequences to the local aquatic life, and humans that may come into contact with water; as a result the natural disinfection process of natural waters is hindered.

The river fish *Gambusia affinis* and the crustacean *Daphnia magna* are severely intoxicated on exposure to phenol derivative concentrations of 40 mg/l for only 15 min. Accordingly, this ecologically deadly concentration would be easily reached by simply dumping 1 l of unprocessed OMWW into 100,000 l of circulating water; hence, for a typical Andalusian olive oil factory relapsing 5000 l of OMWW per hour on average, the collecting waterway should have a flow-rate of at least 100,000 l/s in order to avoid the aforementioned noxious effects (González-López J. et al., 1994). More specifically, OMWW has proved to have an almost immediate toxic action on *Carasius auratus* at concentrations of 10% (Bellido E., 1989a,b), on *Cyprinus carpio* and *Chondrostoma polylepsi* at concentrations of 6.8 and 8.8% (Fiestas Ros de Ursinos J.A., 1977), as well as on heterospecific populations of phytoplankton in the Guadalquivir river (Bellido E., 1989a,b) and the aquatic microbial flora (Martínez J. et al., 1986) at concentrations of circa 10%. OMWW pollution studies have also been done in the rivers Vomano, Saline, and Foro, in Abruzzo, Italy (Di Giovacchino L. et al., 1976).

The impact of OMWW in fluvial environments (rivers) was studied on the Alento river in Chieti, Italy. The results obtained with the application of the biotic index (E.B.I.) and the diversity index[11] (Simpson and Shannon indices) revealed the structural destabilization of the aquatic community with consequent reduction of the river capacity for reducing the effects of polluting substances through internal mechanisms of self-purification (Cicolani B. et al., 1992).

The acute toxicities of 13 samples of OMWW, from traditional and continuous processes collected from different regions of Portugal, were evaluated by Microtox, Thamnotoxkit, and Daphnia tests using three aquatic species: *Vibrio fischeri* (formerly *Photobacterium phosphoreum*), *Thamnocephalus platyrus*, and *Daphnia magna* and correlated with several physical and chemical parameters (Paixão S.M. et al., 1999). The acute toxicity of OMWW expressed in LC_{50} or EC_{50}, ranged from 0.16 to 1.24% in Microtox test, 0.73 to 12.54% in Thamnotoxkit F test and 1.08 to 6.83% in Daphnia test. These values reflect the high toxicity of OMWW to all test species. Statistical analysis of the results shows a high correlation between the two microcrustacean bioassays. Microtox test did not correlate significantly with the

---

[11]A diversity index is a mathematical measure of species diversity in a community. Diversity indices provide more information about community composition than simply species richness (i.e. the number of species present); they also take the relative abundances of different species into account.

other bioassays used. Establishing relationships between toxicity and physico-chemical parameters was difficult, although in microcrustacean bioassays, significant correlations were established between some chemical properties of OMWW and their toxic effects. Polyphenolic compounds presented little toxicity and were not biodegradable, whereas tannins were highly toxic and biodegradable.

OMWW from a Ligurian mill (Italy) was fractionated by ultrafiltration and reverse osmosis techniques and tested for toxicity on aquatic organisms from different trophic levels: the alga *Pseudokirchneriella sucapitata* (formerly known as *Selenastrum capricornutum*); the rotifer *Brachionus calyciflorus*; the two crustaceans, the cladoceran *Daphnia magna*; the anostracan *Thamnocephalus platyrus*. The fraction most toxic to the test organisms was that from reverse osmosis containing compounds of low molecular weight ($<350\,\mathrm{Da}$) and this was especially due to the presence of catechol and hydroxytyrosol, the most abundant components of the fraction (Fiorentino F. et al., 2003).

Relatively small spills of olive-mill effluents into the sewers have appreciable effects on the wastewater treatment plants, as pollution due to $1\,\mathrm{m}^3$ of OMWW corresponds to $100–200\,\mathrm{m}^3$ of domestic sewage. This overload can be dramatic, taking into account that in some areas the polluting load due to the OMWW during the milling period can be up to ten times the domestic sewage load. Therefore, because of this highly variable input, the same design problems of wastewater treatment plants are encountered as in holiday resorts where the resident population can also increase up to one order of magnitude.

Other negative effects of OMWW on sewers are related to the acidity and the suspended solids contents. Because of the high concentration of organic acids (mainly volatile fatty acids), olive-mill effluents are very corrosive to the sewer pipes (Rozzi A. and Malpei F., 1996). Extensive damage to the sewerage systems due to OMWW has been reported in the Apulia Region (Mendia L. and Procino L., 1964), and these corrosion phenomena are the main reasons why direct discharge of OMWW in sewers has formally been forbidden for many years, although in practice illegal dumping of OMWW and sludges in sewers has been a common disposal method for olive-millers.

It is worth noting that flotation/sedimentation tanks for oil recovery in the mills are not affected by acid corrosion even if they are made of limestone. The protection is probably due to a film of lipids which coats those tanks, and which no longer exists when the OMWW is discharged into the sewers, because of the much lower concentration of fats in the waste waters (Mendia L. and Procino L., 1964). However, the lipids in OMWW may form an impenetrable film on the surface of rivers, their banks and surrounding farm lands. This film blocks out sunlight and oxygen to microorganisms in the water, leading to reduced plant growth in the soils and river banks and in turn erosion.

Even though the fraction of pollutants as suspended solids in OMWW is low (of the order of 10–20%), the actual concentration is quite high. Suspended solids settle in the sewers close to the mills' discharge pipes and sediments build up. These obstructions hinder the normal circulation of the sewage, which will also settle.

The sediments, from both OMWW and domestic wastes, undergo anaerobic fermentation with consequent malodors and increase of the acidity contents of the wastewater.

Effects on Atmosphere

Olive-mills generate gas emissions resulting in significant odor complaints. Many of the volatile organic acids and other low-boiling organic substances create characteristic odors that can be detected around the olive-mills.

Fermentation phenomena take place when OMWW is stored in open ponds and/ or discharged on the land or into natural waters (Balice V. et al., 1986). As a result methane and other pungent gases (hydrogen sulfide, etc.) emanate from pond evaporation plants and pollute waters or soil. This leads to considerable pollution by odors even in great distances, especially during the oil production period.

Analysis of the composition of OMWW stored for several months in an open pond showed that almost half of its COD is composed of volatile fatty acids — see Fig. 3.3. Among the volatile fatty acids, butyric acid (18% of COD or 1,36 g/l) and the acids caproic, valeric, and isobutyric are particularly malodorous compounds (Le Verge S. and Bories A., 2004).

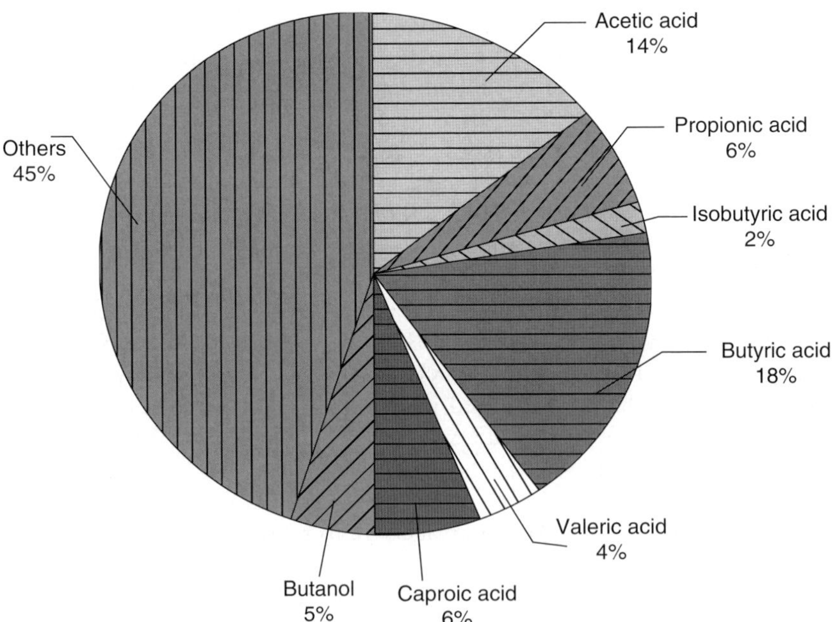

Fig. 3.3. Composition of OMWW (expressed in % COD) after storage in evaporation pond (May 2003) (Le Verge and Bories A., 2004).

An olive cake with high moisture content is also a source of odor nuisance, especially during warm and dry weather. Upon prolonged storage, seepage water also contributes to this odor unless a special drainage system is provided for the olive cake. During the drying of olive cake, an essential preparatory stage in the extraction of residual oil (pomace- or seed- or orujo-oil), extremely pungent odors are released in the waste gases. This causes problems for the functioning of seed-oil extraction plants near residential areas — see also Chapter 10: "Uses", section: "Recovery of residual oil".

Analysis of the condensates from the crude cake dry-distillation showed that the main pollutants were mixtures of organic acids of low molecular weight (8–10 g/l of concentrate) and fatty compounds (fatty acids or their esters, 5–10 g/l of the concentrate), the latter being found in the vapors in the form of air colloids (Papaioannou D., 1988) (See Table 3.1). The presence of short-chain organic acids could be explained as an intermediate product of anaerobic fermentation during the period after the olive cake had left the olive processing plant and before it entered the olive press oil-processing plant.

Table 3.1. Characteristics of the condensate of the wet press-cake dry distillation (Papaioannou D., 1988)

(a) *Water phase*
pH: 3.4–3.8
Color: Pale yellow
Fatty substances (floating): 0.5–1% v/v (of the wet phase)

COD	10,000–11,000 mg/l
BOD_5	7000–8000 mg/l
Acidity	5000–6000 mg/l (as –COOH)
Double bonds	100–120 mg/l (as –CH=CH–)
Total nitrogen	70–80 mg/l
Phosphates	40–50 mg/l
Sulfides	Not detectable (acetic lead qualitative test)

Organic acids (liquid chromatography)

Folic acid	8.2%
Acetic acid	81.4%
Propionic acid	1.5%
Butyric acid	3.6%
Lactic acid	5.3%

(b) *Fatty phase* (when separated from the water phase)

Color	Bright yellow
Acidity	10–15 mg/l (as –COOH)
Double bond	30–40 mg/l (as –HC=CH–)
Color change	Bright violet, after 2 h of aeration

Chapter 4

The Effect of Olive-Mill Technology

Evolution of Production Methods

According to Pliny the Elder[12] (see Fig. 4.1) the extraction of oil from the olive fruit was introduced by Aristaios, son of Apollo and the nymph Kyrene. Aristaios was considered also as the inventor of olive press (Pliny the Elder VII, 199). For this reason he was particularly honored in Sicily by the olive producers (Diodorus[13], IV, 82).

Olive oil production and trade spanned the centuries since the beginning of the Mediterranean civilizations. In Palestine, olive oil was extracted as early as the Chalcolithic period (after the Neolithic and before the Bronze Age, between about 4500 and 3500 B.C.) and this is attested by the discovery of primitive rock-cut installations. In Crete the finding of oil lamps, which show signs of burning, from the Early Minoan period attest to the knowledge of oil extraction, although proper presses were not used at that time (Hadjisavvas S., 1992).

In antiquity, as today, the production of olive oil involved three essential stages: (i) crushing, (ii) pressing, and (iii) separation of oil from water.

---

[12]Pliny the Elder (Gaius Plinius Secundus, A.D. 23–79), Roman naturalist, encyclopedist and writer born in Verona. He served a cavalryman in Germany and from his experiences wrote his first book "On the Use of the Javelin by Cavalry", the beginning of a literary career of enormous output. His famous Natural History (*Historia Naturalis*) was published in the year 77 A.D., two years before his death and is the only work of Pliny to survive. The work in 37 volumes is encyclopedic in coverage and includes information on astronomy, chemistry, geography, natural history, agriculture, medicine, astrology, and mineralogy. A popular translation covers five volumes, each of about 500 pages. Over 400 different authors are cited. Pliny was a compiler and the work is a monumental collection of science, technology, and ignorance. Although Pliny appears overly credulous, his encyclopedic coverage is the best known and most widely referred source book of "classical" natural history. Pliny is also a rich source of agriculture and horticulture.

[13]Diodorus Siculus, late 1st century B.C., Greek historian.

Fig. 4.1. Pliny the Elder.

Crushing

In the most primitive method, the olives are thought to have been placed into a trough and crushed with the use of a large pestle or treaded under the foot. Another simple method of crushing olives in antiquity was by spreading the fruit onto a hard surface and rolling a large cylindrical stone over it. An important change in the production techniques was the invention of the round crushing basin based on rotary motion which enabled the use of animal power. The two classical forms of this equipment, as described by the Roman agricultural writers Cato[14] (see Fig. 4.2) and Columella[15], were the *trapetum* and the *mola olearia*.

---

[14]Cato the Elder or Cato the Censor (Marcus Porcius Cato) (234–149 B.C.), Roman statesman and writer. His *De Agri Cultura*, a treatise on farming, is the oldest surviving prose work in Latin.

[15]Columella Lucius Junius Moderatus (1st century A.D.), Roman writer on agriculture, born in Gades (now Cádiz), Spain. Of his work there remains the 11-volume entitled *De re rustica* (On Agriculture), treating general husbandry, the care of domestic animals, and farm management. The 10th book, modeled on Virgil, is in hexameters. A short essay on trees also survives *De Arboribus* (On Trees). Columella's Latin is facile and elegant, and his information is surprisingly practical and accurate. They are considered to be the most comprehensive and systematic of all Roman agricultural treatises. The works may have been written on request or commission from a certain Publius Sivinus, known only from Columella's references to him. His work is amazingly modern in feeling and devoid of superstition, although the discussion of slaves is disconcerting.

The *trapetum* (from the Greek word "τραπητής") was the olive-mill *per excellence* in his time (Cato, *De agri cultura* 21–23). The *mola olearia* comprises roughly the same elements as the *trapetum* and looks very similar. It is well possible that the *mola olearia* is a later version of *trapetum*. Drachmann A.G. (1932) studied in detail the operations of *trapetum* and *mola olearia* and gave instructive reconstructions based on Cato's and Columella's accounts, respectively — see Figs 4.3 and 4.4.

Fig. 4.2. Cato the Elder.

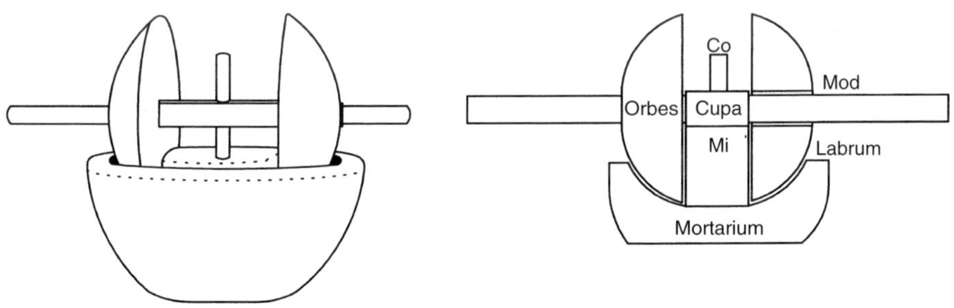

The immovable par of a trapetum was made of lava in the shape of a large cup (mortarium) housing a central pillar (miliarium). The miliarium was a few cm higher than the lip of the cup (labrum). On the top of the miliarium there is a square hole, in which an upright iron pin (columella) was fastened by means of lead. The movable part consisted of a wooden beam (cupa), which fitted over the collumela and rested on the miliarium in a horizontal position; on its

Fig. 4.3. The *trapetum*.

two arms were threaded two willstones (orbes), flat on the side towards the miliarium, but convex on the outer side, so that they dipped into the ring-shaped cup. The orbes were kept in their place by a system of washers and wedges. When properly adjusted they would keep a distance of exactly one Roman inch from the miliarium, from the bottom of the hollow and from its outer, curved side. When the trapetum was filled with olives and the cupa, which projected beyond the orbes to form handles, was turned, the orbes would perform a double rotation, going round the miliarium and at the same time turning on their axles. The result was that the olives were crushed, but the olive stones were not, which was indeed the point of the whole arrangement.

Fig. 4.3. (*continued*).

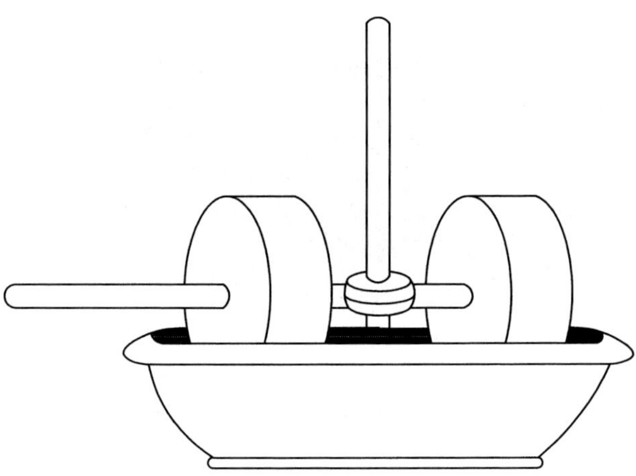

The mola olearia consisted of two cylindrical mill stones rotating on a horizontal axle which was carried by a vertical beam that turned around also, and was placed in the middle of the flat surface on which the grinding took place. The fact that the two mill stones were carried by the short cross piece and did not rest on surface, allowed the mill stones to be adjusted thus preventing the olive stones from being crushed.

Fig. 4.4. The *mola olearia.*

In early modern Europe, the traditional animal- or water-powered olive-mills used a vertical millstone that turned on a metal pivot around a vertical axle and rolled upon another circular stone (the dormant stone), horizontally placed, crushing the fruit by simple pressure.

In Calabria, olives were crushed in the *trappeto*. The *trappeto* — which was different from the Roman olive-mill *trapetum* — used a vertical millstone approximately 1.1 m in diameter and measuring 0.4 m on its edge. The dormant stone was often concave, like a basin, so that the millstone made only partial contact with it. A mule supplied power, while a man fed the olives into the stone with a spade. In Apulia, millstones were bigger (1.5–2.0 m), heavier and rolled on a flat dormant stone. The edge of the millstone was rounded, reducing the contact surface by two-thirds and increasing the force applied at the point of contact. The Apulian olive-mill was carved into the rock, a few meters underground. This type of mill was widespread in southern Apulia because of its relatively inexpensive construction and the insulating properties of stone (warmth being a crucial factor in oil production. By comparison, traditional mills in Provence used smaller millstones and were powered either by a horse (*moulin à sang*) or by waterwheel (*moulin à eau*). Mill of the latter type used a horizontal wheel placed under the dormant stone and turning around the same axle as the millstone (Mazzotti M., 2004).

Pressing

In the early history of olive oil extraction, simple installations consisting of a sloping crushing floor connected to a lower collecting vat were the devices combining the first two stages of olive processing (Frankel 1984, III). The first important technical improvement in oil production was the introduction of the lever and weights press — in the Late Bronze Age in Crete, Cyprus, and Ugarit[16] and in the Iron Age in ancient Israel — which became the most popular type in Antiquity. The mode of operation of the lever and the weights press is illustrated is Fig. 4.5. In its basic form, this press consisted of a long wooden lever, one end pinned in a recess in a wall or between two pillars, while the other could be pulled down to exert pressure on whatever was under the lever or beam — in this case, a bag of olive paste. The sacks used with this type of press were made of vegetable fiber to allow the oil to pass through them.

To cope with the increased output, a series of improvements were brought about in the pressing operation. The culmination of these improvements was the employment of the screw (*cochlea*) in lever press, which became the most popular combination from the time of its introduction up to the middle of the twentieth century — see Fig. 4.6. The introduction of the screw was a major technical improvement in the pressing operation, the second after the lever itself. Its application enabled greater force to be brought in and as a consequence the press bed could be placed anywhere between the anchoring point and the screw. A screw press could easily be operated outdoors as there was no need for a pulley to raise the lever up (Hadjisavvas S., 1992). Although the technical characteristics of the new invention were far more advanced than the lever and weights press, the latter continued to be used up as late as the nineteenth century.

Up to the middle of the eighteenth century, the olive oil extraction was based on the same basic types of machinery used in the antiquity. That is not to say that technology had stagnated. Archaeological and ethnographic research has

---

[16]Ugarit (modern site Ras Shamra) was an ancient cosmopolitan port city, sited on the Mediterranean coast of northern Syria.

One end of the press beam is anchored either in a wall recess or is attached to a cross piece supported by two wooden uprights. A pile of bags is pressed by the beam and the liquid pours into an open large vessel underneath. Two pierced boulders hang from the free end of the beam to pull it down. A worker adds his weight in this respect (Hadjisavvas S., 1992).

Fig. 4.5. Lever press as depicted on an Attic Skyphos, about 520–510 B.C. (Photograph © Museum of Fine Arts, Boston).

documented the continuously changing forms of oil-production machinery. And always, changes in the socioeconomic setting of the oil-producing regions — such as a shift to large-scale production for export-shaped technological developments. However, the basic process of interconnected social, economic, and technological change is in no way unidirectional (Mazzotti M., 2004). For instance, during the late Roman–early Byzantine period with trade declining on the Mediterranean routes, provinces once renowned for their exports began to rely more on local consumption, and producers in those areas turned from comparatively sophisticated technology back to simpler alternatives. The result was the success and diffusion of a rudimentary new milling–pressing system that involved simply rolling a cylindrical stone over the olives. Oxen were preferred over horses not only because they were stronger, but because of their more regular pace (Mazzotti M., 2004).

The modern golden age of olive oil began around 1750. Increasing consumption of oil for cooking and eating was only one factor accounting for growing demand

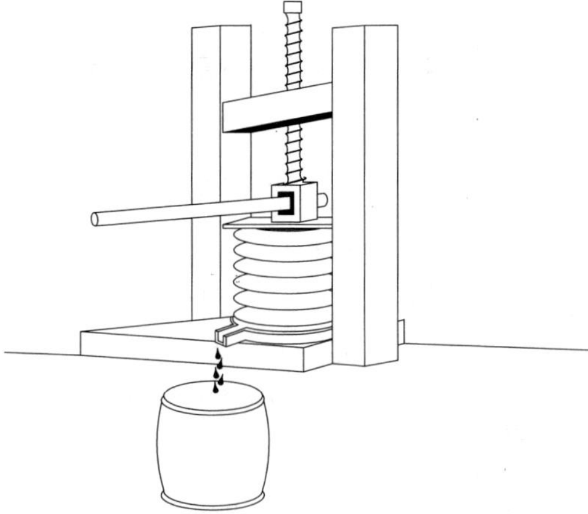

Fig. 4.6. The screw press.

and rising prices; industrial uses, which included lighting, lubrication, and the manufacture of soap and wool, also played an important part. As international demand for olive oil steadily grew, the main concern for producers and traders became not the supply but rather the quality of the product. In the middle of the eighteenth century, the centers of production of high-quality oil were Provence (Aix), the Italian Riviera (Genoa), and Tuscany (Lucca). Among the characteristics that distinguished the oils of these regions from those made in the rest of the Mediterranean basin the most important was their low acidity. They were also more transparent, sweeter, and crucially, easier to preserve.

Four basic types of oil presses were in use around 1750, each with endless local variations. Torsion presses were common in Corsica and southern Italy. In this type of press, very ripe olives were put in a large sack made of goat hair, which was then pushed into place in a wooden trough. Two people then twisted the sack using a pair of sticks, forcing oil from the olives, which collected in the trough and drained into a receptacle. Each sack of olives would be pressed in this way several times, with hot water being used at the end to help extract the last remnants of oil from the fruit.

The lever or beam press (*presse à arbre*) was common in southern France. A more complex variation included the use of a capstan which lifted a counterweight. The lever could also be forced down with the aid of screws. In the first edition of *Encyclpopédie*, Diderot praised a counterbalanced lever-and-screw press common in Provence and Languetoc, called the *pressoir à gran banc* or "Greek press", and recommended its use in the manufacture of olive oil. Another variation was the *pressoir à taissons*, in which the screw was fixed to the ground. The other most common type of press, the screw press, worked by means of the direct action of one or two screws. The double screw press, in which a wooden beam is forced downward by two fixed screws, was found all over Italy, along the Adriatic coasts, and on some Greek islands. One problem with this type of machine was that the upper beam and the two lateral pillars broke down rather easily.

During the second half of the eighteenth century, a group of reformer–entrepreneurs in different parts of Mediterranean Europe employed a mix of enlightenment ideas and advanced technology to rationalize and mechanize olive oil production as a way of meeting the increased demand. The reformers sought to build new machines which would not merely produce more oil but high-quality oil. Prices could be increased and new markets created only by changing the nature of the product and widening the range of possible uses. High-quality oil not only lasted longer than common oil, which made it desirable for long-distance trade; it also burned more efficiently in lamps and made a better lubricant for industrial machinery.

According to Mazzotti M. (2004) the modernized methods of making olive oil did not evolve in some sort of natural development but were rather the consequence of the new meaning attached to oil production by reformer–entrepreneurs. In order to produce low-acidity oil, olives must be processed earlier in the season, when they are less ripe. An early ripe meant paying more for labor, as under-ripe olives were harder to pick. Similarly, the practice of leaving olives to ferment after they had been harvested, common all over the Mediterranean had to be abandoned. Traditional mills included large storage facilities — the *zimboni* in Calabria, the *camini* in Apulia, the *tulhas* in Portugal — where the fruits fermented for weeks or even months before being processed. One reason for this practice was that fermented olives were easier to crush and press, a crucial matter in regions lacking waterpower.

Many traditional mills, designed for fully ripe olives, proved unequal to the task of crushing fresh olives. Redesigning the millstones around the need to use fresh olives, however, precipitated other changes to the structure of the mill. Axles and pivots could not bear the weight of the new stones, and therefore had to be reinforced with metal parts and massive masonry. Also, traditional sources of power could no longer drive the stone effectively. Vertical wheels replaced horizontal ones placed under the basin of the mill, preventing water from cooling the basin and, hence increasing the effectiveness of the crushing action.

To facilitate the flow and to avoid it from becoming dense because of cold weather, the mill temperature was kept constantly high, and loads of hot water were poured on the stacked containers. In time, the same procedure was maintained as presses were introduced that were made totally or partially with metal. Meanwhile, the beam press was rapidly abandoned. The longer levers and additional capstans or counterweights did increase the power of beam presses, but in the late eighteenth century this design reached its structural limits. The consequences of a giant lever suddenly freed by the rupture of a capstan or a rope were spectacular and devastating (Mazzotti M., 2004).

The continuous, regular, and synchronized functioning of the new mills required also a new and more intense kind of work. Technological innovation could succeed only where the local workforce could be effectively disciplined to its new role. Disciplining the workforce served other purposes besides maximizing output.

Fig. 4.7. Olive-mill, *Nova Reperta* (Johannes Stradanus, 1523–1605).

It was also the concrete epitome of a new social order that the bourgeois elite of southern Europe sought to impose upon the rural communities (Mazzotti M., 2004). The traditional design of an olive-mill was a constitutive element of southern European societies — see Fig. 4.7. Modifying it meant modifying traditional landscapes and ways of life as well. Technical innovation succeeded only where the reformers succeeded in reshaping traditional ways of life as well traditional machinery.

At the turn of the twentieth century, hydraulic presses were introduced. The most recent techniques have radically changed the oil extraction concepts and methods. The continuous three-phase centrifugal process was introduced in the 1970s notably to increase processing capacity and extraction yield, and to reduce labor. In the early nineties, the two-phase centrifugal process was introduced, where no process water is used.

Oil Separation

A variety of methods, all based on the principle of gravitation were applied to separate oil from water. The simplest way was by skimming the floating oil by hand or with the help of a ladle. The second method was to draw off the water through a stoppered hole at the base of the receptacle tank. In the third method, the floating oil was conveyed into a lateral tank through an outlet at the rim of the receptacle tank. All the above methods involve a second stage of settlement that was accomplished in settling vats. The fresh oil was "muddy" and time was needed for the impurities to settle to the bottom of the vat. The settling vats were provided with a central concave depression for the collection of the remaining impurities.

Most of the olive-mills were built along the Mediterranean coast or close to rivers, and creeks. The availability of water was a basic factor for the construction of an olive press. Water was necessary in many stages during the complex procedure of turning olives into oil. First, olives were washed before being crushed. Boiling water was poured on the pile of woven bags after the initial pressing, thus washing any remaining oil from them and the press bed. Boiling water was also used before and after each pressing operation to wash old rancid oil, which could spoil the taste of fresh oil. The dregs and other impurities left on the surface of the press were washed away by water and disposed of in the area surrounding the press. At the site of a press excavated in Cyprus were found many olive stones within a layer of blackened soil just below the central outlet of the press indicative of possible soil pollution (Hadjisavvas S., 1992). There is archaeological evidence that olive effluents have been damaging delicate shoreline environments for thousands of years around the Mediterranean.

Olive Oil Production Systems

The process of olive oil production can be subdivided in two main phases: (1) preparation of a homogeneous paste and (2) oil extraction and purification.

In the 1st phase, the olives are processed by means of grinding and mixing pulp and olive stone, followed by a heating process to further break down olive cells and to create large oil droplets. In the 2nd phase, oil is extracted by a press or a decanter. Water and solids are thus separated from the oil and further centrifuged in order to recover residual oil. Oil is purified through clarification by sedimentation or filtration by vibrating screens. OMWW streams are also clarified before disposal. Residual solids from the purification step are mixed with those coming from the extraction step.

In modern olive-mills, extraction from the olive paste is based on the principles of:

- pressing (traditional or classical system);
- centrifugal (continuous system):
  - three-phase;
  - two-phase;
- stone-removing process;
- percolation (selective filtering);
- chemical separation;
- electrophoresis.

The last four methods are hardly used. In the pressing system and the three-phase continuous system the waste is a liquid waste product (OMWW) and a solid waste product; in the olive-mills of two-phases, the waste is a slurry waste (2POMW). The pressing system, as well as the three- and two-phase centrifugal systems are given schematically in Fig. 4.8.

A critical aspect of olive oil production is represented by high energy requirement of the milling process. According to the technology utilized in olive processing, namely pressing or continuous centrifugation systems, the amount of energy consumption, with reference to one ton of treated olives, is 40,000–50,000 and 48,000–65,000 KJ, respectively (Basile P. et al., 1998; Caputo A.C. et al., 2003).

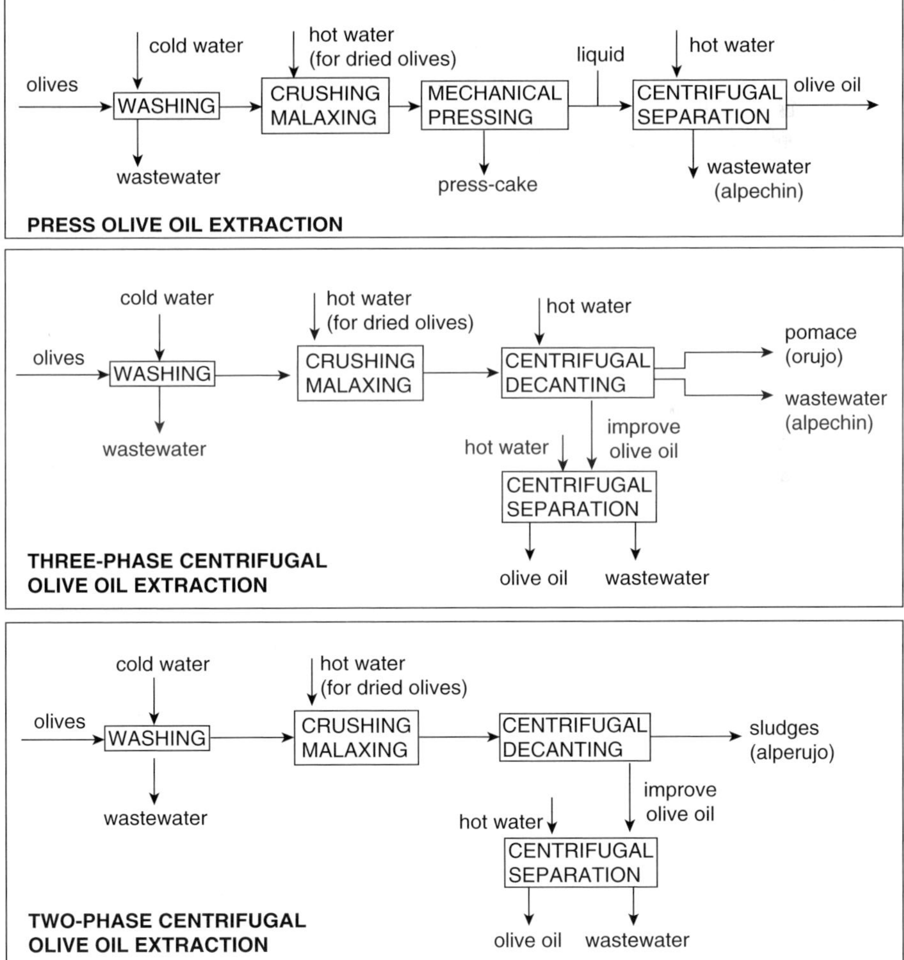

Fig. 4.8. Present olive oil production systems (adapted from Vlyssides A.G. et al., 1998).

The quantity and quality of the produced liquid and solid wastes are strongly influenced by the oil extraction method. The traditional press system produces the "strongest" OMWW, with concentrations of the order of 100–200 g COD/l. The three-phase system produces more dilute OMWW. According to the latest developments in olive oil production, the very emission of OMWW can practically be reduced to nil by transferring it in the spent olive residues. This two-phase technology is considered to be very promising but in fact it simply transfers the problem of disposing of the olive-mill waste from the mill to the oil refineries, where the spent olive residues, prior to oil solvent extraction, must be dried with considerably higher energy requirements than the case for traditional or continuous oil production processes.

The effect of the extraction process on the quality of the virgin olive oil is well documented (Di Giovacchino L. et al., 1994; Stefano G. de et al., 1999; EU project: AIR3-CT93-1355). In fact, the extraction system affects the composition of minor components of oil, including phenolic compounds, which are known to have several functional, sensory, and nutritional properties. For instance, the phenols are correlated to (i) the pungent and bitter taste of olive oil, (ii) the reduction of the oxidative process of fruity flavored aromatic compounds, (iii) the improvement of the olive oil shelf life, and (iv) the health benefits of olive oil consumption in the Mediterranean diet — see Chapter 10: "Uses", section: "Recovery of organic compounds". The phenols are either originally present in the olive fruit or formed during the processing of olive into oil.

Phenols present in olive paste are soluble in water and oil, depending on their partition coefficients (K_p) and temperature. Addition of water to the paste alters the partition equilibrium between aqueous and oil phases and causes a reduction of phenol concentration through dilution of the aqueous phase. A coincident lower concentration of these substances occurs in the oil phase. As a matter of fact, a large amount of the antioxidants is lost with the wastewater during processing — see Chapter 2: "Characterization of olive processing waste", section: "Organic compounds". The partition coefficients (K_p) between oil and water phases of selected phenolic compounds, as determined experimentally and predicted using the UNIFAC model, are given in Tables 4.1 and 4.2, respectively — see Chapter 2: "Characterization of olive processing waste", section: "Experimental techniques", "Phenol characterization". The K_p of the phenolic antioxidant compounds was estimated to be from as low as 0.0006 for oleuropein to a maximum of 1.5 for 3,4-DHPEA-EA (3,4-dihydroxyphenylethanol-elenolic acid). Because the K_p values were very low, some changes in the process were introduced in order to achieve a higher concentration of antioxidants in the oil. A temperature increase could lead to increasing the partition coefficient. Also limiting the quantity of water during oil extraction formed the basis for designing alternative processes for increasing the antioxidant concentration in the olive oil (Rodis P.S. et al., 2002).

The distribution of hydrophilic phenols between the oil and the water phase, as related to their solubility, is not the only mechanism involved in the reduction of the oil phenolic concentration during malaxation: oxidative reactions catalyzed

Table 4.1. Experimentally estimated partition coefficients (K_p) of olive oil antioxidants between oil and water phases (Rodis P.S. et al., 2002)

Antioxidant	Partition Coefficient (K_p)
Oleuropein	0.0006
3,4-DHPEA	0.0100
Protocatechuic acid	0.0390
Tyrosol	0.0770
Caffeic acid	0.0890
3,4-DHPEA-EDA	0.1890
3,4-DHPEA-EA	1.4900

Table 4.2. Partition coefficient (K_p) of olive oil antioxidants among oil and water phases as predicted by UNIFAC (Rodis P.S. et al., 2002)

Antioxidant	K_p (25 °C)	K_p (45 °C)	K_p (65 °C)
Oleuropein	0.0012	0.0034	0.0087
3,4-DHPEA	0.0004	0.0012	0.0026
Protocatechuic acid	0.0250	0.0430	0.0580
Tyrosol	0.0970	0.1510	0.2220
Caffeic acid	0.3600	0.5920	0.7970
3,4-DHPEA-EDA	0.1870	0.2990	0.4130
3,4-DHPEA-EA	11.8000	16.5000	20.0400

by endogenous oxidoreductases such as polyphenoloxidase and peroxidase can promote the phenolic oxidation during processing (Servili M. et al., 2004).

Interaction between polysaccharides and phenolic compounds present in the olive pulp may be involved in the loss of phenols during processing. Polysaccharides may link hydrophilic phenols in the past, thus reducing their release in the oil during crushing and malaxation (Servili M. et al., 2004). In this regard, it has been shown that the use of technical enzymatic preparations containing cell wall degrading enzymes during processing can improve the oil phenolic concentration (Siniscalco V. and Montedoro G.F., 1988; Siniscalco V. et al., 1989; Ranalli A. et al., 2003).

Vierhuis E. et al. (2001) showed that the addition of commercial enzyme preparations reduced the complexation of hydrophilic phenols with polysaccharide, thus increasing the concentration of free phenols in the pastes and their release in the oils and the vegetation waters during processing.

Pressing Process (Traditional or Classical System)

In practice, modern presses operate on a pile of layers of paste laid between mats of nylon. The paste is subjected to hydraulic pressure gradually reaching

$300\text{--}500\,\text{kg/cm}^2$ depending on the characteristics of the olives like maturity and kind of the fruit. Oil and water flow, either from the sides of the pile or down a central pole. In this way the solids are separated from the oil/wastewater mixture. This mixture is separated by sedimentation, and later by centrifugation.

With the double press method the paste is pressed in succession. The first press uses about half the pressure of the second press. Such presses may hold up to 500 kg of paste per pressing. One pressing process takes up to two hours. The pressure given on the olive paste is about $100\,\text{kg/cm}^2$.

The pressure system does not require addition of water to the olive paste. However, if the olives are difficult to process and the oil phase does not separate easily from other phases, or when ripe olives are processed in such a system, addition of small quantities of water ($3\text{--}5\,\text{l}/100\,\text{kg}$ of olives) during crushing, kneading, and washing of the tower after squeezing may be required.

After pressing the pile is dismantled, the olive cake removed and the mats are reloaded with the next batch. Spontaneous decanting can separate the oil that flows out or by centrifugation in vertical centrifuges, which separate the vegetation water from the oil. Depending on the degree of separation, the appearance of the oil can range from perfectly limpid to turbid to meet customer preferences. The presence of particles will lead to a less stable and more prone to rancidity oil.

In general, this process is associated to high quality oil due to the low temperature needed for the extraction; however the resulting oil quality is very dependent on the hygienic conditions during processing. If the press is not kept as clean as possible during the processing, then the oil quality will be lower due to the contact of the oil with old and already oxidized particles. The pressing process is costly from the point of view of manual labor and the need of filtering materials.

Centrifugation

The extraction of oil with centrifugal power started at the end of the nineteenth century and passed through several evolutionary stages. The centrifugal systems use horizontal centrifugal separators, known as *decanters*. The continuous centrifugation involves the steps of: crushing of the olives, mixing the olive paste, and centrifuging with or without water addition according to "three-phase" or "two-phase" mode, respectively.

i) Three-Phase Centrifugation

In 1965, the Alfa Laval firm presented in the market the centrifugal olive cluster "CENTRIOLIVE" and in 1969 the "COSI", while the Pieralisi firm put in the market the cluster "SC" in 1971. Since then, several manufacturers have designed and put in the market a number of centrifugal live clusters. Since then, different manufacturers have designed and put in the market a number of centrifugal olive clusters, all based on the same operation principle. This process exploits the specific weight differences between water and oil. The water-thinned paste is turned at very high speed in a horizontal centrifuge. Traditional continuous processing of olives using centrifuging extractors requires the addition of warm water, and the resulting olive paste is separated into three phases: oil, vegetation plus any added water (OMWW), and olive cake (stones and pulp residue).

The three-phase centrifugal mills are advantageous compared to the traditional ones because it

- requires less human labor;
- has higher olive oil production rates.

Disadvantages of this process include:

- increased amounts of wastewater that is produced due to increased water utilization (1.25 to 1.75 times more water than press extraction),
- loss of valuable components (e.g. natural antioxidants) in the water phase,
- quality characteristics of the oil are not as high as in the case of the press system,
- problems of disposal of the wastewater.

It has been a common practice to recycle the obtained OMWW in its pretreated or untreated form to the process of olive oil extraction. This water-based by-product, which comprises the vegetation water and the water used in the different stages of oil production plus, olive washing water, waters from filtering disks and from washing of equipment, and rooms, could help to reduce the demand for fresh water. However, because of its oxidative nature, it affects negatively the quality of the produced olive oil and it is hardly used anymore.

ii) Two-Phase Centrifugation

The failure to develop a suitable end-of-pipe wastewater treatment technology gave the opportunity to technology manufacturers to develop the two-phase process, which uses no process water, and delivers oil as the liquid phase and a very wet, olive cake (2POMW) as the solid phase using a more effective centrifugation technology. This technology has attracted special interest where water supply is restricted and/or aqueous effluent must be reduced.

Spain was the first country where the two-phase system was used and from there the new technology was spread and installed around the world — see also Chapter 1: "Introduction". The two-phase extraction process has substantially reduced the volume of wastes produced with the traditional three-phase system, which requires the supplementary addition of waters (Alba-Mendoza J. et al., 1990). Nevertheless it has created a new waste material, 2POMW in large quantities (approximately 4,500,000 tons in Spain for the year 2002), which requires new knowledge about how it must be handled (Junta de Andalucía, 2002; Alburquerque J.A. et al., 2004). The depleted product, with a fat content lower than 1% and moisture around 65%, is an inconvenient and very abundant residue (around 80% of the ground olive) impossible to store in the olive-mill.

Decanters based on the two-phase process have been developed by several companies (Alpha Laval, Sweden; Pieralisi, Italy; Oliomio, Italy; Flottweg GmbH; Fethil, Turkey, and others). The performance of such a decanter (Pieralisi, Jesi, Italy) was evaluated in comparison to a traditional three-phase extraction process and was found to produce olive oil in similar yields to the three-phase process, but of a superior quality in terms of polyphenols and diaphanous content. The higher contents of total phenols and *o*-diphenols were attributed to the greater amount of added water used in the three-phase decanter diluting the aqueous phase and thus reducing oil phenol contents as a result of partitioning. Oils produced by the traditional three-phase process were approximately half as stable as the two-phase oils, as determined by the Rancimat method[17]. In addition, the two-phase process did not produce a vegetation water phase during oil extraction.

EP557758[18] (1993) discloses a two-phase system designed and developed by Westfalia Separator A.G. In this process — outlined in Fig. 4.9 — , the washed olive fruit (1) is fed to a mill (2) and, from there to a mixer (3) in which the olive paste is disrupted. When fresh olives are used, the paste is produced without addition of water, whereas, when dried olives are used, a small amount of water is added appropriate for the condition of the olives. The disrupted paste is fractionated in a two-phase helical conveyor centrifuge (5) into oil (6) and a solid/water mixture (7). The solids/water mixture (7) is suitable for subsequent extraction and drying. The oil (6) is fed, with the addition of a small amount of water (9) to the disc centrifuge (8) in which pure oil (10) and separated water (11) are obtained. The water is relatively clean and can be mixed with the circulating water used for washing

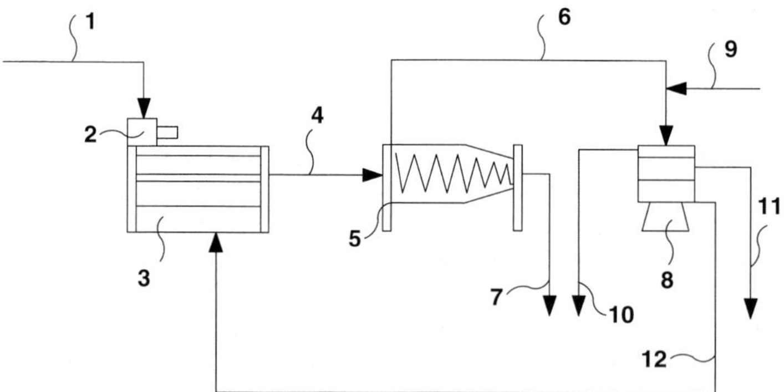

Fig. 4.9. The two-phase centrifuge developed by Westfalia Separator A.G. (EP557758, 1993).

---

[17]Method used to determine the oxidative stability of the oil; see also: Aparicio, Roda, Albi, and Gutiérrez (1999). Effects of various compounds on virgin olive oil stability measured by Rancimat. *J. Agric. Food Chem.*, **47**, 4150–4155.

[18]The patent was revoked on 16.05.2001.

the olives. Solids (12) are discharged from the disc centrifuge at periodic intervals of time.

Several studies have been carried out to ascertain the influence of the two- and three-phase centrifugal decanters employed in olive process on oil yield, composition, and quality of virgin olive oil (Ranalli A. and Martinelli N., 1995; Ranalli A. and Angerosa F., 1996; Angerosa F. and Di Giovacchino L., 1996; Piacquadio P. et al., 1998; Stefano G. de et al., 1999; Koutsaftakis A. et al., 1999; Di Giovacchino L. et al., 2001, 2002). In one of these studies tests were performed in an olive-mill equipped with centrifugal decanters at two- and three-phase on a homogeneous lot of three olive varieties (*Coratina*, *Nebbio*, and *Grosse di Cassano*) at an industrial level (Ranalli A. and Martinelli N., 1995; Ranalli A. and Angerosa F., 1996). The results showed that the two-phase centrifuge frequently yielded higher oil outputs. Furthermore, as the vegetation water was not separated from the stone, the amount of liquid effluent produced was much lower. This goes towards solving the age-old and very difficult problems connected with the production of this highly polluting outflow. However, the obtained olive by-product is in a sludge form with a moisture content of 55–70%, while the traditional olive cake has a moisture content of 20–25% and 40–45% in the press system and the three-phase centrifuges, respectively (Alba-Mendoza J. et al., 1990), making the industrial recovery of the residual oil difficult and expensive. Furthermore, the semi-solid by-product was characterized by higher values of the pulp/stone ratio, as well as the greater weight produced. The effluent was produced in small quantities, besides being more concentrated and thus richer in fat, dry residue, phenols, and *o*-diphenols. The COD and turbidity values were also higher. It was also shown that virgin oils extracted with the two different decanters do not differ in free fatty acids, peroxide value, and ultraviolet absorption. The extraction system did not modify qualitatively the phenolic composition of virgin olive oils obtained with the two different centrifugal decanters. However, virgin oils obtained by the two-phase mode showed a greater concentration of phenolic compounds than the homologous oils obtained by the three-phase mode (Cert A. et al., 1996; Piacquadio P. et al., 1998). In particular, the highest differences were observed for aglycone derivatives of oleuropein, such as 3,4-DHPEA-EA (3,4-dihydroxyphenylethanol-elenolic acid) and 3,4-DHPEA-EDA (3,4-dihydroxyphenyl ethanol-elenolic acid dialdehyde), and ligstroside that are the most concentrated antioxidant phenolic compounds of virgin olive oil (Stefano G. de et al., 1999). Oils processed by the three-phase mode showed a significant correlation between their stability and their phenolic concentration. In general, olive oils obtained by the two-phase mode were of a far higher quality, mainly for their higher oxidative stability and better organoleptic characteristic, so that they are wholly comparable to those extracted by pressing or filtering. In addition, the significant reduction of processing costs, as well as the lower utilization of hot water and electrical energy, must also be emphasized.

High-performance liquid chromatography (HPLC) was applied to evaluate simple and complex olive oil phenols in the streams generated in the two-phase extraction system using *Arbequina* and *Picual* cultivars (García A. et al., 2001a).

The malaxation stage reduced the concentration of ortho-diphenols in oil circa 50–70%, while the concentration of non-ortho-diphenols remained constant, particularly the recently identified lignans: 1-acetoxypinoresinol and pinoresinol. Oxidation of ortho-diphenols at laboratory scale was avoided by malaxing the paste under a nitrogen atmosphere. Phenolic compounds in the wash water used in the vertical centrifuge were also identified. 3,4-DHPEA (hydroxytyrosol), *p*-HPEA (tyrosol), and 3,4-DHPEA-EDA were the most representative phenols in these waters. Hence, phenolic compounds in the wash waters came from both the aqueous and the lipid phases of 2POMW.

Compared to three-phase decanting, the two-phase decanting has the following advantages:

- The construction of the two-phase scroll centrifuge is less complicated and thus, is more reliable in operation and less expensive than the three-phase decanter.
- In two-phase decanting the disk centrifuge for subsequent treatment of the vegetation water is not required.
- During operation of the three-phase scroll centrifuge the separated oil and the water may be remixed; volatile compounds from the vegetation water may cause a sticky deposit on the centrifuge.
- The throughput of the two-phase centrifuge, related to the oil quantity, is higher because no additional water is required to produce the pulp. Energy consumption is also reduced as a result of the lower processing quantity.
- Oil produced by the two-phase decanting is of higher quality; especially it has higher oxidation stability and better organoleptic characteristics. The operating costs are lower, compared to the three-phase decanting process. Water utilization in the olive-mill decreases considerably.

The disadvantages of the two-phase decanting are:

- The two-phase process, although it produces no wastewater as such, it combines the wastewater that is generated with the solid waste to produce a single effluent stream of semi-solid form (~30% by mass). This doubles the amount of "solid" waste (*alperujo*) requiring disposal, and it cannot be composted or burned without some form of (expensive) pretreatment.
- 2POMW has a moisture content of 55–70%, while the traditional olive cake has a moisture content of 20–25% and 40–45% in the press system and the three-phase centrifuges, respectively. This greater moisture, together with the sugars and fine solids that in the three-phase system were contained in OMWW give 2POMW a doughy consistency and makes transport, storage and handling difficult — it cannot be piled and must be kept in large ponds.
- 2POMW is characterized by higher values of the pulp/stone ratio, as well as the greater weight produced. The effluent is produced in small quantities, besides being more concentrated and thus richer in fat, dry residue, phenols, and *o*-diphenols. The COD and turbidity values are also higher. Furthermore, 2POMW is a rather new type of waste, which has not been fully characterized yet.

- This two-phase technology transfers the problem of disposing of the olive-mill waste from the mill to the seed-oil refineries; 2POMW, prior to oil solvent extraction, must be dried with considerably higher energy requirements than the case for traditional or continuous oil production processes, making the industrial recovery of the residual oil difficult and expensive.

OMWW originating from traditional olive-mills based on press system shows, usually a high COD and total solids content. The second type in rank of COD is OMWW from a three-phase extraction system, and finally the lowest values belong to olive-mills with a two-phase extraction system, where the only residue is the water used to wash the oil up. Regarding solids in suspension, the three-phase olive-mills show the highest values, followed by press olive-mills and by two-phase olive-mills. The pH shows an opposite trend, as the vegetation water of two-phase olive-mills show values higher than that of three-phase olive-mills, and press ones (Andres M. et al., 2001).

Stone-Removing Process

This method relates to the production of olive oil without crushing the stones and it was highly appraised in antiquity. According to Columella (XII, 52) and Cato (LXVI) the olive stones were not to be crushed during the pressing since this was considered to spoil the flavor of the oil. The crushing devices used for that purpose were the *trapetum* and the *mola olearia* — see section: "Evolution of production methods" and Figs 4.1 and 4.2.

US4370274 (1983) discloses an apparatus for recovering olive oil from destoned olives. Initially, olives are fed to a pulper that separates the olive stones from the pulp. The pulp is then taken up by an extraction screw that subjects the pulp to an extraction pressure sufficient to withdraw a liquid phase, comprising oil, water, and a minor proportion of olive pulp. The liquid phase is collected in a bin and then sent to a clarifying centrifuge that separates the residual pulp from the liquid phase to obtain a mixture comprising olive oil and vegetation water. A purifying centrifuge then separates the vegetation water and a small proportion of solid matter from the mixture to obtain an olive oil, substantially free of vegetation water that is collected in a tank. According to the inventor, the water can be directly disposed to a sewer-system.

Additional devices that may be used are disclosed in: IT1276576 and IT1278025. As above, these devices can be used to separate the pulp from the stones prior to processing of the crushed olive pulp into oil, water, and solid residues.

EP581748 (1994) describes a process comprising the steps of:

i. kneading the olives in a thermo-regulated room without crushing the stones, obtaining an homogeneous and completely granulated paste at about 40–45 °C, because of the combined action of temperature, mechanical stirring, and motion of metallic surfaces within the paste;

ii. extracting the oil by circulation of hot water at about 40–45 °C, with a ratio water to paste of 3:1;

iii. filtering the paste by gravity where the separation of water and oil occurs.

OMWW produced by the stone-removing process has the following advantages:

- reduction of the pollution load of OMWW due to the removal of the not crushed stones; said waters have, with respect to those produced by the conventional processes, the following features:
  - lesser acidity,
  - reduction of BOD_5 up to 8 times to that of the conventional process,
  - smaller amount of organic compounds refractory to biological digestion,
  - smaller amount of suspended solids.
- reduction of production and undertaking costs of the mill, as the used machines are considerably cheaper than the conventional one for what concerns supplying, installation, and maintenance. With the same production, smaller nominal power of the engines is required and this means that the energy demand and undertaking costs are reduced;
- obtaining oil of high quality as the stones are removed;
- eliminating the olive stones that absorb a considerable part of the produced oil; the production yield is, therefore, increased;
- use of the olive stones as an energy resource, in consideration of the fact that the olive stones have a greater calorific value than common firewood;
- facilitating the easier recovery of "useful" polyphenols such as hydroxytyrosol.

OMWW thus obtained are substantially free of compounds that are found primarily in olive stones, such as tyrosol and other highly polluting monophenolic compounds (WO0004794, 2000; WO0218310, 2002) — see Chapter 10: "Uses", section: "Recovery of organic compounds".

The use of new technologies to extract oil from destoned paste can improve the oil phenolic concentration — see Table 4.3. The phenolic oxidation during processing is catalyzed by the peroxidase, which is highly concentrated in the olive seed. The destoning process, by excluding the olive seed before malaxation, partially removes the peroxidase activity and consequently can reduce the enzymatic degradation of the hydrophilic phenols in the oils processing, thus, improving their concentration and oil oxidative stability (Servili M. et al., 2004).

Percolation (Selective Filtering)

The first studies to build a machine to process olives with this method date back to 1911; in 1951 the "Afin" prototype was built (now called Sinolea). It takes advantage of the different surface tensions of the liquid phases in the paste. To this end a steel plate is plunged into olive paste. When it is withdrawn again, it will be coated with oil because of the different surface tensions. In the past, the percolation

Table 4.3. Qualitative parameters of virgin olive oils obtained from destoned and control (whole fruit) pastes evaluated at time 0 and after 12 months of storage at room temperature (25 °C) (Montedoro G.F. et al., 2001)

	Oils of control olive pastes		Oils of destoned olive pastes	
	Time 0	Time 12	Time 0	Time 12
Free acidity (g oleic acid/100 g oil)	0.29	0.31	0.25	0.30
Peroxide number (meq O_2/kg oil)	6.1	25.4	5.4	21.7
K_{232}	1.922	4.000	1.826	3.250
K_{270}	0.136	0.234	0.110	0.190
Total polyphenols (mg/kg)	345	150	355	195
Orthodiphenols[a] (mg/kg)	250	85	270	100

[a]Evaluated colorimetrically and expressed in mg/kg as 3,4-DHPEA equivalent.

system was coupled with pressure, while at present it is coupled with the centrifugal decanter (Di Giovacchino L. et al., 2002).

Sciancalepore V. et al. (2000) studied the effects of the cold percolation system on the quality of virgin olive oil from two different Italian cultivars (*Coratina* and *Oliarola*). The quality was also compared with that of oil extracted with the current centrifugation system using a two-phase decanter. Tests were performed in an industrial olive-mill equipped with the two extraction systems. The oils extracted with cold percolation system showed, in all cases, lower free acidity, peroxide value, and ultraviolet (UV) absorption (K_{232} and K_{270}) and higher polyphenol contents in comparison to oils obtained by two-phase centrifugation. These results were confirmed by the autooxidation stability of the oils examined.

Electrophoresis

By this method the separation of oil is obtained by electrophoresis. It concerns a method that has been developed only at an experimental stage and comprises the steps of:

- Crushing the olives and kneading the paste.
- Dilution of the paste by hot water with a ratio 3:1 (water/paste) obtaining a homogeneous mixture.
- Separation of the oil by floating; the mixture water/paste is subjected to the passage of direct current that by electrophoresis determines the de-emulsification of the oil that is available after a certain time at the head of the electrophoresis tanks.

In practice, this method has been abandoned.

Chemical Separation

A method already known is the chemical method of separation. The working steps are the following:

- Crushing the olives by millstones, simultaneously crushing the stones and obtaining the olive paste.
- Dilution of the paste with alkali containing water within suitable tanks equipped with heat steamers.
- Standing within said tanks to separate the oil phase that lasted many hours.

Such a method has been abandoned although it has the merit to have first opened a way to research on which the modern method of centrifugation of paste is based.

All the above mentioned methods involve systems and machine arrangements always very expensive, since both the stone crushing and the centrifugation or pressing of the paste involves very high mechanical stresses of the machines that consequently are very heavy and must be realized employing resistant materials and very accurate machining.

Part II

Treatment Processes

Chapter 5

Physical Processes

Physical processes involve the separation of different phases through mechanical means. These phases could be a variety from solid–liquid to liquid–liquid.

The main physical processes are:

- Dilution,
- Sedimentation/Settling,
- Filtration,
- Flotation (dissolved air flotation, gravity flotation),
- Centrifugation,
- Membrane technology (Microfiltration, Ultrafiltration, Nanofiltration, and Reverse Osmosis).

Dilution

Dilution is a simple way to reduce the organic load of OMWW. As dilution water may be used, water coming out of wells or irrigation water or water from nearby streams or brooks, which are in abundance during the winter period the olive-mills operate. Dilution can also be carried out in the sewerage system, either directly within the mill (mainly by adding washing water) or outside the mill, by adding domestic sewage. The ratio between the polluting load from the olive-mill(s) and the resident population in the surrounding area can be an important factor in the selection of the treatment process. When the load due to OMWW is low, compared to the domestic effluents, OMWW can be disposed of in existing or planned sewage treatment plants (Boari G. and Mancini I.M., 1990) As a consequence, appreciable savings can be obtained. This is one of the cases in which dilution is quite beneficial to improve treatability of OMWW by biological systems. When the mill(s) load is high (load ratio > 1), then independent treatment is probably more appropriate.

Dilution is also carried by the continuous process (three-phase). Dilution is a mixed blessing depending on the treatment process which is used. It is obviously detrimental if one plans to use concentration processes such as distillation or to store OMWW in *ad hoc* tanks prior to treatment, while dilution is clearly beneficial when biological treatment is considered — see Chapter 8: "Biological processes", section: "Anaerobic processes". In every case, except concentration, dilution by itself decreases the concentration of OMWW and, therefore, makes it easier to reach the required standards for the final effluent. However, the large quantities of water needed for the dilution process makes it unsuitable for use in areas with limited water resources.

Sedimentation/Settling

Sedimentation (Settling) is the simplest and most widely used physical pretreatment method (Georgacakis D. et al., 1986; Velioğlu S.G. et al., 1987; Al-Malah K. et al., 2000). Much of the organic matter in OMWW is in a suspended form, rather than in solution and removal of the sediment or sludge brings about a large reduction in BOD_5 of OMWW. The sedimentation is a natural process, which, after approximately 10 days, results in two liquid fractions, a low COD supernatant, and a high COD settled sludge (Georgacakis D. and Dalis D., 1993; Georgacakis D. and Christopoulou N., 2002). Sedimentation of OMWW that took place in a $650\,\mathrm{m}^3$ concrete holding basin with conical basin, resulted in a supernatant with an average COD value of 22 g/l, corresponding to 68% of the total OMWW volume and in a sludge with an average COD value of 162.4 g/l, corresponding to the remaining 32% of the total OMWW volume (Georgacakis D. and Dalis D., 1993).

ES2116923 (1998) describes a process of storing OMWW in a large-diameter and shallow-depth cavity (pit) (1) — see Fig. 5.1 — in which there is sedimentation of a

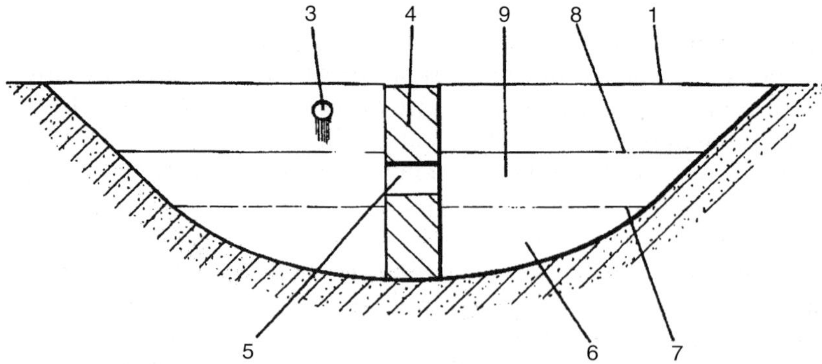

Fig. 5.1. Storage and sedimentation of OMWW in a large-diameter and shallow-depth cavity (pit) (1) (ES2116923, 1998).

high percentage of OMWW to give a semi-sludgy mass (6), above which there floats a mass of water (9); the semi-sludgy mass (6) being removed by any conventional system and transferred to a place for drying and hardening with a view of obtaining a solid product, which can be used as fuel product to replace other products such as wood, coal, and the like. It can also be used as fertilizer for agriculture and even as insulator and protector for maintaining humidity in certain types of plantation.

The sedimentation method is considered slow and it usually requires use of costly flocculants to facilitate the aggregation of small particles into large agglomerates, which can then be removed more effectively from OMWW — see Chapter 7: "Physico-chemical processes", section: "Precipitation/Flocculation". Another disadvantage of this method is that the supernatant and the settled sludge have to be treated further at an additional cost before being disposed of.

Flotation

Flotation is a unit operation, which removes solid or liquid particles from a liquid (such as oil droplets or suspended solids from OMWW). Adding a gas (usually air) to the system facilitates separation. Rising gas bubbles either adhere to or are trapped in the particle structure of the suspended solid, thereby decreasing its specific gravity relative to liquid phase and affecting separation of the suspended particles.

When OMWW is stored for some time (a couple of days) a crust is formed on the surface. If samples of this crust are studied under a microscope, small drops of oil are observed. The oil content of samples taken from the surface with a crust and from the bottom with sludge of a 1 l OMWW having been stored in a basin was found to be 0.2 and 0.03%, respectively. The results show that in order to obtain the greatest quantity of oil from OMWW flotation is preferred over sedimentation (WO9211206, 1992).

Methods of flotation include dispersed- and dissolved-gas flotation. Dispersed-gas flotation, commonly referred to as froth flotation, is not widely used in wastewater treatment. Experiments on formation of solids by flotation were done by Escolano Bueno A. (1975). Curi K. et al. (1980) used gravity separation and dispersed air flotation to evaluate the feasibility of oil recovery from OMWW without much success.

The dissolved-air flotation (DAF) method is also referred to as pressure flotation in which air dissolved in water under pressure is released in the form of small air bubbles by discharge to the atmospheric pressure.

DAF has been evaluated, as potential pretreatment technique, for the removal of suspended solids of OMWW (Mitrakas M. et al., 1996). The pilot unit of Fig. 5.2 was used to investigate the influence of retention time, operating pressure, and chemical addition on the method's efficiency to reduce organic loading and total solids of OMWW as well as the efficiency of flotation to separate oils as foams (Mitrakas M. et al., 1996).

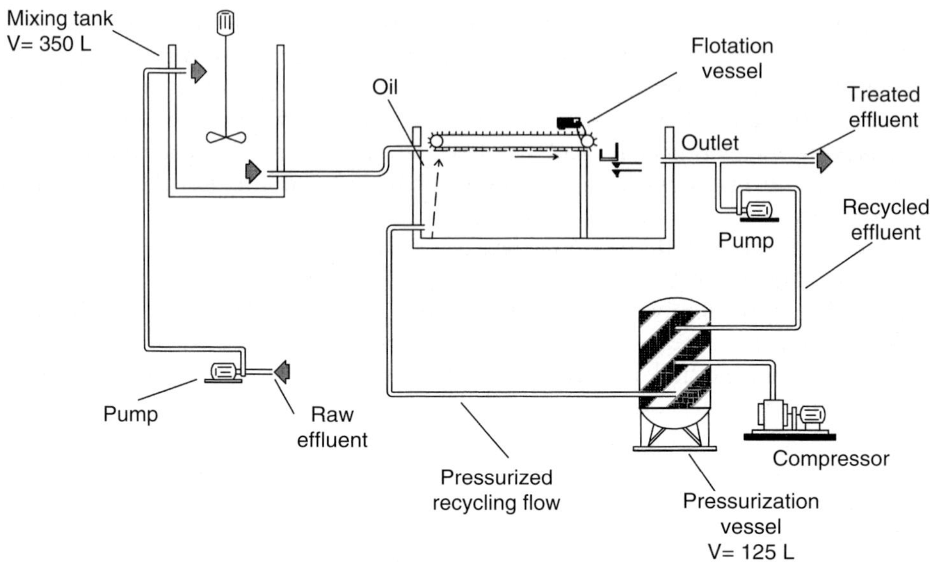

Fig. 5.2. Schematic diagram of the pilot DAF unit (Mitrakas M. et al., 1996).

The high content of OMWW in suspended solids made the DAF technique quite inefficient, since the ratio air to solids was out of the typical working range of 0.005–0.06. DAF can remove COD as well as oils but not as efficiently as centrifugation. Thus, acidified OMWW gave maximum COD reduction of 30% and oil recovery of 30%. DAF performance with raw OMWW was one-half to one-third of these values. An additional disadvantage of dissolved air flotation is that recovered oils (or hydrolyzed oils) should be extracted from the foam of the DAF unit. DAF application in practice was not feasible, despite the relatively high, almost 30%, oil recovery by this process.

Centrifugation

Mitrakas M. et al. (1996) investigated the separation efficiency of this technique and the influence of chemical additions on the effectiveness of the method to reduce COD and to recover oils contained in OMWW. In general, when an OMWW sample is subjected to centrifugation, three segregated phases are formed: a surface layer containing oil, an aqueous layer containing the soluble materials, and a sediment layer where suspended and colloidal matter are concentrated.

Centrifugation proved capable to fully separate suspended solids, which in turn significantly improved COD removal and oil recovery. Removal of COD by centrifugation of raw OMWW reached 70% and oil recovery 30–50%, depending on the origin of the raw OMWW. Changes in the chemical environment had a considerable influence on the centrifugation yield. At pH $=2$ (acidification by H_2SO_4) the highest oil recovery (47%) and a simultaneous high COD decrease (67.8%) were achieved. The sediment obtained from centrifugation, at pH $=2$, was more cohesive, with the lowest volume (15%) and water content (80%). In that aspect centrifugation turned out to be preferable to sedimentation owing to smaller volume of the separated phase. The quality of the obtained oils, however, was low because of their hydrolysis. Addition of lime and precipitation of Ca-salts of fatty acids somehow improved COD removal to about 83%, but oil recovery became very low (12%) and the resulting sediment was jelly-like and bulky.

According to Mitrakas M. et al. (1996) the simplicity of the centrifugation process and the oil recovery represents a serious advantage of this technique, since the oil can be recycled. Considering that typical oil losses in the effluents of small olive-mills reach 1–1.5%, a recovery of 0.3–0.75% (i.e. 30–50%) represents a significant revenue that can quickly offset the capital and operating costs of the process. However, the resulting treated OMWW still contains 50–70 g/l of COD, despite 70% removal of COD in the centrifugation step. This COD results from dissolved organic matter and cannot be removed by physical or even chemical treatment. Only biological treatment of the centrifuged OMWW appears appropriate for further reduction of COD to acceptable levels. The high organic load, however, would make such biological treatment expensive for small olive-mills.

WO9728089 (1997) discloses a method of extracting floated and suspended particles from OMWW during the olive oil extraction process. Prior to reaching the waste drainage stage and immediately after exiting the centrifuging stage of an olive-mill, the particles are separated from OMWW through filtering, sinking tanks, and centrifuging independently, or in conjunction with one another. The floated and suspended particles amount to 3–5% of the OMWW produced within a three-phase centrifugal system. The invention claims to simplify and improve the further process stages for neutralization of OMWW and its adsorption by nature. In addition, the recovered by-product could be used as animal feed.

In general, further COD reduction is not possible by these pretreatment techniques, due to the soluble organics in OMWW, which cannot be removed by these processes.

ES2091722 (1996) describes a process for reducing the moisture content of 2POMW by a combination of settling and centrifuging stages. The process comprises the following steps: (i) an initial stage of preparing 2POMW with the optional addition of additives, specifically water (0–10%) and talc (0–1.5%); (ii) passing the mixture to a settler or a horizontal centrifuge yielding a residue with reduced moisture content and in parallel a liquid; (iii) passing the liquid to a settling stage after which water (5–15%) is added and through this to (iv) a vertical centrifuge from which both oil and OMWW are obtained. The process avoids the main problems associated with the drying of 2POMW — see Chapter 7: "Thermal processes", section: "Physico-thermal processes".

Filtration

Filtration serves the same purpose as sedimentation. Suspended and colloidal solids in OMWW may be removed by filtration. Despite the high solid and colloid contents of OMWW, filter-press equipment is occasionally used in small olive-mills. Pressure filtration has been proposed in the literature as a possible pretreatment method for OMWW with encouraging results (Velioğlu S.G. et al., 1987; Bradley R.M. and Baruchello L., 1980). Using a filter-press Mitrakas M. et al. (1996) found that the physico-chemical characteristics of OMWW (suspended solids, oils, and fats present) led to rapid clogging and the formation of an impenetrable cake of solids and oils, which decreased the yield and made this process practically unsuitable. COD removal, however, was high, almost higher than centrifugation, in agreement with the results of Bradley R.M. and Baruchello L. (1980).

Velioğlu S.G. et al. (1987) designed a treatment scheme on a small-scale unit based on simple physical processes. The treatment scheme consisted of sedimentation, flotation by gravity and straw filtration within a single unit. Some of the conclusions were: (a) organic matter could not be removed effectively (about 25%); (b) substantial amounts of oil and suspended solids (about 80%) could be removed when the system is operated at an overall hydraulic detention time of 1 h; (c) the effective life of the straw filter was about 2 days, after which it should be replaced.

GR1001839 (1995) describes a system for purifying OMWW, which combines a filtration and a dilution step. The filtration step comprises a self-cleaned mechanical filter and three tanks: one for removing the suspended solids and two for separating the residual oils by flotation. The dilution step comprises an active carbon filter for the decolorization of OMWW and two tanks: one for the dilution of OMWW with large quantities of water and one for the oxygenation of the treated OMWW.

ES2087827 (1996) — see also ES2087032 (1996) — describes a process for the decontamination of OMWW by mechanical filtration to remove solids and separation of oil and water fractions. OMWW is passed through regeneratable mechanical filters, which retain 98.8% of the solids. The liquid passing through is collected and the oil separated from the water in a second filter. The water fraction is recycled to the process after a further filtration to remove last traces of oil.

WO2005003037 (2005) describes a system of filters for use in the treatment of OMWW, wherein said system of filters is composed of a combination of sub stratums of natural products selected from a group comprising turf, sand and sawdust, and optionally one or more filters of resins selected from the group consisting of cationic, mixed-bed, and PVPP (polyvinlypolypyrrolidone). Optionally, OMWW is pretreated e.g. by centrifugation, for the removal of solids (~30%). Such a system offers several advantages as compared to filtration systems known in the art. First of all, the system of filters is made of inexpensive physical product filters. Secondly, the system offers a considerable degree of flexibility. As the system is made up of a discrete number of filters, the number of each kind of filter in the system is easily changed. Such flexibility is required in cases of changes in the volume and/or chemical composition of OMWW to be treated. Thirdly, it permits the treatment of large volumes of OMWW and the rate at which OMWW is treated is very fast compared to methods known in the art. Another aspect of the invention relates to a process for retention and recovery of antioxidant phenolic compounds, present in OMWW — see Chapter 10: "Uses", section: "Recovery of organic compounds".

Membrane Technology (Microfiltration, Ultrafiltration, Nanofiltration, and Reverse Osmosis)

This technology is based on the separation of particle sizes that are in the same phase, i.e. all components are in solution. The basic principle manifests itself into different membrane methods, differentiating themselves through the particle size they separate and how they separate them. Those methods that are of interest in the treatment of OMWW are microfiltration, ultrafiltration, nanofiltration, and reverse osmosis.

With the microfiltration particles with a diameter of more than $2\,\mu m$ are separated. Thus colloidal constituents are completely removed.

DE4210413 (1993) describes a membrane for the separation of polydispersions and/or emulsions which could be adapted in a simple manner to the purification of OMWW. This membrane can be used for the microfiltration and deemulsification of the suspended residual oil in OMWW. The membrane is essentially composed of a porous carrier support grid with a membrane layer of bonded powder material. The carrier support grid is a metal felting or woven material, or a glass fiber woven, with a mesh width of 5–$60\,\mu m$. The membrane layer consists of metallic or ceramic powder or a temperature resistant plastic powder with grains of up to $15\,\mu m$ and an organic binder. The powder grains and the binder are structured to give passage openings in the layer to allow the continuous phase to pass through and retain the dispersed or emulsified phase. For hydrophilic characteristics, a ceramic powder (preferably Al_2O_3) and a binder (preferably polyethersulfone) are mixed in a ratio to give a ceramic powder content of minimum $55\,wt.\%$. The membrane layer can also contain pigments. The membrane has high mechanical strength and stability combined with sufficient permeability.

Microfiltration has also been used as a pretreatment procedure prior to the photocatalytic oxidation of OMWW (Vigo F. and Cagliari M., 1999).

Ultrafiltration has a cut-off of $0.1\,\mu m$. With the help of this process, suspended pollutants such as oils or phenolic compounds can be eliminated besides the colloidal constituents (Carrieri C., 1978; Jemmett M.T. et al., 1983; Halet F. et al., 1997; Mameri M. et al., 2000b). However, dissolved components, such as those determined by the sum parameters COD, are only insufficiently removed by means of this process. With the ultrafiltration, only a small amount of waste is produced because the residual moisture in the concentrate is low. Ultrafiltration, even if it allows very high removals of lipids and polyphenols, is affected by poor selectivity (indeed, large amounts of biodegradable COD are also removed). Canepa P. et al. (1987, 1988a,b), who studied the influence of the cut-off of polysulfonated organic membranes to treat OMWW, found that between 50 and 75% of the COD was removed.

During ultrafiltration of OMWW severe fouling of the membrane occurs, thus affecting process performances. Fouling reduces the permeate fluxes and determines both efficiency decrease and variation of membrane selectivity; it also makes the process highly expensive owing to repeated plant shut-down for cleaning and washing the membranes. Permeate flux profiles show, typically, an initial drop from the value obtained with osmotized water, then a smoother but continuous decay until a steady state is reached. That kind of time-dependent profile is caused by both concentration polarization and fouling. While the former is a reversible process caused by an increased transport resistance in the boundary layer, the latter is an irreversible phenomenon comprising the effect of surface fouling, adsorption, gel layer formation, pore blocking or reduction of pore diameters, cake formation, and adhesion of particles on the membrane. Membrane fouling depends on several factors, such as membrane characteristics, feed solution properties, such as molecular size of solutes and their interaction with the membrane, operating conditions (transmembrane pressure, flow rate, and temperature).

Halet F. et al. (1997) and Mameri N. et al. (2000b) presented a membrane technique to treat OMWW using different commercial ultrafiltration membranes: one organic (polysulfone) and two ceramic (ultrafine ZrO_2 pores supported on alumina of coarse porosity) membranes. Before ultrafiltration most of the suspended matter, oil, and fat were removed by centrifugation. The influence of the hydrodynamic parameters (transmembrane pressure and flow rate) and the membrane cut-off on the efficiency of the ultrafiltration process was evaluated, and it was shown that the polysulfone membrane could reduce pollution due to organic matter by decreasing the value of the COD by about 90%. Moreover, the nature of the ultrafine pore membrane appeared to be an important parameter, which may strongly increase or decrease the capacity of the membrane. The membrane cut-off did not have a strong influence on the performance of the process, but if the membrane pores were too large the stability of the dynamically formed membrane decreased at transmembrane pressures greater than 0.2 Mpa.

A semi-pilot plant (capacity $1.5\,m^3$/day) was used to investigate the ultrafiltration of OMWW by a type of membrane described in US4188354[19] (Vigo F. et al., 1981, 1983b). Results, showed severe fouling at feed concentration of 58 g COD/l (40°C, 2 atm, 4 m/s). The problem was alleviated by adding a 3% non-ionic biodegradable detergent solution to the feed, which ensured a constant permeate concentration of

---

[19]Granted to Munari S. and Tecneco S.p.A. on 12 February 1980.

30 g COD/l throughout the 90 h trial. The results are discussed in terms of gel-layer formation. The main parameters affecting the process are rate of flow (0.054 m/s), temperature (highest temperature compatible with the membrane and its support), and chemical treatment (addition of detergent).

Within the reconversion activities of the industrial complex of S. Eufemia Lamezia (Italy) into an environmental service plant system, an ultrafiltration plant for OMWW by polymeric (polysulfone) membrane batteries was started in November 1995 suitable to treat a nominal capacity of 300 m^3/day flowing from olive-mills. The treatment plant is based on the following steps: (i) dirty water storage, (ii) oil removal system, (iii) settling of suspended solids, (iv) tangential filtration on polymeric membrane, and (v) eluate treatment by means of polymeric membranes in a double-step biological process complying with Italian standards. The final reconversion of the industrial complex into an environmental service plant system will be composed of a sewage treatment line, an oil vegetation treatment line, a solid waste incineration line, and a compost solid waste line (Borsani R. and Ferrando B., 1996).

In another approach, OMWW were pretreated by centrifugation and then ultra-filtered in a flat-sheet membrane module (Turano E. et al., 2002). The combination of centrifugation and ultrafiltration allows a COD reduction of about 90%. Moreover, a complete separation of fats, completely rejected by the membrane, from salts, sugars, and polyphenols contained in the permeate is attained. The experimental part was directed to investigate the fluid-dynamic aspects related to the ultrafiltration of OMWW. The complex rheological behavior of OMWW has been preliminary examined and the permeation efficiency was evaluated as a function of several parameters such as the importance of pretreating wastewater, the effects of localized turbulence, promoted by ultrafiltration module geometry, and of the main operating variables (transmembrane pressure and feed flow rate). Ultrafiltration experimental results, obtained in a laboratory-scale flat-sheet membrane module, are interpreted using both the cake-filtration and the resistance-in-series models, thus allowing the evaluation of R_f that represents the effect of fouling on separation efficiency. An estimation of specific cake resistance, α, was, therefore, performed on the basis of the feed concentration of total non-water compounds present in the waste showing that pretreated OMWW gave a lower α with respect to raw OMWW by a factor of about 1000. Moreover, it was found that at the same transmembrane pressure, lower values of α corresponded to a greater R_f and that higher local turbulence implies lower specific cake resistances. The results obtained could give useful indications for a preliminary characterization of pilot and industrial modules utilized for OMWW treatment at a significant COD reduction and a selective separation of valuable compounds that are present in the waste.

Nanofiltration is a form of filtration that uses membranes to preferentially separate different fluids or ions. Nanofiltration is not as fine a filtration process as reverse osmosis, but it also does not require the same energy to perform the separation. Nanofiltration also uses a membrane that is partially permeable to

Table 5.1. Feed and permeate parameters of a DA-5 nanofiltration system (General Electric Company, 1997–2003)

Constituent	Feed	Permeate	Rejection (%)
Bacteria (No/ml)	108	0	100
Suspended solids (mg/l)	1090	0	100
COD (mg/l)	8950	705	92
BOD_5 (mg/l)	5970	500	92
Oil/grease (mg/l)	150	0	100
Dissolved solids (mg/l)	7460	3000	60

Table 5.2. Design and operating parameters of a DS-5 nanofiltration system (General Electric Company, 1997–2003)

Element type	DK8040FJL
Feed spacer	50 mil parallel
Number of elements	100
Pretreatment	200-mesh backwashable screen
Flux	6–13 gfd (10–22 LMH)
Feed pressure	862 kPa (125 psig)
Feed volume reduction	75%
Cleaning	Daily with alkaline cleaner

perform the separation, but the membrane's pores are typically much larger than the membrane pores that are used in reverse osmosis.

A technical paper of General Electric Company[20] (1997–2003) describes a nanofiltration system (DS-5) which, in conjunction with a flash evaporator, can reduce the volume of the OMWW stream by 75% — see Table 5.1. Permeate from the nanofiltration system is reused in the processing plant to reduce incoming water costs. The system described in Table 5.2 incorporates DS-5 nanofiltration elements constructed with wide feed spacers to prevent feed channel plugging by suspended solids. The wider feed channel allows higher feed flow volumes and crossflow velocities, thereby, reducing membrane surface fouling.

Nanofiltration membranes prepared from selected types of poly(amidesulfonamide) have been proven to be effective in removing oil from OMWW. Under an operating pressure of 14–21 kPa (2–3 psi), a constant flux of $5 l/m^2 \cdot h$ and 99.6% retention of a solution of 5000 ppm olive oil could be achieved with said membranes over a period of 430 h. In addition, the superiority of the tested

---

[20]http://www.gewater.com/library/tp/834_Nanofiltration_.jsp.

poly(amidesulfonamide) materials (four homopolymers and four copolymers) characterized by excellent retention and high flux rate, was evident from the results of a study comparing it with polysulfonamide, poly(ether amide), and commercially available regenerated cellulose. (Wing-Hong Chan and Sai-Cheong Tsao, 2003).

Reverse osmosis is a separation process working in the molecular range. Depending on the membrane, the cut-off is 20 up to 1000 mol/l. With reverse osmosis a good water quality is obtained that allows reutilization of the water for industrial production processes (Pompei C. and Codovilli F., 1974; Jemmett M.T. et al., 1983; Rampichini M., 1987). Laboratory scale tests were carried out on the purification of OMWW by reverse osmosis in a DDS (De Danske Sukkerfabrikker, Copenhagen) apparatus. Membrane DDS 999 achieved a 91.4% reduction of COD and 98.2% of BOD_5 from fresh OMWW and 97.4% reductions from OMWW stored for 3 months — resulting in an overall COD and BOD reduction of 99.99% of the fresh OMWW. Even better results could be obtained by maintaining more than 2 mg/l dissolved O_2 during storage (Pompei C. and Codovilli F., 1974). The waste amount resulting from this process, however, is rather high. Up to 20% of the treated water volume occurs as concentrate, which has to be treated before disposal. Reverse osmosis has an efficiency of more than 90% in removing organic matter, but on the other hand it has high operating cost and sludge disposal problems (Fiestas Ros de Ursinos J.A., 1961b).

Earlier studies showed that about 99% of COD have been reduced by combining ultrafiltration and reverse osmosis (Jemmett M.T. et al., 1983; Rampichini M. et al., 1987; Canepa P. et al., 1987, 1988a,b; GR88100368, 1989).

Canepa P. et al. (1987, 1988a,b) built a laboratory scale pilot plant to treat OMWW with a long-term integrated membrane process combining ultrafiltration (polysulfone) and reverse osmosis (polypiperazine) membranes with adsorption on porous polymers on running time — see Fig. 5.3. The process was optimized so that from wastewater entering with a COD content of about 90 g/l it was possible to obtain a COD reduction of about 99% with the recovery of polyphenols to be used in alimentary industries and of a concentrated paste for oil extraction, furfural production, or combustion. The proposed integrated membrane process, without adding any chemicals or thermal energy, permits the treatment of OMWW, obtaining about 70% of fresh water with good characteristics both for recycling and irrigation.

A combination of membrane techniques has also been used for the treatment of OMWW. One of the them comprises the steps of: (i) pretreatment to remove suspended solids larger than 5 mm in size; (ii) micro- or ultrafiltration through membranes which have a minimum separation and maximum porosity of 2000 and 0.8 µm respectively; (iii) passing over an ion exchange resin; and (iv) a purification phase on a reverse osmosis membrane (GR88100203, 1989). A similar process comprises the steps: (i) filtering the OMWW discharged from olive-mills through mechanical filters; (ii) ultrafiltration through semipermeable capillary membranes (preferably polyvinylidene fluoride –PVDF– and polyamides), which are periodically washed with an aqueous solution of hypochlorite and/or industrial detergent;

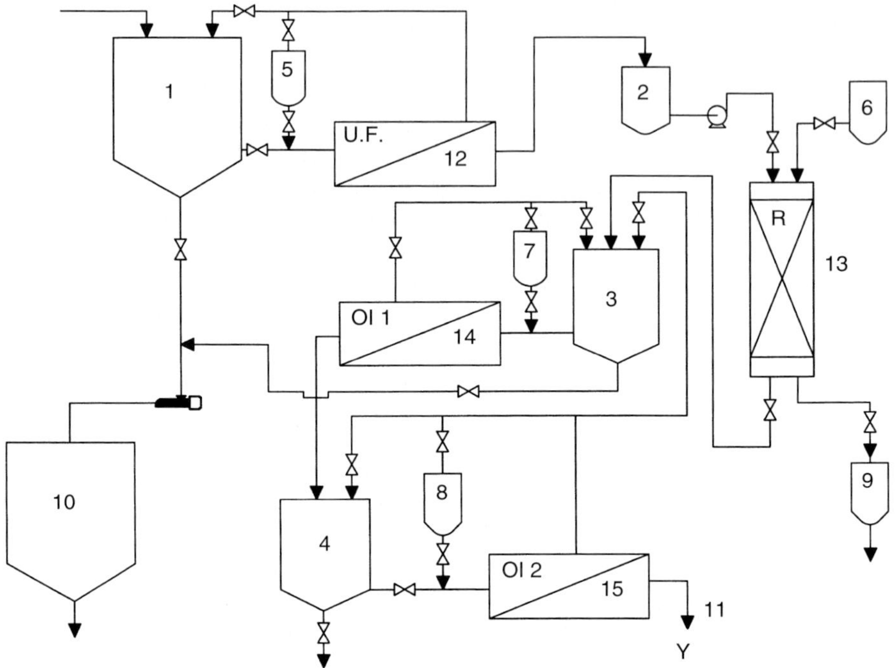

Fig. 5.3. Schematic drawing of integrated membrane pilot plant. (1–4) service tanks; (5–8) washing tanks; (9) polyphenols recovery tank; (10) concentrated recovery tank; (11) fresh water discharge; (12) UF pilot plant; (13) resins column; (14–15) reverse osmosis plants (Canepa P. et al., 1988).

(iii) adjusting the filtrate to neutral pH; (iv) treating by reverse osmosis in semipermeable capillary membrane filters; (v) repetition of stage (iv). The system is claimed to increase the operational flexibility and efficiency of purification of OMWW (GR88100203, 1989).

EP1424122 (2002) describes an installation for recycling wastes generated in the pig farms and olive-mills. The installation mainly consists of three operations:

i. Separation of the liquid phase from the solid phase. In this operation, the wastewater is tipped onto an inclined Johnson screen in which part of the liquid phase is distilled. The rest of the waste in a paste form is compressed in a press where a piston is introduced quickly into a cylinder whose shape is perforated. In the degree that the waste is compressed the liquid phase pours through the grooves in the cylinder and is collected together with the liquid coming from the Johnson screen.

ii. Separation of the fibrous substances in suspension in the liquid phase. In this operation, the liquid obtained from the previous operation is subjected to a process of micro- or ultrafiltration at atmospheric pressure with some lower air diffusers in the body of the filtration.

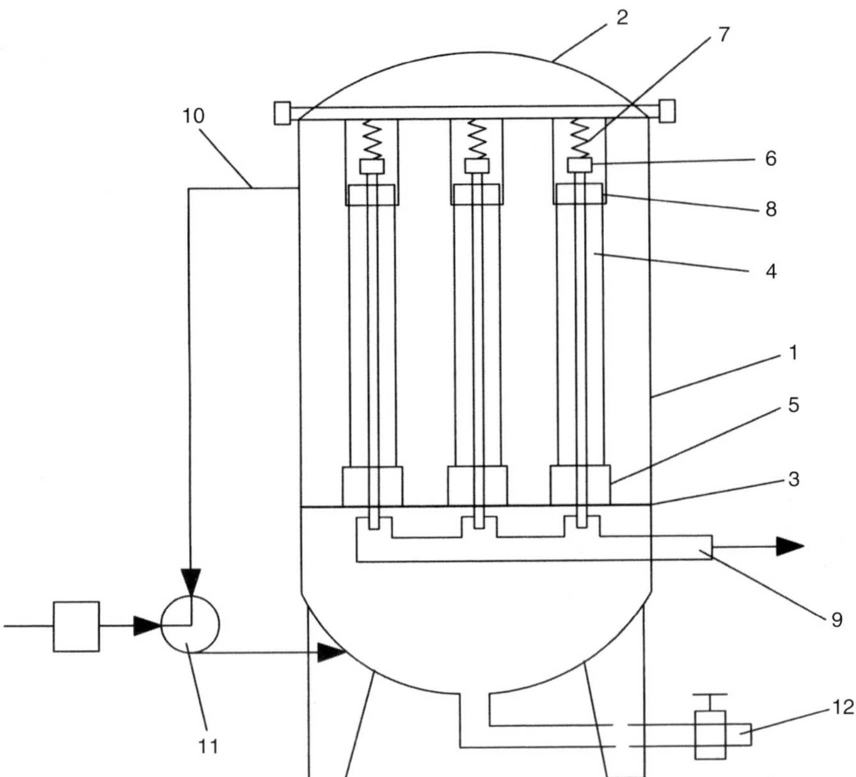

Fig. 5.4. Representative diagram of the inverse osmosis equipment in the third operation (EP1424122, 2004).

iii. Separation of dissolved minerals and the obtaining of drinking water — see Fig. 5.4. In the last operation the water coming from the previous operation is fed to a reverse osmosis device (1); said device is made up of a pressurized tank (10–25 bar) with an upper opening lid (2), on the inside of which there is a support structure (3) for the reverse osmosis filtration membrane elements (4), and which are completely submerged in the liquid to be filtered. Said elements (4) are fixed to the support structure (3), respectively by means of a lower support plate (5) and a liquid duct. The filtration elements (4) are fixed at the top, respectively by a coupling (6) held in position by a spring (7) and at the end of the filtration duct and a cylindrical perforated body (8) with external fixing. Each one of the membrane elements emptying ducts of the filtration is connected to the collector (9) for the drinking water. In turn, the tank (1) has an extraction outlet (10) at the top for the liquid contained and for the recirculation of the same, connected to a pressure pump (11). At the bottom of the tank (1), there is a lower drainage outlet (12) for the complete emptying.

WO2004064978 (2004) describes a filter arrangement for liquids containing solids such as OMWW, the filter arrangement including a substantially cylindrical filter membrane having a longitudinal axis, a brush arrangement within the cylindrical filter membrane, the brush arrangement being adapted to be rotated on an axis co-axial with a longitudinal axis of the cylindrical filter membrane, inlet means to supply liquid to be filtered into an inlet of the cylindrical filter membrane and exit means to extract sludge from an exit of the filter membrane and a filtrate exit — see Fig. 5.5. Pressure may be used within the filter membrane and vacuum outside to assist with filtration. Preferably, the brush arrangement includes a helical brush with bristles. The brush bristles do not touch the filter membrane so that wear of the membrane does not occur and there is not excess load on the motor. The bristles of the brush help to demulsify the olive oil. The mechanism of demulsifying is believed to relate to the action of the tips of the fibers on the emulsified oil. There may be several stages in such a process with the first stage being a microfiltration with pore sizes down to 0.1 μm. A next stage could be an ultrafiltration stage with pore sizes down to 0.01 μm. A next stage could be a nanofiltration stage with pore sizes down to 0.001 μm. A final stage could in effect a reverse osmosis stage with pore sizes down to 0.0001 μm.

The membrane processes are suitable to concentrate the organic substances and allow the recovery of some valuable components. These membrane processes can be used in modular plants and, therefore it becomes easy to plan projects for the decontamination of OMWW either for individual or associated olive-mills. Besides, the application of these technologies depends on the possibility of the economical recovery of the concentrate, whose composition must be controlled and, if necessary, modified according to the final destination. For example, if the concentrate is used as animal feed, the presence of polyphenols can cause some reduction of the protein bioavailability; on the other hand, the recovery of polyphenols can give economical benefits as they can be used as natural antioxidants and pigments.

The ideal membrane for purification/concentration of OMWW should be able to concentrate organic compounds, while the mineral materials are passed through and remain in the clear permeate. It is preferred to remove a significant part of the organic material without simultaneously removing minerals.

Although these processes have been proposed several times for the treatment of OMWW (Pompei C. and Codovilli F., 1974; Carrieri C., 1978; Vigo F. et al., 1981, 1983a,b; Canepa P. et al., 1987, 1988a,b), their advantages in this particular application have still to be clearly demonstrated. It has been claimed (Camurati F. et al., 1984) that membrane processes allow one to separate from OMWW valuable compounds with a high added value such as polyphenols (as antioxidants and flavoring agents) but to the authors' knowledge not a single demonstration plant is in operation today.

The main drawback of membrane processes is related to the limit of the concentration factor. In fact, they can concentrate the waste at a concentration appreciably lower than distillation processes. Moreover, the final products (retentate and permeate) have to be processed prior to disposal. The former is a liquid without

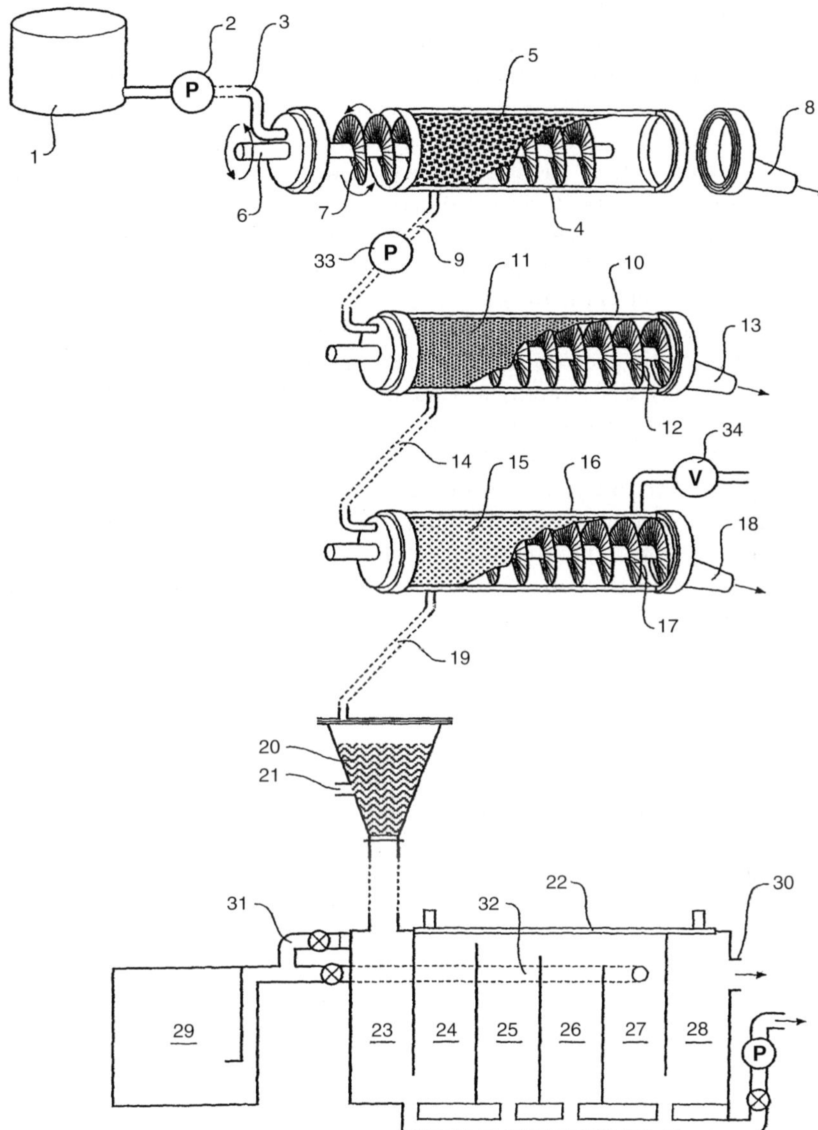

Fig. 5.5. Wastewater treatment arrangement by which both water and solids can be removed from wastewater according to the invention (WO2004064978, 2004). (1) centrifugal separator; (2) pump; (3) pipe; (4) 1st housing cylinder; (5) membrane screen; (6) shaft; (7) helical brush; (8) exit pipe; (9) pipe; (10) 2nd housing cylinder; (11) membrane screen; (12) helical brush; (13) exit pipe; (14) pipe; (15) membrane screen; (16) 3rd housing cylinder; (17) helical brush; (18) exit pipe; (19) pipe; (20) cone-shaped cyclone separator; (21) pipe; (22) overflow tanks; (23), (24), (25), (26), (27), (28) chambers; (29) holding tank; (30), (31), (32) pipes; (33) pump; (34) vacuum system.

much use and the latter must be post-treated because it still contains organics whose COD is far from negligible. Besides, the application of membrane processes involves some technological difficulties related to the presence in OMWW of gelling substances, like pectins, that give rise to fouling phenomena strongly reducing the membrane efficiency (Vigo F. et al., 1981, 1983b; Jemmett M.T. et al., 1983). Therefore, the removal of these substances before the treatment of OMWW with membrane processes seems to be absolutely necessary.

A number of pretreatments have been used in order to avoid the fouling of the membranes. Massignan L. et al. (1985, 1988) applied enzymatic, chemical, and chemical-physical pretreatments in order to overcome the membrane fouling during the reverse osmosis process of OMWW. Tests have been performed using a commercial pectolytic preparation, normally used in the oenological industry, and calcium chloride ($CaCl_2$) solutions. Calcium chloride was selected as flocculation agent because of its capacity to interact with pectic substances and to modify the protein solubility ("salting out" phenomenon). Treatment with pectolytic enzymes allowed a viscosity reduction up to 50% without, however, any water filterability. Better results were obtained by chemical treatment with calcium chloride solution at room temperature. The treatment of 8–10 days aged OMWW with calcium chloride induced an effective separation of the gelling substances. Almost the same results can be obtained by natural aging of OMWW for at least 4 months. Permeation rate of 380–390 $l/m^2 \cdot day$, COD reduction of 98 and 80% recovery factor were observed both for calcium chloride pretreated and naturally aged OMWW.

Prefiltration through a synthetic porous sand base gave better results than chemical clarification or centrifugation as far as fouling of the membranes during subsequent ultrafiltration or reverse osmosis purification of OMWW is concerned (Boari G. et al., 1980).

Chapter 6

Thermal Processes

The processes available in this area are numerous, but usually have one thing in common: they all involve the concentration of the olive-mill waste by reducing its water content and eventually reducing the total amount of the waste. For convenience, the thermal processes are divided into three main categories. The first category involves physico-thermal processes and it comprises evaporation and distillation of OMWW and drying of olive cake and 2POMW. The second category involves irreversible thermo-chemical processes and it comprises combustion and pyrolysis. The third category involves a combination of physical and biological processes and it comprises lagooning. The thermal processes usually form a part of an integrated approach for the treatment of OMWW (ES2024369, 1992; ES2084564, 1996).

Physico-Thermal Processes

Evaporation/Distillation

Physico-thermal processes consist of evaporation and distillation of OMWW where a concentrated solution — "molasses" or concentrated "paste" — and a volatile stream consisting of water vapor and volatile substances are produced. These processes give a large reduction in COD and BOD_5 and require possibly only one more step of treatment, e.g. biological (ES2024369).

Evaporation differs from distillation in that when the volatile stream consists of more than one component no attempt is made to separate these components. In evaporation, OMWW is separated into a residue containing non-volatile organics and mineral salts, and a condensate that consists of water and volatile substances. The evaporation of OMWW reduces its volume by 70 to 75% and brings down the polluting load to more than 90% in terms of COD (Di Giacomo G. et al., 1991).

This operation can be easily done by using the existing industrial evaporators, and it can be optimized up to a level whose value depends on the type, origin, and age of OMWW. The residue produced can be used as animal feed, fertilizer, or put back in the distillation process. The condensed vapor can then be used in the olive washing step in the olive-mill.

Although the evaporation is claimed to reduce significantly the volume of OMWW and its polluting load, there is disagreement about the COD reduction results among several technical reports. According to Annesini M.C. and Gironi F. (1991), the large bias on the results is mainly due to the wide variability of OMWW characteristics. In fact, differences in extraction processes, in olive ripening, in storage time of olives before milling or storage time of OMWW before treatment, cause large differences in the concentration of volatile organic pollutants. Storage time has the most negative effect on the evaporation behavior of OMWW; the residence of OMWW within the storage deposits for few days causes the development of aerobic and anaerobic fermentation volatile compounds such as ethanol and volatile fatty acids, which during the distillation are transferred in the distillate, increasing its concentration in polluting substances which results in the spoiling of the purification process. Experimental evidence indicates that a long storage time must be avoided if OMWW has to be treated by evaporation or distillation because the increase in the volatile pollutant concentration reduces significantly the equipment separation efficiency. These problems can be overcome by addition of a bactericidal compound, which prevents formation of volatiles by spontaneous fermentation of OMWW, and which permits the storage of OMWW for longer periods (GR89100788, 1991). Those compounds are selected among thermolabile, oxidizable, hydrolysable, or in general degradable compounds; the bactericidal compounds are added to the OMWW in concentrations preferably between 500 and 3000 ppm. Suitable bactericidal compounds are:

- thermolabile carbamates, e.g. sodium dimethyl-dithiocarbamate and disodium diethylene-bis-dithiocarbamate,
- thermolabile quaternary ammonium salts, e.g. alkyl-dimethlyl-benzylammonium chloride and benzalkonium chloride,
- thiocyanate derivatives hydrolysable in alkali environment, e.g. methylene-bis-dithiocyanate,
- aldehydes oxidizable in air, e.g. formaldehyde and glutaralaldehyde.

To heat OMWW to such temperatures that the substances begin to vaporize requires a lot of energy. The energy to evaporate the water can be provided either by a man-made heat source or by a natural source (air). Trials have shown that a typical electric energy requirement is $100 \, kWh/m^3$. It can be clearly seen that from an economical point of view, this process is not favorable. It is also evident, when considering that the dry content of OMWW is usually within the range from 5000 to 120,000 ppm, that this solution is quite expensive and unreasonable. Obviously, the more concentrated OMWW is, the more economical is the distillation treatment per unit mass of concentrated COD. Distillation processes, which are already used in the desalination, chemical, and food industries, have been tested on OMWW, namely, vacuum, multiple effect (to reduce energy requirements) and flash evaporation (Rozzi A. and Malpei F., 1996).

In Italy and Spain, several pilot plants have been designed to treat OMWW by evaporation and distillation processes. Such a pilot plant is described in: ES2021191 (1991); the installation consists of an evaporation plant fed with OMWW from a controlling tank by a pump through a preheater. At the outlet from the evaporation plant there is double condenser connected to the heating plant and the clean water supply to the olive-mill. There is then provided a cooling system and finally settler units with a small tank between them for aerobic treatment, yielding as by-products, OMWW concentrate, biological sludge, and wholly purified water. IT1211951 (1989) describes a double distillation process for the purification of OMWW in which the condensed water passes through an active carbon filter.

The main drawbacks of these processes are related to the post-treatment and disposal of the produced emissions. A first problem can be the disposal of the concentrated "paste". Its use as animal feed is limited by the very high concentration of potassium. Otherwise, it can be burnt to feed the boiler which provides the thermal energy to the distillation plant, but its combustion induces air pollution, which has to be dealt with by post-treatment of the gases. A second problem is related to the condensed distillate. The distillate is not made of pure water but carries away an appreciable fraction of volatile compounds found in OMWW such as volatile acids and alcohols. These compounds are the cause of the high COD of the condensate, which can reach 3 g COD/l, and make an additional treatment of the distillate necessary prior to discharge or reuse.

An evaporation process has been developed in Spain, which claims to solve most of the above problems, by exploiting natural evaporation of OMWW in ambient air (Fiestas Ros de Ursinos J.A., 1992; ES2043507, 1993). OMWW is sprayed on specially perforated panels with a very large specific contact surface area. A fraction of the wastewater is evaporated and carried away with the air, which circulates naturally through the panels. The apparatus, as shown in Fig. 6.1, consists of a series of modules, in each of which there are two evaporation chambers (1–1′) and a single condensation chamber (2) formed between the previous two and separated from them by metal partitions (3), which ensure good heat transmission between these chambers. OMWW is sprayed into chambers (1–1′) through nozzles (5) and via the latter streams passes a current of air, which is saturated and heated up subject to the heat generated by a source (12), which takes up the aforesaid air current via a pipe (11) and returns it via a pipe (13) to the intermediate chamber or condensation chamber (2), from where the heat passes to the evaporation chambers via the partitions (3); water evaporation takes place in these chambers (1–1′) with the consequent concentration of OMWW, whilst the evaporated water condenses in the chamber (2) for reuse in the olive-mill. The energy requirements of this process are limited to the recirculation pump. Unless the weather in the selected area is very rainy (which is not likely in most Mediterranean locations), this system

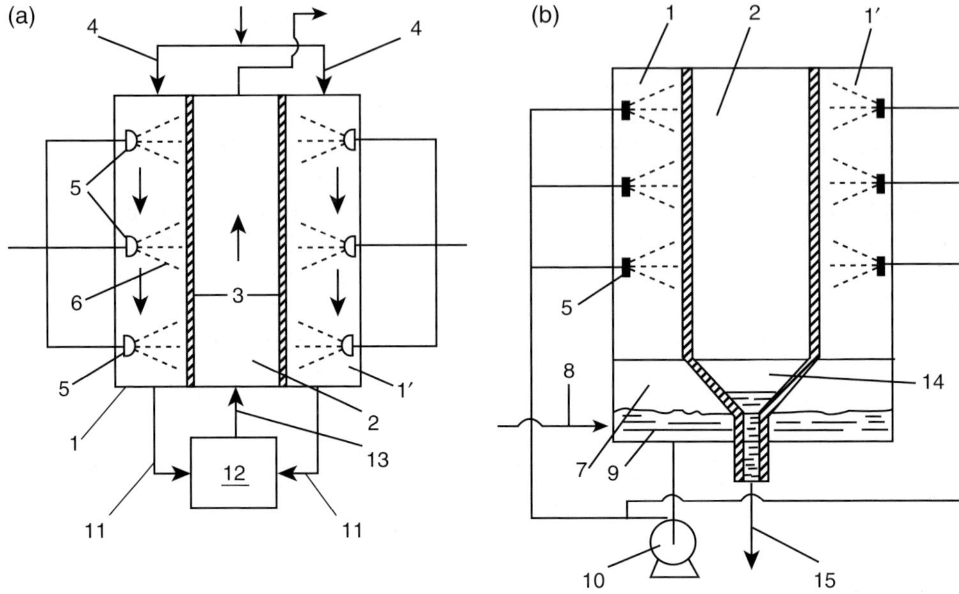

Fig. 6.1. (a), (b) Apparatus for evaporation and concentration of OMWW (ES2043507, 1993).

is fairly efficient. Maximum evaporation rates can reach values of the order of $1 \, m^3$ per m^2 of panel per month.

A possible drawback is related to the odor problems. As the smell of OMWW is normally considered quite strong and unpleasant, the plants based on this process should be located at some distance from residential areas, especially, downstream to the direction of prevailing winds.

EP295722 (1988) describes a process and an apparatus for the purification of OMWW by evaporating it in a stream of unsaturated air from the atmosphere. OMWW is sprayed in a tower and the air is fed through the tower in counter-current flow to OMWW. The concentrate obtained by evaporation has a solids content of 30 to 50% and is fed to a drying stage in order to obtain the organic matter contained in OMWW in powder form. According to the inventor, this is a technologically simple, yet reliable method for the disposal of such wastewater, which requires little capital investment and has low operating costs. However, the general changes in the technology of olive oil production, accompanied by relatively large amounts of wastewater and associated smaller dry matter contents of OMWW, decreases the economic efficiency of the process.

An eolic hydropump has been used for the evaporation of OMWW (Montero M., 1989). In the eolic hydropump, a turbine powered by the wind agitates and pulverizes the surface water of OMWW.

So far, the most often used industrial process for the thermal treatment of OMWW is based on a concentration section with a single or multiple effect evaporator from which a distillate is obtained, whose residual COD value is not superior to 5–7% from the initial one; at the same time 10–20% of the liquid feed leaves the evaporator from the bottom as a concentrated solution characterized by a water content of 30–50% by weight. The recondensed vapor can be subjected either to biological treatment or to distillation in order to regain an alcoholic phlegm, whose commercial value should not be completely neglected. The concentrated solution, although in smaller quantities than at the beginning, is much more difficult to work with and must be destroyed in some way, or made inert and then got rid of. Many systems have been proposed for the treatment of this concentrated solution; for example, one can spread it over the soil, or use it as a component for the production of fodder or of agricultural amendments, or as a fuel.

ES8708149 (1987) describes a process for elimination of OMWW comprising the following steps: (i) mixing OMWW with olive cake; (ii) centrifuging to remove the olive stones; (iii) feeding the effluent into a drying oven to evaporate the water; (iv) recovering the solid residue containing the dried waste, and the valuable organic substances contained in the aqueous phase.

A process was developed to eliminate OMWW by mixing it with exhausted stones and evaporating in a furnace commonly used for stone drying (Lanzani A. et al., 1988). Residual fats, sugars, proteins etc. are concentrated in the stones, and the evaporated water vapor is emitted to the atmosphere without risk of air pollution. The furnace is fuelled by cheap waste products. However, this process has some disadvantages, such as high energy costs, the caramelization of sugars, and the formation of condensates of organic acids.

A process for the purification of OMWW by evaporation uses olive stones soaked with OMWW in a desiccation plant (EP330626, 1989). OMWW is sprinkled on the olive stones to be introduced into a hot air cyclone of a desiccation plant where the water evaporates and the solid parts in suspension deposit on the above mentioned olive stones; these olive stones are sprinkled again with OMWW and reintroduced into the plant and so on until complete evaporation of the water. The novelty is that the olive stones used for carrying OMWW, which must be evaporated in the hot air cyclone, are exhausted olive stones. The process is claiming to be more economical than previous processes and not to cause air pollution.

Another thermal process is making use of the heat emitted by an internal combustion engine which uses its motive power to generate electric current by means of an electric generator in such a way that 52% of energy of the fuel which is not converted into electrical power and which is converted into heat is used to bring an organic product mixed with water, such as OMWW, to boiling point; by evaporation and subsequent condensation 70–85% of the water can be separated and made use of for any purpose, the other small percentage being evaporated off, and the remainder, approximately 10% of the solid product or juice comprises a fraction from which between 5–20% of oil can be extracted, while the remainder of the fraction is an organic material which can be used as fertilizer (ES2101651, 1997).

The disadvantages of the evaporation/distillation processes are summarized as follows:

- Necessity of frequent stoppage, which falls upon the decrease of the stationary efficiency of the boiler generating the living fumes. Besides, upkeep costs are too high due to the necessity of eliminating the incrustations of sugars and lime produced at the evaporators, which are in fact the reason of the plant's stoppages, by means of additions of acids or bases.
- Water obtained through concentration of OMWW, has a pH between 4 and 4.5 and a maximum BOD_5 over 4 g/l even though it has been discolored. The distillate is not made of pure water but carries away an appreciable fraction of volatile compounds found in OMWW such as volatile acids and alcohols. As a result, its pollutant effect on the watercourse is not completely removed; moreover, it cannot be discharged or reused at the olive extraction process because it would increase the degree of acidity of the pure olive oil; an additional treatment of the distillate is, therefore, necessary.
- Due to the acidity of the reclaimed water, and according to the plant operating conditions, every element in contact with OMWW, must be stainless steel or materials resistant to these operating conditions.
- Disposal of the concentrated "paste". Its use as animal feed is limited by the very high concentration of potassium. Otherwise, it can be burnt to feed the boiler which provides the thermal energy to the distillation plant, but its combustion induces air pollution which has to be dealt with by post-treatment of the gases.
- The energy consumption requirements are high.

According to Azbar N. et al. (2004), certain pretreatment methods may help evaporation, increase its efficiency, and reduce the energy consumption. For example, centrifugal separation would be a good pretreatment step before evaporation. This step would bring two advantages; it would increase olive oil production in the mill, and the distillate from the evaporator would be cleaner. Use of other pretreatment alternatives such as chemical precipitation and filtration could make evaporation easier, too. Another pretreatment step before evaporation could be increasing the pH of OMWW. This keeps more of the volatile organic compounds in the solid fraction during evaporation, thus ending in a lower COD distillate. Elimination of volatile organic compounds from this distillate can be considered if it is feasible. This evaporation normally represents 5–6% of the cost of evaporation but assures better distilled water quality for reuse. Alternatively, condensed water can also be reserved for fire fighting when needed or for land irrigation (Azbar N. et al., 2004).

Recently, the solar radiation as a renewable energy source for the distillation of OMWW has been used in a number of applications for dewatering of OMWW in order to reduce the original volume and thus, render the remaining sludge easy for further manipulation and treatment. The use of solar distillation is promising due to reduced energy consumption and relatively high temperatures achieved in the distillation device. Potoglou D. et al. (2003) investigated the efficiency of a solar distillation device on a laboratory scale for the dewatering of OMWW. A quantity of OMWW was left inside the solar distillation for nine days, under outdoor conditions. It was shown that the distillate produced was free from solids, 80% lower in terms of COD and 90% in terms of TKN, while the basin residual was in solid form with only 15% water concentration and without any odor emissions.

The main objective of the EU project: SOLARDIST (EVK1-CT-2002-30028) is the treatment of OMWW by means of a combination of a solar distillation and biological processes. The establishment of a solar distillation wastewater treatment system for small to medium size olive-mills in combination with a biological treatment, like constructed wetlands, it is estimated that can eliminate up to 98% of the OMWW's organic matter content, including phenols and polyphenolic compounds. As well as this, it will allow the reduction of the air pollution and odor generated from uncontrolled evaporation pools, due to the installation of the solar distillation. In addition, the system will allow composting of the organic solid waste generated at the same time in the olive-mill. The SOLARDIST is claimed to be an easy-to-handle, almost maintenance-free and practically cost-free system for the treatment of OMWW.

Evaporation/distillation techniques were also used for the characterization of OMWW. Annesini M.C. et al. (1983) proposed a method for the physico-chemical characterization of OMWW by means of batch distillation runs. The experimental results allow determination of equilibrium ratios and initial concentrations of three pseudo compounds, which can be utilized to describe the behavior of OMWW in an industrial distillation process. The qualitative and quantitative evaluation of the phenolic content and the antibacterial properties of OMWW were studied by Saez L. et al. (1992) during evaporation in simulated evaporation ponds. No antibacterial effect was detectable in subsequent evaporation for 90 days.

Drying

The crude olive cake has a moisture content of around 20–25% in press systems and 40–45% in three-phase centrifuges (Alba-Mendoza J. et al., 1990), while 2POMW has a moisture content of approximately 55–70%. The initial moisture content of both wastes has to be reduced to about 5–8% in order to be able to extract the residual oil and recover their energy content — see Chapter 10: "Uses", sections: "Recovery of residual oil" and "Generation of energy".

For the drying of olive cake, contact, convection, and radiation drying processes can be used. In convection drying, heat is transferred to olive cake by means of hot gases. Water contained in olive cake evaporates and is conveyed by the hot gas flow. Examples of this type of driers are drum driers and fluidized bed driers. The resulting dried olive cake is deoiled with an organic solvent (hexane) and then, can be either incinerated for energy production, reused in agriculture, or land filled, while the air emission must be treated appropriately. The main drawback of this method is the high energy demand. However, this disadvantage is justifiable against the background that the resulting final product can be reused for generation of energy. From an economic point of view, high investment and operating costs are associated to drying plants; moreover, to ensure trouble-free operations a trained and qualified personnel is required.

The drying of 2POMW presents problems because of its high moisture and sugar contents. The classical driers, e.g. rotary kilns (drums) and trays, have a low thermal efficiency due to the poor air–solid contact and can present several problems. The employed driers were designed for three-phase olive cake made up of loose particles of stone and pulp with a homogeneous moisture distribution that can easily be piled up and fed through rotary driers. On the other hand, the high moisture content of 2POMW (55–70%) demands much more energy and the sugars present in it make it sticky and difficult to dry. 2POMW tends to stick to the drier's walls, particularly to the initial part of the drum where the gases are hot, obstructing the gas stream and increasing fire risk (Arjona R. et al., 1999).

A qualitative description of the 2POMW drying process was developed that describes the characteristics of the different phases that make up the whole process (Arjona R. et al., 1999). The drying process was studied at laboratory scale and the drying rate was determined with respect to operating conditions (temperature and air velocity) and agglomerate size. The experimental results show that a constant-rate drying period, which has traditionally been taken into account for designing 2POMW driers, does not exist. The operating conditions — size, temperature, and moisture content — at which 2POMW could ignite and cause a fire in an industrial drier were determined. The loss of volatile matter during the drying process, which modifies the composition of product and may affect the quality of the oil to be extracted, was also evaluated. The results of this experimental work allowed the development of a useful drying model for designing new driers and for assessing the behavior of existing ones.

The most common solution adopted by the industry is to use two rotary driers in line. The first drier is fed with a mixture of fresh and dried 2POMW having a moisture content of approximately 52–55% (wet basis) to avoid stickiness and dries it up to 25–30%. The second drier brings the moisture content of the mixture below 8%. Currently, the driers are operated manually or, at the most, with a simple system to control the inlet gas temperature. Arjona R. et al. (2005) developed, implemented, and tested at a 2POMW industrial drier a control system based on PID controllers[21] (applied to the first drier) that minimize the operational problems and improve the production and the energetic efficiency.

Despite the several investigations being performed to optimize the drying process of 2POMW from an operational and energy-saving point of view, the high energy cost of reducing its moisture content remains a clear drawback.

---

[21]Proportional-plus integral-plus differential controllers or three-term controllers.

Irreversible Thermo-Chemical Processes

The main irreversible thermo-chemical processes are combustion and pyrolysis. These are the most radical and destructive techniques, which eliminate any possibility of further uses of olive-mill wastes. Both processes are mainly used for the decomposition of concentrated solutions of OMWW and/or solid olive-mill wastes (olive cake or 2POMW), complementary to other treatments. As already mentioned, the thermo-chemical processes can be coupled to an evaporation or biological treatment. However, combustion and pyrolysis cause environmental problems arising from the emission of toxic substances in gas form, require high-energy consumption, very expensive facilities, as well as further energy wastes, caused by the transport of olive-mill waste to the incinerating facilities.

The development of appropriated technologies, which avoid the production of pollutants and other problems, while maximizing process efficiency, permits the sustainable disposal of solid olive-mill wastes. There are three main thermo-chemical processes by which this renewable energy source can be utilized, namely gasification, briquetting, and combustion (direct firing) or co-combustion (co-firing) — see Chapter 10: "Uses", section: "Generation of energy".

Combustion

Complete combustion (incineration, direct firing, burning) is the rapid chemical reaction of feed and oxygen to form carbon dioxide, water, and heat. Combustion is a method widely used for the disposal of waste material but the problem with OMWW is that it contains around 80–83% water and consequently is unable to sustain combustion without predrying. Therefore, this process is more suitable for very "strong" wastewaters (highly concentrated OMWW) so that the combustion can be self-sustained. Another problem is that olive-mill waste generation is a seasonal activity, which means that if the incinerator is to be run throughout the year then, other fuels are also required.

A commercial plant for disposal of OMWW by combustion is described by Arpino A. and Carola C. (1978). Fired by fuel oil, the plant could dispose of $20\,m^3$/day at a combustion temperature of 800°C — ejecting smoke at 400°C at a speed of 4 m/s. Technical data, operating costs, and the cost to the olive-mill during two years of operation are analyzed in tables and alternative fuels are recommended: up to 90% fuel savings could be achieved by using exhausted olive stones.

Amirante P. and Mongelli G.L. (1982) described the construction and operation of an Alfa-Laval incineration plant for the treatment of OMWW with the aid of diagrams, heat balance, and cost analysis. The incinerator efficiency with regard to heat recovery from OMWW was high and air pollution was insignificant.

A modular purification plant is described by Baccioni L. (1981), containing a section for purification and recycling of waters and a burner for destruction of waste muds discharged from olive-mills using exhausted stones as fuel. Diagrams show the product flow of OMWW and mud discharged from a continuous plant flow of water in a continuous plant (with recycling), a recycling plant, and a burner for OMWW. In practical operation, about 250 kg/h of exhausted stones or 160–180 kg/h of stones were needed for combustion of 1000 kg mud/h, with possible recovery of 250 kcal/h from the combustion gases.

A system to eliminate OMWW in a continuously processing olive-mill by thermal drying comprises a double combustion and drying air circuit being equipped with a hot-air drier, which can have diffusers for conveying the air through the product to be dried; the air is then passed by a blower to an exchanger. A combustion unit passes combustion gases through the exchanger and then via a venturi to an exhaust gas cleaner before discharge to atmosphere (ES2088340, 1996).

ES2032162 (1993) describes a process consisting of passing OMWW through a combustion chamber, yielding water vapor, combustion gases, and inorganic residues — see Fig. 6.2. The water vapor and the combustion gases pass through a heat exchanger yielding condensate water, which can be used as a diluent in the process. The non-condensable hot gases are passed through another heat exchanger, which raises the general temperature of the purification process. The process allows elimination of contaminating organic material in the effluent, and provides condensate water and hot gas for use in the process, together with usable inorganic residues.

The potential of using OMWW residues and olive stones as fuel was investigated by Vitolo S. et al. (1999). OMWW samples were separated by evaporation into an aqueous liquid (80–90% of the initial volume) and a residue in which approximately 98% of the organic load was concentrated. Pyrolysis and combustion tests on this residue and on olive stones showed that a mixture of the two may be useful as a fuel to provide heat for the evaporation of OMWW. The use of the concentrated solution as fuel is particularly attractive — its high heating value is calculated between 2000 and 3000 kcal/kg; on one hand it contributes to bringing down the polluting load, on the other hand it could also be a helpful low-cost alternative to the ordinary fuel for the evaporator and/or the bottom column boiler in case the distillation is considered. For this reason, in the few existing OMWW treatment plants there are usually special boilers, which should burn the sludge and concentrated solution coming from the aqueous effluent concentration

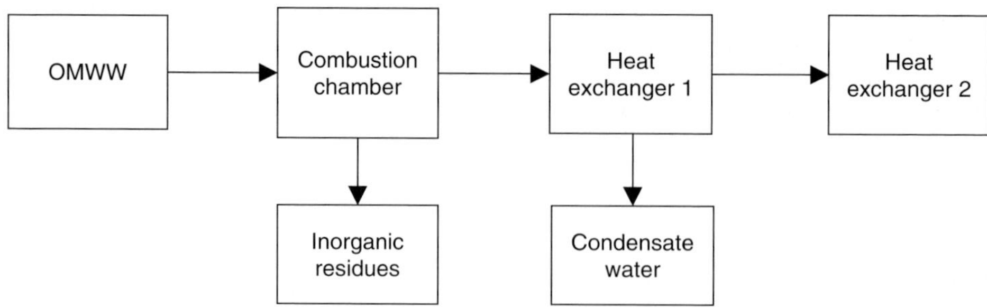

Fig. 6.2. Block diagram of the combustion process described in: ES2032162 (1993).

(Di Giacomo G. et al., 1991). However, it is well known that usually these devices do not work as they are supposed to, since the inorganic salts present in the concentrated solution (5 to 10% by weight) melt during the combustion and their deposits encrust the pipes of the boiler making them inefficient. Therefore, this part of the plant is often left unused and it is replaced by boilers that work with methane or oil making the purification process much more expensive. A process has been described which permits to execute the combustion of the concentrated solution, avoiding the typical problems related to the high salt content — see section: "Pyrolysis" (Di Giacomo G. et al., 1989; Di Giacomo G., 1990; IT1231601, 1991).

A process for obtaining alternative electrical power through the use of OMWW is described in: ES2092444 (1996). OMWW is passed from a storage tank to a boiler heated by olive waste, OMWW reaching the outlet from the boiler at 600°C. OMWW passes through an evaporator and at the outlet from this dissociates into steam and hot liquid at 400°C. The steam is used to drive a turbine whose axis is connected to an electrical power generator, while the steam from the turbine is collected and passed to a cooling tower where it is separated into hot waste liquid and water at 85°C; the hot waste liquid being reused in the actual process.

In general, the combustion process of OMWW has the following disadvantages:

- It is not self-sustained and it has a very high energy cost due to the necessity of evaporating great amounts of water.
- It is highly pollutant because it gives off toxic substances to the atmosphere in gas form produced during the combustion process.
- Neither the water nor the organic material in OMWW can be reused.
- Because OMWW is a seasonal activity, other additional fuels are required for running the incinerator the rest of the year.

Catalytic incineration, using a system similar to the catalytic convertor used for the exhaust gases from cars, was applied on an experimental basis for the treatment of waste gases from the drying of crude olive cake with good results (Papaioannou D., 1988). Catalytic incineration is based on the complete incineration of the pollutant substances in the presence of a catalyst made of platinum, iridium, etc. All the organic substances contained in the waste gases pass through the honey-comb catalyst, at temperatures of 350°C, giving off CO_2 and water. There were no traces of any of the pollutants in the gases once they had passed through the catalyst.

The drawbacks of this method were: (i) the high energy consumption (5 kg of fuel oil per ton olive cake); (ii) the high investment of $1.2\,m^3$ platinum catalyst for a daily production of 200–500 tons olive cake and the short time-life of the catalyst due to the presence of phosphorus in the stream of pollutants.

Pyrolysis

Pyrolysis (retorting, destructive distillation, carbonization) is the thermal decomposition of an organic material in the absence of oxygen. Pyrolysis is mainly used for the decomposition of concentrated solutions of OMWW and/or olive cake.

Petarca L. et al. (1997) studied the pyrolysis of the concentrated solution obtained from the evaporation process of OMWW in a laboratory-scale apparatus to detect the yield and properties of the gases, oil, and carbonaceous residue produced at different temperatures.

As said earlier, during the combustion of concentrated OMWW the inorganic salts that are present in the concentrated solution (from 5 to 10% by weight) melt and encrust the pipes of the boiler rendering them unfunctional. A process has been developed, which permits to execute the pyrolysis of the concentrated solution, avoiding the typical problems related to the high salt content (Di Giacomo G. et al., 1989; Di Giacomo G., 1990; IT1231601, 1991). In particular, the concentrated solution previously mixed with olive stones is pyrolized to separate the inorganic salts, which deposit on the charcoal bed. The heavy organic compounds originally present in OMWW are vaporized and leave the reactor with the gaseous stream, together with water and other volatile products resulting from the partial thermal decomposition of the olive stone present in the charge. Since the thermal decomposition process of the concentrated solution is strongly exothermic, the excess heat can be used to lower the COD of the treated water to values compatible with environmental regulations. Evaporation and pyrolytic tests performed at laboratory scale with solid-like mixtures of olive stones and concentrated OMWW with 50% water have demonstrated the technical feasibility of this process (Di Giacomo G. et al., 1991). The main by-product of this OMWW purification process is the charcoal whose commercial value can be of help in reducing the treatment purification costs. On the other hand, the integration of an OMWW purification process with the carbonization of exhausted olive cake and/or other discharged lignocellulosic residuals can add value to these by-products which, in the modern agricultural management have lost much of their traditional value.

In another process, OMWW was mixed with the residual fly ash produced by coal combustion in thermoelectric power plants and submitted to pyrolysis and activation process in order to obtain an adsorbent material (Rovatti M. et al., 1992). The pyrolysis produced an oily liquid fraction, with a good calorific value, a high hydrogen content gaseous fraction and a carbonaceous matrix dry residue — see Chapter 10: "Uses", section: "Activated carbons".

The process of pyrolyzing olive-mill waste products and other organic materials by means of microwave has been contemplated in: WO8904355 (1998). The process comprises the steps of: (i) preheating the waste organic materials at superatmospheric pressure and a temperature of at least 60°C, substantially without pyrolysis, by means of a hot gas stream; (ii) feeding preheated material directly to a microwave discharge zone having an atmosphere comprising a substantially oxygen-free gas at superatmospheric pressure; (iii) pyrolyzing the preheated material in said zone by means of a microwave discharge in the low gigahertz frequency range (e.g. about 2.4 or about 0.91 GHz) for about 15 to 60 minutes to produce solid fission products containing elemental carbon and gaseous by-products; (iv) recycling at least some of the gaseous by-products to the hot gas stream to effect preheating. The process could possibly be applied to the decomposition of 2POMW or olive cake.

A process for reutilizing and eliminating 2POMW based on a combination of processes comprising separation–extraction–pyrolysis or carbonization–gasifying combustion sequence is described in: ES2150360 (2001). The lignaceous part of stone, present in sludge, is separated during separation stage, while the pulp, already free of stone parts, is separated from oil during the extraction stage. Carbonization of the lignaceous part is conducted in a furnace at 550°C for 20 minutes. During the gasifying stage, the product of carbonization is subjected, in presence of steam, to temperature of 850–1000°C, and gaseous effluents from carbonization are utilized for drying the oil-free pulp, thus, saving energy.

Lagooning

With the use of large lagoons (artificial evaporation ponds or storage lakes), the sun's energy is used to speed-up the process of evaporation and drying of OMWW. Moreover, OMWW is partially degraded by a natural biological route, over very long time periods. This technique for OMWW disposal imposes treatment times of the order 7–8 months, in practice, from one milling season to the subsequent season, depending on the climatic conditions of the area. It has been estimated that for every 2 tons of olive processed, $1\,m^3$ of lagoon volume is required for storage and natural evaporation in Izmir, Turkey (Kasirga E., 1988; Azbar N. et al., 2004).

Most Mediterranean countries dispose OMWW in artificial evaporation ponds, the most developed being evaporation ponds provided with an impervious layer and those that use soil as a receptor medium, for instance, evaporation and infiltration ponds for large amounts of OMWW (Escolano Bueno A., 1975). Actually, lagooning has been one of the first processes to be used for the treatment of OMWW. By the end of the 1970s, the disposal of OMWW had become the main pollution problem in the Guadalquivir river basin (the river was called "the black river" at the time), the area of greatest olive oil production of Spain. For this reason in 1981 the Spanish Government prohibited the discharge of OMWW into rivers and subsidized the construction of ponds for its storage during the milling period and the evaporation of its water during the warm Andalusian summer. About 1000 evaporation ponds were constructed and subsequently the water quality of the rivers of the Guadalquivir basin improved greatly. However, the ponds caused serious negative environmental impacts on nearby areas due to the foul odors, insect proliferation, leakages, infiltrations, and silting with sludges (Rosa M.F. and Vieira A.M., 1995). Nevertheless, the main problem with evaporation ponds was their insufficient capacity because of the progressive change from the classic system, which produces 0.5–1 l of OMWW per kg of olives, to the three-phase continuous centrifugation system, which produces more than double of OMWW (1.3–2 l/kg) (Cabrera F. et al., 1996).

Results of experiments on disposal of OMWW to ponds and lagoons suggest no difficulties in application of this process, especially in rural olive-mills of medium size (Escolano Bueno A., 1975). About $1\,m^3$ $OMWW/m^2$ was a reasonable load. In large lagoons, anaerobic fermentation produced a strong acetic acid smell, no longer perceptible at 100 m distance. No anaerobic fermentation was observed in small ponds.

Lagooning has been used for pollution control and OMWW disposal as fertilizer after solar drying (Leon Cabello R. and Fiestas Ros de Ursinos J.A., 1981; Shammas N.K., 1984) and for storage in order to obtain load equalization during the whole year before the treatment by other processes (Balice V. et al., 1986). Removals of COD ranging from 20–30 to 75–80% have been obtained after 2–4 months.

There has been no attempt to recover biogas from OMWW treatment ponds, which operate mainly under anaerobic conditions, although in principle it could be possible to cover the lagoons with suitable gas-proof films and extract the biogas. This procedure would reduce emissions of methane in the atmosphere, which contribute to the greenhouse effect (Rozzi A. and Malpei F., 1996).

In a study in Portugal, vacuum evaporation panels consisting of plates with $30°$ inclination were fixed into a lagoon in order to separate solid and liquid phases. The liquid phase was evaporated while the solid phase remained on the plates and was taken out to be used as fertilizer (Duarte E.A. and Neto I., 1996).

The design of an evaporation pond must take into consideration, among other things the following factors:

  i) volume of OMWW produced by each of the olive-mills to be serviced,
  ii) climate of the region,
  iii) hydrology of the ground,
  iv) proximity of natural waters,
  v) distance from dwelling areas.

All these considerations allow the determination of the height of the pond.

The excavation costs comprise digging operations and removal of unearthed soil. The estimation of the excavations costs (between 7 and 20 €) is difficult because they depend on the type of the soil and the distance from the disposal site.

The following costs have been proposed for the purchase and the placing of the lining material (Le Verge S. and Bories A., 2004):

anchoring trench: 7.5 €/m;
draining geotextile with anti-piercing characteristics: $6 €/m^2$;
geomembrane of high density polyethylene (HDPE) with a thickness of 1.5 mm:
   $7 €/m^2$;
draining geotextile with anti-piercing characteristics: $6 €/m^2$;
layer of intermediate material (e.g. coarse gravel, flintstones, or cobbles) 0/31.5:
   $2 €/m^2$;
removal cost of the unearthed soil: layer of pebbles 0/31.5: $2 €/m^2$.

In addition, the cost of sealing a pond of $1000 \, m^2$ is estimated at 20,000 €. This cost is reduced at 16,000, if the cleaning of the pond is made with the help of a ditch cleaning machine.

The rate of evaporation differs from one type of OMWW to another. A study carried out by the Institute of Olive Tree and Subtropical Plants of

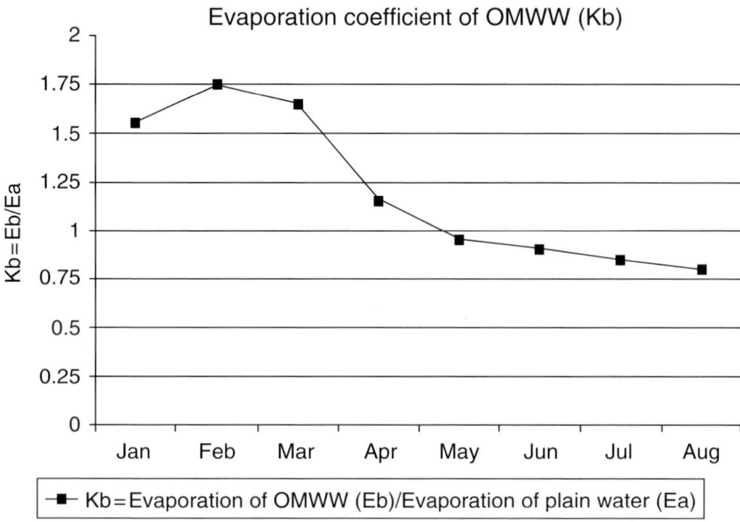

Fig. 6.3. Rate of evaporation of OMWW (Michelakis N. et al., 1999).

Chania-NAGREF (National Agricultural Research Foundation) recorded the evaporation mode of OMWW in a time period ranging from January to August (Michelakis N. et al., 1999) in terms of its coefficient of evaporation (Kb); Kb is defined as the ratio of evaporation of OMWW to the evaporation of plain water — see Fig. 6.3. The evaporation of OMWW is higher than that of the water up to the month of April (Kb > 1). Actually the dark color of OMWW enhances the absorbance of solar light increasing consequently, its temperature.

Rainfalls can cause an elevation of the OMWW level in the pond. For this reason the solid residue must be removed before the start of the new harvesting season. In addition, the oily film formed on the surface of the pond must also be removed because it obstructs substantially the evaporation of OMWW. In order to be sure that the dimensions of the pond are sufficiently large to tolerate the inflow of raining water it is necessary to know the annual rainfall pattern of the region. Areas with frequent and intense rainfalls require large evaporation areas.

In countries with a shortage of suitable large surface areas, the installation of a plastic coverage above the pond allows to increase the pond's height and reduce its surface.

To summarize, lagooning presents the advantages of low investment and maintenance cost.

On the other hand, lagooning has the following drawbacks:

- Threat of leakage of OMWW through the soil and into the groundwater. Preventative measures such as lining the lagoon and suitable maintenance is vital for the proper functioning of the lagoon.

- This method requires the availability of large collecting basins far from inhabited areas due to the unpleasant smell of OMWW and the presence of insects. The lagoons may have to be located 1 or 2 km away from the olive-mill and appropriate pipes will be needed to transport OMWW safely, i.e. without possible leakage into the soil. Considering the large volumes of OMWW produced each year, and the necessary time for the disposal thereof, this solution is affected by the drawback that large, adequate surface areas have to be available for long periods (about $1\,m^2$ for each $2.5\,m^3$ of OMWW); consequently, these large land surface areas are rendered useless for active agriculture. Rising costs for successive enlargements of the occupied surface area are required.
- The end product is useless as fertilizer, or for irrigation. Anaerobic fermentation produced a strong acetic acid smell, perceptible at long distance.

In conclusion, although this process is widely used, from the environmental point of view its use must be carefully implemented to avoid previously mentioned cautions.

Chapter 7

Physico-Chemical Processes

This type of processes involves the use of additional chemicals for the neutralization, flocculation, precipitation, adsorption, chemical oxidation, and ion exchange of OMWW.

The suspended and colloidal matter of OMWW is olive fines and juice, a part of which is biodegradable (pectins, proteins, etc.) and another is not biodegradable (tannins, oils, etc.). A significant fraction of the colloidal matter corresponds to pectin substances contained in olive juices, in the form of negatively charged hydrophilic colloids.

Neutralization

Neutralization is the restoration of the hydrogen (H^+) or hydroxyl (OH^-) ion balance in solution so that the ionic strength of each are equal. The neutralization technique can be used as a pretreatment procedure for the removal of the suspended or colloidal matter of OMWW and it is performed either by reducing pH to the point of zero charge (pH$\sim$2–4) via the addition of acids (e.g. H_2SO_4, HCl, HNO_3) or by increasing it (pH $= 11$) via the addition of caustics (e.g. $CaCO_3$, $Ca(OH)_2$, $NAOH$). By increasing the hydrogen ion (H^+) concentration or by adding specifically absorbed ions (Ca^{2+}) the negative surface charge of the suspended hydrophilic colloids is reduced and this leads to their neutralization and destabilization. The increase of pH as a pretreatment step before evaporation of OMWW helps in keeping more of the volatile organic compounds in the solid fraction during evaporation, giving a distillate of lower COD. The reduction of pH as a possible pretreatment of OMWW has attracted little attention so far (Mitrakas M. et al., 1996). Apart from colloids destabilization, pH reduction is also expected to contribute to the acidic hydrolysis of oils to fatty acids, which can be easily separated from effluents.

A process, disclosed in: ES8706800 (1987), consists of (i) filtering OMWW in a self-cleaning filter to remove large suspended solid particles; (ii) treating the filtrate with strong inorganic acid, e.g. H_2SO_4, in sufficient quantity to give pH 2–2.5; (iii) decanting by gravity into a lower part consisting of flocks and sludge and an upper aqueous part containing 0.5% olive oil; (iv) pumping the latter to a neutralizer; (v) addition of NaOH to give pH 7–7.5; and (vi) injection into the cycle as process water.

PT85790 (1987) describes a process consisting of subjecting the mixture of liquid and solid residues obtained simultaneously in the olive press to a treatment with an alkali carbonate, drying, and extraction with solvent, to obtain neutralized oil and a cake with good nutritional characteristics for feeding animals.

The strongly acidic composition of the pollutant gases generated during the drying of olive cake permits, as a principle, their retention in a solid alkaline filter bed, creating alkaline salts — see Table 3.1. Papaioannou D. (1988) proposed a method for reducing the pollutants and the odor in the waste gases based on a solid-bed filter with a filling material of ash and a fibrous filter, with a pore diameter of 6 μm, of fiber glass to trap the fatty substances — see Fig. 7.1. Of the possible alkaline beds, one made of ash was selected since this is a substance which has strong alkalinity and which is in plentiful in the seed-oil extraction plants. Dry lime hydrate or amorphous calcium carbonate were added occasionally in order to improve the alkalinity and porosity of the bed. The exhausted ash of the alkaline bed can be discarded after use. Its almost neutral composition makes it possible to be reused as a natural fertilizer, since it is a valuable source of mineral salts for plants. The fibrous filter can also be reused after it has been rinsed with hexane, exactly as in the process of extracting oil from dry olive cake.

Precipitation/Flocculation

Precipitation is the technique whereby a precipitate-inducing agent is added to the wastewater to transform dissolved chemicals into an insoluble solid form through a chemical reaction, so that it precipitates out.

Flocculation is an agent-induced aggregation of particles suspended in liquid media into larger particles. Essentially, it can be described as the destabilization process of a stable colloidal dispersion by the addition of a chemical known to effect destabilization.

It has been shown that during storage, OMWW undergoes natural self-purification due to a spontaneous flocculation/denaturization of the proteins (Annesini M.C. and Gironi F., 1991; Carlini M., 1992; Riccardi C. et al., 2000). The suspended material already present in the wastewater interacts with the help of long chain proteins acting as flocculating agents. The individual particles combine together to form a flock, which becomes denser than the surrounding medium and then settles. Flocculation happens also upon heating. It is, thus, expected that other results will be obtained with fresh OMWW directly from the mills than with cold and old OMWW.

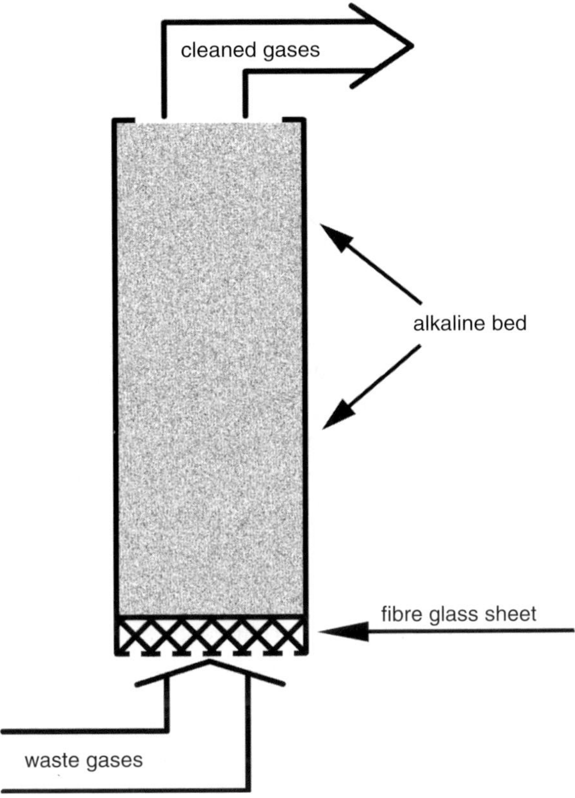

Fig. 7.1. Flow diagram of the solid-bed filter (Papaioannou D. 1988). The bed container has a diameter of 10 cm and a height of 30 cm. The fiber glass mesh filter has a thickness of 2 cm, a diameter of 10 cm, a specific weight of $0.25\,g/cm^3$ and a pore diameter of 6 μm.

Escolano Bueno A. (1975) and Raimundo M.C. and Oliveira de J.S. (1976) were among the first to use flocculation/coagulation[22] processes to remove oil, suspended solids and BOD_5 from OMWW. A maximum reduction of 40% for COD can be achieved while a precipitate that has to be disposed of is a large disadvantage of this method. In another study a preliminary treatment of OMWW by flocculation/coagulation or electro-coagulation allowed to remove about one-third of its COD and led to a significant decrease of the polyphenols content (Jaouani A. et al., 2000). The process is not very efficient in reducing the concentration of pollutants in OMWW because most organics found in OMWW are difficult to precipitate

---

[22]The terms flocculation and coagulation are both used in connection with formation of aggregates, frequently interchangeable and sometimes with distinctions that vary among professional disciplines. Although no distinction is made in this review, the more common types of distinction appearing in literature are enumerated in the "Glossary".

(e.g. sugars and volatile acids). The main disadvantages of flocculation can be summarized as follows:

- the results are only partial, not quantitative, because the separated fraction is only a fraction of the initial content;
- the precipitated material has then to be disposed of.

The flocculation/precipitation techniques may be used as pretreatment procedures for the removal of organic matter from OMWW. With the tendency towards anaerobic biological processes for final treatment of the pretreated OMWW, pretreatment with lime and iron flocculants caused no inhibitory effects on the methanogenic activity (Zouari N., 1998). The flocculation process can be also used as post-treatment to remove residual pollutants and suspended solids after biological processes (Fiestas Ros de Ursinos J.A., 1992).

The flocculants currently in commercial use are conveniently classified as organic or inorganic, and can also be of the anionic or cationic type.

Inorganic Flocculants

The inorganic flocculants of the cationic type include ferrous chloride, ferrous sulfate, ferric chloride, ferric sulfate, chlorinated ferric sulfate, aluminum sulfate, chlorinated basic aluminum sulfate, calcium chloride (Massignan L. et al., 1988), magnesium chloride, and magnesium sulfate. Flocculants can also be of the anionic type such as sodium aluminate or calcium aluminate. Inorganic flocculants reported in the literature for the purification of OMWW are described in the following paragraphs.

Ferric Chloride

When chlorides of trivalent iron ($FeCl_3$) are added to the water, they usually produce flakes of iron hydroxide, which make the impurities coagulate and simultaneously adsorb to the hydroxides (co-precipitation). Tests with iron chloride for flocculating OMWW proved that the trivalent iron immediately was reduced to divalent because of the reductive capacity of OMWW. At large dilution with oxygen-containing water flakes of iron oxide and a clear liquid phase were produced. This method requires such large quantities of water and energy for aeration that iron proved to be useless as a flocculant (WO9211206, 1992). In addition, iron chloride should not be used for flocculation/precipitation if the precipitated material (sludge) is to be used as feed for animals.

Ferric Sulfate and Aluminum Sulfate

Ferric sulfate and aluminum sulfate are commonly used as efficient flocculants of complex organic compounds in certain wastewaters. Tests with aluminum sulfate showed that no flakes of aluminum hydroxide were produced when the compound was added to undiluted OMWW. Prolonged aeration did not give any formation of flakes. Aluminum sulfate did not thus prove to be a suitable flocculation material for the same reasons as indicated for ferric chloride. The precipitated material could also not be used as feed for animals (WO9211206, 1992). Aluminum sulfate hydrate $[Al_2(SO_4)_3 \cdot 18H_2O]$ was tested along with other chemical substances (lime and hydrogen peroxide) as part of an integrated process for the decolorization of OMWW (Flouri F. et al., 1996). The aluminum sulfate was converted to an equivalent amount of $Al(OH)_2$ and was used in this form. Although all chemical substances exerted a clear decolorizing effect yet the least effective was aluminum hydroxide, followed by lime and hydrogen peroxide. Aluminum hydroxide in concentrations of $4\,g/l$ yielded an average percentage decolorization of 25%. At higher concentrations the percentage decolorization dropped even further.

In another process OMWW, having an initial $COD = 240\,g/l$, was first treated with sulfuric acid ($0.4\,cm^3/100\,cm^3$ OMWW) under agitation and filtration; the filtrate ($COD = 40\,g/l$) was neutralized with lime, $Ca(OH)_2$, and then treated with aluminum sulfate $[Al_2(SO_4)_3 \cdot 18H_2O]$ ($35\,g/l$) or limestone ($CaCO_3$). The obtained filtrate had a reduced organic load ($COD = 12\,g/l$) and could be treated further biologically or with reverse osmosis (IT1191528, 1988).

It has also been reported that only $10\,g/dm^3$ of $(NH_4)Fe_2 \cdot (SO_4)_3 \cdot 12H_2O$ was sufficient to precipitate almost 45% of the initial COD and color (Zouari N., 1998). The maximum amount of COD removal that could be attained was close to 70%. The complexing effect of iron was complete after $3\,h$.

Sodium Silicate

Sodium silicate (Na_2SiO_3) has been tested as a flocculant for OMWW (WO9211206, 1992). Added in large amounts the whole solution flocculated to a gel, which could be filtered after stirring. When using smaller amounts the flocculation took place after some time. The gel could only be sedimented or flocculated further after thorough stirring. The silicates made the sludge unsuited as feed for animals. In addition, gel materials were difficult to process further, among other things, because such material consistencies plugged filters and membranes and prolonged and expensive purification processes were necessary to avoid this problem. The conclusion was that sodium silicate was useless as a flocculant for OMWW.

Lime

Lime stabilization is a recognized means of treating municipal sludge prior to land application. Lime is used primarily for pH control or chemical precipitation in wastewater treatment. At the same time it assists flocculation, frequently functioning as a co-flocculant. The major types are quicklime CaO or mixed MgO and slaked or hydrated lime, which may be $Ca(OH)_2/MgO$, or $Ca(OH)_2/Mg(OH)_2$. The purity of quicklime depends on the type and the efficiency of the treating kiln, but generally it contains more than 90% of calcium oxide which when hydrated generates calcium hydroxide.

Lime precipitation has been employed as a minimal pretreatment procedure for the removal of organic matter content (GR870652, 1987; Lolos G. et al., 1994). The sludge produced upon addition of 0.5–3% quicklime (CaO) reduces the concentration of suspended solids in OMWW by 27.6%, whereas 77.1% of the oil and grease are distributed in the precipitate fraction. The optimum lime dose for flocculation of OMWW was found to be 2.5% w/v. The COD as well as the pollution load in terms of phenolic compounds are not strongly affected by the addition of CaO. The COD removal was found to depend strongly on the level of suspended solids of the untreated OMWW and was independent on the lime dose when the lime added was in the range between 5 and 30 g/l. Treatment of OMWW with CaO proved ineffective in removing color, when used in concentrations ranging from 4 to 35 g/l, yielding a maximum decolorization of only 15% (Flouri F. et al., 1996). The liquid fraction corresponds only to 25–43% of the total waste volume, contains less recalcitrant compounds and in general has a lower COD value. The organic material extracted by the lime addition and received as sludge, appears to have an economic interest due to its high energy potential (gross calorific value of total solids is 22,830 kJ/kg).

OMWW samples were analyzed for concentration of total, fixed, volatile and suspended solids, COD, oil fat, polyphenols, volatile phenols, nitrogen, and reducing sugars before and after treatment with lime (addition of lime until a pH of 12 was reached in OMWW). Lime treatment reduced levels of all investigated pollutants by 63–95%, with the exception of volatile phenols (average reduction 28%). Additional experiments conducted on phenol mixtures revealed that adsorption efficiency on lime varied widely according to compound structure; compounds with two phenolic groups in the middle (e.g. catechol) were adsorbed completely, compounds containing both phenolic and carboxyl groups (e.g. vanillic acid) were adsorbed partially, while compounds having only one phenolic or carboxyl group (e.g. vatic acid) were not adsorbed at all. More efficient filtration of lime treated OMWW compared with untreated OMWW was observed. It was concluded that lime treatment is an effective, low cost means of reducing pollutants in OMWW (Aktas E.S. et al., 2001).

It has been reported that 10 g/l of hydrated lime, $Ca(OH)_2$, was sufficient to precipitate more than 50% of the initial COD and remove 50% of the initial color within a short contacting time (Zouari N., 1998). The removal efficiency increased with increasing lime concentration and is maximized at $pH = 11 \pm 0.5$ (Tsonis S.P. et al., 1987). With lime treatment, 55% of COD and 70% of color removal may be reached, but for economical and biological considerations, treatment with 10 g/l calcium dihydroxide was sufficient. The effect of lime was complete after 12 h.

The above results can be explained by the fact that the pectin substances present in OMWW in the form of negatively charged colloids can be destabilized, either by increasing $[H^+]$ concentration, or by adding Ca^{2+} ions. It should also be noted that the relatively high COD removal with addition of $Ca(OH)_2$ is attributed to sweeping flocculation — sedimentation that the hydroxide causes to colloids (Tsonis S.P., 1987).

The volume of the resulting sludge is large and could not be adequately handled by sedimentation, requiring that it be dewatered. Straining was found an inexpensive means for handling this sludge.

The addition of lime temporarily halts biological activity. However, lime renders organic molecules more accessible to microorganisms.

A process for purification of OMWW, — applied also to effluents from wineries, paper-making plants, distillers for producing alcohol from sugarcane, abattoirs and municipalities — consists of : (i) adding $10\,g\,Ca(OH)_2$ to each liter of wastewater and agitating for several seconds; (ii) adding $10\,ml$ 50% sodium hypochlorite (NaOCl) solution per liter of wastewater, together with $1\,ml$ organic flocculant solution (0.025%) and agitating for several seconds; and (iii) leaving the flocculate to settle and separating by decanting and filtering (ES2009267, 1989).

Tests with lime were also conducted to recover the remaining oil in OMWW. An amount of 70.6–96.4% of the oil was recovered when OMWW was treated with lime whilst air was passed through the mixture, then filtered using a muslin cloth disk, dried, and extracted with organic solvent. It is concluded that recovering the oil phase from OMWW can reduce pollution and regain an economically important by-product.

In general, lime precipitation results in a 40–50% reduction of the organic matter but produces large quantities of sludges (Mendia L. and Procino L., 1964). Moreover, the effluents after precipitation, as well as the chemical–organic sludges that are produced, have all the pollution load of the initial OMWW leading to serious disposal problems (Fiestas Ros de Ursinos J.A., 1991).

Lime has also been used for the treatment of odors emanating from evaporation ponds. The emitted gases are degradation products of the anaerobic fermentation of OMWW and are mainly composed of volatile fatty acids (butyric, caproic, valeric, and iso-butyric). The addition of lime neutralizes the volatile acids in their salts which are non-volatile and non-malodorant. On the other hand, the neutral volatile (alcohols, aledhydes, esters, etc.) or basic compounds are not eliminated.

The addition of lime in an evaporation pond during the phase of intense emission of odors (May 2003) up to pH 10, or $33\,m^3$ of hydrate lime (30%) in an evaporation pond of volume $1200\,m^3$ reduced the volatile fractions of the acids butyric and valeric by 86% and the caproic acid by 88% (analysis by GC and SPME). The effect of neutralization of the volatile acids was immediately evident and was prolonged for two months (June–July, 2003), a period characterized by particularly elevated temperatures. In general, a maximum amount of $1\,m^3$ of hydrated lime is recommended for every $40\,m^3$ of OMWW (Le Verge S. and Bories A., 2004).

Odor control in evaporation ponds was used for measuring the effect of $Ca(OH)_2$ on OMWW (Lagoudianaki E. et al., 2003). Different amounts of $Ca(OH)_2$ were added in 21 beakers containing 1l of OMWW. The mixture was stirred for 45 min and left to settle. The Odor Threshold was used for determining the effect of the treatment in the odors of the beakers three and 30 days after. Both sets of measurements indicated important reduction in OMWW pollutants and odor emission when $10\,g/l$ $Ca(OH)_2$ were added. In order to evaluate these results in more realistic conditions, plastic containers were filled with 6 l of OMWW, relevant amounts of $Ca(OH)_2$ were added, the mixture was stirred manually and left to settle in the open. Again the same odor reduction was noticed.

Washing with an alkaline solution (lime) has been used for the treatment of waste gases generated during the drying of crude olive cake. Washing was based on a system in which the waste gases were sprayed with a lime hydrate solution. All the pollutant substances were absorbed into the washing liquid, and the lime hydrate neutralized the acids and saponified the fatty substances. The liquid could be recycled for reuse, and eventually discarded when it was overloaded (Papaioannou D., 1988).

The drawbacks of this method were: (i) the high water consumption ($1.5\text{–}2\,m^3$ per ton olive cake); (ii) the high pollutant load in the liquid waste, equivalent to a 20,000–30,000 population from a production of 200–250 tons olive cake per day; (iii) the high energy consumption (about $2\,kWh$ per ton of olive cake).

Miscellaneous Inorganic Flocculants

Natural environmental-friendly clay minerals such as bentonite have been used for flocculation and sedimentation of suspended pollutants in OMWW. A remarkable characteristic of the bentonite, except its strong swell and adsorption capacity, is its sheet-shaped or also rod-shaped structure, which has an exceptional large surface area. The inside of a bentonite particle is fissured, layered, and full with voids. The areas of the various walls of these voids and layers form the so-called inner surface of these porous materials; the inner surface area of a gram of bentonite can amount to several hundred square meters. Additionally, the bentonite has deposited cations, which can move freely and which can be replaced by other cations, found in OMWW in the form of polluting particles. The bentonite particles also function as ion exchangers, where the exchange process takes place either only on the external surface of the particles or also by the cations deposited among the inner layers of the bentonite.

The process and apparatus described in: DE19529404 (1997) use bentonite enriched in montmorillonite for the treatment of OMWW. The bentonite and other reactant agents are continually added to the OMWW in a fine stream and homogeneously mixed into OMWW in strict relation to the pollution load. The clay particles have a large inner surface and act as an ion exchange medium, which together with the other reactant agents bind the pollutants by adsorption, followed by flocculation. The flocculant particles are separated as sediment in a subsequent settlement basin. The water skimmed from the settlement basin is discharged to the public drain. The sludge is continually discharged to a sludge tank and is subsequently dewatered, compressed, and dried.

CZ9401911 (1996) describes a process for the disposal of OMWW by using an intensive grinding and mixing of OMWW with natural bentonite, which is thus activated. OMWW, with a density between 800–1000 kg/m^3, is continuously mixed, at a ratio of upto 300 kg (preferably 200 kg) of natural bentonite per 1 m^3 of OMWW. The mixture is further mixed up with a solid phase of waste, i.e. press cakes and with vegetable waste, while inoculation material is being added. The mixture is inoculated and subject to forced aeration. Advantage of this way of disposal of OMWW is claimed to be the use of simple and cheap equipment, which, for example, can only consist of the tank and the grinding and mixing pump. Another advantage is the absence of chemicals and availability of all necessary raw materials as well as energy and especially that of inoculation material. The latter can easily be prepared from a soil containing microorganisms being for a long period modified through the action of olive tree products and, thus provided with specific biodegradation properties. Resulting product is a loose, lumpy mass suitable for exploitation in agriculture. This material neither decomposes further nor smells bad even in a humid environment.

Studies on the effect of the various inorganic flocculants on OMWW purification, reported in the literature, are summarized in Table 7.1. These investigations have employed lime 3–54 g/l, aluminum sulfate 0.12–160 g/l, ferric chloride 2–160 g/l, and other aluminum or iron salts, alone or in combination.

A major disadvantage of using inorganic chemicals for OMWW conditioning is that it significantly increases the sludge mass. The use of inorganic flocculants, and especially lime, should not be considered without provision for the handling and disposal of the resulting sludge. In addition, most of the inorganic flocculants proved ineffective in reducing the pollution load of OMWW and did not justify their use in the majority of the olive-mills (Bradley R.M. and Baruchello L., 1980; Shammas N.K., 1984).

Organic Flocculants

Organic flocculants are water-soluble polymers with weight-average molecular weights ranging from about 10^3 to greater than 5 × 10^6 and include natural and synthetic flocculants. Of organic flocculants there are many different types and depending on the properties, which are necessary for such materials there is a distinction between anionic, cationic, and non-ionic agents. If some subunits of the polymer's molecule are charged, it is termed a polyelectrolyte. Polyelectrolytes containing both positive and negative charges in the same molecule are termed polyampholytes. Although the non-ionic water-soluble polymers do not fall within the definition of a polyelectrolyte, they tend to be placed in the same category in the flocculant literature. The organic flocculants are more expensive on a unit-weight basis than the inorganic flocculants in general use, but the required dosage is much lower.

The organic flocculant for OMWW must have cationic properties on account of the charge on the surface of the colloid particles; it must also have non-toxic

Table 7.1. Effect of various inorganic flocculants on OMWW purification (compiled by Tsonis S.P. et al., 1987)

Chemical	Dose g/l	Operation/ process	Raw waste characteristics		Treatment efficiency		Remarks	Reference
			Parameter	Value g/l	Parameter	% Removal		
Calcium oxide CaO	50	1 h mixing and filtration			VTS	56		Fiestas Ros de Ursinos, 1953
Ferric chloride or ferrous sulfate	2				BOD_5	40		Fiestas Ros de Ursinos, 1977
Aluminum sulfate	2				BOD_5	<37		,,
$Ca(OH)_2 +$ $Al_2(SO_4)_3$	18 + 0.12 54 + 0.12	Successive addition of chemicals, flocculation	Oxydability COD	28.8 118.8	Oxydability	22.4 21.4		Raimundo et al., 1976
$Al_2(SO_4)_3$	10–160	1 min mixing (100 rpm), 40 min flocculation (40 rpm) and 8 h sedimentation	Oil	5–6	Oil	Max 96	Optimum dose 150 g/l	Curi et al., 1980

$FeCl_3$	10–160	"	"		"	Max 88	Optimum dose 10 g/l	"
$FeCl_3$ + $Ca(OH)_2$	(5–25)+ (25–5)	"	"		"	Max 88	Optimum dose 5 g/l $FeCl_3$ and 25 g/l $Ca(OH)_2$	"
Alum	3				COD	25	In presence of polyelectrolyte	Bradley R.M., 1980
$Ca(OH)_2$	3–12	Jar test flocculation and 1 h sedimentation	TC	37.5	TC	3.7–20	Capillary suction time of sludge >26 sec, Specific resistance $>10^{10}\,s^2/g$	Boari G. et al., 1980
$FeCl_3 \cdot 6H_2O$	10–18	"	COD	112	"	20–30.5	"	"
Poly(aluminum chloride)	8–24	"	TS SS	77.85 1.4	"	4–38.5	"	"
$Ca(OH)_2$	10 20 30	45 min mixing and 24 h sedimentation	BOD_5	18.9–20.2 27.0–30.3 28.7–29.1	BOD_5	7.3–9.0 24.6–29 43–50	Good sludge dewaterability	East Cretan Section Technical Chamber of Greece, 1980

and non-polluting properties to be used later as inter alia feed for animals or fertilizers.

ES820395 (1982) and ES8307286 (1983) describe a flocculation process comprising the steps of: (i) treating OMWW in a stirred tank containing a coil, for stabilization and heat treatment, (ii) transferring OMWW to a second stirred tank for dispersing and blending of an anionic polymer flocculant (polyacrylamide), then, (iii) transferring OMWW to a third stirred tank for dispersing and blending of a cationic polymer flocculant (polyamine), and finally, (iv) bringing OMWW to a sedimentation and thickening tank for the separation of the suspended solids and the residual oil. The process is claimed to decrease the BOD_5 by 50%.

A process is characterized in that OMWW is subjected to a controlled flocculation by means of an aqueous solution of cationic surfactants of fatty nitrogen derivative type with a weight ratio of OMWW/fatty derivatives of 0.2–0.7%. The treatment is carried out in a tank with mechanical stirrer and measuring pump for the flocculants. After filtering, decanting, or centrifuging, a non-polluting liquid phase and an organo-mineral solid residue are obtained (ES2011366, 1990). In a modified process, the liquid is subjected to repeated treatment with the same flocculants or with condensation products of formaldehyde and dicyandiamine with or without additives and partly recycled to adjust its concentration and finally passed through biological filters. The sludges are combined, mixed with pruning from olive or other tress, or cereal straw to produce compost by a fermentation process (ES2028497, 1992).

WO9211206 (1992) describes a process and a plant for purifying OMWW where the waste material resulting from the processing of olive fruit, (grinding, pressing, etc.), is subjected to an initial separation of solid, water-insoluble material, and water-soluble material by a flushing method. The obtained suspension is stable based on the surface charge of the particles. Chitosan is added to the aqueous phase to flocculate the colloidally floating particles in the aqueous phase. Chitosan is a biopolymer which has in the above mentioned aqueous system been shown to precipitate and flocculate the dissolved and floating organic materials and salts in the aqueous solution form. Chitosan is a water-soluble cationic polysaccharide made from chitin [poly($1\rightarrow4$)-acetamido-2-deoxy-β-D-glucose], by partial deacetylation with alkali. Chitin is derived from crustacean shells (fishery waste), and, therefore, the raw materials supply is limited at present. Chitosan, as natural product, is suited for feed or fertilizer on account of its nutritional value and has additionally adsorbing properties for emulgated oil. The principle for the present purification method is that the added chitosan neutralizes the charge on the surface of the particles so that they coalesce (flocculate) into larger units (aggregates) and may thus be separated from the water. The chitosan was admixed to OMWW in an amount of 100–200 g/m^3 OMWW. The chitosan was added in the form of an acetic acid solution for adjustment of the pH to a value in the interval 5.5–7.0, preferably pH = 6.5. In addition, a water-soluble calcium compound such as calcium hypochlorite or calcium nitrate may be added in an amount of up to 200 g/m^3 as well as other optional adjuvants for flocculation and aggregation of the flakes,

Table 7.2. Results of flocculation tests (WO9211206, 1992)

Flocculant	Organic carbon	Purifying effect
Silicate	27.3 mg/l	15.5%
Chitosan	20.5 mg/l	36.5%

which have been produced with chitosan to larger and more solid flakes. Such adjuvants may, e.g. be organic polymer materials such as Prästol™ and/or Zetag™, which may also be added to OMWW in an amount of 50–100 g/m³.

The effect of chitosan as a flocculant on OMWW purification was compared to that of silicate — see Table 7.2. The results of Table 7.2 show that chitosan gave a better flocculating effect. The precipitation was measured to 17 g dry matter per liter OMWW. As mentioned before, silicate as a precipitation agent will result in the forming of a gel representing a significant problem at a subsequent filtering of the aqueous phase after the flocculation step. Conversely, chitosan gives no such gel formation.

Adsorption

Adsorption is a physico-chemical process, consisting of the attachment of dissolved compounds (adsorbate) from polluted waters to the surface of a solid substance (adsorbent). Adsorption not only takes place at the visible surface of the solid, but also in its pores. The attachment takes place in two steps: transportation of the adsorbate to the surface of the adsorbent and the attachment itself. There are two methods for intensification of the transport process. During the fluidized-bed process, the adsorbent powder is stirred with the wastewater, while during the fixed-film process the wastewater flows along the grainy adsorbent in reactor.

Organic compounds (adsorbates), which can be removed from OMWW are coloring substances (mainly tannic acids), hardly or non-biodegradable pollutants, bactericidal, or inhibiting compounds. The adsorption is usually used in combination with other treatments (GR870652, 1987; EP324314, 1989).

One of the most widely used adsorbents is activated carbon. Activated carbon is especially suited because of its large inner surface (500–1500 m²/g) and its high adsorptive capacity, but unfortunately it cannot be reused. However, its calorific value is very high so that it can be incinerated without problems. Curi et al. (1980) used adsorption on activated carbon to investigate the dark color removal of OMWW, but they did not report any values regarding the change in COD nor phenols content of the treated samples. It has been estimated that between 60–80% of the organic constituents from OMWW can be adsorbed by activated carbon (EU project: FAIR CT96–1420 "IMPROLIVE"). Strong contamination has negative effects on the workability of the plant so that OMWW should be pretreated, for example in an activated sludge tank.

Generally, methods of employing activated carbon to remove contaminants from OMWW have met with only limited success due to either the limited adsorption capability or the high costs of the adsorbents. The use of activated carbon has been severely inhibited by associated processing difficulties and the inherent high initial cost of the material. In addition, the high attrition and regeneration losses, which occur when activated carbon is employed results in high running costs. Activated carbons in powdered form, although available at relatively low initial cost since they are produced largely by the partial incineration of waste liquors from paper manufacture, are difficult to remove from the treated water because of their highly subdivided state, which results in very low settling rates. Thus, when using powdered activated carbon, each contact stage requires a subsequent settling having a long residence time and the use of expensive organic polymers as flocculants. Furthermore, after removal from the water, no practical techniques have been developed for regenerating activated powdered carbon for use. Thus even, if the initial unit cost of the powdered activated carbon is relatively low, the overall operating cost becomes exorbitant since the material can be used only once and then must be disposed of at an additional cost.

Granular activated carbons, such as those produced from coal, are expensive adsorbents because they require a multistep process for their manufacture in order to produce them with uniform particle size and acceptable hardness. Even though these materials have greater hardness and attrition resistance than the so-called "soft" activated carbons produced from other materials such as wood, nut shells and the like, the attrition resistance of granular activated carbon is lost due to attrition in the handling and use of material. This may occur, for example, when the granular activated carbon is removed from the wastewater contacting bed(s) and regenerated in a device such as a multiple-hearth furnace, the regenerated granular activated carbon then being recycled to the contacting bed(s). Not only does this represent a high operating cost due to the make-up with fresh granular carbon material, but the fines produced by said attrition are difficult to remove from the treated water and, therefore, represent a source of contamination. Furthermore, because of the fragility of the granular activated carbon, the wastewater treating processes that have been devised employing such carbon have been severely limited, since such processes must necessarily inhibit the motion of the carbon granules in the process in order to minimize attrition.

Activated carbons can be obtained from the olive stones and solvent-extracted olive pulp (Mameri M. et al., 2000a; Moreno-Castilla C. et al., 2001; Galiatsatou P. et al., 2001, 2002). Approximately, 90% of the olive pulp consists of woody material, which is rich in lignocellulosic precursor. Use of this precursor for the preparation of activated carbons not only produces a useful material for purification of contaminated environments from phenolic derivatives, but also contributes to minimizing the solid wastes. The activated carbons were proved to be efficient adsorbents for the removal of phenols and COD decrease in OMWW — see also Chapter 6: "Thermal processes", section: "Pyrolysis"; Chapter 10: "Uses", section: "Activated carbons".

Bentonite and other clays are used sometimes as low-cost adsorbents. Promising results were obtained by adding $Ca(OH)_2$ (up to pH 6.5) and 10–15 g/l of bentonite, and then feeding the mixture to a laboratory-scale continuous anaerobic reactor without providing an intermediate phase separation (Beccari M. et al., 1999b, 2000, 2002). Preliminary biotreatability tests performed on the pretreated OMWW showed high bioconversion into methane at very low dilution ratios (1:1.5). The results confirm the double role played by bentonite: adsorption of lipids (the most inhibiting substances present in OMWW) and release of the adsorbed biodegradable matter in the anaerobic reactor (Beccari M. et al., 2002).

Activated clay is a new low-cost adsorbent, which has been tested for treating OMWW. OMWW conditioned with a series of pretreatments steps composed of settling, centrifugation, and filtration was then subjected to a post-treatment process, namely adsorption on activated clay (Al-Malah K. et al., 2000). The dynamic response of phenols concentration, pH and COD, using different concentrations of activated clay showed a peak at which adsorption capacity was achieved. The maximum adsorption capacity for the tested concentrations of activated clay was reached in less than 4 h. It is thought that adsorption of phenols and organics is reversible mainly due to hydrophobic interactions. The maximum removal of phenols was about 81%, while it reached about 71% for organic matter.

Separation of complex organic compounds from OMWW by means of adsorption on specific resins is an economical alternative. DUOLITE® XAD 761™ is an aromatic resin adsorbent which has been investigated for OMWW decolorization (Zouari N., 1998). The DUOLITE® XAD 761™ resin is used industrially for the adsorption of mono- and poly-aromatic compounds. It removes color, protein, iron complexes, tannins, hydroxymethyl furfural, and other ingredients responsible for off-flavors (Technical sheet of the Duolite Company). The degree of adsorption tends to increase with molecular weight in a given homologous series and has more affinity for aromatic than aliphatic compounds. The aromatic adsorbent resin retained more than 50% of the coloring compounds (chromophores) corresponding to removal of more than 60% of the initial COD after treating three bed volumes of crude OMWW. The efficiency depended on the volume treated.

Advantages of the adsorption process are:

- low space requirements,
- no water pollution,
- no odor emissions, and
- low costs for adsorbent.

Among its disadvantages are:

- limited purification efficiency,
- running plant costs, and
- qualified personnel are required to ensure trouble-free operation.

Chemical Oxidation Processes

The oxidizing agent used is chosen from the group formed by oxygen, oxygen derivatives (e.g. hydrogen peroxide or ozone), chlorine, chlorinated derivatives (e.g. chlorine dioxide, sodium hypochloride, calcium hypochlorite, potassium hypochlorite, sodium chlorite, sodium chlorate, or bleach) or potassium permanganate. A mixture of oxidizing agents can also be employed.

Chlorinated derivatives (e.g. chlorine dioxide, sodium hypochloride, calcium hypochlorite, potassium hypochlorite, sodium chlorite, sodium chlorate, or bleach) or potassium permanganate have been used at various stages of OMWW purification (ES8607039, 1986; GR88100203, 1989; ES2009267, 1989; WO9211206, 1992; Bellido E., 1987, 1989a,b).

Ozonation is usually adopted for water disinfection, but it also has a high potential as pretreatment method. The characteristic of ozone is that it is rather selective towards double bonds. Theoretically, it would leave intact the proteins and the sugars of OMWW, which are biodegradable anyway, and attack selectively the double bonds of unsaturated fatty acids and phenols. In this way, the total COD would vary to a lesser extent, because the toxic compounds are present in minor concentration and the biomass potential to feed an anaerobic reactor would not be lost.

Ozone or hydrogen peroxide, possibly combined with UV radiation, on the one hand is used because of the high oxidation potential of these oxidizing agents, on the other hand it is possible to operate under the condition of atmospheric pressure and ambient temperatures without problematic decomposition products of the oxidizing agent. In principle, the reaction mechanisms of H_2O_2 and O_3 are identical, differences only occur in provision and reactivity. The utilization of H_2O_2 has turned out to be environment friendly because this oxidizing agent has no negative effects. However, since the H_2O_2 quickly undergoes decomposition, the storability is limited. The HO· radicals formed during the H_2O_2 decomposition have negative effects. Using suitable agents (e.g. titanium dioxide) or UV radiation the development of OH· radicals can be considerably enhanced — see "Advanced oxidation processes (AOPs)".

Ozone has to be produced on site according to demand. In oxidation systems using O_3 it is possible: (a) to convert inorganic components into higher oxidation stages; (b) to cleave also hardly biodegradable organic compounds; (c) to destroy bacteria; and (d) to destroy especially odorous, taste-causing and coloring substances. With the help of UV radiation the effect of both oxidizing agents can be increased additionally by radical formation.

The opportunities offered by O_3 for the abatement of organic pollutants in OMWW were explored by means of experimental investigations pointing to the chemical and kinetic characterization of ozonation processes of model compounds belonging to different chemical classes. The research has recently been focused on phenolic compounds present in OMWW, such as *p*-coumaric acid, vanillic acid, and 3,4-dihydroxybenzoic acid. Tests were performed also on OMWW, at different pH (EV5V-CT93-0249).

A purification procedure was developed that is implemented synchronously and automatically in a continuous-flow fashion with the factory production and involves an operational sequence including drastic oxidation with sodium hypochlorite (NaOCl), decantation, filtration, active carbon catalysis and aeration-ozonation — see Fig. 7.2 (González-López J. et al., 1994; Bellido E., 1987, 1989a,b). Prior to injecting ozone, and in order to enhance its oxidizing power, OMWW was subjected to treatment with activated carbon in order to simultaneously accomplish the catalytic decomposition of the chlorinated compounds and the adsorption of the derivatives and other organic substances. The efficiency of the proposed procedure in reducing the initial concentration of polyphenols in OMWW was tested by the same authors. The results obtained in the oxidation experiments show that the efficiency of the process is circa 97.75% (38 OMWW samples), i.e. somewhat lower than those achieved by other authors (99.9%) by using a sequence of aerobic and anaerobic treatments on a laboratory scale (Maestro-Durán R. et al., 1991) and variously prediluted OMWW from a factory equipped with a continuous extraction system (Borja-Padilla R. et al., 1991b,c). As regards the operativeness of the process, its implementation on a plant scale involves automatic performance of the oxidizing steps, and the overall duration of the process is slightly over 3 h. According to the above results, the sequential chlorination, catalysis-adsorption on activated carbon and final ozonation involved in the proposed purification procedure, the individual effects of which are quite well known in the treatment of OMWW (Fiestas Ros de Ursinos J.A., 1977; Janer del Valle M.L., 1980; Ranalli A., 1991) result in a high purification efficiency. The procedure is claimed to neutralize the acidity and negative redox potential, ensure the oxidation of organic and inorganic compounds,

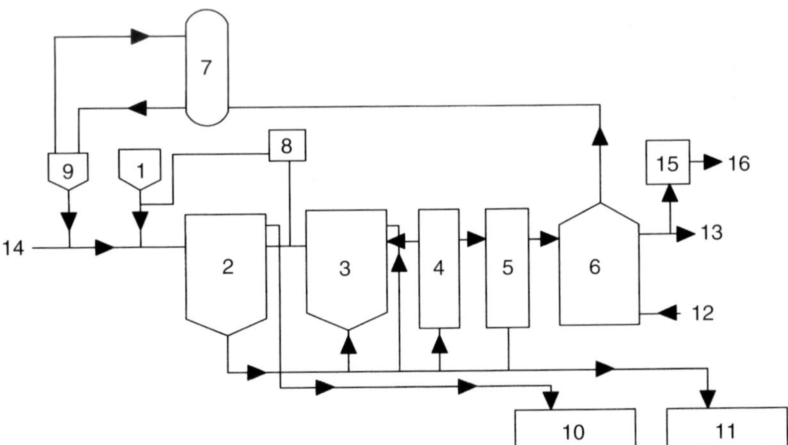

Fig. 7.2. Block diagram of the purification process (González-López J. et al., 1994) (1) sodium hypochlorite; (2) reactor; (3) settler; (4) inert filler; (5) active charcoal filter; (6) degasifier; (7) absorber; (8) reaction control; (9) absorbent; (10) oil; (11) slurry; (12) air; (13) effluent; (14) OMWW; (15) demineralizer; (16) effluent.

reduce the amounts of residual oils, suspended and extractable solids, sterilize the microbial flora, and eliminate its toxicity.

Advanced Oxidation Processes (AOPs)

In an attempt to optimize oxidation processes, recent research and development work indicate that oxidation rate limitations may be removed and lowered if conventional oxidants are replaced by combination of oxidants as well as combinations of oxidants with ultraviolet radiation. Such mixed oxidation systems have been labeled advanced oxidation processes (AOPs). AOPs are characterized by the production of the highly oxidative hydroxyl radical (HO$^{\cdot}$) at ambient temperature. The HO$^{\cdot}$ radical may be generated by a number of photochemical and non-photochemical pathways. Due to its strong oxidative nature, which is much greater than other traditional oxidants, the HO$^{\cdot}$ radical is able to completely transform organic carbon compounds to CO_2. Common AOPs, such as H_2O_2/UV, O_3/UV, and H_2O_2/O_3/UV, involve UV photolysis of O_3, H_2O_2, or both to generate OH$^{\cdot}$ radicals. But these radicals can also be generated with a semiconductor (photocatalysis), which absorbs UV radiation when this is in contact with water. The latter process is of special interest since it can use (solar) UV, if the semiconductor used has an appropriate energetic separation between its valence and conduction bands, which can be surpassed by the energy content of a solar photon ($\lambda \geq 360$ nm). Titanium dioxide particles (TiO_2) have demonstrated to be an excellent catalyst for this application.

The principal AOPs used in OMWW treatment are:

- O_3/H_2O_2
- Photolysis of O_3
- Photolysis of H_2O_2
- Photocatalysis
- Fenton reaction.

Photolysis of O_3

The effect of the combined O_3/UV treatment on the oxidative degradation of four phenols (caffeic, *p*-coumaric, syringic, and vanillic acids), which are major pollutants in OMWW, was evaluated and compared to single photolysis and single ozonation (Benítez F.J. et al., 1995, 1997b). The combined use of O_3 and UV slightly increased the rate of phenol oxidation compared with the rates for the single oxidations performed; the combination of processes generated OH$^{\cdot}$ radicals which increased degradation rates.

Compared to other conventional AOPs, the O_3/UV system exhibits a lower degree of mineralization but faster kinetics. Similarly a limitation on the use of O_3 in OMWW treatment is the generation and mass transfer of sufficient O_3 through the water to efficiently oxidize the organic contaminants.

Photolysis of H_2O_2

Photolytic processes of water pollutants abatement by hydrogen peroxide are being studied as a profitable way to oxidative degradation of scarcely reactive substrates. The high effectiveness of H_2O_2 photolysis for the treatment of wastewater is in fact provided by reactions involving OH˙ radicals' generation. Application of H_2O_2 photolysis processes can be normally recommended for wastewater treatment where no special concern to process selectivity is required.

The H_2O_2/UV system has been used for the chemical degradation of several model phenolic compounds present in OMWW (Benítez F.J. et al., 1996a, 1998). When the degradation is promoted by the combination of H_2O_2/UV, a reaction rate equation is proposed that includes two contributions: the single photooxidation and the increase in the process by the action of OH˙ radicals, which are generated from H_2O_2 by the presence of UV radiation.

The H_2O_2/UV is efficient in mineralizing organic pollutants, but exhibit slow kinetics compared to O_3/UV.

A disadvantage of conventional AOPs, such as O_3/UV and H_2O_2/UV, or their combination, is that they cannot utilize abundant solar light as the source of UV light because the required UV energy for the photolysis of the oxidizer is not available in the solar spectrum.

Photocatalysis

In the photocatalytic oxidation, TiO_2/UV, a titanium dioxide semiconductor absorbs UV light and generates OH^- ions. The overall process, taking place in the photocatalytic mineralization of organic pollutants at a semiconductor surface, is summarized in Table 7.3.

Under illumination electrons (e^-) in the valence band (VB) of the semiconductor are excited in the conduction band (CB) (2). The electron holes (h^+), which are produced by the excitation, are powerful oxidizing agents — calculations show that their oxidation potential is sufficient for complete oxidation of nearly any contaminant. This complete destruction is termed mineralization.

Under proper conditions, the photochemical electrons (in the conduction band of the semiconductor) and the photoexcited holes (in the valence band of the semiconductor) can be made available for redox reactions. The photogenerated holes in the VB must be efficiently positive to carry out the oxidation of adsorbed OH^- ions or H_2O molecules to produce OH˙ (the oxidative agents in the degradation of organics) (5, 6). The photogenerated electron usually reacts with oxygen (7).

Photocatalytic oxidation was used to treat OMWW, previously clarified by microfiltration, and then illuminated in the presence of anatase-type TiO_2 using both natural and artificial light (Vigo F. and Cagliari M., 1999). TiO_2 in the anatase crystal form is the most commonly used and catalytically active photocatalyst applied to OMWW treatment. Compared to pure anatase or rutile a nanocrystalline titania catalyst with high surface area ($68-100\,m^2/g$) containing varying amounts of anatase

Table 7.3. Basic reactions of the photocatalytic oxidation

Heterogeneous Photocatalytic Mechanism
Promotion of electrons from valence to conduction band Creation of electronic vacancies on the catalyst surface Radical degradation of the organic reactants

$$\text{Organic pollutant} + O_2 \rightarrow CO_2 + H_2O + \text{mineral salts} \tag{1}$$

$$\text{Semiconductor} + h\nu \rightarrow h^+_{VB} + e^-_{CB} \tag{2}$$

$$e^-_{CB} + D \text{ (donor)} \rightarrow D^{\cdot +} \tag{3}$$

$$H^+_{VB} + A \text{ (acceptor)}^- \rightarrow A^{\cdot -} \tag{4}$$

$$h^+_{VB} + OH^-_{ads} \rightarrow OH^{\cdot}_{ads} \tag{5}$$

$$h^+_{VB} + H_2O \rightarrow OH^-_{ads} + H^+_{ads} \tag{6}$$

$$e^-_{CB} + O_2 \rightarrow O^{\cdot -}_2 \tag{7}$$

and rutile phases has shown a significantly higher catalytic activity during the photocatalyzed degradation of *p*-coumaric acid — a pollutant found in OMWW (Basca R.R. and Kiwi J., 1998).

Marques P.A.S.S. et al. (1996) investigated the effect of photocatalysis on the detoxification of diluted OMWW (initial TOC 80–90 mg/l). The catalyst used was titanium dioxide (TiO_2) in several concentrations, normally used as a paint pigment and which has the additional advantage of low cost. Oxygen was used as the oxidant agent for the photocatalytic degradation of organic compounds. Sodium persulfate ($Na_2S_2O_8$) was also used as an additional oxidant. A TOC degradation of 98% was obtained. These initial results suggest that this method can be very appropriate to degrade organic toxic compounds such as polyphenols. However, industrial and extensive research on photocatalytic degradation of OMWW and the application of sunlight has yet to be carried out.

The interest for TiO_2 photocatalysis processes is also rapidly growing due to the capability of exploiting solar UV for chemical reactivity promotion. Photocatalysis using solar energy is a promising and cost-effective method of OMWW treatment — compared to the use of rather expensive UV-lamps — considering that the major olive oil producing countries benefit from high intensity solar irradiation throughout the year. From the commonly known and frequently applied AOPs only two can be powered by sunlight; heterogeneous photocatalysis with TiO_2

and homogeneous photocatalysis by the photo-Fenton reaction ($Fe^{2+}/H_2O_2/UV$) — see section: "Fenton reaction". These two photocatalyzed processes were investigated by the EU project: FAIR5-CT97-3807 "LAGAR" for the purification of OMWW by using solar radiation captured by simple, inexpensive, and efficient non-concentrating solar collector technology, which is considered to be the best technological solution to solar detoxification systems. The efficiency of the process has been demonstrated by Gernjak W. et al. (2004), where OMWW was treated in various types of pilot-plant photo-reactors using solar light photocatalysis over TiO_2 or solar light coupled with Fenton. To enhance process efficiency, a pretreatment step such as flocculation and/or decantation was employed to remove suspended solids as they obstructed light from entering the liquid.

This technology is opposed to the most conventional techniques where the number of pollutants can only be degraded very slowly, incompletely, or not at all. In addition, it is cheap (solar energy, low cost catalysts). The oxidation process is quite fast and occurs under ambient conditions, i.e. room temperature and pressure, with oxygen as the oxidant. A wide spectrum of organic pollutants can be converted to water and carbon dioxide. No chemical reactants must be used and no side pollutants are produced. TiO_2 is inexpensive, non-toxic and has long catalyst life.

Four different cinnamic acids (ferulic acid, caffeic acid, *p*-coumaric acid, and cinnamic acid) have been used as probes in order to study the effect of solar light catalyzed by 2,4,6-triphenylpyrylium hydrogen sulfate on phenolic compounds present in OMWW. The parent cinnamic acid underwent no photodegradation under the employed reaction conditions. The ferulic and caffeic acids reacted faster than *p*-coumaric acid (Miranda M.A. et al., 2000, 2001). Methylene blue has also been used as a photocatalyst for the photodegradation of *p*-coumaric acid, but it resulted in slower degradation. Other advanced oxidation processes (O_3/UV) have been tested as well; as expected *p*-coumaric acid abatement is much faster (100 times), but O_3 and UV are dangerous and expensive for industrial uses. In contrast with other phenolic acids, O_3, and UV do not show an important synergistic effect in *p*-coumaric acid oxidation. This could be due to differences in the absorption spectra. Major *p*-coumaric acid oxidation intermediates have been identified and quantitated by HPLC (Amat A.M. et al., 1999).

Poulios I. et al. (1999) investigated the photocatalytic degradation of proto-catechuic acid, a biorecalcitrant phenolic compound typically found in OMWW in aqueous heterogeneous solutions containing semiconductor powders ($Ti O_2$, ZnO) as photocatalysts, both in the presence of artificial and natural illumination. It was observed that, ZnO is more efficient as a photocatalyst, both in respect of degradation as well as mineralization. The photocatalytic treatment converted the protocatechuic acid into compounds, which can be more easily attacked by microorganisms by a following biological treatment An integrated photocatylitic-biological system, under solar exposure, for the destruction of biorecalcitrant phenolic compounds, seems a logical choice for the treatment of OMWW.

Fenton Reaction

The Fenton oxidative process is a method of chemical oxidation and coagulation of organic compounds present in wastewater streams. The Fenton reagent is a mixture of hydrogen peroxide (H_2O_2) and ferrous ion ($FeSO_4$). The process is based on the formation of reactive oxidizing species able to efficiently degrade the organic content of the wastewater. Although the chemistry of Fenton's systems involves a rather complex mechanism, its theoretical background could be described by the following set of reactions:

Under acidic conditions, in the presence of H_2O_2, Fe^{2+} and organic substrate, the following redox reactions take place:

$$Fe^{2+} + H_2O_2 \rightarrow Fe^{3+} + HO^- + HO^{\bullet} \tag{1}$$

$$HO^{\bullet} + Fe^{2+} \rightarrow Fe^{3+} + HO^- \tag{2}$$

$$HO^{\bullet} + RH \rightarrow H_2O + R^{\bullet} \tag{3}$$

$$R^{\bullet} + Fe^{3+} \rightarrow R^+ + Fe^{2+} \tag{4}$$

Reactions (1) and (2) are initiation and termination reactions, while reactions (3) and (4) are propagation reactions. The hydroxyl radical $HO^{\bullet}$ can attack and break down the organic compound RH or be captured by Fe^{2+}. The radicals $HO^{\bullet}$ can attack the organic compounds in a minimal time and non-selectively, removing hydrogen atoms or added to unsaturated carbon-to-carbon bonds. In addition, the following secondary reactions are possible:

$$HO^{\bullet} + H_2O_2 \rightarrow H_2O + HO_2^{\bullet} \tag{5}$$

$$2R^{\bullet} \rightarrow R - R \tag{6}$$

Because of reaction (5), H_2O_2 captures and deactivates by itself the $HO^{\bullet}$ radicals and thus, an increase in its concentration does not always lead to an increase of the efficiency of the oxidation (3). To the dimerization reaction (6) can be attributed the capability of the Fenton reaction to decolorize the organic compounds.

The desired reaction for the organic radicals ($R^{\bullet}$) is given by the reaction:

$$R^{\bullet} + O_2 \rightarrow O_2R^{\bullet} \tag{7}$$

The organic radicals react with the dissolved oxygen and are transformed to $O_2R^{\bullet}$, while the lack of oxygen leads to the undesirable recombination of the organic radicals according to reaction (6) and the breaking down of the organic material is slowed down. The consumption of H_2O_2 can be considerably diminished by adding oxygen or air to the waste. It is desirable that the quantity of Fe^{2+} be as small as possible, so that reaction (2), which is using $HO^{\bullet}$ is not favored.

The Fe^{3+} ions produced react with the hydroxyl ions to form complexes:

$$[Fe(H_2O)_6]^{3-} + H_2O \leftrightarrow [Fe(H_2O)_5OH]^{2-} + H_3O^- \tag{8}$$

$$Fe(H_2O)_5OH]^{2-} + H_2O \leftrightarrow [Fe(H_2O)_4(OH)_2]^- + H_3O^- \tag{9}$$

In the pH range from 3.5 to 7, the above complexes have the tendency to polymerize, while successive hydrolytic reactions with a greater hydroxyl number:

$$2[Fe(H_2O)_5OH]^{2-} \leftrightarrow [Fe_2(H_2O)_8(OH)_2]^{4-} + 2H_2O \tag{10}$$

$$[Fe_2(H_2O)_8(OH)_2]^{4-} + H_2O \leftrightarrow [Fe_2(H_2O)_7(OH)_3]^{3-} + H_3O^- \tag{11}$$

$$[Fe_2(H_2O)_7(OH)_3]^3 + Fe(H_2O)_5OH]^{2-} \leftrightarrow [Fe_3(H_2O)_5(OH)_4]^{5-} + 7H_2O \tag{12}$$

Therefore, with a series of hydrolytic reactions, which are possibly accompanied by dehydration reactions, complexes of Fe^{3+} are formed. As the charge of the Fe^{3+} is decreasing due to the increase of the number of hydroxyl ions, the repulsion between ions is reduced and their tendency to polymerize is increased. Some colloidal polymers of hydroxyl, and finally insoluble accretions of ferric oxide hydrate, which precipitate are possibly produced. It is to these accretions that the coagulation action of Fenton reagent is due. Organic molecules as well as suspended particles are entrained by the accretions and precipitate. A considerable percentage of the decrease of the waste's COD value, after the Fenton process, is due to this action of coagulation-aggregation. During the chemical oxidation a large number of small aggregates (flocks) is produced, which have a low rate of precipitation. The rate of precipitation increases considerably by the addition of poly-electrolytes. When calcium hydroxide is used as the aggregating reagent, a reaction with the ferrous and ferric ions, as well as with the sulfate and carbonate radicals takes place, yielding insoluble ferrous hydroxide, calcium sulfate, and carbonate that constitute the coagulation nuclei, according to the following reactions:

$$Ca(OH)_2 + H_2CO_3 \leftrightarrow CaCO_3 \downarrow + H_2O \tag{13}$$

$$FeSO_4 \cdot 7H_2O + Ca(HCO_3)2 \leftrightarrow Fe(HCO_3)_2 + CaSO_4 + 7H_2O \tag{14}$$

$$Fe(HCO_3)_2 \leftrightarrow Fe(OH)_2 + 2CO_2 \tag{15}$$

Further addition of calcium hydroxide gives:

$$Fe(HCO_3)_2 + 2Ca(OH)_2 \leftrightarrow Fe(OH)_2 + 2CaCO_3 + 2H_2O \tag{16}$$

Thereafter, the ferrous hydroxide [$Fe(OH)_2$] produced is oxidized by the dissolved oxygen to ferric hydroxide [$Fe(OH)_3$]:

$$4Fe(OH)_2 + O_2 + 2H_2O \leftrightarrow 4Fe(OH)_3 \downarrow \qquad (17)$$

The oxidation is favored by the high pH values, which results from the addition of calcium hydroxide.

$$Fe_2(SO_4)_3 + 3Ca(HCO_3)_2 \leftrightarrow 3CaSO_4 + 2Fe(OH)_3 \downarrow + 6CO_2 \qquad (18)$$

while its combination with calcium hydroxide gives:

$$Fe_2(SO_4)_3 + 3Ca(OH)_2 \leftrightarrow 3CaSO_4 + 2Fe(OH)_3 \downarrow \qquad (19)$$

The insoluble precipitate $Fe(OH)_3$ that is formed in both cases entrains the suspended colloidal particles.

The Fenton process combines oxidation and aggregation and also increases the concentration of dissolved oxygen. For satisfactory decolorizing of soluble colorants to be achieved, the pH must be below 3.5. With a pH higher than 4, the Fe^{2+} ions are unstable and are easily transformed to Fe^{3+} ions, which have the tendency to form complexes with hydroxyl. With a pH more than 9, these complexes form $[Fe(OH)_4]^-$. In addition, H_2O_2 is unstable under alkaline pH and loses its oxidative power because of its break down to oxygen and water. For those reasons, the Fenton system (Fe^{2+}/H_2O_2) system loses its oxidative action with increasing pH.

Rivas F.J. et al. (2001b) established that typical operating variables such as reagent concentration ($H_2O_2 = 1.0$–$0.2\,M$; $Fe^{2+} = 0.01$–$0.1\,M$) and temperature ($T = 293$–$323\,K$) exerted a positive influence on COD and total carbon removal. The optimum working pH was found to be in the range 2.5–3.0. The exothermic nature of the process involved a significant increase of the temperature of the reaction media. The process was well simulated by a semiempirical reaction mechanism based on the classic Fenton chemistry. From the model, the reaction between ferric iron and hydrogen peroxide was suggested to be the controlling step of the system. Also, the simultaneous inefficient decomposition of hydrogen peroxide into water and oxygen was believed to play an important role in the process.

It is claimed that the combination of Fenton oxidation under low pH, followed by coagulation under high pH can result to a considerable overall decolorization of OMWW. With the Fenton procedures soluble colorants are decolorized, while with the coagulation procedure insoluble colorants are removed (EP1157972, 2001).

The Fenton's reagent has been used as part of an integrated treatment for the degradation of OMWW — see Chapter 9: "Combined and Miscellaneous Processes". Fenton's reagent treatment moderately reduced COD and to a greater extent the phenolic compounds (Beltrán-Heredia A.J. et al., 2001c).

The reaction between H_2O_2 and Fe^{2+} to produce hydroxyl radicals can be a valuable mean in the oxidative treatment of OMWW. The treatment can take advantage of the fact that no complicate reactor is needed. However, the use of catalysts in liquid phase (for instance Fe^{3+} sulfate solution, $Fe_2(SO_4)_3$) leads to the formation of precipitates which are difficult to eliminate (mainly hydroxides), causing severe environmental problems. Furthermore, the use of this type of catalyst requires the control of the addition of the catalyst, which adds additional complications to the process. Moreover, this process does not achieve color reductions greater than 70%, under normal pressure and temperature conditions.

The Fenton reaction can be enhanced efficiently in the presence of UV radiation. The combination of Fenton reaction and UV radiation is known as photo-Fenton — see also section: "Photocatalysis". By irradiation of light with wavelengths below 580 nm, the generated Fe^{3+} in reaction (1) is reduced to Fe^{2+} according to reaction:

$$Fe^{3+} + H_2O \rightarrow Fe^{2+} + HO^{\bullet} + H^{+} \qquad (20)$$

Thus, as can be seen from reactions (1) and (20), two $HO^{\bullet}$ radicals are generated per photocatalyzed cycle of the ferric/ferrous system.

EP1097907 (2001) describes a process for the treatment of OMWW by means of catalytic oxidation with H_2O_2, using heterogeneous catalysis in the presence of metal pairs with anodic and cathodic characteristics (for instance, iron and copper). One of the main advantages of this process over the use of liquid catalysts is the self-regeneration of the metals by means of a process of oxidation/reduction of their ions, which leads to only a trace of the metals in solution, therefore overcoming the usual need for a step involving their recovery and/or elimination from the peroxidized effluent (for example, due to iron oxide precipitates). Another advantage of the use of heterogeneous catalysts is the fact that it is not necessary to control the addition of the catalyst, contrary to homogeneous catalytic oxidation processes that use a liquid catalyst (e.g. Fe^{3+} added in the form of $Fe_2(SO_4)_3$) of, where the control of the catalyst addition is absolutely necessary.

The process allows an organic content reduction of around 50%, a reduction in toxicity of more than 80% and a reduction in color greater than 75%, with low costs in terms of reagents and energy under normal pressure and temperature conditions.

Wet Oxidation

The oxidation of organic substances in the liquid phase using oxygen is called wet oxidation. The process takes place at increased pressure (10–220 bar) and temperature (120–330°C). With increasing pressure the temperature rises, which leads to an increasing degree of oxidation. With far-reaching material conversion only the inorganic final stages CO_2 and water (and possibly other oxides) are left. However, the application of this technology to the wastewaters from olive oil and table olive industries showed that the oxidation is not strong enough because the organic concentration remains very high, 75% of the initial value (García-García P. et al., 1989). With incomplete degradation the original components (which are non-degradable) are decomposed to biodegradable fragments so that it is useful to install a biological treatment stage downstream of the wet oxidation stage.

Besides oxygen, oxygen derivatives are also used so that even hardly degradable constituents of OMWW can be destroyed or attacked. Possible oxidizing agents are ozone (O_3) or hydrogen peroxide (H_2O_2), optionally combined with UV.

Treatment systems by wet oxidation with H_2O_2 were described by Chakchouk M. et al. (1994). They studied the wet oxidation and subsequent biodegradation of OMWW and reported that the poor biodegradability of OMWW (mainly due to the presence of polyphenols and tannins) was significantly improved after wet oxidation pretreatment since the oxidized mixture was easily biodegraded under aerobic conditions. This was so due to the oxidation of the original compounds to lower organic acids such as formic, acetic, oxalic, and succinic acids. However, these systems require fairly high temperatures and pressures.

Mantzavinos D. et al. (1996a) studied the wet oxidation of *p*-coumaric acid — one of the biologically recalcitrant phenolic compounds present in OMWW — and identified many of the reaction intermediates formed and their evolution over time. The literature suggests that all of the detected intermediate compounds are in fact biodegradable at either aerobic and/or anaerobic conditions, while *p*-coumaric acid itself is not easily broken down. Assuming that all the oxidation intermediates are biodegradable, Mantzavinos D. et al. (1996a,b) proposed an integrated chemical–biological treatment for OMWW comprising a brief period of wet air oxidation to transform the recalcitrant polyphenols to intermediates followed by a biological treatment stage in which these intermediates are further oxidized biologically — see Figs 7.3 and 7.4. This approach has the potential to dramatically reduce the volume of wet oxidation reactor necessary to achieve the treatment objective, which is to remove TOC from the wastewater.

A principal disadvantage of the wet oxidation process is that a long reaction time is needed for an efficient oxidation. Under ecological aspects wet oxidation has to be regarded critically, considering the strong air emissions and the high-energy demand. The process may have negative effects on air and natural resources, therefore, it does not meet the demands made by environmental legislation. Other disadvantages of this process are limited plant reliability and the resulting running costs for the plant. Moreover, qualified personnel are required to ensure trouble-free operation.

Electro-Chemical Oxidation

The electro-chemical oxidation of organic pollutants present in OMWW is a promising process for substances, which are recalcitrant to biological degradation. The electrooxidative processes (taking place at the anode of the electrolytic cell) for the oxidation of recalcitrant organic substances have been extensively studied since the early eighties. Electro-chemical processes have been successfully applied in the purification of several industrial wastewaters as well landfill leachate and domestic

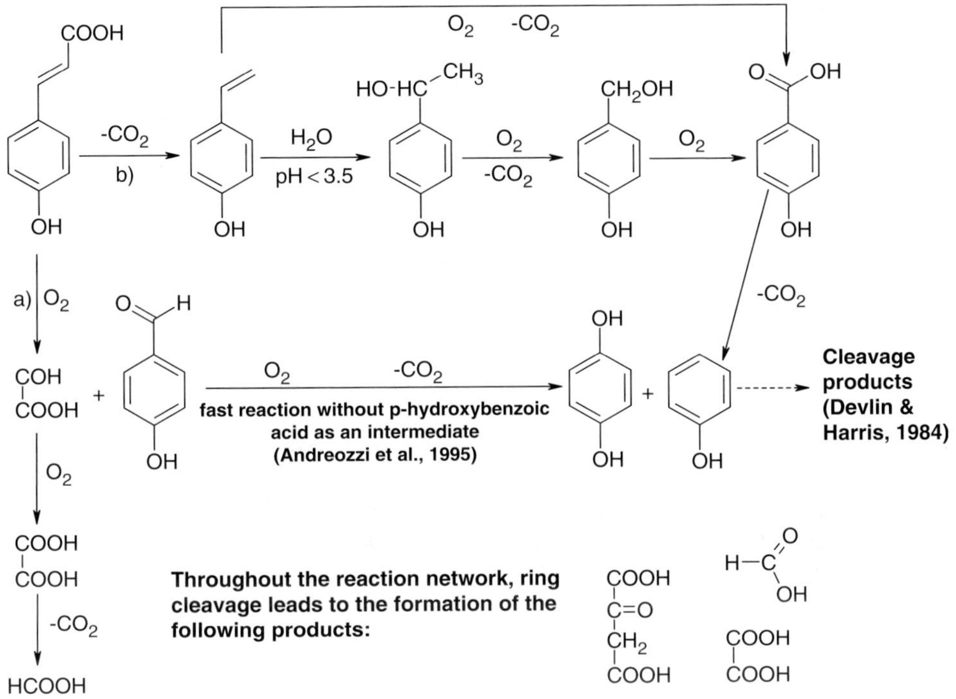

Fig. 7.3. The proposed reaction network for the oxidation of *p*-coumaric acid (Mantzavinos D. et al., 1996a).

sewage. Their competiveness against other counterparts (chemical or photochemical processes) depends mainly on the electrode material and on the type of the electrolytic cell employed (with or without membrane, with or without recycling, with bi- or three-dimensional electrodes, etc.). SnO_2, PbO_2, Pt, or Pt–Ti are usually used as anode and steel as cathode. NaCl, Na_2SO_4, H_2SO_4 are used as electrolyte salt in concentrations ranging from 0.2 to 2 N.

An electro-chemical oxidation process is outlined by Vigo F. et al. (1983a), based on the action of direct current on its organic components, with added sodium chloride (NaCl) and current density of $1–4\,A/cm^3$. Under optimum conditions, with 10 g NaCl/l, 2–6 V, and 3000 W, more than 95% of the organic substances were destroyed (COD reduction from 20,000 to 500 mg/l). For an oil mill producing $4\,m^3$ OMWW/day, electrical power installed should be around 40 kW and cooling water $80\,m^3$/day.

High-temperature oxidation with sodium persulfate ($Na_2S_2O_8$) and/or electro-oxidation in the presence of NaCl were used for the treatment of OMWW. With both solutions, the oxidation of the phytotoxic and biotoxic compounds is only partial and has a high-energy demand. Furthermore, complex management problems have to be solved. The end product is unsuitable for use for watering purposes and in agriculture, as fertilizer (Vigo F. et al., 1990).

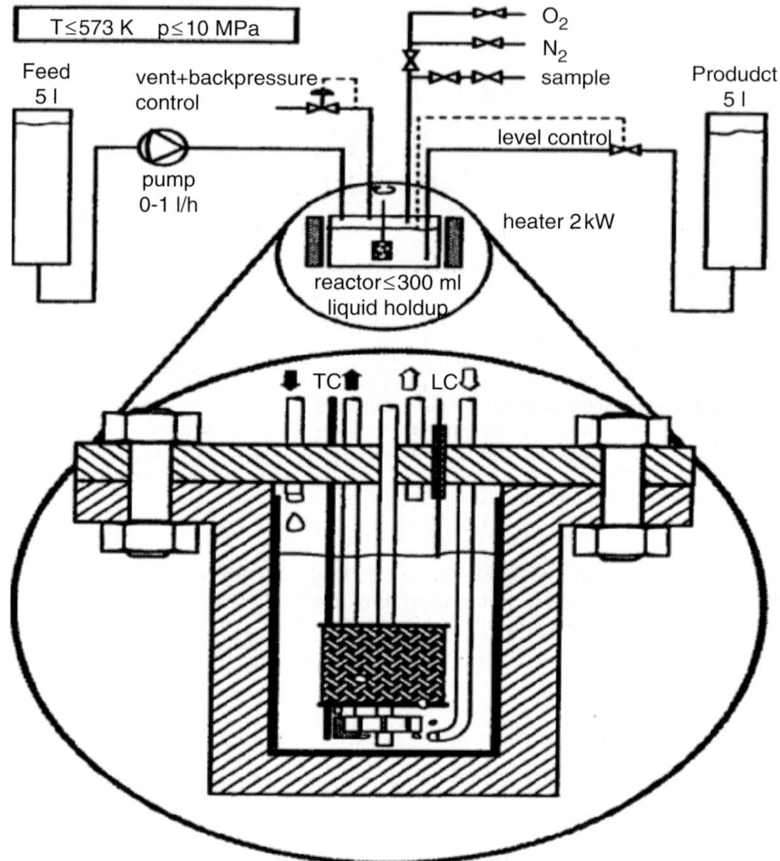

Fig. 7.4. Schematic diagram of the wet oxidation reactor system (Mantzavinos D. et al., 1996a,b).

Israilides C.J. et al. (1997) treated OMWW with the use of an electrochemical method using Ti/Pt as anode and Stainless Steel 304 as cathode — see Fig. 7.5. The apparatus has the following components: (i) electrolytic cell; (ii) recirculation reactor; (iii) OMWW input; (iv) pH control, and (v) cooling system. In this technique, NaCl 4% (w/v) as an electrolyte was added to OMWW and the mixture was passed through an electrolytic cell. Due to the strong oxidizing potential of the chemicals produced (chlorine, oxygen, hydroxyl radicals, and other oxidants) the organic pollutants were wet oxidized to carbon dioxide and water. A number of experiments were run in a batch, laboratory scale and pilot plant. After 1 and 10 h of electrolysis at 0.26 A/cm^2, total COD was reduced by 41 and 93%, respectively; TOC was reduced by 20 and 80.4%, respectively, VSS were reduced by 1 and 98.7%, and total phenolic compounds were reduced by 50 and 99.4%, while the mean anode

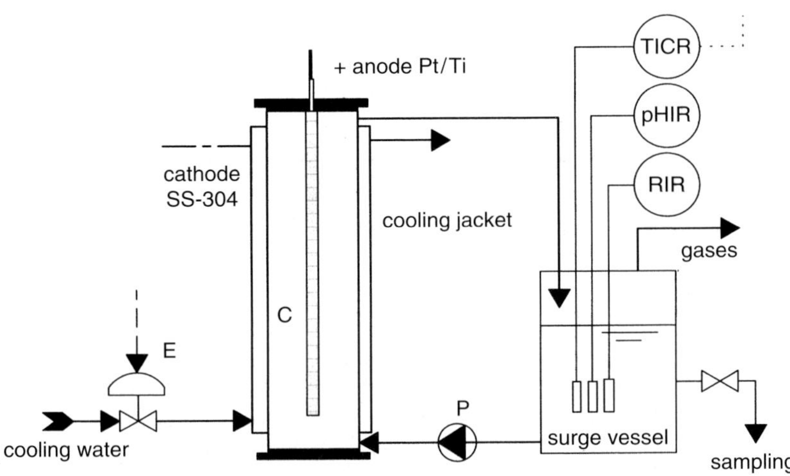

Fig. 7.5. Experimental laboratory pilot-plant (Israilides C.J. et al., 1997).

efficiency was 1960 and 340 g/h·A·m^2, respectively. The mean energy was 1.273 kWh per kg of COD removed and 12.3 kWh per kg of COD removed for 1 and 10 h, respectively. These results strongly indicate that the application of electrooxidation for complete oxidation of OMWW is not feasible. However, it could be used as an oxidation pretreatment stage for the detoxification of OMWW; the utilized energy is then 4.73 kWh per kg of COD within the first three hours.

In one of the embodiments of: DE3804573 (1989) OMWW is subjected to electro-flotation (anoxidation and flotation by hydrogen gas). The pH drops to 4–5 where the proteins are separated and by means of the small hydrogen bubbles float and removed from the solution. One part of the purified water can be reused for irrigation. The other part is subjected to anodic oxidation with an electrolytic cell. With this treatment the COD value is reduced to less than 150 mg/l.

Electro-chemical oxidation experiments were carried on OMWW samples at a dimensional stable anode (DSA) in the presence of NaCl (Polcaro A.M. et al., 2002). The results obtained indicate that the rate of degradation of phenolic compounds (*p*-hydroxybenzoic and protocatechuic acids) is high, provided that chloride ions are present in solution. Oxidation of phenolic compounds is faster than that of biodegradable substances, such as sugars or aminoacids. Moreover, investigation on the trend of toxicity, during the treatment, seems to exclude that toxic intermediates persist in solution when phenolic compounds are removed. When phenolic compounds are completely removed, the toxicity of OMWW is very low and the initial dark color of OMWW is nearly completely disappeared.

The salt concentration that is required in order to achieve low enough electrical resistance between the electrodes and consequently reduce operating costs is higher than permitted by environmental legislation in several countries (e.g. Italian law 152/99). The cost of the post-treatment required to remove the salt introduced in advance will likely be unacceptable. Discharging the treated effluent into the sea might represent the only exception.

A pilot study was carried out to investigate the electro-chemical oxidation of *p*-coumaric acid, which is a biorefractory organic pollutant of OMWW, over Pt–Ti anodes at electrolyte salt concentrations as low as 0.02 N Na_2SO_4 (Saracco G. et al., 2000). The operating test conditions were compatible with direct discharge of the after-treatment effluent in natural water basins or rivers as regards the electrolyte salt content. The study indicated that the kinetics are enhanced by: (i) increasing the temperature, (ii) enhancing current density, (iii) alternating electrode potential, (iv) lowering the initial pH, and (v) dissolving Fe^{3+} ions in the anolyte. This last feature was enabled by the presence of *in situ* generated hydrogen peroxide. H_2O_2 acted as an oxidant through reaction mechanisms parallel to those occurring at the anode surface. Experimental and analytical indications suggested that bulk and electrode oxidation pathways possibly co-exist; this might lead to comparatively high specific abatement efficiencies.

The possibility of oxidizing at a PbO_2 anode the phenols and polyphenols, present in OMWW, has been studied as a pretreatment for the submission of such wastewater to the traditional biological treatments (Longhi P. et al., 2001). The results obtained by operating at current densities ranging from 500 to $2000\,A/m^2$ show that it is possible to reduce the concentration of the phenolic components, which interfere with the biological treatments, down to low values without decreasing too much the total organic content of OMWW.

The disadvantages of electro-chemical oxidation of OMWW can be summarized as follows:

- high energy cost;
- possible formation of toxic organochlorinated by-products that need to be removed from OMWW prior to its disposal (Giannes A. et al., 2003);
- post-treatment to remove the electrolyte salt used in high concentrations. Conversely, salinity of the treated OMWW is not a problem, if it is to be discharged in seawater;
- skilled personnel are required.

Therefore, the electro-chemical oxidation, although effective, does not seem to be feasible, except if it is to be used as a pretreatment for detoxification, for example to maximize the effectiveness of a biological post-treatment. Alternatively, electro-chemical oxidation may be coupled with another AOP such as the Fenton's reagent to achieve improved rates (i.e. the electro-Fenton process).

Ion Exchange

Ion exchange is the substitution of ions in solution using a chemical. This substitution is ideal to remove heavy metals, earth-alkali metals as well as chloride, nitrate, or sulfate ions. Available materials for this problem are chelate-producing, semi-acid cation-exchangers. These materials can be regenerated with highly dissociating inorganic acids, like sulfuric acid. Another possibility is the elimination of phenols and polyphenols. In this case the use of a semi-acid anion-exchanger must be employed, which can be regenerated with methanol. Usually this technique is used for the purification of the lye-wastewaters generated in the black table olives industry.

Ion exchange resins may be used for the treatment of OMWW. Such a process is described in Chapter 5: "Physical processes", section: "Membrane technology" where an ion exchange resin forms a part of an integrated treatment of OMWW.

The bentonite particles function also as ion exchangers, where the exchange process takes place either only on the external surface of the particles or also by the cations deposited among the inner layers of the bentonite — see section: "Miscellaneous inorganic flocculants".

Chapter 8

Biological Processes

Biological processes employ the use of microorganisms to break down biodegradable chemical species present in olive-mill wastes. The actual type of microorganism that is involved depends on the conditions in which the olive-mill waste is treated, i.e. aerobic or anaerobic. Anaerobic process is used for removing organic matter in higher concentration streams, and aerobic process is used on lower concentration streams or as polishing step to further remove residual organic matter and nutrients from the wastewater.

An appropriate solution for the decontamination of OMWW could be the biotechnological application of microorganisms able to metabolize the toxic compounds of this waste product. The biological approach requires deep knowledge of the catabolic routes used by the microorganisms for the different compounds of OMWW in order to select the most appropriate species or "design" new strains that effectively degrade the wide variety of these substances. In most cases it has been found that biological processes are more economic and efficient than physical/chemical processes, especially anaerobic processes.

Anaerobic Processes

Anaerobic biodegradation (digestion) consists of a series of microbiological processes that convert organic compounds to methane and carbon dioxide. While several types of microorganisms are implicated in aerobic processes, anaerobic processes are driven mostly by bacteria. The anaerobic process has three major steps: hydrolysis, acidogenesis, and methanogenesis. During hydrolysis, consortia of anaerobic bacteria break down complex organic molecules (proteins, cellulose, lignin, and lipids) of the influent matter into soluble monomer molecules such as amino acids, simple sugars, glycerol, and fatty acids. The monomers are directly available to the next group of bacteria. Hydrolysis of the complex molecules is catalyzed by extracellular enzymes such as cellulases, proteases, and lipases. Acidogenesis includes fermentation and anaerobic oxidation (β-oxidation), which are executed by fermentative acidogenic and acetogenic bacteria, respectively. Fermentative acidogenic bacteria convert sugars, amino acids, and fatty acids to organic acids (e.g. acetic, propionic, formic, lactic, butyric, or succinic acids), alcohols and ketones (e.g. ethanol, methanol, glycerol, acetone), acetate, carbon dioxide, and hydrogen. Acetate is the main product of carbohydrate fermentation. Acetogenic bacteria convert fatty acids (e.g. long chain fatty acids) and alcohols into acetate, hydrogen, and carbon dioxide, which are used by the methanogens. In the methanogenesis step, acetate, hydrogen, and carbon dioxide are converted into methane. This is done by methanogenic bacteria composed of both gram-positive and gram-negative bacteria with a wide variety of shapes.

Anaerobic biodegradation is affected by temperature, retention time, pH, H_2 partial pressure, chemical composition of wastewater, and the presence of toxicants. Anaerobic biodegradation can take place under psychrophilic ($< 20°C$), mesophilic (25–$40°C$), or thermophilic (50–$65°C$) conditions. Biodegradation under thermophilic conditions is the most common. It allows higher loading rates and is also conductive to greater destruction of pathogens. One drawback is its higher sensitivity to variabilities in operational parameters and toxicants. Because of their slower growth, as compared with acidogenic bacteria, methanogenic bacteria are very sensitive to small changes in temperature, which leads to a decrease of the maximum specific growth rate, while the half-saturation constant increases. The hydraulic retention time (HRT), which depends on wastewater characteristics and environmental conditions, must be long enough to allow metabolism by anaerobic bacteria in digesters. Digesters based on attached growth have a lower HRT (1–10 days). The retention times of mesophilic and thermophilic digesters range between 25 and 35 days, but can be lower. pH and the H_2 partial pressure have a strong influence on β-oxidation and the methanogenesis. Acidogenic bacteria produce organic acids, which tend to lower the pH of the bioreactor. Under normal conditions, this pH reduction is buffered by the bicarbonate that is produced by methanogens. Under adverse environmental conditions, the buffering capacity of the system can be upset, eventually lowering the pH and stopping the production of methane — a condition known as "souring" of the anaerobic bioreactor. Alkalinity is important to buffer the decreasing pH.

The anaerobic process is the most widely investigated technique for the decontamination of OMWW. The most important reasons for the choice of anaerobic biodegradation as a treatment method are the feasibility to treat wastewaters with a high organic load and the techno-economical structure of the olive-mills. In addition, it offers the advantages of low energy consumption, production of an energy-rich gas (methane) that may be amenable to further uses after some preparation and relatively small amount of sludge that must be subjected to subsequent treatment. On the other hand, this same low rate of sludge accumulation is responsible for the high sensitivity of anaerobic systems to the recalcitrant components of the inflowing OMWW and for the increased discharge of microorganisms.

The biological disposal of OMWW by anaerobic biodegradation has been investigated by several researchers for the production of methane (Fiestas Ros de Ursinos J.A. et al., 1982; Aveni A., 1983, 1984; Boari G. et al., 1984; Rigoni-Stern S. et al., 1988; Rozzi A. et al., 1989a; Dalis D., 1991; Martín-Martín A. et al., 1991; Georgacakis D. and Dalis D., 1993; Tekin A.R. and Dalgiç A.C., 2000) or for the recovery of valuable materials, such as coloring compounds and polysaccharides (Iniotakis N. et al., 1989, 1991) — see Chapter 10: "Uses", sections: "Recovery of organic compounds" and "Biogas production". The treated OMWW can also be used as a liquid fertilizer (Vassilev N. et al., 1998) — see also Chapter 10: "Uses", section: "Use as fertilizer/soil conditioner". However, inhibitory effects towards methanogenic bacteria caused by high concentration of aromatic compounds (phenols) and lipids have been reported (Boari G. et al., 1984; Hamdi M., 1991a,b, 1993a; Beccari M. et al., 1998). Preliminary laboratory and pilot scale experimentation on diluted OMWW showed that the anaerobic contact process was able to give high organic removal efficiency (80–85%) at 35°C and at an organic load (lower than 4 g COD/l d (Aveni A., 1984); however, especially at high feed concentrations, the process proved unstable due to the inhibitory effects of substances such as polyphenols and potassium. Moreover, additions of alkalinity to neutralize acidity and ammonia to furnish nitrogen for cellular biosynthesis were required. Consequently, although anaerobic degradation of OMWW is feasible, and quite appealing from an energetic point of view, the presence of phenolic and lipidic inhibitors decelerates the process, hinders removal of part of the COD and detracts from its economic viability (Fiestas Ros de Ursinos J.A. et al., 1982; Boari G. et al., 1984).

The seasonal nature of OMWW production means that an anaerobic bioreactor treating this waste must have the ability for easy start-up operations every year. Tsonis S.P. (1991) used a 3-unit large laboratory scale anaerobic system with 200 l digesters as well as a pilot anaerobic unit of 5.5 m^3 active volume to study the restart-up of seasonally fed anaerobic units digesting OMWW. It was found that the restart-up of units kept for prolonged periods of time under non-feeding conditions could be effected in a reasonable period of time lasting less than 30 days. The initial start-up of the anaerobic biodegradation of OMWW, when using cow manure as seeding material was also studied in a 5-unit laboratory system and was found to necessitate unacceptable periods of time for a treatment system operating on a seasonal basis.

Therefore, pretreatment of OMWW in order to remove such recalcitrant compounds prior to anaerobic biodegradation is a key step. The various pretreatments employed for improving anaerobic degradation of OMWW (dilution, differential distillation, preculture with yeast, acidogenesis, and aerobic fermentation with *Aspergillus niger*) have been reviewed by Hamdi M. (1996). One of the most popular pretreatments is dilution — see Chapter 5: "Physical processes", section: "Dilution". The most promising results were obtained on UASB reactors, both at laboratory and pilot scale (tank capacity 15 l and 5 m^3, respectively), fed on diluted waste (COD = 13–18 g/l) (Boari G. et al., 1984). Volumetric loading rates (16–21.5 COD g/ l · day) and 70% removal efficiencies were obtained with these digesters. Start-up of

UASB reactors fed on OMWW is a delicate step, which still has to be fully controlled and optimized. The best results were obtained by diluting the waste (COD = 5 g/l) and increasing its concentration of available nitrogen by addition of urea. Granulation of the sludge, as achieved in Dutch UASB digesters fed on sugar beet wastewater, was not obtained, but, even so, the settleability of the sludge was very good. Hamdi M. (1991a) studied the effects of agitation and pretreatment on the batch anaerobic degradation of OMWW. Agitation decreases methane formation in anaerobic degradation of unmodified OMWW. Acidified OMWW is less toxic than is raw waste. Pretreatment of OMWW by fermentation with *A. niger* decreases the toxicity for methanogenic bacteria and facilitates anaerobic degradation. Moreover, agitation did not affect gas production.

Various reactor types have been utilized for anaerobic biodegradation of OMWW (anaerobic bioreactors or digesters). In fact, most of the bioreactors developed thus far within the waste management field have been tested at laboratory-, pilot-, and/or full-scale studies. Anaerobic bioreactors and technologies currently available and studied with OMWW as substrate include:

- up-flow anaerobic sludge blanket (UASB) reactor,
- anaerobic baffled reactor (ABR),
- continuous-flow stirred tank reactor (CSTR) (anaerobic contact reactor),
- anaerobic filter reactor (up-flow and down-flow),
- expanded or fluidized bed reactor,
- combinations and comparisons among various reactors.

Up-Flow Anaerobic Sludge Blanket (UASB) Reactor

One of the early studies of using a UASB reactor for the anaerobic treatment of OMWW is attributed to Aveni A. (1985).

Zouari N. and Ellouz R. (1996b) used UASB reactors to digest OMWW anaerobically. Trials revealed that COD was removed by growth of active biomass, and by adsorption in and on the sludge, inferring irreversible inhibition of methanogenic bacteria. Acclimatization of the sludge stopped because of adsorption of colored olive compounds on bacteria. Removal of 50% of the initial color from OMWW through resin treatment (DUOLITE® XAD 761) showed that the resultant OMWW was more susceptible than crude OMWW — see also Chapter 7: "Physico-chemical processes", section: "Adsorption".

The anaerobic treatability of OMWW was investigated using a laboratory scale UASB reactor operated for about six months (Ubay G. and Ozturk I., 1997). The effects of various operating conditions including pH, feed strength, and hydraulic retention time on the performance of the anaerobic treatment process were determined. In the first part of this study, the reactor was operated with feed COD concentrations from 5 to 19 g/l and a retention time of 1 day, giving organic loading rates from 5 to 18 kg COD/$m^3 \cdot$ day. Soluble COD removal was around 75% under these conditions. In the second part of the study, feed CODs were varied from 15 to

22 g/l while retention times ranged from 0.83 to 2 days; soluble COD removal was around 70%. A methane conversion rate of 0.35 m^3 per kg COD removed was achieved during the study. The average volatile solids (VS) concentration in the reactor had increased from 12.75 to 60 g/l by the end of the study. Sludge volume index (SVI) determinations performed to evaluate the settling characteristics of the anaerobic sludge in the reactor indicated excellent settleability with SVI values of generally less than 20 ml/g. Sludge granules ranging from 3 to 8 mm in diameter were produced in the reactor. The second order substrate removal kinetic model was applied by assuming that hydraulic conditions in the UASB are approximately completely mixed and the model fitted well to the steady state operating results.

The influence of Na and Ca alkalinity on UASB treatment of OMWW was studied by Rozzi A. et al. (1988). Laboratory scale studies were carried out on treatment of OMWW in an UASB digester with a view to minimizing the amount of added alkali needed and characterize the role played by added alkaline salts. Efficiency of removal of total organic carbon (TOC) was studied at different levels of addition of Na_2CO_3, $NaHCO_3$, or $Ca(OH)_2$; the influence of addition of $Ca(OH)_2$ with $NaHCO_3$ was also studied. High levels of alkalinity (both as Na and Ca) were needed for stable anaerobic treatment of the effluent diluted to 3–8 kg TOC/m^3. Bicarbonate (or carbonate) buffers the system, and lime mainly enhances sludge settling characteristics. In a later study, anaerobic-process control based on bicarbonate alkalinity addition and regulation during start-up and overload operation was described (Rozzi A. et al., 1994). The experiments were performed with laboratory scale hybrid digesters fed on OMWW. An automatic instrument monitored bicarbonate concentration and controlled its value in each reactor by alkali addition. Two different start-up procedures were tested: step-increases of substrate concentration at constant feed-flow rate and step-increases of flow rate at constant feed-concentration, keeping in-reactor bicarbonate alkalinity constant by automatic control. Two overload experiments were carried out. In the first case, the volumetric loading rate was increased up to 8 kg TOC/m$^3 \cdot$ day keeping both reactors under automatic control. No major differences were observed in the performance of the reactors. In the second case, overload was induced by a 100% step-increase of the substrate concentration as acetic acid and alkalinity was controlled in one reactor only. As expected, the system without control became unstable and sour conditions developed.

Anaerobic Baffled Reactor (ABR)

A prototyped anaerobic baffled reactor (ABR) at a laboratory scale was built by Schaelicke D. (1995) to treat OMWW — see Fig. 8.1. The ABR reactor has been successfully used in treating a great variety of wastewaters (distillery, swine waste, molasses, etc.). This process uses a series of vertical baffles to force the wastewater to flow under and over them as it passes from the influent to the effluent. The bacteria within the reactor tend to rise and settle with gas production, but move horizontally at a relatively slow rate. The wastewater can therefore come into contact with a large

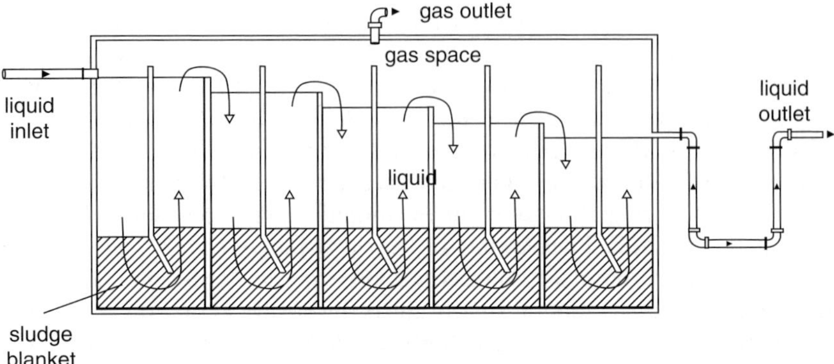

Fig. 8.1. Prototyped anaerobic baffled reactor (ABR) at a laboratory scale for the treatment of OMWW (Schaelicke D., 1995).

active biological mass as it passes through the reactor and the effluent is relatively free of biological solids. Preliminary results have shown that the ABR reactor was quite successful in treating diluted OMWW, while a hybrid ABR reactor could be used to treat raw OMWW.

Recently another study was reported on the use of an ABR reactor for the treatment of OMWW (Khabbaz M.S. et al., 2004). The laboratory scale unit operated at influent COD concentrations of 1–5 g/l at a hydraulic retention time (HRT) of 48 h and a constant temperature of 36°C. Maximum COD removal was achieved at an influent COD concentration of 3 g/l. The substrate loading removal rate was compared with predictions made from the Kincannon-Stover and the Monod model. Analysis of data indicated that the Kincannon-Stover model could produce the best fit with the experimental results.

Combinations and Comparisons Among Various Reactors

A technique based on the controlled mesophilic anaerobic degradation of both the supernatant and the sludge from settled OMWW was investigated by Georgacakis D. and Dalis D. (1993). The settling is a natural process which, after approximately 10 days, results in two liquid fractions, a low COD supernatant and a high COD settled sludge. Two different types of anaerobic digesters were used, a fixed-bed type for the supernatant and a plug-flow type for the sludge. Concentrated aqueous ammonia and sodium carbonate were added to adjust the C/N ratio and the pH of the solution in each digester. In both digesters, biogas production and COD reduction exceeded the rates mentioned in the literature for diluted raw OMWW; the combined values were 2.28 l biogas/l working volume digester, COD reduction of 94.02%, and a final COD value of 4 g/l OMWW. The results indicated that a total reactor volume of 9.21 l/l OMWW was required, 4.08 l for the fixed-bed and 5.12 l for the plug-flow digesters. In this way, the anaerobic-biodegradation system suggested is of reduced size and becomes cost-effective compared with other digestive system suggested in the literature for such wastes.

A comparison between an anaerobic filter reactor and an anaerobic contact reactor for the digestion of prefermented OMWW is given by Hamdi M. and García J.L. (1991). These two reactors differ as regards the means of retaining microorganisms in the fermenter: in the contact reactor, this depends on settling and sludge return, while in the fixed film reactor it depends on attachment of micro-organisms on surfaces. During the anaerobic degradation of OMWW prefermented by aerobic growth with *A. niger*, a stationary state was reached more quickly with the anaerobic contact process than with an anaerobic filter, but was more stable with the anaerobic filter. The daily methane production and COD removal recorded with the anaerobic filter were greater than those obtained in the anaerobic contact reactor. The anaerobic filter yielded a biogas with a higher percentage of methane and effluent with a lower volatile fatty acid and volatile solid content than the anaerobic contact reactor. The immobilizing of volatile solids (VS) in an anaerobic filter fermenter is a means of reducing the inhibition of methanogenic bacteria by the residual phenolic compounds present in prefermented OMWW. A yield of 0.15 and 0.33 l methane/g COD removed was obtained with the anaerobic contact and anaerobic filter reactor, respectively. Additional advantages of fixed film over contact reactors include the elimination of mechanical mixing and sludge settling and return. A similar study reached the same results using OMWW prefermented with *Geotrichum candidum* (Borja-Padilla R. and González, A.E. 1994).

Dalis D. et al. (1996) evaluated the anaerobic biodegradation of OMWW in a two-stage pilot plant with an up-flow type and an anaerobic filter (fixed-bed type reactor) working in series. The pilot plant system operated in the mesophilic range ($35 \pm 1°C$) during approximately 390 days, and with organic loading levels that ranged between 2.8 and 12.7 g COD/l·day. Concentrated aqueous ammonia was added to the total raw OMWW to adjust the C/N ratio to the optimum value of 20/1, and this also achieved stabilization of the pH values in the digesters with a range about neutrality. In a series of seven consecutive experiments, for the first stage with up-flow bioreactor, optimum values of specific biogas production rate were stabilized at 2.1 l/l digester·day with an 83% COD reduction (with a volumetric load of 11 g COD/l·day). During the second stage with fixed-bed bioreactor, the biogas production rate was stabilized at 0.22 l/l digester·day with an 8% COD reduction (with a volumetric load of 0.19 g COD/l·day). Phenols were reduced during the anaerobic degradation process in both digesters with a concentration reduction, which reached 75% in the up-flow digester; with the use of the second stage (fixed-bed reactor) a further reduction of 45% was obtained. On the basis of these results, it is suggested the employment of an up-flow digester, especially in combination with a fixed-bed-type reactor used as a complementary treatment, as an economical and effective treatment for reducing the organic load of total raw OMWW.

Borja-Padilla R. et al. (1996a) investigated the feasibility of using a hybrid anaerobic reactor combining a filter and an UASB reactor for the anaerobic degradation of wash waters derived from the purification of virgin olive oil. The reactor was operated under mesophilic conditions at different influent substrate concentrations. Hydraulic retention times ranged from 0.20 to 1.02 days under normal operating conditions. COD removal efficiencies of more than 89% were achieved at an organic loading rate of $8.0\,kg\ COD/m^3 \cdot day$. The organic loading rate was gradually increased from 2.6 to $7.1\,kg\ COD/m^3 \cdot day$ within 16 days, but anaerobic reactor performance did not significantly change. The system was able to tolerate an organic loading rate of up to $17.8\,kg\ COD/m^3 \cdot day$ with an average COD removal efficiency of 76.2%. Although the reactor was fed by diluted influent, with an average COD of $1.030\,g/l$, at very high hydraulic loadings (HRT $= 4.8\,h$), COD removals were more than 75%.

Previous works on the anaerobic treatment of OMWW have shown that lipids, even if more easily degraded than phenols, were potentially capable of inhibiting methanogenesis more strongly. The surface active/detergent activity of long chain fatty acids (LCFAs) present in OMWW (oleic and linoleic acid) can cause microbial cell lysis, and acetate-utilizing methanogens used during the methanogenesis are particularly vulnerable (Ching-Shyung Hwu and Lettinga G., 1996). Beccari M. et al. (1998) investigated the anaerobic degradation of OMWW and concluded that saturation of LCFAs is the key factor to prevent inhibition of methanogenesis — see also Chapter 9: "Combined and miscellaneous processes". The mechanism of degradation of unsaturated LCFAs (oleic and linoleic acid) is thought as a saturation followed by β-oxidation — see Fig. 8.2. The experiment was carried out in a semi-continuous laboratory scale two-reactor system fed with diluted OMWW. Phase separation (acidogenesis and methanogenesis) was not complete in the two reactors. Indeed, a moderate methanogenic activity was allowed to be established in the first reactor. This scheme proved to be suitable to obtain an almost quantitative biotransformation of unsaturated LCFAs to palmitic acid in the first reactor, thus drastically lowering lipid inhibition on methanogenesis in the second reactor.

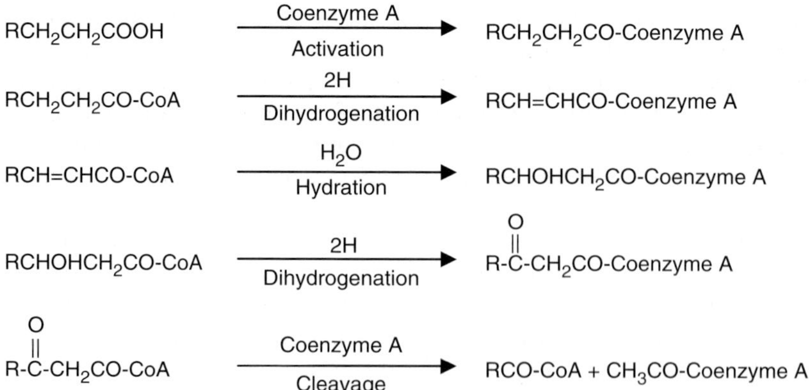

Fig. 8.2. Steps of β-oxidation (Novak J.T. and Carlson D.A., 1970).

The addition of an easily biodegradable substrate such as glucose did not improve biodegradation of polyphenols. A continuous flushing of N_2/CO_2 through the "acidogenic" reactor did not promote the β-oxidation of palmitic acid. The two-reactor system with partial phase separation could find useful application for the anaerobic treatment of any lipid-containing wastewater. But other studies (Novak J.T. and Carlson D.A., 1970) showed that the degradation of unsaturated LCFAs is as fast as or even faster than the degradation of saturated LCFAs. The β-oxidation and not the saturation of LCFAs was assumed as the rate limiting step, so that just the total amount of LCFAs affect the degradation independently of the kind of LCFAs.

Bioremediation of effluents containing LCFAs, including dairy and olive-mill effluents, can be affected by the inhibitory action of oleic acid present in wastewater. The effects of temperature on oleate toxicity to acetate-utilizing methanogens were studied by Ching-Shyung Hwu and Lettinga G. (1996). Acute toxicity tests were performed at three temperatures (30, 40, and 55°C) with four different anaerobic sludges. Inhibitory effects of oleate showed a significant dependence on temperature and were also more pronounced in flocculent than granular sludges. Oleic acid concentration causing 50% inhibition was 0.35–0.79 mol/m^3 at 55°C, 0.53–3.37 mol/m^3 at 40°C, and 2.35–4.30 mol/m^3 at 30°C. Overall, oleic acid was more than 12-fold more toxic to thermophilic flocculant sludge than to mesophilic granular sludge. It is concluded that mesophilic (30–40°C) bioremediation treatment is more suitable for fat/oil/grease-containing effluents, due to the temperature-related toxicity effects of oleic acid.

EP324314 (1989) describes a reactor for the anaerobic biodegradation of wastewater, especially suitable for OMWW (dry solid matter up to 10%). The wastewater is broken down in anaerobic conditions by the action of micro-organisms charging the multiple fragments of active carbon on support bodies rotating on a shaft in a body of the liquid. Carbon granules are preferably embedded in a layer of adhesive on a plastic support. The supports may project radially from the shaft, with intermediate gaps through which the liquid flows for stirring and enhanced contact with the granules. The container for the liquid is preferably divided into compartments each containing a set of support bodies mounted on a common shaft. Partition extend between the compartments alternately from above and below, imposing a meander-like course on liquid traveling between inlet and outlet at opposite ends of the horizontal drum container. Treatment time is relative short and granules are effective for long periods without regeneration.

AquatecOLIVIA 3w GmbH developed a two-stage process and installation — the so-called AquatecOLIVIA technology — for the treatment of OMWW (DE19829673, 2000). In the 1st stage OMWW is subjected to acidification to effect biogenic flocculation and separation of the suspended constituent. In the 2nd stage the pretreated OMWW is subjected to anaerobic mesophilic treatment with a suspended and substrate-immobilized biomass. Moving-bed biofilm immobilization is employed, using hollow cylindrical supports acting as a carrier material.

Table 8.1. Mean concentration values and parameters downstream of the individual treatment stages of OMWW (DE19829673, 2000)

Parameter	Raw OMWW	Acidification	Flocculation	Methanation
COD (g/l)	70	40	26	3
COD degradation rate (%)	0	43	63	96
DM content (g/l)	71	39	15	2.8
Phenol (g/l)	3	0.6	0.5	0.08
pH	5.4	4.2	4.6	7.1
HRT (days)	–	2.5	–	8

The use of a biofilm by sessile microorganisms minimizes unwanted discharge of active biomass and increases the process stability. In addition, it can achieve short start-up times with a seasonal mode of operation of the installation. In experiments on the two-stage anaerobic treatment of OMWW a purification of 92 to 97% could be achieved, with a largely odor-stable effluent. Mean concentration values and parameters downstream of the individual treatment stages are given in Table 8.1. The sludge production, downstream of the solids separation, is approximately 0.2–$0.25\,m^3/m^3$ OMWW at a DM content of approximately $36\,g/l$. The production of biogas having a methane content of 65–60% by volume by the methanation of OMWW runs at 7–$8\,m^3/m^3$ OMWW and 350–$400\,l/kg$ COD removed.

Within the scope of the EC program: LIFE-Environment, a semi-pilot plant based on the AquatecOLIVIA process was installed in 1999 on the island of Crete, Greece and has been operated successfully since then. The system operates roughly five months a year and purifies OMWW from an olive-mill that produces approximately 400 tons of olive oil per year (three-phase decanter).

Co-Digestion

Mixing and digesting OMWW with other effluents offers several advantages such as: (i) reduction of feed COD and total phenols concentration; (ii) no need to add nutrients (i.e. nitrogen and phosphorous if OMWW is mixed with effluents rich in nutrients); (iii) the possibility of running a year-round treatment plant based on the co-digestion of seasonally generated effluents. The concept has been demonstrated in a number of studies (Carrieri C. et al., 1986, 1992; Gavala H.N. et al., 1996; Angelidaki I. and Ahring B.K., 1996, 1997a,b; Angelidaki I. et al., 1997, 2002).

Carrieri C. et al. (1986, 1992) investigated the anaerobic treatment of sewage sludges (primary and secondary) mixed to concentrated OMWW in a laboratory scale anaerobic contact digester. Experimental results indicate that it is possible to increase appreciably (100%) the volumetric loading rate of anaerobic contact digester by adding of soluble substrates and keeping stable operating conditions. The proposed treatment is economically very attractive if the load due to high strength wastewaters is of the same magnitude as the sludge load, as it does not require additional reactor volume.

Gavala H.N. et al. (1996) developed a mathematical model for the co-digestion of OMWW, piggery effluents and dairy effluents to predict the response of a digester subject to seasonal feed variations. An organic loading rate of 3.84 g COD/l· day was found to be safe for a digester operating on a year-round basis, fed sequentially with piggery, piggery-OMWW, and piggery-dairy wastewaters.

Angelidaki I. et al. (1997) developed a mathematical model for describing the combined anaerobic degradation of complex organic material, such as manure, and a lipid-containing additive, such as OMWW, based on a model previously described (Angelidaki I. et al., 1993[23]). The model has been used to simulate anaerobic co-digestion of cattle manure together with OMWW and the simulations were compared with experimental data. Simulation data indicated that lack of ammonia, needed as nitrogen source for synthesis of bacterial biomass and as an important pH buffer, could be responsible for the problems encountered when anaerobic degradation of OMWW alone is attempted. It was shown that the amount of nitrogen needed to obtain a stable degradation of OMWW could be provided by manure during co-digestion of OMWW and manure.

In another study Angelidaki I. et al. (2002) investigated the combined anaerobic degradation of OMWW with swine manure. In batch experiments it was shown that for anaerobic degradation of OMWW alone nitrogen addition was needed. A COD/N ratio in the range of 65:1 to 126:1 was necessary for the optimal degradation process. Furthermore, it was found that methane production rates during digestion of either swine manure alone or OMWW alone were much lower than the rates achieved when OMWW and manure were degraded together. Admixing OMWW with manure at a concentration of 5 to 10% OMWW resulted in the highest methane production rates. Using UASB reactors, it was shown that co-digestion of OMWW with swine manure (up to 50% OMWW) was success-ful with a COD reduction up to 75%. The process was adapted for degradation of OMWW with stepwise increase of the OMWW load to the UASB reactor. The results showed that the high content of ammonia in swine manure, together with content of other nutrients, make it possible to degrade OMWW without addition of external alkalinity and without addition of external nitrogen source. Anaerobic treatment of OMWW in UASB reactors resulted in reduction of simple phenolic compounds such as mequinol[24], phenyl ethyl alcohol, and ethyl methyl phenol. After anaerobic treatment the concentration of these compounds was reduced between 75 and 100%. However, the concentration of some degradation products such as methyl phenol and ethyl phenol were detected in significantly higher concentrations

---

[23]Angelidaki I., Ellegaard L., and Ahring B. K. (1993). A mathematical model for dynamic simulation of anaerobic digestion of complex substances: Focusing on ammonia inhibition. *Biotechnol. and Bioeng.*, **42** (2), 159–166.

[24]Synonyms: 4-methoxyphenol; hydroxyquinone; monomethyl ether; 4-hydroxyanisol.

after treatment, indicating that the process has to be further optimized to achieve satisfactory removal of all xenobiotic compounds.

Marques I.P. (2001) and Marques I.P. et al. (1997, 1998) studied the co-digestion of OMWW and piggery effluents (either raw or anaerobically digested) in an up-flow anaerobic filter. Successive volumetric increases of OMWW ranging from 8 to 91% were mixed with the piggery effluent and treated through an up-flow anaerobic filter to promote gradual adaptation of the microbial consortium — see Fig. 8.3. The response of the digester was positive, and only at 91% OMWW the reactor performance was lower, suggesting that a concentration on the order of 83% was the highest in terms of the efficiency and stability of the reactor. At this inlet composition, the mixture loading rate varied from 5.0 to 5.7 kg COD /$m^3 \cdot$day and the total COD removal was 73 to 75%, with a gas production of 1.7 to 2.1 $m^3/m^3 \cdot$day (66 to 68% CH_4). The corresponding volume of the piggery effluent (17%) was enough to maintain an influent ammonium–nitrogen (NH_4^+–N) concentration of 0.17 to 0.19 kg NH_4^+–N/m^3, which was practically spent. Nevertheless, as some polymerized phenolic compounds are not readily biodegradable, no significant decrease in the black color of OMWW was achieved. However, since those lignin residues are innocuous and oxidizable into humic matter, the produced effluent appears as environmentally safe and suitable for agricultural irrigation.

Conclusions Regarding Anaerobic Bioreactors

Most of the results from the anaerobic treatment of OMWW were obtained on a laboratory scale and serious operational problems were encountered when passing from the laboratory to the plant scale. According to the conclusions of the EU project: AIR3-CT94-1987 "BIOWARE" it is possible to transfer from laboratory to pilot plant an anaerobic process for treatment of OMWW (anaerobic filter technology), if the following conditions are fulfilled:

- In start-up large amounts of inoculum must be introduced gradually into the digester in order to facilitate microbial colonization.
- The suspended solids in OMWW must be reduced by filtration using a filter with smaller pores.
- Olive-mills must be large enough to feed a digester of 250 to 500 m^3 with a daily production by the mill of 60 to 170 m^3. Although there are many small olive-mills in the south of Europe, over 90% of OMWW is produced by a few very large plants. The size of these mills is much greater than the minimum required to show an economic benefit from applying the anaerobic degradation technology developed. Hence, it was suggested that the technology is restricted to these big mills.
- The treated OMWW can be spread on soil without further treatment and in higher concentrations than untreated OMWW. This would be possible due to the lower phenol content and stabilization of the treated OMWW. Soil bacteria

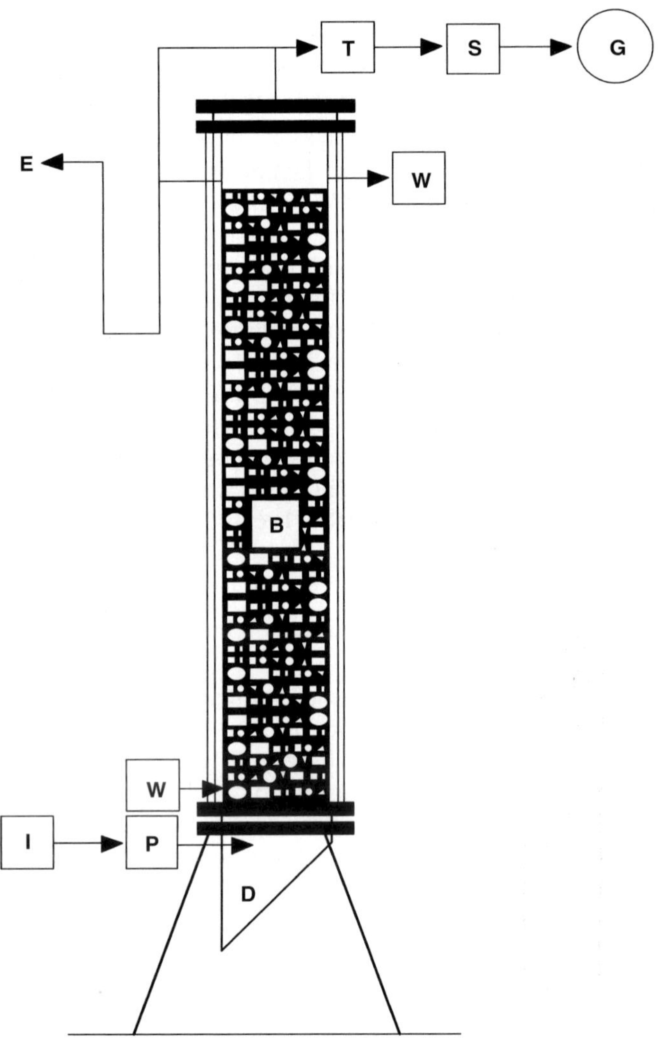

Fig. 8.3. Up-flow anaerobic filter scheme: B, packed bed; D, settling bottom; E, effluent storage, G, gas meter; I, influent storage; P, pump: S, gas sampling; T, settling tube, and W, heating water bath system (Marques I.P. et al., 1998).

and fungi further reduce the residual water-soluble phenols due to absorption into the humic acid fraction and by break down. It was found that the yield of barley grown in soil irrigated with treated OMWW from the anaerobic plant was higher than in the control, suggesting that the effluent contained a useful amount of organic matter as well as micro- and macro-nutrients. The treated OMWW can, therefore, be disposed of by spreading on land without

harmful environmental effects and with positive effects on crops grown in the treated soil, after one month from application.
- The plant must be installed in areas where the climatic conditions are favorable avoiding the risk of frost.

Anaerobic treatment as only process is not suited for 2POMW because of its low water content compared to OMWW; problems with mixing and clogging may arise during treatment. Moreover, anaerobic treatment requires further treatment measures, which leads to additional costs. Another problem is the long starting-up of the process after a longer shutdown period. These problems were also the reason for the breakdown of anaerobic plants in Greece. In the meantime, these plants have been shut down. An economically reasonable solution would be a joint treatment in existing fermentation plants. For this purpose however, the local situation has to be suited, i.e. the fermentation plant should have free capacities and be situated near the olive oil production to avoid high transportation costs and beginning digestion of 2POMW. The obtained biogas can be used for energy production (EU project: FAIR CT96-1420 "IMPROLIVE").

Landfills

Landfills can be considered as anaerobic bioreactors with very little control by the operators. A landfill in the methanogenic stage could act as an anaerobic filter and reduce the pollution load of OMWW while also acting as a temporary storage tank. Sanitary landfills of municipal solid waste might be used to reduce the storage volume required at plants giving year-round treatment of OMWW. In a study, a lysimeter in pilot scale was used to simulate a cell of a sanitary landfill. It was filled with municipal solid waste screened by an 80 mm mesh sieve mixed to municipal sludge. Results showed that when OMWW was spread on the top of the lysimeter at a loading rate not exceeding 0.4 kg COD/l of reactor steady methanogenic activity was maintained in the layers of refuse and a 70% removal of COD was obtained in the OMWW leachate collected. Higher loading rates reduced methanogenic activity and COD removal efficiency. Nevertheless, OMWW collected from the bottom of the landfill was more easily treated by anaerobic degradation than was the raw OMWW (Boari G. et al., 1993). In a similar study, the results, obtained from lysimeter filled with municipal solid waste, showed very high gasification and reduction of the organic load of OMWW, with no inhibitory effect on waste degradation processes, providing OMWW is added to an active methanogenic system or buffering the landfill system during the acid phase (Cossu R. et al., 1993).

In very dry climates, which involve negative hydrological balances for the landfill systems, OMWW can be added in order to maintain the right moisture, which promotes anaerobic degradation of the organic fraction of municipal solid wastes and includes the degradation of OMWW (Rozzi A. and Malpei F., 1996).

Obviously, this form of disposal requires storage of the effluents during most of the year, but it is a viable solution in those areas in which the volume of OMWW produced is relatively small.

Aerobic Processes

Aerobic process is the process that relies on microorganisms that thrive under aerobic conditions i.e. where plentiful of oxygen is available and a sufficient amount of food is present. Two issues must be made clear in discussing aerobic processes. First, one must specify the type of microbial fauna utilized making the distinction between mono- and poly-culture. Second, one must specify the technology utilized in order to achieve the desired goal (i.e. waste treatment). Aerobic technologies currently available and studied with OMWW as substrate include:

- attached-growth (biofilm, fixed-film);
  - trickling filter,
  - packed-bed reactor,
  - rotating (disk) biological contactor (RBC).
- suspended-growth;
  - activated sludge,
  - sequencing batch reactor (SBR).
- aerated lagoons (stabilization ponds);
- controlled wetlands.

Attached-Growth (Biofilm, Fixed-Film)

In an attached-growth system, sessile microorganisms grow on the surface of a carrier in the bioreactor creating a slime layer called biofilm. The microorganisms covering the surface of the carrier media use components of the wastewater as food source. The surface area of the media supporting the growth of organisms is the effective part of the system. The diffusion processes in biofilm plants are more important than in activated sludge plants because unlike activated sludge flocs the biofilm plants are shaped approximately two-dimensional. But while on one hand diffusion is necessary to supply the biofilm with substrate and oxygen, on the other hand the final metabolic products must be removed from the biofilm. Biofilm processes are used when the aim is very far-reaching retention and concentration of the biomass in a system. This is especially the case with the slowly reproducing microorganisms in aerobic or anaerobic environment. Due to seasonal production of olive processing wastewaters and to the rather slow growth rates of the microorganisms, the biofilm processes are less suited for the treatment of OMWW.

Aerobic biofilm reactors which have been reported for the treatment of OMWW include trickling filter, packed-bed reactor and rotating biological contactor (RBC), fluidized bed reactor, moving-bed reactor and biological aerated filter.

Trickling Filter

The trickling filter is a container filled completely with highly permeable filling material to which microorganisms are attached. The wastewater is distributed by means of a rotary sprinkler on top of the material and then trickles through it. The filling material (e.g. stones, lava slag, or plastic bodies) serves as a carrier. Biological growth and activity depend on a constant supply of dissolved oxygen. The effluent from the filter carries with it living and dead organisms and waste products of the biological reactions. Effluent sludge flocs are indicators of the efficient functioning of the trickling filter and are separated from the water in settling banks. If the wastewater is not free of solid matter, it should be prescreened to reduce the risk of clogging (Cortinovis D., 1975).

Packed-Bed Reactor

A packed-bed reactor system was used for the degradation of two fractions of pretreated OMWW (Bertin L. et al., 2001); both fractions, one deriving from natural OMWW through reverse osmosis treatment and containing low-molecular weight organic molecules, and the other obtained from an anaerobic laboratory scale treatment plant fed with OMWW, were rich in monocyclic aromatic compounds. Two aerobic fixed-bed biofilm reactors were developed by immobilizing the cells of a co-culture of two bacterial strains (*Ralstonia* sp. LD35 and *Pseudomonas putida* DSM1868) on Manville silica beads and on polyurethane foam cubes. Both supports were found to give rise to a microbiologically stable and active biofilm. Two identical glass columns with an external jacket in which water at 30°C was continuously recycled were used as the bioreactors — see Fig. 8.4. The inlets for OMWW were at the bottom of the columns, whereas the outlet for exhaust air was at the top. A recycle line continuously carried the contaminated water from the top to the bottom of the reactors. The two biofilm reactors were found to be similarly capable of rapidly and completely biodegrading the components of a synthetic mix of nine monocyclic aromatic acids, typically present in OMWW and the low-molecular weight aromatic compounds occurring in the anaerobic effluent in batch conditions. Under the same conditions, the silica bead-packed reactor was found to be more effective in the removal of high-molecular weight phenolic compounds from the anaerobic effluent with respect to the polyurethane cube-packed reactor. The co-culture of the two bacterial strains was

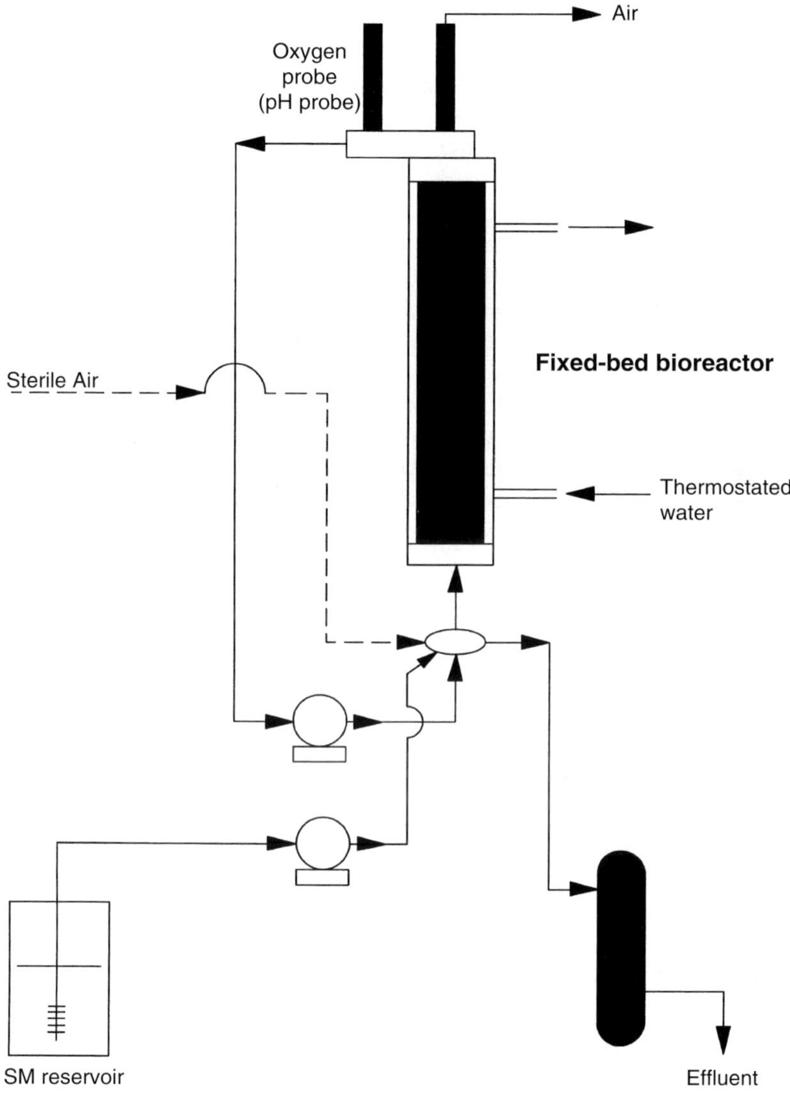

Fig. 8.4 Schematic diagram of an aerobic fixed-bed biofilm reactor (Bertin L. et al., 2001).

able to biodegrade seven of the nine components of the tested synthetic mixture, while protocatechuic acid (2,6-dihydroxybenzoic acid and 3,4,5-trimethoxybenzoic acid were the two non-degraded compounds — see also section: "Bacteria". Packed-bed reactors are simple to construct and operate but can suffer from blockages and from poor oxygen transfer.

Rotating Biological Contactor (RBC)

Rotating biological contactor (RBC) is a biofilm reactor similar to trickling filter in that organisms are attached to support media. In the case of the RBC, the support media are slowly rotating discs that are partially submerged in a tank with wastewater having a neutral pH. Oxygen is supplied to the attached biofilm from the air when the film is out of the water and from the liquid when submerged, since oxygen is transferred to the wastewater by surface turbulence created by the discs' rotation. The biomass that grows in the tank of the contactor is of dual form: suspended and attached to the filling material. The attached biomass grows on the particles of the filling material. The suspended biomass grows inside the wastewater volume of the reactor. The wastewater volume in the reactor is low and almost equal to the half volume of the statically rotating contactor. As a result, the volume remaining for the growth of the suspended biomass is also low, as is the volume of the wastewater to be treated. This method with such a layout for the contactor results in huge capital and operating cost, since the process requires a multiple contactor system for the bioconversion of large wastewater volumes.

WO9935097 (1999) brings about an improvement to the RBC technique used for OMWW treatment. The invention features an additional transport (linear or circular) motion either of the contactor, implemented by placing the contactor in a trolley-frame, or of the treated wastewater with the rotating contactor remaining fixed, implemented by kinetic energy addition to the liquid by means of mixers. The invention is applicable for reactors with aerobic microorganisms, whose breathing time can be practically utilized for the additional suggested motion. The time required for a full lap of the trolley or the wastewater must be less or equal to the time needed for the microorganisms for their next inhalation. With the additional motion the same RBC inoculates, mixes, and oxygenates a multiple tank volume and, thus, a greater OMWW volume. This occurs because during the transport motion, the RBC continuously inoculates the whole active tank volume with microorganisms, thus, increasing the population density of the suspended biomass, continuously mixes the whole active tank volume (suspended biomass) bringing the lower microorganisms to the upper — rich in oxygen — layer of the tank and oxygenates the whole active tank volume (suspended biomass), since the lap time step is less than the breathing time step of the microorganisms.

Suspended-Growth

In a suspended-growth system, microorganisms are maintained in suspension in the wastewater. Suspended-growth reactors which have been reported for the treatment of OMWW and TOWW include activated sludge and sequencing batch reactor.

Activated Sludge

The activated sludge system (aeration and sedimentation tanks) is the main representative of the suspended-growth aerobic system. The activated sludge is the most widely used method to bring about stabilization in wastewater having organic matter constituents. The method depends on establishing and maintaining a population of degrading microorganisms and providing close contact of the degrading microorganisms and a supply of dissolved oxygen. The microorganisms feed and grow upon the oxidizable material in the wastewater and form a suspended floc of "activated sludge" in the water. Air bubbled through the water or absorbed by constantly renewing the air–water interface (by agitation) replenishes the oxygen needed for the biological oxidation. The mixture of wastewater and activated sludge, known as "mixed liquor", is then settled to separate the activated sludge solids from the treated (i.e. reduced BOD_5) water. Part of the settled activated sludge is usually mechanically returned (by pump) to the aeration site (usually a tank or vessel).

The solids in an activated sludge system tend to build up due to accumulation of inert material and the growth of microorganisms. To control the amount of solids during aeration, the excess solids, i.e. "excess sludge" are wasted from the system regularly. Typically, the influent wastewater is mixed with about 20–30% by volume of activated sludge and approximately the same weight of suspended solids, which enter the treatment system each day, must be wasted as excess activated sludge.

OMWW containing large amounts of organic substances and non-biodegradable substances cannot be treated on biological plants (Di Giovacchino L. et al., 1988; Mascolo A. et al., 1990). However, a biological degradation of OMWW by the activated sludge process could be provided, if said OMWW is previously and suitably diluted with easily biodegradable wastewaters, such as, e.g. municipal sewage. After such a dilution, the biotoxic substances contained in OMWW are, by now, in low concentrations and no longer capable of deactivating the bacterial fauna, which controls the biodegradation process. However, the necessary value of the ratio of OMWW/municipal wastewater is very small, owing to OMWW biotoxicity. The use of this technique to dispose OMWW requires a large number of low-capacity biological treatment facilities sited in areas with low dwelling density. Unfortunately, in these areas the production of municipal sewage is small and consequently insufficient in order to feed such facilities. Therefore, this solution is only a partial one, and constrained to particularly favorable local situations. But, in this case too, OMWW must be transported — consequently, with additional costs — to their treatment facilities (Perrone S., 1983; EP520239).

DE2640156 (1978) describes a process for the purification of OMWW comprising two aeration stages with dilution by recycled activated sludge water. OMWW passes from a storage tank to a first activation stage where it is diluted at a ratio of 1:50 to 1:400 with recycled activated sludge water. After intermittent aeration at a space loading of $1–8\ BOD_5/m^3 \cdot day$ another dilution at the same rate follows in a second activation stage. Separation from the activated sludge in a settling basin is followed by denitrification and final purification. After aeration, a constant quantity of activated sludge is recycled. No make-up clean water is needed for the dilution and recycled water from the aeration tanks has proved satisfactory. The system results in very high loaded plants with excellent economy in operation. In the outgoing fluid BOD_5 and COD are kept below 25 and 150 mg/l, respectively.

Velioĝlu S.G. et al. (1992) have shown that effective BOD_5 and residual oil removals could be realized using a completely mixed activated system with sludge recycle. The overall system behavior and removal efficiencies were observed and evaluated by using the solids retention time as the major controlling parameter. Kinetic coefficients, oxygen utilization rates and other relevant parameters are determined to serve as a basis of design of such systems. Although, the obtained data indicated that activated sludge treatment of OMWW effectively removed BOD_5 and oil, further treatment of the effluent would be required prior to final disposal.

Borja-Padilla R. et al. (1995a) investigated the feasibility of using a completely mixed activated sludge system for the treatment of wash waters derived from the purification of virgin olive oil in a two-phase extraction process. Monod and multi-substrate activated sludge models were applied to data derived from laboratory scale units operating at four different input COD concentrations and four different solids retention times — see Table 8.2. The data in the Table indicate that more than 93% of the input COD concentration can be removed by the various activated sludge systems, ranging from $\theta_x = 4$–15 days and $S_i = 700$–200 mg COD/dm^3. Thus, the results of using a completely mixed activated sludge system

Table 8.2. Effluent COD concentration (S_e, mg/dm^3), mixed-liquor suspended solids concentration (X, mg/dm^3), and sludge volume index (SVI, cm/g) of various completely mixed activated sludge operational methods (Borja-Padilla R. et al., 1995a)

| Solids retention time, θ_x (days) | Initial influent COD, S_i (mg/dm^3)* | | | | | | | | | | | |
| | 700 | | | 1200 | | | 1700 | | | 2200 | | |
	X	S_e	SVI	X	S_e	SVI	X	S_e	SVI	X	S_e	SVI
4.0	1680	56	190	–	–	–	4140	108	220	–	–	–
6.0	2065	47	160	3460	74	170	4740	97	185	5970	122	190
8.0	2355	41	125	–	–	–	–	–	–	7020	97	180
8.5	–	–	–	–	–	–	5985	76	170	–	–	–
9.0	–	–	–	4025	63	155	–	–	–	–	–	–
10.0	–	–	–	4410	60	145	6210	72	140	7925	85	160
12.0	2605	37	125	–	–	–	–	–	–	–	–	–
15.0	–	–	–	4760	54	125	–	–	–	8460	83	130

*Values are averages of four determinations; the differences between the observed values were less than 3% in all cases.

to treat this tertiary wastewater from washing of virgin olive oil to achieve a low COD residue are promising and compatible with those reported in the literature for other food wastewaters. Results showed also that effluent COD concentration (S_e) correlated with input COD concentration (S_i) and was proportional to the product of input COD concentration and specific growth rate (μ). The multi-substrate model fitted the data well and could be used to predict effluent COD from a wide range of input substrate concentration.

Sequencing Batch Reactor (SBR)

A sequencing batch reactor is an activated sludge type wastewater treatment system that can carry out various treatment operations in one tank. A specific volume of wastewater, called a batch, is first screened to remove larger particles within the water. The reactor is a tank into which air is pumped to ensure that a sufficient supply of oxygen is present for aerobic biochemical processes to occur. The addition of oxygen allows microorganisms to consume dissolved organic matter in the wastewater that are not removed by a screening or settling process. After a specified period of aeration, the wastewater in the reactor is allowed to settle. The sludge that settles on the bottom now primarily consists of the microorganisms that have fed on the organics in the wastewater. Sequencing batch reactors utilize an activated sludge treatment process. After the treated effluent is discharged, all but a small portion of the sludge, which is rich in microorganisms, is removed from the reactor. This helps quickly reestablish a population of microorganisms within the next batch of wastewater delivered to the reactor, reducing the amount of time necessary for treating each batch. Usually more than one reactor is needed so that while one batch of wastewater is being treated, additional flow can be directed elsewhere. The number of reactors ultimately depends on the expected volume of wastewater flow and the amount of time allowed for treatment of each batch in the reactor. A longer retention period produces less sludge and cleaner effluent.

The main advantage of sequencing batch reactors is that they produce effluent low in organic compounds and thus can be used to meet strict effluent standards. The system can be effectively used as part of a larger system when the removal of the nutrients nitrogen and phosphorus are required. Other advantages are that it can be located on a small area of land, and it is relatively easy to expand this system by adding additional reactors. However, the operation of this system is more complex than others. The system does tend to be more costly to construct and operate than most others, yet it usually has fewer maintenance problems over its lifetime.

A study on a bubble-column reactor for treating OMWW is described by Hamdi M. and Ellouz R. (1992a). The reactor design serves for the growth of the mold culture *A. niger*, with the use of carrier bodies, which is said to promote fermentation on OMWW by eliminating hardly degradable constituents. The study is directed toward the pretreatment of OMWW by the use of a special mold structure and not towards reduction of pollutant load.

Conclusions Regarding Aerobic Bioreactors

The traditional aerobic systems of OMWW treatment do not produce acceptable results, as OMWW contains phenols, which are enzymatic inhibitors, and prevents the spontaneous development of aerobic bacteria (Ragazzi E. and Veronese G., 1982; Olori L. et al., 1990). Ragazzi E. and Veronese G. (1989) reported that this antimicrobial activity is produced by the phenolic compounds tyrosol and hydroxytyrosol. Besides polyphenols, potassium content could also hinder the aerobic treatment.

In addition, the aerobic treatment systems have a number of inherent drawbacks:

- Trickling filters take up too much space, and tend to cause secondary pollution such as odor and flies. Moreover, due to seasonal production of OMWW and the slow growth rates of the microorganisms, these processes are less suited for the treatment of OMWW.
- Activated sludge processes generate large amounts of biosolids, and require careful monitoring because they are susceptible to shock caused by sudden changes in loading.
- Rotating biological contactors are harder and more compact, but they are expensive and prone to mechanical problems.
- Reactors using fixed submerged media perform well at low loadings, but they are easily plugged by excessive build up of biomass.
- Sequencing batch reactors are more complex to operate than others and tend to be more costly to construct and operate than most others.

In general, aerobic biological processes are less attractive for the treatment of OMWW because of:

- high use of energy,
- high use of nutrients (to reach a ratio $BOD_5{:}N{:}P = 100{:}5{:}1$ from $BOD_5{:}N{:}P = 100{:}1{:}0.5$),
- very high production of secondary sludge which has to be disposed of,
- high capital cost — see also Chapter 9: "Combined and miscellaneous processes", Table 9.1.

Use of Specific Aerobic Microorganisms

Several investigations have been carried out using specific microorganisms capable of growing aerobically on diluted OMWW in order to reduce the initial organic load and phenolic content and obtain proteins and vitamins (Fiestas Ros de Ursinos J.A., 1961a, 1966, 1967; Montedoro G.F. et al., 1986; Amat A.M. et al., 1986, 1987; Hamdi M., 1991a; Hamdi M. et al., 1991a,b; Hamdi M., 1993a; Gharsallah N. et al.,

1998; Borja-Padilla R. et al., 1995d,e; Zouari N. and Ellouz R., 1996a). On the other hand, there is limited information on the use of microorganisms for the bioremediation of 2POMW (Jones C.E. et al., 2000; EU project: FAIR CT96-1420 "IMPROLIVE").

Fungi

The bioremediation of OMWW has been attempted with a large array of fungi aiming at neutralizing its heavy pollutant effect, for converting it into new value-added products or for rendering it susceptible to further degradation treatment (review by Zervakis G. and Balis C., 1996). The use of filamentous fungi for OMWW pretreatment has been shown to reduce toxicity and improve the biodegradability in anaerobic degradation (Hamdi M., 1991a, 1996; Borja-Padilla R. et al., 1998b). In particular, the pretreatment of OMWW with higher fungi, — see Table 8.3 — which produce polyaromatic hydrocarbon-degrading enzymes, has been used to detoxify and decolorize OMWW. However, their use on a large scale is difficult compared to bacteria. The application of fungi in a large scale is limited by the difficulty of achieving continuous culture because of the formation of filamentous pellets and mycelia. Moreover, COD reduction and color removal obtained after OMWW biotreatment varied, even with the same microorganism and operating conditions.

The structure of the aromatic compounds present in OMWW can be assimilated to many of the components of lignin (Sanjust E. et al., 1991). Only few microorganisms, mainly white rot basidiomycetes, are able to degrade lignin by means of oxidative reactions catalyzed by phenol oxidases and peroxidases. Both the

Table 8.3. Some of fungal species used to detoxify OMWW

Aspergillus niger	Hamdi M., 1991a; Hamdi M. et al., 1991a,b; Hamdi M. and Ellouz R., 1992a,b; Hamdi M. and García J.L., 1993; García-García I. et al., 2000
Aspergillus terreus	Martínez-Nieto L. et al., 1993; Borja-Padilla R. and González A.E., 1994; Borja-Padilla R. et al., 1995d,e, 1998b; García-García I. et al., 2000
Coriolus versicolor	Yesilada O. and Fiskin K., 1996; Yesilada O. et al., 1998
Funalia trogii	Yesilada O. et al., 1995, 1998
Geotrichum candidum	Borja-Padilla R. et al., 1992b,e,h, 1995c, 1998b; Martín-Martín A. et al., 1993; Assas N. et al., 2000, 2002; Fadil K. et al., 2003
Lentinus edodes	Vinciguerra V. et al., 1993, 1995; D'Annibale A. et al., 1998, 2000; García-García I. et al., 2000
Phanerochaete chrysosporium	Sayadi S. and Ellouz R., 1992, 1995; Gharsallah N. et al., 1999; García-García I. et al., 2000; Kissi M. et al., 2001
Phanerochaete flavido-alba	Pérez J. et al., 1998; Hamman O. et al., 1999; Blánquez P. et al., 2002
Pleurotus ostreatus	Sanjust E. et al., 1991; Flouri F. et al., 1996; Zervakis G. et al., 1996; Setti L. et al., 1998; Kissi M. et al., 2001, Aggelis G. et al., 2003

low degree of specificity, which characterizes these enzymes, and the structural relationships of many aromatic pollutants with the natural substrates of enzymes, have suggested the use of ligninolytic organisms and their enzymes for the treatment of these kinds of substrates (Tomati U. et al., 1995).

Laccase as well as other lignin-modifying enzymes (Sayadi S. and Ellouz R., 1995) were evaluated for the treatment of OMWW. Laccase is produced in significant amounts by the white rot fungus *Lentinus edodes* (strain SC-495). Laccase (E.C.1. 10.3.2 para-diphenol: oxygen oxidoreductase) and is a multi-copper oxidase able to catalyze the one-electron oxidation of a wide array of substrates, such as phenols, aromatic amines, benzenethiols, hydroxyindoles, and phenothiazinic compounds, with simultaneous reduction of oxygen to water. The low substrate specificity exhibited by laccase and its ability to oxidize priority pollutants has attracted interest for its use in OMWW treatment. Laccase presents the distinct advantage that it does not require the addition of hydrogen peroxide like peroxidases, and it generally exhibits broader substrate specificity than tyrosinase (D'Annibale A. et al., 2000). However, enzymes are proteins and one of the main drawbacks of using them to detoxify OMWW is their instability towards thermal and pH denaturation, proteolysis, and inactivation by inhibitors. Immobilization of enzymes to solid supports often supports stability and allows their reuse. Several supports have been made to immobilize laccase. D'Annibale A. et al. (1998) reported the immobilization of *L. edodes* laccase on a polyurethane-sponge and its use in the biodegradation of OMWW. Throughout three consecutive treatment cycles of the effluent, significant abatement of its polluting characteristics was attained. In fact, its contents in total organic carbon, total phenols, and total ortho-diphenol were dramatically reduced. In addition, an extensive effluent decolorization and apparent depolymerization of the high molecular weight fraction were observed. The study provided evidence that the depolymerization of the high molecular weight fraction as well as the maximum extent of phenol removal in the effluent requires the simultaneous presence of laccase and manganese-peroxidase (MnP), although the degradation of this fraction is not necessarily associated with the extent of effluent decolorization. Another study reported the immobilization of *L. edodes* laccase on EUPERGIT® C (Röhn Pharma, Weiterstadt, Germany), an epoxy-activated polyacrylic matrix (D'Annibale A. et al., 2000). This support exhibits an array of interesting features: (i) wide pore distribution; (ii) good hydrodynamic properties; and (iii) improved stability of the immobilized proteins. The immobilization of *L. edodes* laccase on EUPERGIT® C increased pH, thermal, and proteolytic stability with slight modifications in laccase oxidation efficiency. The use of immobilized laccase in the removal of OMWW phenolic inhibitors could be a viable form of pretreatment to improve the process efficiency in the anaerobic degradation of OMWW for methane production, since one of the main obstacles of this process is due to the inhibition of methanogenic bacteria exerted by OMWW phenolic compounds (Hamdi M., 1996; D'Annibale A. et al., 2000).

Other studies performed on OMWW with the free mycelium basidiomycete *L. edodes* demonstrated its effectiveness in the degradation of the effluent and revealed that the fungus was able to carry out a significant decolorization in the presence of a readily available carbon source like glucose and without air or oxygen purging (Vinciguerra V. et al., 1995). Moreover, it was shown that ortho-diphenols were degraded earlier than other phenolic constituents (Vinciguerra V. et al., 1995) such as the monophenol tyrosol, whose biotransformation products have been also investigated (Vinciguerra V. et al., 1997). In this study tyrosol was converted by cell-free preparation of the fungus into a dimeric tetracyclic ketone. Conversely, the alcohol corresponding to the above ketone was isolated after seven days from whole cells of *L. edodes* incubated with tyrosol.

Among the white rot fungi *Pleurotus* species presented the potential to degrade and convert both OMWW and olive cake into mushrooms and fodder. In parallel, remediation is achieved through biomass production with a simultaneous decolorization and decrease in phytotoxicity. The white rot basidiomycete *Pleurotus ostreatus,* which presents the advantage of being an edible mushroom, has been shown to degrade phenolic compounds from OMWW. In *P. ostreatous* the enzyme responsible for phenolic compounds and aromatic amines oxidation, by reducing molecular oxygen to water, is the laccase induced by OMWW or other substrates (Tomati U. et al., 1991; Martirani L. et al., 1996). Although, treatment of OMWW with purified laccase showed a significant reduction of phenolic content, no decrease of its toxicity was observed when tested on *Bacillus cereus* (Martirani L. et al., 1996).

Sixteen strains belonging to six different species of *Pleurotus* were investigated for the ability to grow and decolorize OMWW (Flouri F. et al., 1996). The tests with *Pleurotus* isolates were carried out on plates using different concentrations (25, 50, 75, and 100%) of sterilized OMWW solidified with 1.5% agar. For all strains tested, decolorization proceeded more slowly than radial growth. Among the six *Pleurotus* species, *P. ostreatus* (ATCC 34675), and *P. cornucopiae* (ATCC 38547) were the most efficient. On the basis of these results the tested *Pleurotus* strains can be divided into five groups.

A: highly effective (*P. ostreatus* ATCC 34675, *P. cornucopiae* ATCC 38547),
B: effective (*P. ostreatus* ATCC 38538, *P. ostreatus* LGAM P58, *P. dryinus* CBS 44977),
C: medium effective (*P. ostreatus* LGAM P15, *P. ostreatus* LGAM P62, *P. pulmonarius* LGAM P46, *P. pulmonarius* ATCC 36050, *P. pulmonarius* LGAM P26),
D: low effective (*P. cystidiosus* LGAM P50, *P. eryngii* LGAM P63, *P. eryngii* CBS 10082),
E: no decolorization (*P. cystidiosus* ATCC 28597, *P. cystidiosus* CBS 61580).

These observations are in agreement with the findings of previous workers (Galli E. et al., 1988; Sanjust E. et al., 1991).

Tsioulpas A. et al. (2002) also studied the ability of several *Pleurotus* spp. strains to remove phenolic compounds from OMWW, with respect to their laccase activity. All strains tested in this work were able to grow in OMWW without any addition of nutrients and any pretreatment, except sterilization. High laccase activity was measured in the growth medium, while 69–76% of the initial phenolic compounds were removed. The black color of OMWW became yellow-brown and brighter as the strains grew. The lowest phenolic concentrations were reached after 12 to 15 days. A decrease of the phytotoxicity, as described by the parameter Germination Index, was noticed in OMWW treated with some *Pleurotus* spp. strains, although this decrease was not proportional to the phenolic removal. A new parameter, namely Phenol-toxicity Index, was introduced. Using this parameter it was found that the remaining phenolics and/or some of the oxidation products of the laccase reaction in the treated OMWW were more toxic than the original phenolic compounds.

The prospect of exploiting OMWW for mushroom cultivation was examined by Zervakis G. and Balis C. (1996) — see also Chapter 10: "Uses". At a preliminary stage, two *Pleurotus* species, i.e. *P. eryngii* and *P. pulmonarius,* were tested for their ability to colonize an olive cake substrate supplemented with various dilutions of raw OMWW. Some important cultural characters related to mushroom production (earliness, yield, biological efficiencies, and quality of basidiomata) were estimated. The outcome revealed different cultural responses for each *Pleurotus* species examined; the *P. pulmonarius* strain showed better earliness values and *P. eryngii,* although it was a slow growing fungus, produced basidiomata in high yields and of a very good quality. On the other hand, the olive cake substrate supplemented with low concentrations of OMWW (12.5% v/w) behaved satisfactorily as regards the fungal colonization rates and mushroom yield, but when the addition of higher rates of raw, untreated OMWW (75–100% v/w) was attempted, then the *Pleurotus* strains were completely unable to grow. The optimal concentration of OMWW for *Pleurotus* mycelial growth was assessed through measurements of the biomass produced in liquid nutrient media and was found to lie within the 25–50% range, depending on the *Pleurotus* species and on the properties of the substrates examined. Furthermore, the phytotoxic effects that the spent liquid medium possessed were examined in comparison with the phytotoxicity of the raw liquid waste.

The white rot fungus *Funalia trogii* (Malatya) was used for the phenol removal and decolorization of OMWW (Yesilada O. et al., 1995, 1998). The aerobic degradation of OMWW was studied in static and agitated cultures. The white rot fungus *F. trogii* showed 31 and 38% color removal and 77 and 72% phenol removal in static and agitated cultures; 40% COD reduction was also obtained.

The white rot fungus *Coriolus versicolor* was also investigated for the decolorization and total phenol removal of OMWW (Yesilada O. and Fiskin K., 1996). OMWW decolorization and phenol removal occurred during the primary phase of growth. *C. versicolor* removed 80% phenol and 50% color without any additional inorganic and organic sources. No positive effect of inorganic and organic sources was determined. This fungus can also be used to reduce COD (53%) and perhaps become a source of single cell protein (SCP) for supplementation of animal feeds.

The white rot fungi, *C. versicolor* and *F. trogii,* produced laccase on media with diluted OMWW and vinasse. Addition of spent cotton stalks enhanced the laccase activity with a maximum after 12 days of cultivation (Yesilada O. et al., 1998; Kahraman S. and Yesilada O., 1999). Adding glucose, sulfate, or nitrogen had no effect on biodegradation. During growth in optimum conditions, *C. versicolor* removed approximately 63% COD, 90% phenol, and 65% color within 6 days and *F. trogii* removed approximately 70% COD, 93% phenol, and 81% color of the OMWW used. The fungi also excreted large amounts of extracellular laccase into the medium. High biodegradation yields were also obtained by fungi immobilized in calcium alginate gels (Yesilada O. et al., 1998).

Martínez-Nieto L. et al. (1993) described a process that uses mushrooms for the biological elimination of polyphenols. *Aspergillus terreus* gave the overall best results in OMWW at approximately 80% concentration, degrading organic material by 53%, expressed as COD, and 67% expressed as BOD_5. Degradation of the total phenol content, which included the great majority of phenolic compounds, reached 69%.

Initial experiments with aerobic treatment using *Aspergillus niger*, resulted in a decrease in the concentration of tannins and aromatics and correlated decrease in the inhibitory effect of these polyphenols towards methanogenic bacteria (Hamdi M., 1991a; Hamdi M. and García J.L., 1991). This is a considerable advance but even so, *A. niger* did not attack the high molecular weight polyphenols, the biodegradation of low molecular weight aromatics was insufficient and the color was not removed.

OMWW with added N and Mg was used as a medium in a shake-flask, repeated-batch fermentation process with a passively immobilized and acid-producing strain of *A. niger*. The latter reduced the phenolic content of the waste material to 59% of its initial amount and lowered the pH of the medium. Rock phosphate added to OMWW medium was solubilized to a maximum amount of 0.5 g/l during the fourth batch cycle with a corresponding productivity of 10.6 mg P/l·h (Vassilev N. et al., 1998). In another study, free cells of *A. niger* were grown on OMWW supplemented with rock phosphate in a 5 l air-lift bioreactor for 8 days at 30°C in batch and repeated-batch processes. The fungus grew well, reducing COD by 35 and 64% in batch and repeated-batch processes, respectively. A 60% reduction in total sugars was also achieved, but total phenol levels were largely unaltered (Cereti C.F. et al., 2004).

The white rot fungus *Phanerochaete* spp. was used as an alternative culture for the degradation of phenolic compounds in OMWW (Sayadi S. and Ellouz R., 1992, 1993, 1995). It was suggested that in the case of *Phanerochaete chrysosporium* the lignin-degrading system and particularly lignin peroxidase (LiP) and manganese peroxidase (MnP) were the major one responsible for the degradation of phenolic compounds in OMWW (Sayadi S. and Ellouz R., 1992, 1995). The pretreatment of OMWW with the white rot fungus *P. chrysosporium* decreased the COD from 107 to 55 g/l (Sayadi S. and Ellouz R., 1992). A *P. chrysosporium* strain isolated from Moroccan OMWW and its ability to degrade OMWW in different culture conditions was investigated and compared to that of *P. ostreatus* (Kissi M. et al., 2001). The results indicated that *P. chrysosporium* isolate is more efficient than *P. ostreatus* in decolorizing and detoxifying OMWW in the presence of added nutrients. *P. chrysosporium* is able to remove more than 50% of the color and phenols from OMWW within 6 days of incubation, whereas *P. ostreatus* needs more than 12 days to reach similar results in the same conditions. Many factors affecting the treatment of diluted OMWW (20%) by *P. chrysosporium* were studied, including the effects of added nutrients, initial pH, temperature, and inoculated biomass. Once the optimization of 20% OMWW biodegradation process had been set up, higher OMWW concentrations (50%) were tested. The results show that the fungus is capable of reducing all parameters analyzed (color, phenol content, and COD) by at least 60%, after only nine days of growth.

In *Phanerochaete flavido-alba*, enzymes involved in OMWW decolorization process were the manganese-peroxidase (MnP) and the phenol oxidase (laccase), whereas lignin peroxidase (LiP) was not detected in the growth environment (Pérez J. et al., 1998). Hamman O. Ben et al. (1999) attempted to identify optimum culture conditions for the decolorization of OMWW by *P. flavido-alba* for subsequent use in bioremediation assays. Of several media tested, nitrogen-limited *P. flavido-alba* cultures containing 40 mg/l Mn(II) were the most efficient at decolorizing OMWW. Decolorization was accompanied by a 90% decrease in the OMWW phenolic content. Concentrated extracellular fluids alone (showing manganese peroxidase, but no lignin peroxidase activity) did not decolorize the major OMWW pigment, suggesting that mycelium binding forms part of the decolorization process.

Decolorization of fresh and stored-black OMWW by *Geotrichum candidum* was investigated in an aerated batch bioreactor (Assas N. et al., 2000, 2002). During storage of OMWW, autooxidation and subsequent polymerization of phenolic compounds and tannins, gives rise to darkly colored phenolic compounds which are not readily biodegradable. *G. candidum* growth on fresh OMWW decreased pH and reduced COD by 50% removal during the first 3 days and subsequently by a further 15%. In contrast, 75% of the color was removed during the last 3 days of culture because *G. candidum* hydrolyzed phenolic compounds with high molecular weight and removed many simple phenolic compounds. *G. candidum* growth on the stored-black OMWW was rapidly inhibited resulting in low reduction COD (25%) with no decolorization because phenol polymerization was amplified by the increased pH and oxygen. The addition of oxygen to enable *G. candidum* growth and biodegradation of phenolic compounds is critical in order to avoid the polymerization of phenolic compounds and tannins.

Growth and polyphenol biodegradation by three fungi, namely, *Geotrichum* sp., *Aspergillus* sp., and *Candida tropicalis* were studied on OMWW (Fadil K. et al., 2003). These three microorganisms were selected for their tolerance to the polyphenols. The biodegradation process of OMWW was investigated in batch regime by conducting experiments where the initial concentration of COD was varied. Furthermore, some test performed to determine the most important nutrients necessary for aerobic degradation of OMWW. Average COD removals were 55.0, 53.5, and 62.8% in OMWW fermented with *Geotrichum* sp., *Aspergillus* sp., and C. *tropicalis*, respectively. The maximum removal of polyphenols was 46.6 (*Geotrichum* sp.), 44.3 (*Aspergillus* sp.), and 51.7% (*C. tropicalis*). In addition, significant decolorization was evident.

In another study, immobilized *C. tropicalis* YMEC14 under metabolic induction was used for the biodegradation of polyphenols (Ettayebi K. et al., 2003). The process was enhanced by directing yeast metabolism towards biodegradation pathways using hexadecane as co-metabolite and by immobilizing yeast cells in calcium alginate beads. Under immobilization conditions, *C. tropicalis* YMEC14 grown at 40°C in OMWW supplemented with hexadecane resulted in 69.7, 69.2, and 53.3% reduction of COD, monophenols, and polyphenols, respectively.

Scioli C. and Felice B. de (1993) examined the growth of five different yeasts on OMWW in shaker-flasks, for the potential to reduce COD levels and produce biomass. The yeast demonstrating the best growth potential on this medium, without chemical or physical pretreatment, was *Yarrowia lipolytica* ATCC 20255. In a later work, Scioli C. and Vollaro L. (1997) showed that the yeast was capable of reducing the COD level by 80% in 24 h, when grown in a 3.5 l fermenter and to produce useful biomass of 22.45 g/l and the enzyme lipase. After processing, the waters had a pleasant smell and did not exhibit the initial oily aspect and intense smell. The fermentation effluent examined by gas chromatography showed the presence of methanol and ethanol, which were responsible for the pleasant smell. During the process, most of the organic and inorganic substances were consumed and only aromatic pollutants were still present in the fermentation effluents. Therefore, a phenol degrader was used, namely *Pseudomonas putida*, to reduce phenolic compounds in the fermentation effluents after removing *Y. lipolytica* cells. *P. putida* was effective in reducing phenols in only 12 h (Felice B. de et al., 1997).

Studies have also been made of the growth of three yeasts *Candida krusei*, *Saccharomyces chevalierie*, and *Saccharomyces rouxii* on OMWW (Gharsallah N., 1993). These three yeasts have been selected on their ability to tolerate the polyphenols. The cultures were made in a shaker-flask culture and in a fermenter in order to select organisms which can produce large quantities of biomass. These yeasts can be used to reduce BOD_5 (40–50%), and perhaps become a source of SCP for supplementation of animal feeds. Concentration of protein of 3.35 g/l and yield of 0.45 g of biomass per gram of glucose based on glucose consumption were obtained using *S. rouxii* strain.

Chtourou M. et al. (2004) investigated the ability of an isolated yeast, identified as *Trichosporon cutaneum*, to degrade phenolic compounds extracted from OMWW. The yeast was adapted to the OMWW by an enrichment culture. The results of this biotransformation were a decrease in the phenolic content and hence a reduction in the phytotoxic effects of the effluent after the yeast treatment. The kinetic growth of the isolated yeast on phenol was studied over a range of concentrations (0.3–3.0 g/dm^3). The ability of the strain to assimilate simple monomeric phenols and alkyl phenols, at a concentration of 1 g/dm^3, in a synthetic liquid medium used as the sole carbon source was investigated in a batch culture. The aromatic ring cleavage pathway occurred in the yeast through catechol oxidation. Using various concentrations of ethyl acetate extract from OMWW as the sole carbon source, the yeast exhibited growth on the substrate up to 7 g/dm^3 equivalent of phenols. A significant reduction of COD after the treatment of the OMWW extract by the yeast isolate was noticed. The removal of phenol and COD exceeded 80% of the original loading after eight days of treatment, for extracts containing initial COD in the range 19–72 g/dm^3.

Bacteria

It has been shown that *Azotobacter chroococcum* can grow in the presence of various OMWW concentrations, but optimally in 15% OMWW, while preserving its nitrogenase activity (García-Barrionuevo A. et al., 1992). Moreno E. et al. (1990) and Rubia de la (1987) have also shown that the *A. chroococcum* can grow on substrates containing aromatic compounds, which, it can readily metabolize (Hardisson C. et al., 1969).

Azotobacter vinelandii is a free-living N_2-fixing bacterium that has been shown to degrade phenolic compounds and use them as a carbon and energy source. *A. vinelandii* strain A was isolated from soil repeatedly treated with OMWW (Balis C., 1994) and was shown to be particularly active N_2-fixer when grown in sterile OMWW (Balis C. et al., 1996; Papadelli M. et al., 1996). This strain was, therefore, used as an inoculum for aerobic biological treatment aiming to detoxify OMWW (Chatzipavlidis I. et al., 1996).

The use of OMWW as substrate for *A. vinelandii* growth, and application of the treated OMWW to cultivated soils as fertilizer has been proposed — see Fig. 8.5 — (Chatzipavlidis I. et al., 1996; Ehaliotis C.C., 1999; Piperidou C.I., 2000).

Pasetti L. et al. (1996) produced biomass from OMWW by using *A. vinelandii*. 16 l of OMWW diluted to 5% organic matter were inoculated in a fermentor ($T = 30°C$, airflow $= 16$ l/min, stirring $= 100$ rpm) with a strain of *A. vinelandii*. After 2 weeks the bacterial biomass was separated by centrifugation and capsular polysaccharide and exopolysaccharides were extracted. The apparent molecular weight of capsular polysaccharide was determined by gel filtration. The capsular polysaccharide was entrapped in polyvinyl alcohol membranes, which were used to adsorb cadmium and lead ions from a liquid stream.

Fiorelli F. et al. (1995) studied fertility-promoting metabolites produced by *A. vinelandii* grown on OMWW. OMWW diluted to 5% organic matter (d.w.) was

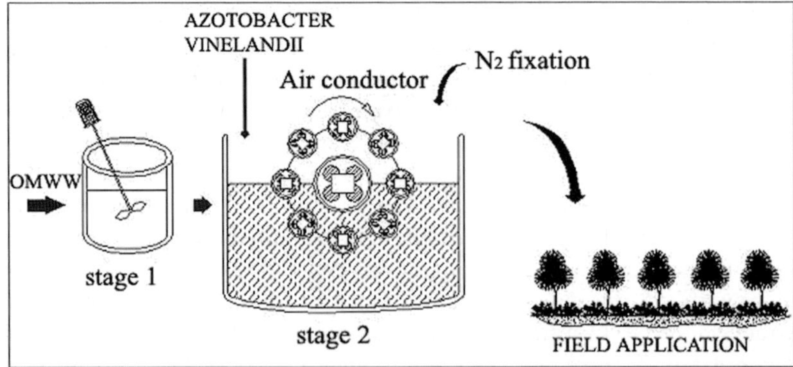

Fig. 8.5. Schematic diagram of diazotrophic bioremediation of OMWW (Chatzipavlidis I., 1996).

inoculated in a 2 l bioreactor ($T = 30°C$, airflow $= 1.4$ lmin^{-1} stirrings $=100$ rpm) with a strain of *A. vinelandii* isolated from soils heavily treated with OMWW. Microbial growth and auxin production were followed during the first week. Exo- and capsular polysaccharides were determined after 2 weeks. Microbial growth, assayed as oxygen consumption, reached after 4 days. Auxin biosynthesis became evident as nitrogen fixation decreased. A two- to three-fold increase in auxin production was recorded when tryptophan was added to OMWW. Exo- and capsular polysaccharides were respectively present in amounts of 1 and 4 mg/l.

Yesilada O. and Sam M. (1998) studied the toxic effects of biodegraded and detoxified OMWW on the growth of *Pseudomonas aeruginosa*. Detoxification and biodegradation of OMWW and toxicity (antibacterial effect) of untreated and treated (detoxified with *Trametes versicolor*) OMWW on a soil bacterium, *P. aeruginosa* were determined. *T. versicolor* biodegraded and detoxified OMWW and can be satisfactorily used for the biodegradation of phenol, COD, and color content. The inoculation of OMWW with *T. versicolor* reduced the toxic effects of *P. aeruginosa*. This research showed that *T. versicolor* could be satisfactorily used for biodegradation and detoxification of this waste.

Two aerobic bacterial strains, a chlorophenol-degrading bacterium characterized as *Ralstonia* sp. LD35 on the basis of the sequence of the gene encoding for 16S ribosomal RNA and *Pseudomonas putida* DSM1868, capable of metabolizing 4-methoxybenzoic acid, were tested for their capacity to degrade monocyclic aromatic acids responsible for the toxicity of OMWW (Di Gioia D. et al., 2001a,b, 2002; Bertin L. et al., 2001). *Ralstonia* sp. LD35 was found to metabolize 4-hydroxybenzoic acid, 4-hydroxyphenylacetic acid, 3,4-dihydroxycinnamic acid, and cinnamic acid, whereas *P. putida* DSM 1868 was capable of metabolizing 4-hydroxy-3-methoxybenzoic acid, 3,4-dimethoxybenzoic acid, and 4-hydroxy-3,5-dimethoxybenzoic acid, as well as 4-hydroxybenzoic acid and 4-hydroxyphenylacetic acid. In addition, the two strains were capable of growing on and extensively biodegrading a synthetic mixture of nine monocyclic aromatic acids commonly found at high concentrations in OMWW. Then, due to the complementary activity exhibited by the two strains, a co-culture of the two bacteria was tested under growing-cell conditions for degradation activity of the same synthetic mixture. Finally, the degradation activity of the co-culture on two fractions of pretreated OMWW was studied. Both fractions, one deriving from natural OMWW through reverse osmosis treatment and containing low-molecular weight organic molecules, and the other obtained from an anaerobic laboratory scale treatment plant fed with OMWW, were rich in monocyclic aromatic compounds. The co-culture of the two strains was able to biodegrade seven of the nine components of the tested synthetic mixture (2,6-dihydroxybenzoic acid and 3,4,5-trimethoxybenzoic acid were the two non-degraded compounds). In addition, an efficient biodegrading activity towards several aromatic molecules present in the two natural fractions was demonstrated — see also section: "Trickling filter".

There is not enough information on the bioremediation of 2POMW and the list of bacteria isolated from 2POMW is short. Responsible microorganisms would have to be capable of tolerating the low water activity (a_w) (due to the high organic and mineral content) of 2POMW. Reduced a_w elicits two major adaptive responses in bacteria: the accumulation of organic (compatible) solutes and alterations in membrane composition. Whilst these responses have been well characterized for a number of bacteria, it has been shown that for those, which can be isolated from OMWW the response may be atypical (Cummings S.P. and Russell N.J., 1996). The a_w of 2POMW is considerably lower than that of OMWW. Recently, six phenotypically distinct groups of bacteria were isolated from Spanish and Greek sources of 2POMW. These different bacteria isolated from 2POMW showed different growth and osmoregulatory responses to conditions of reduced a_w, and there was a correlation between the ability of isolates to withstand low a_w and grow on 2POMW. One isolate (1A), which grew particularly well both on 2POMW and in nutrient broth containing either 10% NaCl or 30% sucrose, was identified as being most closely related to *Bacillus amyloliquifaciens* using biochemical tests and partial 16S rDNA gene sequence analysis. *Bacillus* sp. strain 1A was found to display an atypical membrane lipid response at low a_w since the major change was an increase in the zwitterionic phosphatidylethanolamine rather than an anionic phospholipid such as phosphatidylglycerol. In addition, instead of the expected decrease, there was an increase in the average lipid fatty acid chain length at low a_w without any other compensatory fluidizing change (Jones C.E. et al., 2000).

(Micro-)Algae

OMWW treated with two microalgae, *Chlorella pyrenoidosa* and *Scenedesmus obliquus*, produced a biomass of microalgae and at the same time reduced the amounts of certain components, essentially sugars and salts, thereby diminishing the pollutant effect of this residue — see Chapter 10, "Uses", section: "Use in animal feeding" (Sánchez-Villasclaras S. et al., 1996).

OMWW treated with two phenol resistant microalgae, *Ankistrodesmus braunii* and *Scenedesmus quadricauda*, showed a limited reduction of phenol content after 5 days of treatment, irrespective of algal concentration. Otherwise, cultures of both algae, grown in the dark, degraded over 50% of the low molecular weight phenols contained in OMWW, but they were not completely removed, but were bio-transformed into other non-identified aromatic compounds (Pinto G. et al., 2003).

EU project: ICA-3-1999-00010 "MEDUSA-WATER" proposes to apply novel biotechnological processes to treat OMWW and, where possible combine with urban sewage, and reusing the resulting waters in horticulture. A new system of linear matrix multicellular photoreactors (LMMP) using microalgae cultures for the treatment of the final effluent will be designed, constructed and tested *in situ*. This innovative process intends to achieve an adaptation of the technology to the specific environmental conditions of the regions involved, which will be very low energy consuming and consequently with high economical benefits.

Aerobic–Anaerobic Processes

To enhance the anaerobic digestion of OMWW, an aerobic pretreatment stage may be favorable in reducing the amount of total phenolic compounds and associated toxicity. A preliminary aerobic treatment with specific microorganisms turned out to make shorter the residence time required for the anaerobic process (Borja-Padilla R. et al., 1991c; Maestro-Durán R. et al., 1991). Similarly, a significant increase in methane production rate has been reported for OMWW previously fermented aerobically with *A. niger* (Hamdi M. et al., 1991a) or *G. candidum* (Martín-Martín A. et al., 1993). OMWW, which has been previously fermented with *A. chroococcum*, was readily degraded anaerobically with COD fraction removal higher than 73%. The specific rate of methane production was substantially higher than that obtained in the anaerobic degradation of untreated OMWW. In addition, no inhibition phenomena were seemingly involved since the biotoxicity of the waste is reduced by 30% upon treatment (Borja-Padilla R. et al., 1993a).

Borja-Padilla R. et al. (1995b–d, 1998b) studied the effect of aerobic pretreatment, using three different microorganisms (*A. terreus, A. chroococcum,* and *G. candidum*), on the subsequent anaerobic degradation of OMWW. The anaerobic degradation process was carried out in a bioreactor containing microorganisms immobilized on sepiolite as support so as to assist in the separation of biomass during the sedimentation process. The pretreatment of OMWW with these three different microorganisms was capable of reducing COD and total phenols concentration of the waste as well as toxicity by about 63–75%, 65–95%, and 59–87%, respectively, for the various cultures used. The fact is shown by an enhancement of the kinetic constant for the anaerobic degradation process, and a simultaneous increase in the yield coefficient of methane production. In this context, the use of fungi characterized by proved lignolytic efficiency and, therefore, by a high capability of degrading lignin-related compounds, such as polyphenols, represents a promising perspective (Sayadi S. and Ellouz R., 1992).

Fountoulakis M.S. et al. (2002) found that pretreatment of a thermally processed OMWW with *P. ostreatus* was capable of enhancing the performance of subsequent anaerobic digestion. Aerobic treatment for 21 days led to about 65% phenols removal, which was enough to remove inhibition against methanogenic bacteria.

FR2620439 (1989) describes a process for the degradation of wastewaters of the olive oil and table olive industries comprising the following steps: (i) subjecting the wastewater to aerobic fermentation with yeasts until the leaving effluent contains a quantity of oil which is lower than its process inhibition threshold; (ii) separating the yeasts from the effluents; (iii) subjecting the resulting liquid to methane fermentation; (iv) discharging the separated effluent into the environment directly or after a finishing treatment. Unlike prior art the present process does not necessitate dilution of the wastewaters or physico-chemical deoiling and enables large volumes to be treated (especially to 20 kg COD/m^3).

Notwithstanding the potential interest of the aerobic pretreatment the above mentioned processes fail to completely remove the COD. In addition the aerobic pretreatment is affected by several problems (need of optimizing mycelium growth conditions, disposal of remarkable amounts of excess biomass due to the high aerobic yield coefficients) that have not yet found a satisfactory solution in terms of achieving the desired full-scale performance.

Composting

Composting is a controlled microbial bio-oxidative process that involves a heterogeneous organic substrate in the solid state, which evolves through a thermophilic stage and the temporary release of phytotoxins, leading to the production of carbon dioxide, water, mineral salts, and stabilized waste containing humic-like substances. Composting of olive-mill wastes has been examined as a potential bioremediation treatment of these wastes (Vlyssides A.G. et al., 1999; HR20010028, 2002). By using this method, it is possible to transform either fresh OMWW or sludge from pond-stored OMWW mixed with appropriate plant waste materials (carriers) into organic fertilizers (composts) with no phytotoxicity to improve soil fertility and plant production (Paredes C., 1998, Paredes C. et al., 1996a). Composting can be put into effect by means of a mixture of solids with agricultural waste, essentially as cereal straw, sawdust, or the remaining solid waste from the olive-mill.

OMWW contains on average about 6% of organic matter and 0.4% of mineral salts suspended or dissolved in an aqueous medium. Therefore, their bioremediation through composting must be achieved by adding other materials having a high absorbing capacity, such as agricultural lignincellulose residues. The latter are very poor in nitrogen, usually present in an organically bound form, so a rapidly available nitrogen source is necessary to assure the C/N ratio required for microbial development. A composting process developed by the EU project: ETWA-CT92-0006 used OMWW-wheat straw mixtures. Both physico-chemical aspects of the process and the quality of the end-product were analyzed in a number of following papers (Galli E. et al., 1994, 1997; Tomati U. et al., 1995). Tomati U. et al. (1995) used chopped wheat straw and urea to compost OMWW containing ~7% solids in a forced-aeration static pile. The urea was added to ensure a C/N ratio of ~35. A rapid increase of microorganisms and bio-reactions occurred at the beginning of the process, which led to an increase of the temperature and pH and a decrease of total organic carbon. Degree of humification, the humification rate, and the humification index, respectively, reached the values of 78, 37.8, and 0.28% after two months. A lignin degradation of ~70% was assayed at the end of the thermophilic phase.

No phytotoxicity was recorded on the end product, the chemical and physical properties of which suggest its possible use as fertilizer. Galli E. et al. (1997) studied also the composting of an OMWW-wheat straw mixture. Two aspects of carbon compound metabolism — lignin degradation and bioconversion of phenols — were particularly investigated. Lignin is one of the main components of the mixture and the most resistant fraction in composting materials. It is closely associated with cellulose fibers and hampers the degradation of polysaccharides. Moreover, the aromatic units released during its degradation are essential building blocks for the biosynthesis of the humic substances. Phenols are assumed to be responsible for phytotoxicity and their bioconversion is very important for humic acid biosynthesis. Oxygen consumption, microbial growth and urease activity were greatly enhanced during the thermophilic phase, reaching their maximum in about three weeks. Casein-hydrolyzing protease showed a high initial activity, which sharply decreased after two weeks. The high initial value of protease and the rapid increase of urease activity indicate that nitrogen sources are promptly utilized by the growing microflora. The development of the thermophilic microorganisms, particularly fungi, allows the degradation of lignin. The degradability of the OMWW-wheat straw mixture is made evident by the great oxygen consumption. At the end of the thermophilic phase both phenols and lignin were reduced by about 70%. Composting enhanced diazotrophic microflora as indicated by nitrogenase activity which increased at the end of the thermophilic phase.

The evaporative capacity of an intensive composting process was employed to treat OMWW. A mixture of extracted olive press cake and olive tree leaves was used as the solid substrate for composting. OMWW was added to the composting mass to replenish the water loss during processing in a pilot-scale open static container reactor. The salinity content of the compost was the factor restricting the treatment of OMWW by the process. The rate of OMWW treatment achieved in this study was 2.1 l/kg starting solid substrate (dry weight). The cumulative moisture and volatile solids content reduction during the temperature-induced aeration period of the process was 19 and 45%, respectively (Papadimitriou E.K. et al., 1997).

Della Monica M. et al. (1980) processed OMWW in a tank filled with soil. The effect of the treatment is an enrichment of the soil with readily assimilable nutrient sub-stances to the extent that the soil pollutant mixture becomes soil-compost. Neither sludge nor solid residual products were formed in the process, since they undergo degradation too. The treatment of OMWW and sludge is completed on parcels of land underlain with a waterproof base. The waterproof floor prevents filtration of polluting substances in the treated wastewater from percolating into the underlying soil.

IT1244520 (1994) describes a process and plant for the treatment of OMWW, in which OMWW is poured onto a layer of agricultural earth contained in a tank, the pollutant substances contained in OMWW undergo a degradation process by means of the said agricultural earth, and finally this earth, transformed into soil-compost with fertilizing characteristics, is subjected to washing out with water in order to remove and recover the soluble salts.

Negro Alvarez M.J. and Solano M.L. (1996) evaluated the quality of different products obtained through the composting of the solid residue that results from the flocculation of OMWW. To facilitate composting, the residue of flocculation was mixed with different lignocellulosic residues (straw, vine shoots, olive branches, and olive stone). The composting was carried out in a climatic chamber in PVC containers having a capacity of 5 l. Samples were periodically taken which were characterized and analyzed. Except for the mixture of the residue of flocculation with olive stone, in the rest of the mixtures assayed, an important degradation of organic matter as well as cellulose was observed. In addition, a decline of phytotoxicity, which the initial product presented, was observed. The results obtained show that the composting of this residue, when mixed with others of lignocellulosic character, is an effective manner of resolving the problem, while generating quality products from the point of view of its agricultural utilization.

Co-composting of olive cake and OMWW has been investigated as a potential bioremediation treatment for these wastes. Experimental results from a demonstration plant using olive cake as a bulking material and OMWW in a continuous feed have been reported by Vlyssides A.G. et al. (1996) — see Fig. 8.6. Composting temperature was controlled at 45–65°C and OMWW addition was fed in as necessary to maintain moisture content of 45–60% and to replenish the carbon

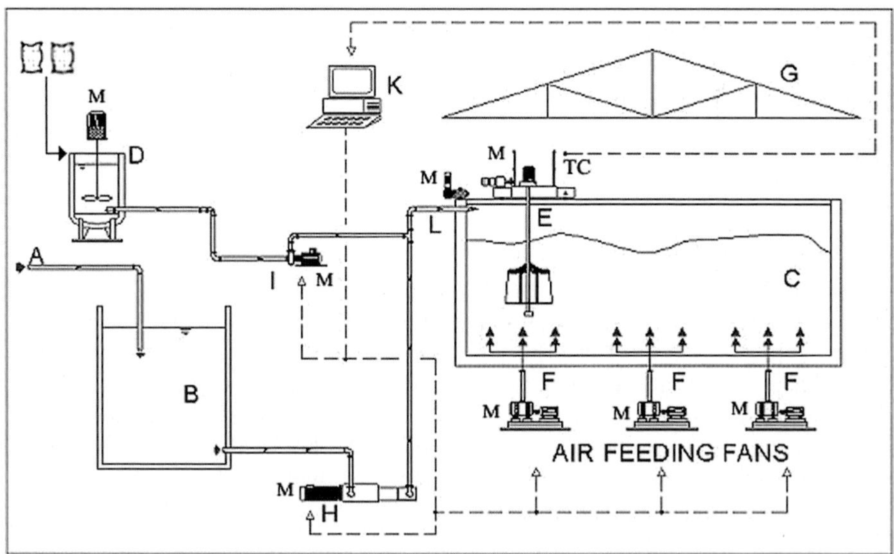

Fig. 8.6. Flow diagram of the demonstration plant (Vlyssides A.G. et al., 1996). A, OMWW feed; B, feed storage tank; C, co-composting bioreactor; D, urea feed system; E, agitator; F, air feeding fans; G, roof to prevent access of rainwater; H, mono-pump for OMWW dosing; I, proportional pump feeding urea solution; K, computer for controlling and data collection; L, traveling bridge for the agitator; M, motors; TC, temperature controller.

substrate level. During 23 days of operation at thermophilic temperature, a total of $263\,m^3$ of effluent was treated and an estimated total of 90,00,000 kcal of total bioenergy was generated. The 23-day thermophilic period was followed by a 3-month mesophilic stabilization period. The resulting composted product is suitable as a high-quality soil conditioner.

Filippi C. et al. (2002) evaluated also the possibility of co-composting olive cake and OMWW. The pH, E.C., total C and N, humic substances, phenolics, volatile acids, lipids, P and K values plus yeast, fungi, heterotrophic, cellulolytic and nitrifying bacteria, and phytotoxicological parameters were monitored during a 120-day stabilization process. Performance of the composting system adopted, together with physico-chemical characteristics of starting material and final product, are reported. Co-composting was found to induce a high level of organic matter change, with decrease of organic carbon, total nitrogen, and C/N ratio, as well as of the easily biodegradable lipids. Good metabolic activity of the microbiological population, with the starting material was also observed. The results obtained suggested that co-composting might be an adequate low-cost strategy for the recycling of olive-mill by-products.

Paredes C. et al. (1996b, 2000, 2002) studied the influence of a bulking agent on the degradation of OMWW during its co-composting with agricultural wastes. Two different piles prepared with OMWW sludge and either maize straw or cotton waste as bulking agents were composted by the Rutgers static pile system in a pilot plant, with the aim of ascertaining the most suitable conditions for degrading the OMWW sludges through composting. The use of maize straw, instead of cotton waste, as a bulking agent led to the following effects on the composting process of the OMWW sludges: (1) a lower mineralization of the organic matter at the end of the active phase of the process; (2) lower total-nitrogen losses by NH_3-volatilization; (3) a higher biological nitrogen fixation, and (4) production of a stabilized organic matter with less humic characteristics. The phytotoxic effects in the pile with maize straw lasted for a longer time, probably due to its slower rate of organic matter mineralization. However, no phytotoxic effects were observed in both mature composts.

The sea grass *Posidonia oceanica*[25] has been used for the production of organic compost or compost for agriculture, with co-composting of organic waste of agricultural, animal, or industrial units — see Fig. 8.7. The procedure applied comprises the collection and transfer of the sea grass to the treatment unit and then mixing with various organic wastes such as OMWW (*Posidonia oceanica* 67%, goat manure 20%, and OMWW 13%) and olive cake (*Posidonia oceanica* 67%, olive cake 23%, grape pomace 6%, olive leaves 1.5%, and sheep manure 2.5%), so that the C/N ratio is approximately 30:1 in the product of mixing. These ratios favor the growth of microorganisms, which control the biological composting process and help sea grass,

---

[25] *Posidonia oceanica* is not an alga, it is in fact a marine plant (phanerogam) which produces flowers. *Posidonia* meadows can only be found in the Mediterranean sea. Its role is incredibly important for the local ecosystems since many other species find their nutrients and housing in *Posidonia* meadows.

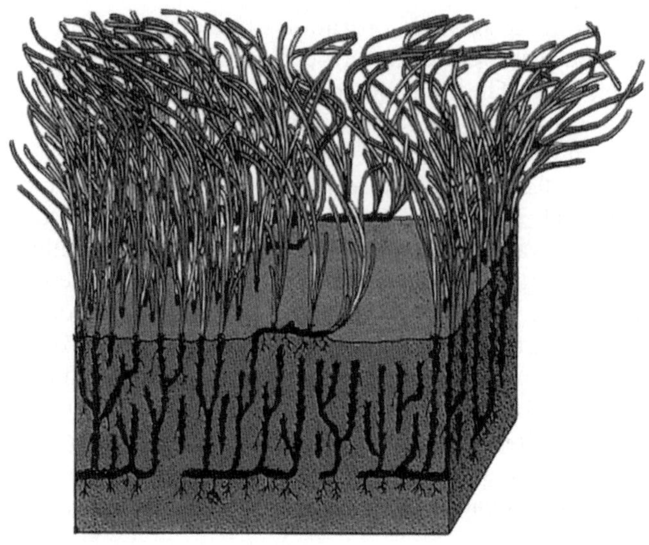

Fig. 8.7. *Posidonia oceanica.*

which is slow to break down naturally, to decompose and release its nutrients. The whole procedure lasts 9–12 months, in two phases. The end product is used as a means of plant growth with fertilizing properties, as a means to improve and enrich soil, as a means against soil erosion and exhaustion, as a product for land reclamation, as a product suitable for reforestation, as a crop-protective agent, and as organic material suitable to be mixed with metal compounds and minerals from industrial units (GR1003611, 2001).

In general, composting seems to be a feasible method to eliminate the toxicity of olive-mill wastes and to turn them into a valuable product (Cegarra J. et al., 1996a; Paredes C. et al., 1996a,b, 1998, 1999a, 2000, 2001 and 2002; Filippi C. et al., 2002). Furthermore, it produces no liquid waste, has a low fixed cost and the final product could be marketed as a high-quality soil conditioner (Vlyssides A.G. et al., 1989). A drawback of composting is the fact that the quantity of (semi-) solid olive-mill wastes is not sufficient for all the waste produced and hence either an additional woody substrate or condensation of the waste is required. In the latter case the condensation of the toxic compounds from the waste will hinder the process of composting and decrease the soil-enhancing quality of the final product. Another drawback is the high increase of pH produced during the composting of olive-mill wastes (Cegarra J. et al., 1996a; Paredes C. et al., 2000) which may limit its agricultural use, not only when used as soil-less substrate but also as soil amendment in high pH soils. The addition of elemental sulfur during the maturation phase of the composting process was considered a recommended method for decreasing the pH of the composts under the organic agriculture regulations — see Appendix II of EEC Council Regulation 2092/91, where elemental sulfur appears as an allowed soil fertilizer (Roig A. et al., 2004). The decrease of the pH reflected the formation of H^+ as result of sulfur oxidation. Sulfur is oxidized to H_2SO_4 by sulfur-oxidizing microorganisms (actionomycetes and filamentous fungi and *Thiobacillus* bacteria) according to the following mechanism:

$$S^o + \tfrac{1}{2}\,O_2 + CO_2 + 2H_2O \leftrightarrow CH_2O + SO_4^{2-} + 2H^+$$

2POMW has been composted by Sciancalepore V. et al. (1994, 1995, 1996). The quality of cured compost obtained by a mixture of crude olive cake, 2POMW and fresh olive tree leaves inoculated with cow manure, after six months of composting has been evaluated. The composting process brought about the total disappearance of phytotoxicity encountered in raw materials. The development of enzymatic activities was positive and no pathogen was found. The compost can, therefore, be satisfactory used as amendment for agricultural crops.

Roig A. et al. (2004) studied in a laboratory scale incubation experiment the effect of different variables (moisture, temperature, and sulfur concentration) on the oxidation rate of elemental sulfur, added to an organic compost prepared with 2POMW and sheep litter. An addition of 0.5% in sulfur (dry weight basis) and a moisture content of 40% were proposed as the optimum conditions to decrease the compost pH by 1.1 units without increasing the electrical conductivity to levels that could reduce the agricultural value of the compost. Compost treated with elemental sulfur did not show any potential phytotoxic effect as far as germination index is concerned. Although temperature was not an important factor for the oxidation rate, the control of moisture was considered to be decisive for the correct development of the process.

A number of laboratory studies assessed the suitability, as a vermicomposting substrate, of exhausted 2POMW either alone or mixed with cattle manure and/or municipal biosolids (Nogales R. et al., 1999). Cattle manure alone was used as a substrate for comparison. Five earthworms (*Eisenia andrei*) were added to 300 g of substrate and incubated for 17 weeks. Substrates examined were: exhausted 2POMW, cattle manure (CM), mixtures of exhausted 2POMW and cattle manure (2POMW:CM 8:1, 2POMW:CM 2:1), mixtures of exhausted 2POMW and biosolids (2POMW:BS 16:1, 2POMW:BS 8:1) and a 16:1:1 mixture of 2POMW, manure and biosolids (2POMW:CM:BS). Where biosolids were added, a preincubation was required to remove substances toxic to earthworms. All substrates supported earthworm growth and reproduction, with growth occurring for 4–8 weeks. Earthworm growth was considerably greater in the manure only substrate than in the exhausted 2POMW only substrate. The addition of manure or biosolids to the exhausted 2POMW enabled similar earthworm growth to that in the manure only. After 17 weeks, the earthworms inoculated at the beginning of the experiment had similar biomass, in all substrates. Larger weights of newly hatched earthworms were obtained in the substrates containing exhausted 2POMW. For a range of reproductive parameters including, among others, cocoon production and hatching success, all substrates were satisfactory with the 2POMW:CM 8:1, 2POMW:CM 2:1, and 2POMW:CM:BS being the most favorable for reproduction. For all substrates with 2POMW, vermicomposting reduced the organic carbon content, appreciably reduced the C:N and reduced the pH. A bioassay indicated that the final products were not phytotoxic.

A further study examined the feasibility of vermicomposting to stabilize exhausted 2POMW, for use as a soil amendment, using cattle manure (CM), anaerobic sewage sludge (ANS) and aerobic sewage sludge (AES) co-composting agents (Nogales R. et al., 1998; Sainz H. et al., 2000). Different ratios of 2POMW to co-composting agent were examined. Earthworm (*Eisenia andrei*) growth, clitellum development, and cocoon production were monitored over 35 days. Exhausted 2POMW alone was an inadequate substrate for vermicomposting on account of slow earthworm growth and infertility. The most effective ratios were: 2POMW:CM of 2:1 and 1:1, 2POMW:ANS of 16:1, 12:1 and 8:1, and 2POMW:AES of 16:1 and 12:1. Vermicomposting for 35 days reduced the dry weight of the substrates by 21–28%, and appreciably decreased their C:N. All final products had low contents of heavy metals. The above study demonstrated that exhausted 2POMW is a suitable medium for vermicomposting when combined with N-rich materials such as cattle manure and sewage sludge in appropriate ratios.

The characteristics of 2POMW are an obstacle for its correct aeration as a composting substrate, because such a process must be carried out in favorable conditions (appropriate moisture, nutrient balance, structure, and air distribution) to obtain a useful product (Alburquerque J.A. et al., 2004).

Phytoremediation (Wetlands)

Phytoremediation technology (also known as phytodepuration) exploits the capacity of plants and their associated microorganisms, such as mycorrhizal fungi and bacteria, to remove pollutants from contaminated water or soil. Phytoremediation utilizes the natural mechanisms of microbial aerobic and anaerobic degradation, as well the plants' ability to stimulate, through the root exudates, the rhizospheric population, that is the microbial community which colonizes the soil area next to the roots.

Wetland treatment system is a form of phytoremediation, that uses living plant systems (hydrophytes such as reeds and other marshal plants) to solve a variety of water pollution problems. Natural wetlands have been used as wastewater discharge sites for a long period of time and the ability of wetland plants to remove pollutants from wastewater is fairly known. Constructed wetlands are man-made structures designed for wastewater treatment and typically have a relatively impermeable bottom and a layer(s) of soil, muck, gravel, or other media to support the roots of aquatic plant species. Two types of constructed wetlands are currently used for wastewater treatment: free-water surface (FWS) and subsurface flow systems (SFS) — see Fig. 8.8.

Wetlands, constructed or natural, are commonly used for treating many types of wastewater, such as domestic sewage, urban and agricultural runoff, industrial and mining wastes. Wetlands can support and provide the necessary biochemical processes needed for the transformation, reduction, and immobilization of pollutants. The passive nature of wetland treatment technologies makes them cost-effective compared to more traditional engineered wastewater treatment systems.

In 1997 in an article from BBC News[26] came reports that scientists from Staffordshire University were using reed beds to clean up pollution from olive-mills in Tunisia. Skerratt G. and Ammar E. (1999) found that reed beds can be used to cultivate bacteria, which break down the pollutants in OMWW, making it harmless. The roots of the reeds provide oxygen to bacteria, which are capable of breaking down the toxic compounds. Other bacteria, which do not need oxygen, can be grown between the roots. Lagooning, which has been used in the past in Tunisia, presented several problems such as leakage of OMWW through the soil and into the groundwater, malodors, and need for available space (about $1\,m^2$ for each $2.5\,m^3$ of OMWW) — see also Chapter 6: "Thermal processes", section: "Lagooning". Central valley olive-mills use modified lagooning techniques to process OMWW.

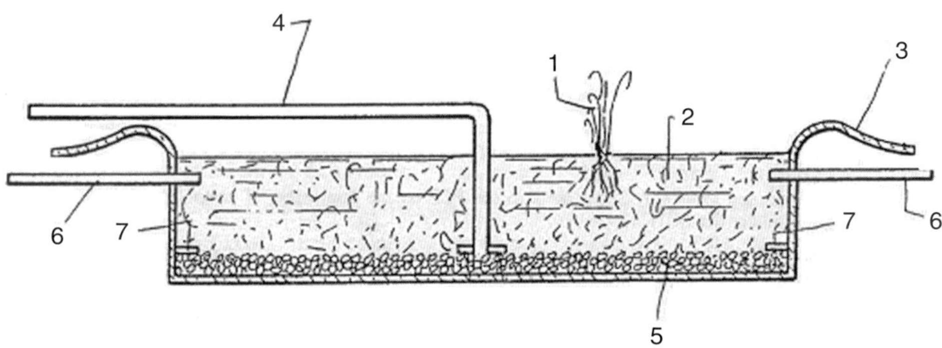

A wastewater treatment system, wherein the plants (1) are planted in a porous substrate (2) enclosed by a water impervious boundary (3). The roots of the plants extend into the substratum to form a root zone. An inlet means (4) is arranged such that wastewater passes towards the rootzone via cobble layer (5), enters the system below the rootzone and flows upwardly through the substratum. The wastewater is removed by outer means (6) located adjacent the rootzone. Flanges (7) are provided around the outlet of the pipe and at the inside wall of the structure, respectively.

Fig. 8.8. Constructed wetland system (adapted from WO9002710, 1990).

---

[26]http://news.bbc.co.uk/1/hi/sci/tech/287309.stm.

This may be a cheap and natural solution for poorer countries around the Mediterranean, which cannot afford expensive waste processing plants.

Reed beds is a natural wastewater treatment system utilizing the bacterial and mineral treatment capabilities also harnessed in part by the more common "constructed wetlands". However, the reed bed is a more highly engineered construction, usually completely isolated from the natural water system and allowing the treatment of highly contaminated effluents.

The most common type of reed planted for water treatment is *Phragmites australis* (the common reed). This is a robust species and grows rapidly being able to tolerate a range of climatic conditions and many types of wastewater. Reed beds are simple and cheap to operate but a significant disadvantage is the time lag between the planting of the reed bed and its ability to effectively treat wastewater. Immature systems can suffer from low porosity, but in due course, the reeds become established and simultaneously, reestablish a root structure, which reintroduces porosity into the substrate. Typically, this process may take two to three years to fully develop. Furthermore, although reed beds have been successful in many applications they are usually prepared on a large site, which is initially dug out of the earth ground and lined with, typically, a low density polyethylene to isolate the system from its surrounding environment. The system suffers also from the drawback of requiring an adjacent site of sufficient size to develop the reed bed system. However, there are many olive-mills which are not close to such sites and which may benefit from the reed bed technology.

A process of OMWW phytoremediation is described in EP1216963 (2002). The process is executed in an absorbing tank, consisting of two separate sections, a lower section for draining and an upper section for tree growing. The lower section constituted by a drainage layer, which takes up from 70 to 30% of the plant height, is made of inert material (for example coarse gravel, flintstones, or cobbles, etc.) characterized by high macroporosity and optionally containing porous material (expanded clay, pumice, etc.). Inside this layer there is a piping for homogeneous inflow and down flow of OMWW. Here, the waste settles until the beginning of the following olive oil session, when new OMWW will be introduced in the plant through the existing pipeline. The upper layer taking up from 70 to 30% of the plant height, made of local soil, appropriately is mixed with manure, in case the cultivar needed it. On this substrate trees are planted in mono or diversified culture, according to the agronomic rules of the specific cultivation.

Plants proved to be tolerant against the waste toxic action and indispensable to the process of phytoremediation are of the arborenous kind. In particular, they belong to the following families: *Betulaceae, Platanaceae, Magoliaceae, Aceraceae, Mirtaceae, Yuglandaceae, Caprifoliaceae, Labiateae, Tiliaceae, Apocynaceae, Salicaceae, Pinaceae, Fagaceae,* and *Cupressaceae.* They can be either singly chosen or not. Among those families, the following ones showed high attitude of adaptability and good physiological growth in the presence of OMWW: *Salicaceae, Pinaceae, Fagaceae,* and *Cupressaceae.* In detail, the more efficient genus for the process of phytoremediation are: *Pinus, Quercus,* and *Cupressus.*

The absorbing tank receives OMWW in wintertime, during the olive milling season (November to February). The waste flows to the drainage layer, where it undergoes degradation and mineralization caused by microorganisms activity; the trees stay, at this time of the year, in a vegetative rest, and they coexist in this state with the waste. At the resumption of the vegetative functions of the trees, the waste will have been already partly degraded and will have put at their disposal bio-assimilable elements. These components can be utilized as nutrients also from the rhizospheric organisms. Using the plants from the above list, the soil microbial population receives an indirect aimed stimulation. Degradation of phenolic and other organic compounds and transformation in humificated matter and assimilable (from the plant) components are achieved — see Table 8.4.

The phytoremediation system in absorbing tanks presents the following benefits:

- High disposal efficiency, even in wintertime (COD and phenols removal, pH and conductivity return to levels consistent with the plants' life).
- Lack of bad smell and of infecting insects, which on the opposite, characterize systems like fertirrigation or lagooning and all the methods where OMWW lays in the open air.
- Zero or limited energy utilization.
- Placing by the single olive-mill, with chances of direct piping from the olive-pressing plant to the phytoremediation system.
- Absence of waste transport and, consequently, of connected expensive and environmental hazardous operations.
- Annual plant reutilization as a natural cycle renewal.
- No need of specialized workforce; only required common farming maintenance (land annual processing, pruning, soil irrigation, etc.).

Table 8.4. Measurements of OMWW phytoremediation during the whole year (EP1216963, 2002). Lower layer made of agriperlite; upper layer planted with a tree belonging to *Qercus Ilex* genus

	November**		May		July		October	
	H_2O	OMWW	H_2O	OMWW	H_2O	OMWW	H_2O	OMWW
pH	7.4	4.97	7.65	6.19	7.24	7.05	7.22	6.9
Electrical conductivity (mS/cm)	0.09	2.73	0.44	2.5	0.16	0.44	0.24	0.63
COD (ml)	109	167,000	6800	7200	8920	6720	5264	5151
Total phenols (mg/g)*	6.2	26.8	7.85	7.05	N.D.	N.D	N.D.	N.D.

*A gram of dry weight from the analyzed sample.
**Analysis was referred to a mixture of agriperlite plus water and agriperlite plus OMWW before construction of the upper layer.
N.D.: Not detectable.

- Low costs for plant building, running and maintenance.
- Economic return linked to trees production industry (ornamental plants, wood fuel, etc.).
- Zero or limited mud production.
- Elimination of environmental damage risks strictly connected with the uncontrolled waste spreading in the ground and/or different waste streams.
- Low environmental impact helped from the plant appearing as a specialized arboretum gifted with highly aesthetic appearance.
- Increase of competitiveness of the oil in the international market, thanks to lower OMWW remediation costs.

One of the objectives of the EU project: ICA3-CT1999-00011 "WAWAROMED" was the comparison (in terms of cost and effluent's quality) of aerobic in-plant treatment and biological treatment by the Epuvalisation technology and a constructed wetland system. Epuvalisation, name which comes from the contraction of two French words: "epuration" and "valorization", is a biologic wastewater treatment technique which uses plants. Based on the nutrient film technique (NFT), this technique has the advantage, not only to purify, but also to produce plants. The effluent, which needs to be purified flows in small channels occupied by the plants and, therefore, is in close contact with the plants' roots on which a constantly growing bacterial flora proliferate, just like a constantly growing trickling filter. In fact, the whole surface of the channels and accessories, in close contact with the effluent, is used by the bacteria as a support mean. The roots are also working like a physical filter, which holds the suspended matters. The plants which were used for the process of epuvalisation were: *Apium graveolens*, *Phragmites australis*, *Ageratum mexicanum*, *Armundo donax*, and *Cyperus* sp. However, the initial results were not encouraging. Preliminary tests showed that the total phenols remained about unvaried and the total COD was not reduced substantially (initial COD: 3 g/l, after a week: 1.8 g/l) (Chartzoulakis K., 2002).

Within the framework of the EU project SOLADIST (EVK1-CT-2002-30028) was developed an easy handling solar distillation plant combined with constructed wetlands. OMWW is heated by the sun and the distillate runs through the constructed wetlands with the effect of a 98% removal of the undesirable organic matter. Operational costs could be reduced by 90% in comparison to state of the art treatment plants. By using the sun as a renewable energy source, the system can run nearly independently from any supply and maintenance, which can lead to high acceptance in the olive-mill industry — see Chapter 6: Thermal processes", section: "Evaporation/distillation".

Irrigation of Agricultural Land/Land Spreading

Irrigation is the process where OMWW is spread across the land and, especially in the olive groves themselves to provide the soil with nutrients and water. OMWW is allowed to percolate through the soil, which acts as a natural biological cleaning agent, breaking down the substances present in OMWW — see also Chapter 3: "Environmental effects", section: "Effects on soil".

Direct irrigation of soil with raw OMWW to save water and fertilizer has long been proposed (Morisot A., 1979; Morisot A. and Tournier J.P., 1986; Fiestas Ros de Ursinos J.A., 1986b), normally using doses less than $800\,m^3/ha$. Moderate doses of OMWW have beneficial effects, increasing soil fertility and microbial population, especially N_2-fixing bacteria, and improving the stability of the soil aggregates and in some instances crop yield — see Chapter 10: "Uses", section: "Use as fertilizer/soil conditioner". ES2051242 (1994) describes a system for the stabilization of soil by spreading OMWW as a continuous film on the surface, covering the soil granules. OMWW is incorporated into soil with optimum moisture content dependent on the type of soil and machine used.

When soil and land characteristics and climatic conditions are favorable, land spreading using high doses (e.g. as high as $5000\,m^3/ha$) of OMWW may be a solution for its disposal. For instance, land spreading of OMWW needs a stretch of flat land (infiltration field) close to the mill, where the soil has an adequate porosity, permeability and hydraulic conductivity, thus allowing infiltration of OMWW, and avoiding stagnancy and runoff. A deep-water table protected by an impervious soil layer is required to prevent groundwater pollution. Also, low rainfall and high evaporation are recommended. The land areas needed, in case OMWW is used directly for irrigation, constitute only a fragment of the total surface cultivated with olive trees. The recycle of the total yearly Italian production of OMWW (about $1,600,000\,m^3$) on the soil, as fertilizer and irrigation, asks only a 2.5–3% of the total Italian surface cultivated with olive trees. In fact, in Italy about 1 million hectares are cultivated with olive trees and to spread $1,600,000\,m^3$ of OMWW it needs about 30,000 hectares, where it is possible to spread $50–80\,m^3/ha$ as the Italian law 574/1996 permits (Di Giovacchino L., personal communication 2004).

Calcareous soils are very effective in reducing organic and inorganic pollution of OMWW (Cabrera F. et al., 1995). Experiments carried out in lysimeters filled with two clayey soils (*circa* 40% $CaCO_3$; *ca.* 40% clay) showed that a 2 m layer of soil almost completely removed the organic and inorganic components of OMWW when it was applied in doses of $5000–10,000\,m^3/ha \cdot y$. This efficiency was maintained for at least two years. In field experiments, the application of OMWW to one of these soils during three successive years at an annual rate of up to $6000\,m^3/ha$ caused changes in some chemical properties of the soil, especially in the upper layer (0–50 cm). Concentrations of soil organic mater, Kjeldahl N, soluble NO_3^- and available P increased soil fertility. On the other hand, soil electrical conductivity and sodium adsorption ratio also increased, but below the levels representing salinization or solidification hazard for the soil. The increase in soil fertility would be expected to allow the agricultural use of the soil, especially with salt-tolerant plants. The low effects of OMWW salts on germination in the soil were attributed to a neutralizing action of Ca, which suggests the possibility of cultivating land between periods of treatment. Furthermore, leaching of mobile species such as Na^+ and NO_3^-, is likely to occur below the 1 m layer, which could lead to the salinization and pollution of the water table. Therefore, special attention must be paid to the hydrogeological conditions of the utilized land area.

Zenjari B. and Nejmeddine A. (2001) reported the results of laboratory experiments carried out to determine the pollution removal capacity of local soil irrigated with OMWW and the effect of successive OMWW treatment on chemical properties of the soil profiles. The study showed that the clay soil has a very effective absorption/adsorption capacity. Over 99% of nutrients and 99% of phenols were removed after the first infiltration with OMWW. On the contrary, after the second infiltration the soil capacity to absorb/adsorb the anions was exhausted, while the phenol concentration was increased in the leachates which can present a risk of contamination of the groundwater. Spandre R. and Dellomonaco G. (1996) reported a link between OMWW spreading and local high concentrations of phenolic compounds in groundwater. When applied to the soil, OMWW induced its enrichment by fertilizers as well as negative effects, like fast filling-in of the soil and contamination by phenols. The latter biodegrade with difficulty, especially those immobilized in deep layers. Alteration of soil physical properties by swelling of soil clayey particles is attributed to the presence of salt. In order to overcome these problems the authors suggest minimizing the quantity of salt used for the conservation of olives, which is responsible for the high content of sodium in OMWW and to increase the time between irrigations so that the soil can recover its purifying capacities. In the same way, special attention should be paid to the amount of irrigation, which depends on the physico-chemical characteristics of OMWW used, and on the hydrogeological conditions of soil irrigated in order to avoid the possible contamination of groundwater (Andreoni N. et al., 1996).

Land spreading of OMWW has the disadvantage that it cannot be extended beyond the two years on the same ground. This problem has been partially solved by an improved process for the purification of OMWW — and effluents from the sugar industries — by irrigation of soil and infiltration followed by drying, with annual digging (ES2041220, 1994). The process comprises infiltration of the residues into the soil to the desired depth by irrigation, followed by natural drying. In the case of OMWW, this is followed by one or several irrigations with clean water. The process is carried out in a controlled infiltration area divided out in halves, which are used alternately for irrigation, and drying. Before commencing the annual infiltration, the sub-soil is cultivated, and after each cycle the surface is scarified.

Although the land application of OMWW at an adequate dose and time has been reported to be beneficial (Fiestas Ros de Ursinos J.A., 1986b; García-Ortíz R.A. et al., 1993) this practice is not very popular. Actually, there is no need for irrigation during the winter season, when olive-mills are in operation. Its main drawback is the high salinity of OMWW, and the low pH, which may both cause a very high concentration of salts and acidity in the ground. Other drawbacks include the dispersion in the environment of substances that are foul smelling and possibly pathogenic. Besides, its abundance of polyphenols may bring a phytotoxic action on plants' roots. In fact, higher ratios of disposed OMWW result in anomalous fermentations of the dispersed organic substances, which damage existing grass and tree crops — see Chapter 3: "Environmental effects", section: "Effects on soil biological properties".

In general, land spreading of OMWW is not universally applicable and it is limited to cases where there is suitable soil (of low permeability) in the proximity of the olive-mills. OMWW suitable for land irrigation must fulfill certain criteria (Cabrera F. et al., 1996), namely:

- the waste must be biodegradable, in whole or at least in part;
- the microorganisms indigenous to the soil will survive and function at reasonable and practicable application rates of the waste;
- the long-term toxic effects of accumulated residues and possible ion adsorption on soil can either be prevented or mitigated;
- reasonable and practical loading rates will neither cause pollution of the groundwater by hazardous constituents nor allow toxic substances to enter the food chain, so that the land treatment site will remain environmentally safe;
- the cost-effectiveness of land treatment in relation to other treatment disposal alternatives is within reasonable limits;
- the land treatment will leave the soil in virtually the same (or even higher) productive conditions as originally.

Land treatment sites may also combine some pedological, climatic, and hydrogeological characteristics to be compatible with the nature, rate, and schedule of application of the waste.

Treated OMWW, which has been deprived of its recalcitrant compounds, meets most of the above criteria and for this reason it has been proposed for irrigation purposes (EP520239, 1992; ES2084564, 1996; FR27249222, 1996; Marques I.P., 2001).

In Italy, land spreading of wastes arising from the processing of olive is specifically regulated under the Law no. 574 of 11/11/1996 on OMWW and olive cake. The prescriptions are listed in Table 8.5. However, the prescriptions of the law have been criticized as they make the inspections quite difficult as the regional and provincial authorities, from which the inspection authorities depend, do not know the exact dates and places of the spreading (Burali A. and Boeri G.C., 2003).

Land spreading of crude olive cake or 2POMW presents olive-mills with considerable organization problems in that the mills have to promptly handle effluents with a high moisture content which are prone, even during short storage periods, to noxious-smelling anaerobic fermentation. Controlled land spreading, in addition to the tight restrictions applied on such methods (for Italy law 574/1996), also presents other problems of a technical nature relating to the percolation of the mass and the requirement for a specific machine which distributes the residue uniformly, particularly in the case of 2POMW. Further, the spreading of these olive wastes even when they have been briefly stored under anoxic conditions, causes the release into the atmosphere of unpleasant odors which are an inconvenience for people living in the vicinity of the olive-mills and in the rural areas where spreading takes place (EU project: LIFE00 ENV/IT/000223 "TIRSAV").

Table 8.5. Prescriptions of the Italian law no. 574 on land spreading of OMWW and olive
cake (European Commission — Directorate-General for Environment, 2001)

Agronomic use:
Olive-mill wastewater: Olive-mill wastewater without pretreatments.
Olive cake: Olive-mill wastewater plus stone fragments and fibrous part of the fruit can
  be used in agriculture and are not subject to Fertilizer Law no. 748.

Quantity
Olive-mill wastewater: From traditional press at $50\,m^3/ha \cdot year$ or from centrifugation
  at $80\,m^3/ha \cdot year$.

Authorization
Spreading operations must be notified to the mayor 30 days before.
Communication must include:
* Type of soil, spreading system, spreading time, hydrological condition.
* The mayor can stop spreading operations if there is a chance of damage to the
  environment.

Spreading systems
* Distribution must be uniform and by-products must be ploughed in.
* During spreading operation run off must be avoided.

Prohibition
Spreading is forbidden, where:
* Distance is less than 300 m to the groundwater draining areas.
* Distance is less than 200 m to the built up areas.
* Soil is used for growing vegetables.
* Soil with a water table depth of less than 10 m.
* Soil where percolation water could reach the water table.

Storage
* Storage period max 30 days.
* Storage must be in a water-proof container.
* The mayor must be notified of storage location.

Chapter 9

Combined and Miscellaneous Processes

Complete abatement of OMWW pollutants can be hardly achieved by the adoption of a single process. Table 9.1 presents an approximate evaluation of the costs and the energy as well the drawbacks of each process. Combustion and concentration by distillation are reliable, but quite expensive and energy consuming. Aerobic processes are not advisable because of: i) mechanical energy; ii) high consumption of nutrients (to reach a ratio BOD_5:N:P $= 100$:5:1 from BOD_5:N:P $= 100$:1:0.5); iii) very high production of seasonal sludge, which has to be disposed of; iv) high capital cost. Anaerobic degradation is quite appealing from an energetic point of view, but it has a long start-up and requires dilution of OMWW with water. Combination of various processes is often the way to optimize the overall process.

The first treatment, if properly chosen, will facilitate the second one, thus leading to a much more effective treatment of the waste. Laboratory scale experiments were carried out in order to identify pretreatment type and conditions capable of optimizing OMWW anaerobic degradation in terms of both kinetics and methane yield. Ultrafiltration, even if it allowed very high removals of lipids and polyphenols, was affected by poor selectivity (indeed, large amounts of biodegradable COD were also removed). Centrifugation turned out to be preferable to sedimentation owing to smaller volumes of separated phase. It has been reported that dilution, acidification, and aerobic pretreatment (by means of a fungal strain, *Aspergillus niger*) were good solutions to the inhibitory problems and consequently, to a better fermentation (Hamdi M. and Ellouz R., 1993).

The difficulties in the anaerobic treatment of OMWW are mainly connected with the presence of biorecalcitrant and/or inhibiting substances, essentially lipids and polyphenols (Beccari M. et al., 1996, 1998, 1999a). The lipids, although are more easily degraded than phenols they are potentially capable of inhibiting methanogenesis more strongly. In batch experiments using synthetic substrates it was shown that an addition of soluble calcium salt reduced the inhibitory effect of long chain fatty acids (LCFAs), provided that the anaerobic culture had not been exposed to LCFAs

Table 9.1. Process evaluation for OMWW treatment (Boari G. et al., 1984)

Treatment	Capital cost[a] (US$/ $m^3 \cdot$ day)	Energy		Concentrate ashes/sludge (kg/m^3)	Drawbacks
		Electric (kWh/m^3)	Thermal (kWh/m^3)		
Combustion	5×10^3	8	670	2 TS	Destruction of recoverable organics
Single effect distillation	1.1×10^4	20^b	730	90 TS	Post-treatment of distillate (2–3 kg $COD/m^3 \cdot$ day)
Activated sludge	2×10^4	30	–	30 VSS	Dilution water. Nutrients addition. Sludge disposal
Trickling filters	1×10^4	15	–	20 VSS	As activated sludge
Anaerobic degradation	4×10^3	<1	200^c 240^d	10 VSS	Dilution water. Long start-up

[a]Per m^3 waste produced per day assuming $BOD_5 = 50\,kg/m^3$.
[b]Energy required by cooling tower fan.
[c]Heating energy, taking into account dilution water, without heat exchanger.
[d]Energy recovered from methane.

before calcium addition (Hanaki K. et al., 1981). Experiments using lauric acid as the model long chain fatty acid confirmed that the decrease in inhibition was due to precipitation of the acid as a relatively insoluble calcium salt (Koster I., 1987); the solubility constants of LCFAs as calcium salts have been calculated (Roy F. et al., 1985). Experiments performed directly on OMWW (Lolos G. et al., 1994) showed that addition of lime removed 77% of lipids whereas the phenolic compounds were not affected. The use of bentonite for cleaning vegetable oils suggested its application to reduce lipid inhibition on thermophilic anaerobic degradation (Angelidaki I. et al., 1990); bentonite was added to a synthetic substrate (glyceride trioleate) and turned out to stimulate methane production probably by binding the substrate on its surface and thus lowering glyceride trioleate concentration in the liquid phase.

These difficulties in the anaerobic treatment of OMWW suggest the use of a physico-chemical pretreatment for the removal of biorecalcitrant and/or inhibiting substances (essentially lipids and polyphenols) as selectively as possible before anaerobic digestion. This way follows a general trend towards integration between physico-chemical and biological processes for wastewater characterized by difficult biotreatability (Scot J.P. and Ollis D.F., 1995).

An integrated treatment of OMWW comprising a sequence of operations is described by Beccari M. et al. (1999b, 2000, 2002); (i) a pretreatment based on the addition of $Ca(OH)_2$ (up to pH 6.5) and bentonite (10-15 g/l) removes lipids almost quantitatively; (ii) the mixture (OMWW, $Ca(OH)_2$, and bentonite) fed to an anaerobic treatment without providing an intermediate solid/liquid separation gives way to high biogas production even at very low dilution rates (1:1.5);

(iii) an eight-day activated sludge post-treatment. The results show that a very high percentage (about 80%) of the phenolic fraction below 500 Da is removed by the methanogenic process whereas the phenolic fractions above 1000 Da are adsorbed on bentonite; the activated sludge post-treatment allows an additional removal of about 40% of total filtered phenolic compounds. The complete sequence of treatments was able to remove more than the 96% of the phenolic fraction below 500 Da (i.e. the most toxic fraction towards plant germination). Preliminary respirometric tests show low level of inhibition exerted by the effluent from the methanogenic reactor on aerobic activated sludges taken from full-scale municipal wastewater plants (Beccari M. et al., 2002).

As for polyphenols removal, the available physico-chemical technologies (precipitation with organic or inorganic flocculants, extraction with selective solvents) can remove up to 75% of polyphenols (Montedoro G.F. et al., 1986). However, high percentages of the initial COD are usually separated together with polyphenols, thus subtracting a useful carbonaceous source from the stream destined for anaerobic degradation. Selective chemical oxidation is also to be considered; in this regard, the potential of ozone for enhancing anaerobic biodegradability of model phenolic compounds has been already reported (Wang Y.T., 1992). However, direct experiments of OMWW ozonation (Andreozzi R. et al., 1998) have shown that the products of ozone attack on lipidic and phenolic compounds can be more inhibitory than the parent compounds themselves.

Andreozzi R. et al. (1998) investigated the possibility of ozonation pretreatment coupled with successive anaerobic fermentations of OMWW. Preliminary tests showed that both total phenols and unsaturated lipids are reduced to about 50% in 3 h of ozonation, and that the total COD remains about unvaried. Nevertheless, ozonated OMWW exhibit in general a longer lag phase and a lower yield in methane than untreated OMWW. These effects are more evident at higher OMWW concentrations. Methanogenic tests were also conducted on OMWW samples to which oleic acid or *p*-hydroxybenzoic acid and their ozonation products were added. Results indicate that the ozonation products of oleic acid are more inhibitory than the original substrate. The inhibitory effects of *p*-hydroxybenzoic acid and its ozonation products show a different dependence upon concentration: inhibition of ozonation products is remarkable at lower concentration, but it increases more slowly as concentration increases. No effect of ozonation is observed on the acidogenic step of fermentation.

Degradation of OMWW has been attempted by the combination of chemical oxidation processes (Fenton's reagent and ozonation) and their consecutive treatment with aerobic microorganisms (Beltrán-Heredia A.J. et al., 2001c). Fenton's reagent treatment moderately reduced COD and to a greater extent the polyphenolic compounds. Ozonation contributed to low conversion of COD and moderate reduction of polyphenols. The aerobic biological treatment reduced COD and polyphenolic compounds to values higher than 70 and 90%, respectively.

Chakchouk M. et al. (1994) used wet air oxidation with the addition of hydrogen peroxide (H_2O_2) for the integrated treatment of OMWW, but its effect was evaluated only as an increased aerobic degradability with no mention of the fate of poly-phenols and lipids or to anaerobic degradability. Moreover, the marked COD decrease observed after treatment has a lowering effect on the energy recovery expected by the successive anaerobic step.

EP520239 (1992) describes a process according to which OMWW is treated with hydrogen peroxide in the presence of enzymes (peroxidase). The addition of the reactants is carried out after the pH of OMWW has being adjusted at values 6.5 ± 0.5, at room temperature. According to the inventor, the use of hydrogen peroxide makes it possible a very effective purification of OMWW to be carried out, by means of the selective removal of biotoxic and phytotoxic principles, while the other substances, useful for the agriculture, are left unchanged. The use of enzymes in combination with hydrogen peroxide promotes the decomposition of the noxious substances into biodegradable substances. The above process can possibly be associated with some techniques known from the prior art, in order to integrally exploit the biological potentialities inherently displayed by OMWW; therefore, said technique is essential as a preliminary treatment, in order to facilitate the disposal of OMWW and shorten to a meaningful extent the time required by the subsequent disposal, e.g. in activated sludge biological facilities. The process does not require any particular plants and/or equipment pieces. It, furthermore, displays the advantage that the efficiency of the process is independent from the volume of OMWW to be treated, so large collection centers are not necessary; it can be carried out also on a local basis, on any desired volume of OMWW, therefore, also with the elimination of energy expenditure due to transport. The process is claimed to be cheap and environmentally compatible, because it takes place at room temperature, and with the use of reactants, which do not leave environmentally toxic residues and therefore, do not cause any secondary induced pollution problems.

Zouari N. (1998) attempted to decolorize OMWW by a physico-chemical treatment prior to anaerobic degradation. Physico-chemical treatment of OMWW was performed using iron and lime as complexing agents, DUOLITE® XAD 761 resin as phenolic adsorbent and hydrogen peroxide as oxidant. It was shown that $10\,g/l$ of ammonium iron(III) sulfate $((NH_4)Fe_2(SO_4)_3 \cdot 12H_2O)$ or calcium dihydr-oxide $(Ca(OH)_2)$ were sufficient to precipitate more than 50% of the initial COD and remove 50% of the initial color within a short contacting time. The aromatic adsorbent resin retained more than 50% of the coloring compounds (chromophores) corresponding to removal of more than 60% of the initial COD after treating three bed volumes of crude OMWW. The efficiency depended on the volume treated. Hydrogen peroxide removed the substituents of the aromatic rings, which resulted in a decrease in length of the coloring compounds in OMWW. However, they were not completely degraded, leading to shorter wavelength absorption. This chemical treatment was efficient in color removal but only 19% COD removal was possible. In all cases, simple aromatics were reduced, as determined by gel permeation chromatography (GPC) analysis. The physico-chemical decolorization of OMWW was efficient in reducing the toxic effect of recalcitrant compounds. The resultant

OMWW by each of these alternative treatments was readily degradable through anaerobic degradation.

A scheme for the treatment of OMWW is given by Shammas N.K. (1984). The scheme is based on emulsion breaking with 1 mg/l H_2SO_4, oil separation in a separation basin, anaerobic lagooning of the water, and sludge drying in drying beds.

EP421223 (1991) describes a process for the disposal of OMWW comprising the following steps: (i) oxygenation at an acidic pH value by feeding 5–10 l of air per liter of OMWW; (ii) oxygenation at neutral pH with pH value 6.8–7.2, by feeding the same air quantity; and (iii) oxygenation in presence of at least one enzymatic substance by feeding 10–20 l of air per liter of OMWW using stirring condition to obtain a substantial humidification of OMWW. The cited oxygenating steps in acidic and neutral conditions for a minimum preselected time ($\sim$24 h) are two critical steps of the process, because they catalyze the fermentation reactions that degrade the organic molecules during acidic oxygenation and lead to the creation of humic acids during oxygenation in neutral conditions. The synthesis of humic acids is then accelerated and substantially completed during the final step of oxygenation in the presence of said enzymatic substances ($\sim$24 h). The pH of OMWW is adjusted by addition of acidifying or basifying liquid ammonium humate. The enzymatic substances are used preferably as a mixture of enzymes selected from amylase, cellulase, lactase, lipase, pancrease, protease, betalactamase, and invertase. The process is claimed to be simple, effective, and inexpensive and gives odor-free products usable as agricultural raw material.

EP441103 (1991), which is an improvement of IT1211951 (1989) — see Chapter 6: "Thermal processes, section: Physico-thermal processes", describes a process and a apparatus, as outlined in Fig. 9.1, in which the condensed water, derived by the double distillation of OMWW, is biologically treated by aerobic oxidation and collected in a tank (2) below the honeycomb tower (6) wherein the oxidation takes place, preferably with frequently repeated circulation (4), accompanied by air induced by a blower at the tower head (7). Organic particles in the water tend to agglomerate, forming a slurry or sludge. This charged water is delivered by a pump (8) to a separate filter tank (9) whose charge of silica sand and/or active charcoal biologically purifies the water for return to the olive-mill or for discharge with the drainage system. Water, which in untreated state has excessive soiling capacity for the drainage system is brought within acceptable limits.

ES2110912 (1998) describes an integrated treatment for the purification of fresh or fermented OMWW comprising a biological treatment with pectolytic enzymes and amylase, followed by filtration and evaporation. Residual olive oil is recovered as well as a concentrate which is enriched in humic acids and mineral salts. The concentrate can be used in agriculture, as liquid fertilizer, or as soil conditioner, or if made of fresh OMWW, in food for humans or animal feeding. In another embodiment said integrated treatment is used for the purification of 2POMW as illustrated in Fig. 9.2.

GR870652 (1987) describes a combined physico-chemical process for the treatment of OMWW comprising the steps: (i) precipitation with an alkali compound,

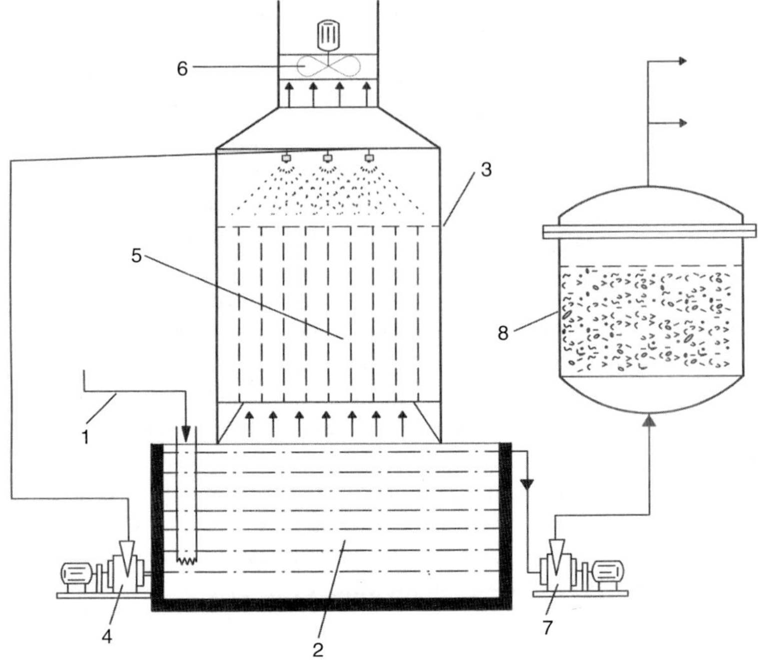

Fig. 9.1. Apparatus for the aerobic oxidation of the condensed water, derived by the double distillation of OMWW (EP441103, 1991).

preferably lime. The calcium hydroxide transforms the organic acids as well as the phenols to salts, which precipitate as calcium salts; (ii) oxidation, preferably with agitation, in one or several stages, which reduces the COD value; (iii) treatment with an acid, preferably sulfuric acid, in presence of a solid decolorizing agent especially, activated carbon. The sulfuric acid completes the oxidation of the dissolved molecules, which are responsible for the coloration of OMWW and promotes their adsorption on the active carbon. In addition, with the help of the sulfuric acid a large part of the calcium remaining in solution precipitates as $CaSO_4$. The resulting water has a minimum contaminant content and can be disposed within legal requirements without further treatment.

A technological process for the recycling of olive-mill effluents for agronomic purposes developed within the framework of the EU project: LIFE00 ENV/IT/ 000223 "TIRSAV" makes it possible to intervene on all types of olive-mill plant currently on the market, whether traditional or continuous-cycle, with either two- or three-phase extraction systems. The action of the technological system on the effluent, which is typically performed in-line in the mill, is divided into three consecutive phases: destoning, mixing, and packaging.

The 1st phase involves treating the effluent (OMWW or 2POMW) to separate the stones which are, therefore, immediately available for use as a fuel. In the 2nd phase,

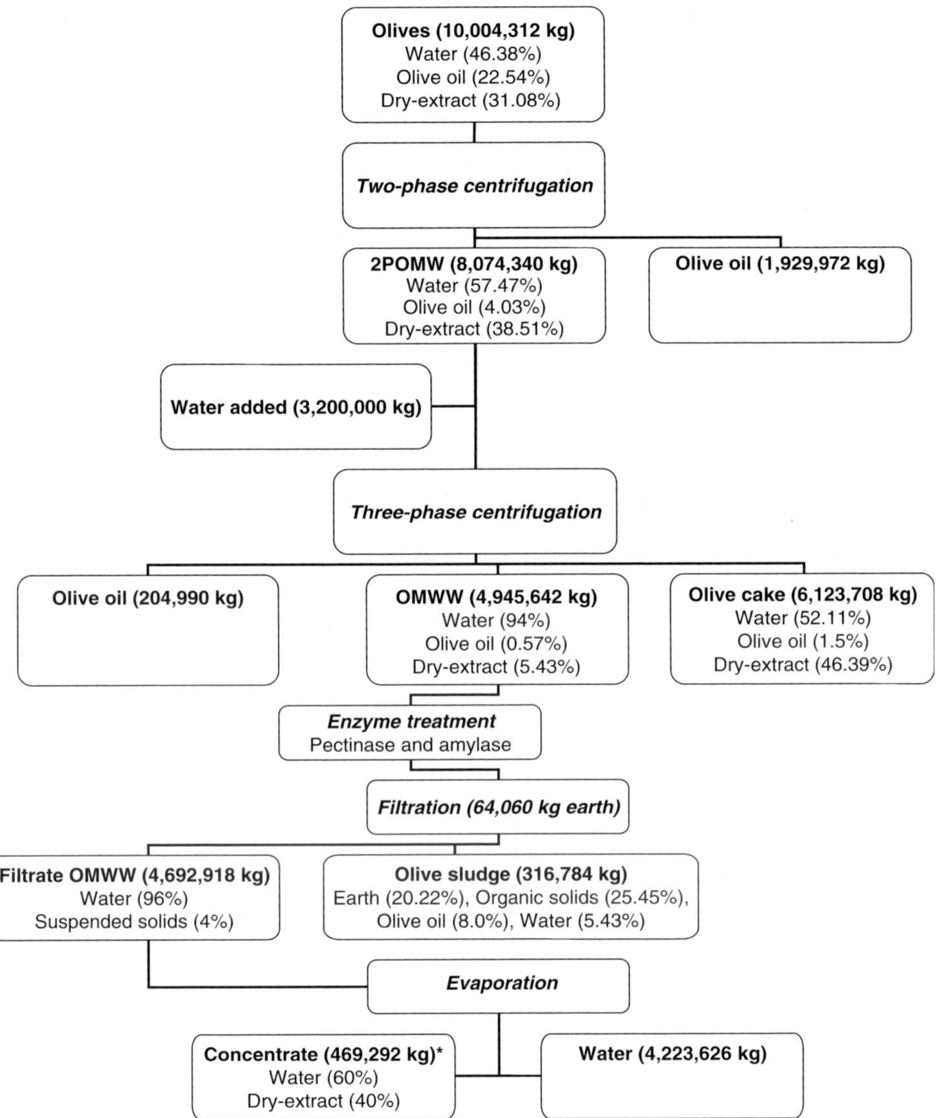

Fig. 9.2. Purification of 2POMW by enzymatic treatment, filtration, and evaporation (ES2110912, 1998).

*The obtained concentrate is composed of; 24.0% humic extract; 2.7% total N; 6.7% K_2O; 1.6% P_2O_5. Elements: Ca, Mg, Fe, Mn, Cu, Zn.

the destoned effluent is collected and stirred in a special tank to prevent the physical separation of its liquid and solid elements, then fed into another tank where it is mixed with additives which:

a) reduce the moisture of the destoned effluent, so that the end product is non-percolating; suitable additives include wood shavings and sawdust from non-treated timber, straw from graminaceous plants, raw wool waste, etc.
b) allow good circulation of air in the end product; preferred additives are straw, olive leaves, and twigs obtained from trimming the olives in the olive-mill upstream of the oil extraction process.
c) reduce the C/N ratio of the initial effluent in order to cause more rapid microbial degradation in the soil and, minimize competition with agricultural crops for the nitrogen contained in the solution circulating in the soil; additives experimented comprise raw wool waste or raw wool itself, both readily available and inexpensive. Along with an organic nitrogen content that varies approximately between 4.5 and 6%, these materials present high hydroscopic capacity which derives, among other factors, from their low moisture content. Raw wool waste and wool have a content of organic matter of about 76% and about 9% ash (mainly composed of calcium, potassium, while iron is the principal constituent of micro-elements). The choice and quantity of additives depends on the type of the olive residue and the end product to be obtained.

The 3rd phase consists of the automatic packaging of the end product in sealed net bags (20–30 kg) to facilitate transport and storage. The bags are stored in static layered piles where the end product undergoes aerobic maturation due to the action of yeasts and bacteria, improving this way its chemical and physical properties. At the same time, the moisture level of the biomass falls, which in practical terms means less weight to be handled and increased agronomic efficiency due to the concentration of nutrients.

The end product can be used as a soil amendment and/or organic fertilizer in olive culture and crop cultivation in general — see Chapter 10: "Uses, section: Use as fertilizer/soil conditioner". The so-called M.A.T.Re.F.O. technology (Method and Apparatus for the Treatment of Oil Mill Effluents) has been patented (application number: IT2004RM000084).

A treatment made up of physical and physico-chemical processes, the realization of which in a rational manner is claimed to achieve the maximum reduction in the organic load with lower energy costs is described in: ES2019830 (1991). According to this process OMWW is treated in a series of steps: (1) Primary reduction of organic load, "oil removal". OMWW is thoroughly mixed with a solvent, so as to obtain: an aqueous phase olive wastewater-I with the oil removed, and another phase formed by a fatty phase — with a solvent, which is used again after distillation. (2) Secondary reduction of organic load, "flocculation". This consists of the separation of organic matter contained in the olive wastewater-I from the preceding process, by means of a coagulation or flocculation, and the flocculated matter is separated from the olive wastewater-II by sedimentation, filtration, or centrifugation. (3) Tertiary reduction of organic load, "formation of carbonated organic compounds". The olive wastewater-II is subjected to the combined action of alkaline-earth hydroxides and carbon dioxide, so as to form organic carbonated compounds, which under certain conditions are insoluble and can be separated by filtration, sedimentation, or centrifugation.

A simultaneous combustion process — as described in: ES2032162 (1993) — is applied to the previous treatment (ES2019830, 1991). Such an improved system for the treatment of OMWW, comprising also biological and thermal processes, is described in: ES2024369 (1992). The combustion process generates the heat energy necessary for continuous regeneration of the reagents used in the process, eliminates contaminating organic matter, and provides the appropriate temperature for optimum operation of the biological process. Thermal dissociation, calcination, activation, and distillation units are included in the system, together with supplementary purification devices for the elimination of OMWW and sludges. The volatile products generated by biological processes, which produce additional calories, are used in the process as an additional fuel for joint thermal processing.

ES2108658 (1997) describes a process for the treatment of highly contaminated and/or toxic wastewaters including those from olive processing. The purification is carried out through bacteria adapted to the residue externally to the reactors involved in the process, said bacteria being fixed to support means which are immersed in the reactors. The process comprises: (a) first and second preliminary physical treatments including screening and/or centrifugation for the separation of solid matter; (b) physico–chemical treatment including sedimentation and/or flocculation and/or coagulation to remove suspended solids; (c) treatment in at least one aerobic biological reactor; (d) treatment in at least one anoxic biological reactor arranged upstream of the first aerobic reactor; (e) treatment in at least one clarification pool to separate the biomass; (f) at least one recirculation of a portion of the slurry generated in the biological reactors to at least one of them; (g) backfeeding from at least one of the aerobic biological reactors to the same; (h) a second refeeding phase from the outlet of the first aerobic reactor to the interior of the anoxic reactor, and (i) treatment in a refining reactor which operates alternatively in anoxic and aerobic conditions. The use of preadapted bacterial strains eliminates or reduces the adaptation time required in the reactor. COD and toxicity levels in the effluent water are reduced to well below acceptable levels for discharge into water-courses. This process is distinguished by a large number of process steps and by the addition of oxygen, which can be assessed as highly energy intensive and, depending on the application, as uneconomic.

ES2084564 (1996) describes an integrated process for the purification and total exploitation of liquid and solid waste product produced at an olive-mill through a combination of a set of physical, chemical, and thermal processes as shown in Fig. 9.3. OMWW is subjected to an accelerated separation of solids (Fig. 9.3, e), which may be done by coagulating and flocculating of the solids held in OMWW and/or by ultrasound emission. The resulting precipitate is isolated from the rest of the solution through decanting; and the obtained mud or solids are mixed with

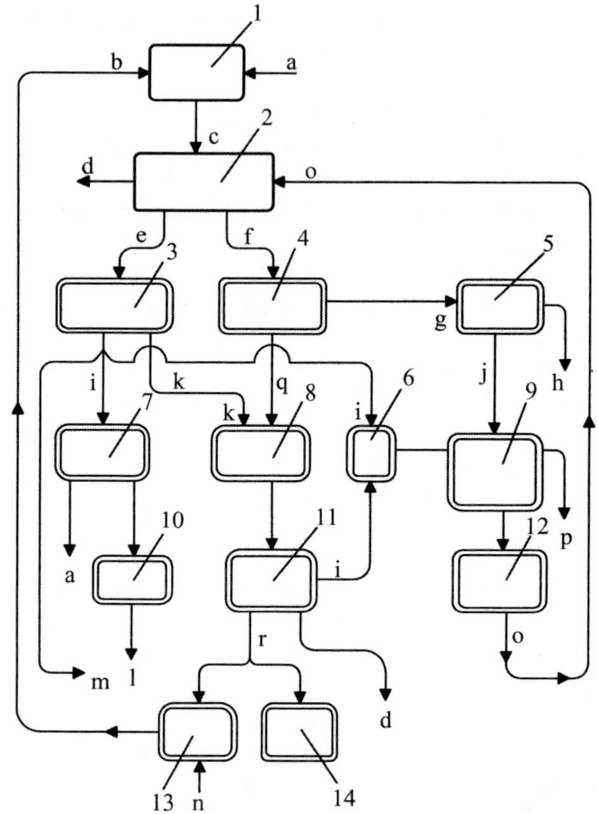

1   Olive grove
2   Olive-mill
3   Accelerated solids separation
4   De-stoning
5   Boiler
6   Roughness elimination
7   Liquids conditioning
8   Solids conditioning
9   Evaporator
10 Liquid fertilizers preparation
11 Oil physical reclamation
12 Interchanging
13 Composting plant
14 Fodders manufacturing

a) Irrigation
b) Organic fertilizer
c) Olives
d) Oil
e) OMWW
f) Olive cake
g) Stones
h) Ashes
i) Liquids from OMWW
j) Heat
k) Solids
l) Liquid fertilizer
m) Evaporation pools
n) Vegetable waste from the area
o) Hot water
p) Concentrate
q) Pulp
r) Solids without oil

Fig. 9.3. Integrated process for the purification of OMWW (adapted from ES2084564, 1996).

the solid waste product resulting from the process of obtaining olive oil at the olive-mill.

The liquid of OMWW (Fig. 9.3, i) may follow three ways: one to the evaporation pools (Fig. 9.3, m) where it is evaporated faster because it lacks solids and oils, the second one to a conditioning of liquids (Fig. 9.3, 7) to prepare liquid fertilizers (Fig. 9.3, 10), and the third one to an evaporation phase (Fig. 9.3, 9). The evaporation is under atmospheric pressure. Through a stage of sudden partial evaporation, previous to the evaporation itself, most of the organic volatiles from the water are removed. All the liquids coming into the evaporator (Fig. 9.3, 9) are subjected to roughness elimination (Fig. 9.3, 6) making them circulate inside an intense magnetic field, which polarize the salt molecules and avoid their deposition on the pipes and elements they circulate through. The calcium carbonate remains in suspension and is evacuated together with the solid waste, which has been removed at the accelerated solids separation stage.

Operating continuously allows to put away the concentrated dissolution (Fig. 9.3, p) formed at the evaporator, to use it as raw material for animal fodders, reclamation of polyphenols, and/or the extraction of chemical products. When the fumes are being condensed, the residual heat is used for the thermal necessities of the olive-mill (Fig. 9.3, 12) and, finally, it supplies hot water (Fig. 9.3, q) to the olive-mill. As for the olive cake (Fig. 9.3, f), the stones are separated (Fig. 9.3, g) from the pulp (Fig. 9.3, q), the stones are used as heater's fuel (Fig. 9.3, 5), the ashes (Fig. 9.3, h) come back to the ground as mineral fertilizer, the pulp (Fig. 9.3, q) is mixed with the solids from OMWW (Fig. 9.3, 8) and are subjected to a process of centrifugation (Fig. 9.3, 11) to extract the oil (Fig. 9.3, d), which has been kept almost completely in the OMWW-solids and in the olive cake. The solids without oil (Fig. 9.3, r) obtained at the oil reclamation stage (Fig. 9.3, 11), are useful in a process of composting (Fig. 9.3, 13), to make fodder (Fig. 9.3, 14), or as boiler fuel. The process is claiming to have zero waste and most of the oil lost in OMWW and solid waste from the mill is recuperated. No chemical additives are used and the oil is pure.

Atanassova D. et al. (2005b) applied ultrasonic irradiation to reduce the antioxidant activity of OMWW and 2POMW. This process comprises cyclic formation, growth, and subsequent collapse of microbubbles occurring in extremely small intervals of tie, and release of large quantities of energy over a small location. Sonochemical degradation in aqueous phase involves several reaction pathways and zones such as pyrolysis inside the bubble and/or at the bubble–liquid interface and hydroxyl radical-mediated reactions at the bubble–liquid interface and/or in the liquid bulk. Sonication of diluted samples was conducted at ultrasonic frequencies of 24 and 80 kHz, an applied power varying between 75 and 150 W and liquid bulk temperatures varying between 25 and 60°C. At the specified conditions, the reduction in antioxidant activity was found to increase with decreasing temperature and increasing power and frequency. Addition of NaCl in the samples also appeared to enhance reduction.

A three-step process comprising adsorption–concentration, catalytic hydrogenation, and regeneration on a fixed bed of adsorbent-catalyst was investigated for the removal of polyphenols from OMWW (Richard D. and Delgado-Nuñez M. de Lourdes, 2003). Tyrosol was taken as representative of the polyphenols present in OMWW. A ruthenium-activated carbon catalyst was used to evaluate the catalytic hydrogenation of the model phenol, tyrosol, under various temperatures (313–353 K) and pressure (0.4–4.0 MPa) conditions, and the Langmuir-Hinshelwood model was used to account for the results. At 353 K and 1 MPa, total hydrogenation of 0.042 mol/kg tyrosol was achieved after 3 h.

WO03000601 (2003) describes a process for purifying several types of wastewater generating at the various stages of the olive processing comprising the following steps:

  i.   homogenization of the process waters such as OMWW, washing water, rinsing water, etc. (agitation or aeration);
 ii.   homogenization of waste lye-water and/or waste oxidation water used for debittering and blackening olives (agitation or aeration);
iii.   degreasing of the homogenate;
 iv.   neutralization (optional addition of H_2SO_4, HCl, NaOH, etc.);
  v.   flotation (poly(aluminum chloride) and cationic polyelectrolyte);
 vi.   sand bed filtration;
vii.   active carbon filtration;
viii. sand bed filtration;
 ix.   ozonation (50–500 g O_3/m^3 depending on the COD value the residue);
  x.   desalination.

The purification system claims to produce completely disinfected, clear, odorless water that complies with all legislative standards. Moreover, the proposed system is economical, fast, effective, easy to manage, and perfectly compatible with the environment.

Boari G. and Mancini I.M. (1990) studied the problem arising from OMWW in Apulia where, during the olive-milling season, organic pollution exceeds that from domestic use by a factor of three. Preliminary research allowed the estimation of organic load per ton of milled olives, and the comparison of the effectiveness of feasible treatment processes (in particular, the anaerobic process and the effect of sedimentation, coagulation followed by aeration). The results have been utilized in the Water Reclamation Plan of the Apulia Region (WRP). This Plan permits the discharge of OMWW into public sewers only when its contribution is less than 20% of municipal wastewater's organic load, or provides the transport of a controlled amount to treatment facilities, over a period ranging from 100 to 300 days. Results of full-scale and pilot biological treatment plants for combined municipal wastewater and OMWW are reported, together with the main project parameters. Anaerobic processes are more economical but their successful operation needs to be confirmed on full-scale plants.

A bleaching process combining clayey soil (7%) and hydrogen peroxide (0.5%) allows the decolorization of OMWW and the elimination of polyphenols. At pH = 6.7, the bleaching led to about 87% reduction of polyphenols and 66%

reduction of COD. The structure of clay and its concentration in iron salts has an effective adsorbent and catalytic effect on the removal of the majority of polyphenols (Oukili O. et al., 2001).

Tsonis S.P. (1997) used OMWW as carbon source in post-anoxic denitrification. A study was undertaken to evaluate the efficiency of applying OMWW as a non-nitrogenous external carbon source in the second anoxic stage of a five stage modified Bardenpho system for nutrients removal in order to assure consistently very low concentrations of total nitrogen (well below 3 mg/l) in the treated effluent. Addition of OMWW was found acceptable only up to 50 mg COD of mill waste/l of wastewater fed to the system because at higher additions color problems in the treated effluent were encountered. The required dosage of OMWW was found to be in the range 4.6–5.4 mg COD/mg $N–NO_3$ removed. Operation with OMWW gives at the same time higher removal of phosphorus. Addition of physico-chemically pretreated OMWW with lime to the second anoxic tank at a rate of 22–45 mg COD/l of municipal type wastewater fed (ratio of volume of OMWW added to the volume of the municipal type wastewater fed 1:1000–1:2000) resulted in a treated effluent with total nitrogen below 3 mg/l and soluble phosphorus well below 1 mg/l.

Another minor process is cryogenesis (Franzione G., 1986). Except the high cost, cryogenesis has the problem that ice crystals of the water of OMWW trap the dissolved phenols and salts.

Part III
Utilization

Chapter 10

Uses

Olive fruit processing produces large amounts of by-products, including liquid and solid wastes arising from olive oil extraction and the production of table olives. The disposal without any treatment of the wastewaters, arising mainly from the olive-mill (OMWW) and to a lesser degree from the table olive industries, is known to cause serious environmental problems. A wide range of technological processes are available nowadays for reducing the pollutant effects of OMWW and for its transformation into valuable products, the most suitable procedures being found to involve recycling rather than detoxification of this waste. Moreover, antipollution legislation has been forcing the utilization of OMWW as an alternative to disposal. Thus, in view of the current need for upgrading by-products at all stages of the olive oil industry (Demichelli M. and Bontoux L., 1996), increasing attention has been paid to discovering a use for OMWW.

OMWW is composed of vegetation water, soft tissues of the olive fruit, and water used at the different stages of oil production. The vegetation water in the olive fruit represents 45–50% of the weight of the fruit and a volume of up to 7 million m^3 is produced each year, in addition to the water added during the olive oil extraction process. The organic matter content is 15–18%, which implies an annual production of 1–1.2 million tons of substance that may be used as raw material either to recover valuable natural constituents/by-products or as a culture from which to develop microorganisms for new products (Fiestas Ros de Ursinos J.A. et al., 1996). Since the early 1970s, the pressure of pollution has promoted studies on conversion of OMWW to useful products like fertilizer, animal feed, a medium of fermentation for single cell protein (SCP) and enzymes, production of alcohol (ethanol, butanol, mannitol), biogas, etc. Lately, several techniques have been developed for the efficient and economic extraction of antioxidants[27]. Fig. 10.1 shows schematically some fields for the end products of treated OMWW.

---

[27]The applications of OMWW have been reviewed by Ramos-Cormenzana (1986) and widely discussed at the *Granada Olive Oil Conference* held in 1995, in Granada, Spain.

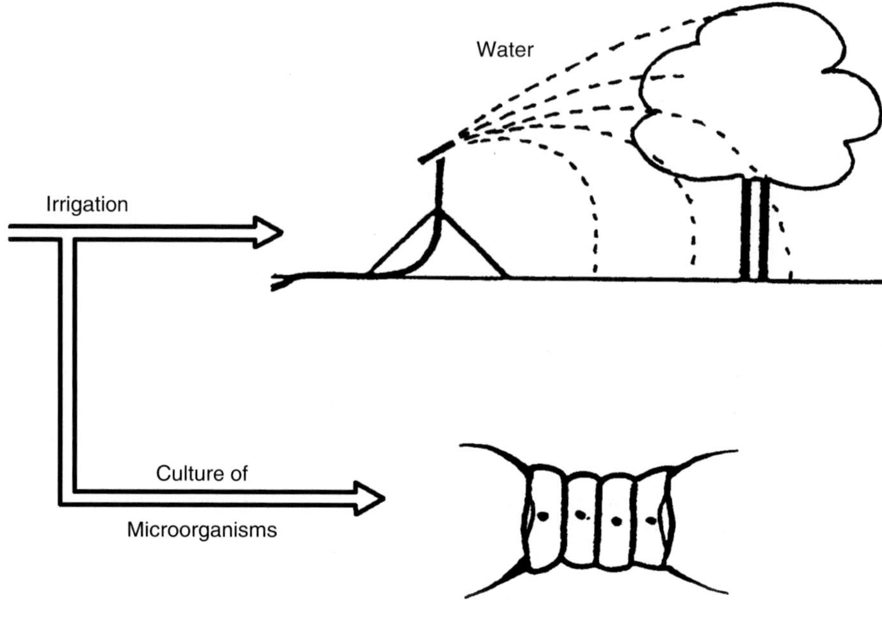

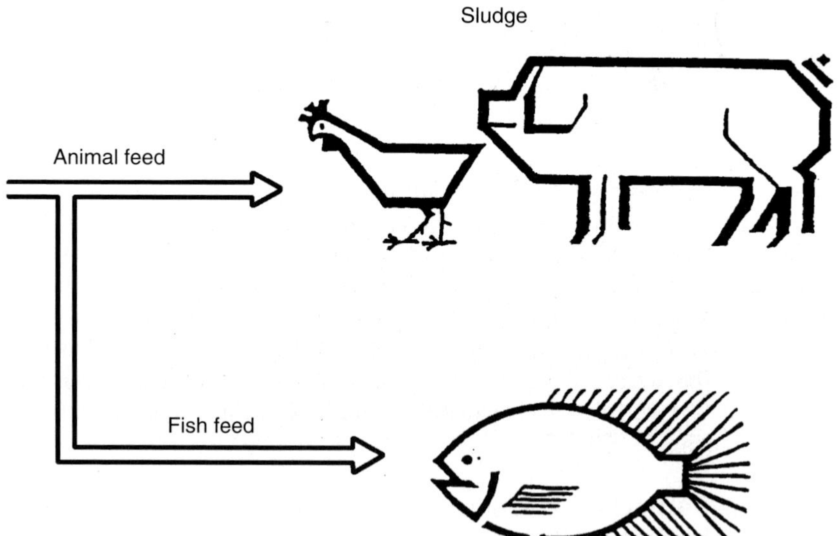

Fig. 10.1. Usage for treated OMWW and sludge OMWW (W9211206, 1992).

The use of olive-mill liquid wastes is known since antiquity. A predecessor of OMWW is amurca. Amurca or olive oil lees, the watery bitter-tasting liquid residue obtained when the oil is drained from compressed olives, had many uses in agriculture. Amurca has been described by several ancient authors as having a universal remedy against insects, weeds, and plant diseases (Columella, Pliny the Elder, Cato, and others). Amurca has also been used for smoothing out plaster floors, oiling leather, etc.; however, many of these uses are not exactly applicable for the modern day[28].

Solid olive-mill wastes have been traditionally used as fuel, both domestic and industrial, and animal feed — see Fig. 10.2. Other uses include use as fertilizer and construction material.

The use of the two-phase processing technique generated a new by-product that is a combination of liquid and solid wastes (2POMW). In Spain, a massive change from the traditional three-phase to the new two-phase process has taken place, and large volumes of this waste (~4.5 million tons per year) are already produced (Junta de Andalucía, 2002; Alburquerque J.A. et al., 2004). The massive production of this by-product intensified the efforts to find possible uses and diminish its environmental impact (EU project: FAIR CT96-1420 "IMPROLIVE").

Fig. 10.2. Piles of olive residue in a warehouse at the excavation site of Karanis in the Fayoum region of Egypt. This olive residue, which was once thought to be bread, was used for animal feed and fuel. (http://www.umich.edu/~kelseydb/Exhibits/Food/text/farm2.html#oil).

---

[28]http://www.oliveoilsource.com.

Olive tree culture by-products include unpicked fruit, pruning, and harvest residues. Historically pruning brush was used for energy on small landholdings, but with availability of alternate energy sources and higher labor costs, only larger wood is still used for burning. Smaller wood and foliage can be chopped and incorporated into the soil or burned. Ashes can be spread on the fields to release potassium and trace elements. Increased production of olives has not been shown to justify expenses involved with such procedures (Amirante P. and Pipitone F., 2002). Using pruning brush for fuel or animal feed seems more promising.

It is also worth mentioning the wastes generated from olive oil use. The main by-product of olive oil use is waste cooking oil. Used cooking olive oil constitutes a waste which is included in the group of urban and municipal wastes. Its main use at present is in animal feed (although controversial) and, in a much smaller proportion, in the manufacture of soaps, biodegradable lubricants (although not recyclable), combustion (recovery of energy in industrial plants), or even fuel for engines (Dorado M.P. et al., 2004). An unusual by-product of the cosmetic use of olive oil in the antiquity was *gloios* — see Fig. 10.3.

Use as Fertilizer/Soil Conditioner

OMWW contains a high organic load, substantial amounts of plant nutrients (3.5–11 g/l of K_2O, 0.06–2 g/l of P_2O_5, and 0.15–0.5 g/l of MgO) and is a low cost source of water, all of which favor its use as a soil fertilizer or organic amendment to the poor soils that abound so much in the countries where it originates (Cegarra J. et al., 1966a,b; Catalano L. and Felice M. de, 1989; Nunes J.M. et al., 2001). Direct application of OMWW to soil has been considered as an inexpensive method of disposal and recovery of their mineral and organic components (Di Giovacchino L. et al., 1990, 1996, 2001, 2002) — see Chapter 8: "Biological processes", section: "Irrigation of agricultural land/Land spreading". The amurca of the ancients was recommended as a fertilizer for olive trees (Cato, XCIII), vines, and fruit trees (Columella, XI, 2; "Geoponika"[29], II, 10), although these latter sources suggested that amurca used for this purpose must be free from salt.

Furthermore, OMWW has a beneficial effect on soil aggregation, soil structure stability, and the hydrodynamic properties of sandy soils and it could be used as a soil conditioner for reducing evaporation in arid and semi-arid areas. Commercial soil conditioners are expensive and the application of such products in developing

---

[29]This book is a compilation of agricultural writings collected and published in the 6th or 7th century A.D. by Cassianus Bassus. For the most part little is known about the individual authors of the various sections except that many lived during the period 200 B.C. to 200 A.D. As with much of the writing of this time, the "Geoponika" is an undiscriminating collection of earlier works, many of which have been lost and can no longer be examined. The content is of great interest to anyone studying agricultural history.

countries such as Tunisia was not always successful (Mellouli H.J., 1996, 1998; Mellouli H.J. et al., 2000).

However, the use of OMWW as a fertilizer is controversial. Many authors have observed phytotoxic effects in plants when this waste is used directly as an organic fertilizer and have, therefore, warned against its direct application (Zucconi F. and Bukovac N.J., 1969; Jelmini M. et al., 1976). Such negative effects are associated with its high mineral salt content, low pH and the presence of phytotoxic compounds, especially phenols — see Chapter 3: "Environmental effects",

Fig. 10.3. Standard hygiene kit of the ancient athlete depicting an *aryballos* (type of oil container) and three *strigils* for scrapping off the *gloios* (mixture of dust, olive oil, and sweat) from the body (The National Archaeological Museum of Naples, Italy) (photograph © Schäfer-Schuchardt H., 1998).

Olive oil was used by Greeks and Romans as a cleaning agent. Olive oil was rubbed on the body to mix with oils and dirt present on the skin. This mixture was then scraped off with a *strigil*, a sickle-like instrument. The resultant scraping (mixture of dust, oil, and sweat) was called *gloios* (γλοιός) and it was also used for cosmetic and medicinal purposes. In fact, the *gloios* that came from athletes was especially prized for its curative properties. Galen*, believed in the efficacy of *gloios* as a medicine. He claimed to have seen *gloios* reduce "cancerous tumors". About a century or so before Galen, Dioscourides** and Pliny wrote about the same substance, which Romans called *strigmentum*. The inscription on a column found at Beroea in Macedonia, Greece (SEG 27.261, about 175–170 B.C.) describes a law concerning the gymnasium and contains one clause outlining provisions for the sale of *gloios* from the bodies of those who were allowed to exercise (Gauthier Ph., and Hatzopoulos, 1993)***. The contractor for the revenues from the *gloios* is to provide a *palaistrophylax* (a guard), who is to serve under the orders of the gymnasiarch and is to be subject to whipping if he misbehaves. The *gloios* was then heated up and impurities like pieces of skin, hair, or sand were taken out. The precious substance will be sold to doctors for use in potent medicine.

*Galen, Medical writer of the Roman Empire who came originally from Asia Minor (modern Turkey) and lived around the 2nd century A.D.

**Dioscourides (also known as Pedanius Dioscorides) Greek writer, probably lived between 40 B.C. and 90 B.C. in the time of the Roman Emperors Nero and Vespasiano. For almost two millennia the medical botanist Dioscourides was regarded as the ultimate authority on plants and medicine. In his treatise *"De Materia Medica"* more than six hundred vegetable, animal, and mineral remedies are described. The plant descriptions were often adequate for identification, including methods of preparation, medicinal uses, and dosages laid the basis for pharmacology.

***Gauthier Ph. and Hatzopoulos (1993) *La loi gymnasiarchique de Beroia*, MELETHMATA 16, (232 pages, 1 map), Athens: Published by K.E.P.A. ISBN 960-7094-82-4.

Fig. 10.3. (*continued*)

section: "Effects on soil". For this reason it would be necessary to carry out a previous treatment in order to utilize it (Fiestas Ros de Ursinos J.A., 1986b).

Several studies have been carried out on the direct application of OMWW to soil as fertilizer either as a fresh liquid or sludge and their effect on soil characteristics and crop production (Morisot A., 1979; Morisot A. and Tournier J.P., 1986; Fiestas Ros de Ursinos J.A., 1986b; Saiz-Jiménez C. et al., 1986; Briccoli-Bati C. and Lombardo N., 1990; García-Rodríguez A., 1990; Di Giovacchino L. et al., 1990, 1996, 2001, 2002; Saviozzi A. et al., 1991, 1993; Levi-Minzi R. et al., 1992; Riffaldi R. et al., 1993). Riffaldi R. et al. (1993) performed a laboratory experiment during a 40-day incubation period in order to evaluate changes in organic and inorganic compounds of soil amended with two doses of sludge obtained from OMWW. Differences between the amounts of organic components of the amended soil and those of the control, although related to doses and sampling time, disappeared at the end of the experimental period. On the contrary, the inorganic anion content was still different for the various processes, which suggest, especially for NO_3^- and SO_4^{2-}, a transient inhibition in the soil–sludge system. A germination test, carried out on the soil amended with different doses of sludge, indicates that after about 20 days even the soil containing the highest dose of sludge did not show toxicity any longer.

Saiz-Jiménez C. et al. (1986) studied also the fertilizing properties of the sludge from OMWW. The humic acid fraction of sludge obtained from OMWW after disposal in isolated lagoons consisted of polysaccharides, proteins, lignins, and relatively high amounts of C_{16}- and C_{18}- fatty acids. Although the composition of this material is different from soil humic acids, it was concluded that the sludge has good soil fertilizer properties.

In general, if soil characteristics are appropriate, OMWW is probably best used as fertilizer since this is an inexpensive method of disposal, and important advantages may be derived from soil fertility, among which the following can be cited (Fiestas Ros de Ursinos J.A., 1986b; Tomati U. and Galli E., 1992):

- the effective use of plant nutrients contained in the waste, mainly K, but also N, P, and Mg;
- a low cost source of water, taking into account the increasing scarcity of water resources for irrigation and;
- supply of organic matter, which enhances microbial activity and improves the physical and chemical properties of soil.

As disadvantages, the following may be cited:

- the high content of mineral salts and the presence of organic compounds, such as fatty acids and polyphenols, both factors being detrimental to soil fertility and;
- the difficulty of storing and disposing of the large amounts of this liquid waste which is produced in short, and often rainy, period of time.

According to the above considerations the following recommendations were made (Fiestas Ros de Ursinos J.A., 1986b):

- OMWW should be applied at a certain distance from trees,
- doses should not exceed $30\,m^3/ha\cdot y$ (those obtained by the three-phase centrifugation system),
- applications should be made in a stepwise fashion,
- at least one month should elapse between the application and the sowing for seeding yearly crops and,
- OMWW must never be added when the crops are in sprouting period.

In particular, OMWW can improve the productivity of the olive trees, if it is used at doses less than (i) $50\,m^3/ha \cdot y$ from a three-phase centrifugation system and (ii) $25\,m^3/ha \cdot y$ from a traditional press system, provided that there is a delay of at least 45 days between the application of the waste and the resumption of sap rising in olives (Le Verge S., 2004).

The use of OMWW for two years out of three can replace a large part of the employed fertilizers. With reference to a traditionally used mineral fertilizer, the use of OMWW can bring an annual saving on fertilizer's costs of 50 €/ha in non-irrigated and 70 €/ha in irrigated olive orchards. To this, one must add the saving on the costs of spreading. In fact, the employment of OMWW permits the reduction of the mineral fertilizers by 550 kg/ha in non-irrigated and 800 kg/ha in irrigated olive orchards, which corresponds to a saving of 55 €/ha and 80 €/ha, respectively (the calculation is based on an hourly cost of 8.5 € for the material and 11.5 € for labor) (Le Verge S., 2004).

OMWW could also be used for the reduction of the contamination of ground-water by pesticide drainage (Cox L. et al., 1997) — see also Chapter 3: "Environmental effects", section: "Leaching".

By using composting technologies, it is possible to transform either fresh OMWW or sludge from pond-stored OMWW mixed with appropriate plant waste materials (carriers) into organic fertilizers (composts) with no phytotoxicity to improve soil fertility and plant production, the process involving the microbial degradation of the polluting load of the wastes — see Chapter 8: "Biological processes", section: "Composting". Results of field and pot experiments using OMWW-composts to cultivate horticultural and other crops have shown that yields obtained with organic fertilization are similar, and sometimes higher, to those obtained with a balanced mineral fertilizer. A comparison between the macro- and micro-nutrient contents of plants cultivated with organic or mineral fertilizers did not generally reveal important differences. However, the cases of iron and manganese are worth mentioning as their bio-availability may be linked to the soil humic complexes originated by the OMWW organic fertilizers (Cegarra J. et al., 1996a). Tomati U. et al. (1996) produced a high quality compost from OMWW, characterized by a considerable presence of nutrients, mainly organically bound nitrogen (1.5–3%), a good level of humification (degree of humification = 78%; humification index = 0.28), and by the absence of phytotoxicity. The agronomic value of a compost thus obtained was assayed both by the "crop test" and following the plant-soil system as influenced by compost supply. Field experiments performed on maize showed that compost, when supplied before sowing in amounts of 60–90 tons/ha (equivalent to a manuring on the basis of organic matter) is able to reduce the need for chemical fertilization. The same quantity supported the nutritional need of rye-grass and horticultural plants. Compost supply enhanced both soil oxygen consumption and nitrogen fixation in the open field. An improvement of activities in the plant-soil system was made evident by pot trials. However, there is limited research on OMWW composting and the use of such compost for crop production (Vlyssides A.G. et al., 1989; Cabrera F. et al., 1993; Amirante P. and Di Renzo G.C., 1991; Montemurro F. et al., 2004).

Composted olive-mill waste could be considered as an appreciable low priced organic ingredient for pot ornamentals growing media, while simultaneously disposal of this waste in a friendly-to-the-environment way could be obtained. In olive oil producing areas usually ornamental horticulture is also developed because of the climate type. Pot ornamentals are mostly grown in soilless growing media that contain the rather expensive peat mixed with an inorganic material. Papafotiou M. et al. (2004) evaluated the amount of peat that could be replaced by olive-mill waste compost in the commonly growing medium of *Euphorbia pulcherrima* (poinsettia), that is peat with perlite. The results of this study suggest that olive-mill waste can replace up to 25% of the peat in the medium with perlite. The quality of the plants produced in this medium was as good as that of the control. Increasing replacement of peat by olive-mill waste compost induced a gradual decrease of the plant height, bract number, and node number where the first bract was initiated. All the above parameters of growth were significantly reduced even when 25% of peat was replaced, except the bract number that was significantly reduced in case of 50%. The decrease of plant height with no simultaneous effects on bract number and flowering could be particularly interesting, as it could contribute to the reduction of the amounts of plant growth retardants that are routinely employed for height control in commercial cultures of poinsettia, and to the decrease of production cost. Higher concentrations of olive-mill waste compost are not recommended for poinsettia production, as they induced late flowering expressed as decreased bract number at the time that control plants and plants grown in 12.5% of olive-mill waste compost were ready for the market.

The co-composting of OMWW and olive cake can yield very good organo-humic soil fertilizer (Vlyssides A.G. et al., 1996). Bouranis D.L. et al. (1995) studied the effectiveness of an organic soil conditioner produced from the co-composting of OMWW with olive cake — see also Chapter 8: "Biological processes", section: "Composting". The concentration of 25% w/w of this material into the conditioner-soil mixtures appears to be the maximum level for the cultivation of tomato plants. The plants grown on this conditioner-soil mixture were 1.52–8.5% times larger than those grown on a sandy loam soil. The pure conditioner cannot be used as a substrate for the growth of tomato plants. The water-holding capacity of the conditioner was almost two times higher than that of the pure soil and remained almost stable for temperatures between 8–40°C. The apparent density of the conditioner was 0.5 times smaller than that of the pure soil. With increased application rate of the conditioner to the soil, there was a decrease in the pH, an increase in the specific conductivity and an increase in the ammonium-nitrogen and phosphorous concentration of the mixture.

ES2037606 (1993) discloses a fertilizer prepared by mixing OMWW with an organic substrate (cellulose paste) or mineral substrate (sepiolite or vermiculite) with optional addition of N, P, and K ingredients to increase its fertilizing value. The fertilizer mixture is milled and fermented in heaps with control of the temperature inside the heap.

ES2002555 (1988) discloses a manure made by the fermentation of mixtures of (i) animal manure (liquid and solid excrement from cattle, sheep, pigs, goats, poultry, rabbits, etc.), (ii) earths which have been used for decolorizing vegetable oil, and (iii) waste from olive processing. The mixture undergoes aerobic and anaerobic fermentation to produce the fertilizer. Organic fertilizers have also been made from agricultural, forestry, industrial, and urban wastes, e.g. vegetable residues, organic fractions of solid urban or industrial residues, sludge from purification of effluent waters, or olive sludge, by adding a liquid fraction consisting of OMWW; after mixing and centrifuging, both fractions are fermented under aerobic conditions, subjected to final milling and sieved (ES8402554, 1984).

2POMW could also have a role as fertilizer, providing that it can be detoxified, for example through bioremediation by breaking down the toxic phenolic compounds.

The effects of the application of 2POMW on plant growth and in physical and chemical characteristics of soil have been evaluated by Jones N. et al. (1998). Considering the results of this study, the utilization of 2POMW as organic amendment appears possible. This practice could provide an interesting solution for this by-product leading to a sustainable development of olive-mill industry.

ES2103206 (1997) discloses an industrial process for the treatment, recycling, and conversion of OMWW and/or 2POMW into pure organic fertilizers. In the 1st stage the contaminating wastes are subjected to a biochemical treatment, passing subsequently into a tank where the OMWW/2POMW waste is absorbed or retained by lignocellulosic materials and undergoes anaerobic degradation; the 2nd stage comprises degradation by bacterial fauna, to give rise to phenolic-lignoprotein substances and stable acids. When the anaerobic stage is complete, the organic materials are extracted from the tank, formed into longitudinal stacks, into which air enters from the sides, creating a draught and initiating the rapid growth of thermophilic aerobic microorganisms (3rd stage) which raise the temperature of the OMWW/2POMW waste held into the lignocellulosic materials to 75°C, bringing about their thermal evaporation (4th stage). The stacks are periodically turned, oxygenating them to reactivate the process of bacterial thermogenesis (5th stage). Once the process is complete, the OMWW/2POMW waste has been converted into pure organic fertilizers.

The degradation product of OMWW or 2POMW by the so-called M.A.T.Re.F.O. technology (EU project: LIFE00 ENV/IT/000223 "TIRSAV") can be used as a soil amendment and/or organic fertilizer in olive culture and crop cultivation in general — see Chapter 9: "Combined and miscellaneous processes". The end product provides a series of beneficial effects on the land including: increased microporosity, hence improved oxygenation of the surface profile of the soil in which root growth and microbial activity occur (cultivation layer); increased stability of aggregates — an important factor preventing erosion phenomena and the formation of surface crusts due to rain action; better hydraulic retention of the land due to its increased microporosity; greater bio-availability of micro-elements for vegetal nutrition due to the known chelating and/or complexing capacity of the organic fraction of the soil which, due precisely to its chemical characteristics, tends to inhibit chemical insolubilization processes on such micro-elements — a phenomenon which generally occurs in calcareous soil. Lastly, the application of the organic substance makes it possible to reach the objective of contrasting desertification in progress in increasingly larger areas of the Mediterranean, especially where olive culture and mill processing activities are more widespread.

The fertilizing effect of microbially treated OMWW (by anaerobic and/or aerobic processes) has been tested on a number of plants. Five types of OMWW with and without rock phosphate, microbially treated with *A. niger* or not, were tested in a soil-*Trifolium* system for their fertilizing ability. The beneficial effect of microbially treated OMWW was more evident during the first crop cycle. Best plant growth response and P uptake were observed in mycorrhizal plants grown in soil amended with fungal treated OMWW with rock phosphate (Vassilev N. et al., 1998).

In another study, the degradation products of OMWW by batch and repeated-batch processes with a passively immobilized strain of *A. niger* were tested for glasshouse-grown durum wheat (*Triticum durum* Desf). Plant growth and yields were compared to those achieved using untreated wastes, and the greatest effects (seed biomass, spike number, kernel weight, and harvest index) were observed using the effluent treated by the repeated-batch process (Cereti C.F. et al., 2004).

OMWW, by treatment with appropriate bacteria in an appropriate environment, can be converted into a natural organic conditioner for enriching natural and chemical fertilizers of any kind (biological and mineral, liquid, solid, and biocompost-based fertilizers), which obviates the drawbacks of conventional conditioners. The known natural organic conditioners for agricultural use consist of humic extracts — frequently obtained from peat by chemical methods; such conditioners, rich in humic and fulvic acids, are used increasingly owing to the extraordinary increase in effectiveness of the agricultural treatments performed in their presence, for example in leaf spraying, herbicidal treatments, wet dressing of seeds (i.e. the application of manure or fertilizers to seeds), in hydroponics, or the like. However, conventional conditioners, owing to their origin, are biologically inert, in that chemical extraction damages and reduces the above mentioned active principles, and once they are spread on the soil, they are consumed by the plants; accordingly, the humic acids become depleted due to consumption and the effectiveness of the product decreases in the course of time. On the other hand, a liquid additive produced from OMWW is biologically active over time, in the sense that the humus-forming process continues on the field after spraying, increasing the effectiveness in a quantitative sense and terms of its duration. ITBO950012 (1996) discloses a liquid additive obtained by aerobic bacterial proliferation in OMWW containing active ferments of high bacterial count (from 500 million up to 2.5 billion bacteria per milliliter) as well humic and fulvic acids. The process for converting OMWW into a liquid additive was improved by the daily addition of organic nitrogen-containing compounds and minerals required for the vital needs of the bacteria. The time required for the conversion of OMWW into a product in which 30% of the organic carbon content is contained in humic material is typically 96 h. Such product applied to the soil through spraying or injection, provided continuity in the conversion of the organic matter into humic substances through the action of such bacteria. The liquid additive, by virtue of its richness in aerobic bacteria, is also useful in the agronomic bacteriological restoration of ecosystems, in producing selection in the soils, in purification plants instead of the expensive lyophilized bacteria, and in the prophylaxis against enterobacteria and pneumobacteria (ITBO950012, 1996).

Use as Herbicide/Pesticide

Since antiquity, the olive-mill liquid wastes were known to have herbicide and pesticide properties and amurca is considered a precursor to modern pesticides in the classical period (Smith A.E. and Secox D.M., 1975; Levinson H. and Levinson A., 1998).

Both Cato (XCI) and Varro (I, 51) recommended that threshing floors be made from a mixture of soil and amurca so that weeds would not grow. Columella, Palladius[30], and the "Geoponika"[31] also made mention of this practice. In this instance, the amurca and soil mixture seems to have dried to a hard plaster-like finish which was thus impervious to weeds. However, Varro specifically noted (I, 51) that amurca was poisonous to weeds, ants, and moles. Varro also observed (I, 55) that where the amurca flowed from the olive presses onto the fields the ground became barren, and he went on to state that amurca was poured around olive tree roots and "wherever noxious weeds grow in the fields". This latter use must be one of the earliest references to a specific weed killing preparation. Theophrastus (IV, 16) wrote that pouring olive oil over their roots could kill trees, young trees being more susceptible to this treatment than mature ones.

The treatment of seeds with amurca was considered to be useful against insects and other animal pests. Virgil[32] (I, 90) recommended that all seeds be soaked in a mixture of amurca and native soda before planting so that greater yields would be forthcoming. Columella (II, 10) held that this latter treatment was successful also in reducing attack by weevils in the mature seed.

Amurca was also used for insect control. Thus, it is mentioned (Columella, II, 9) that unsalted amurca when applied to the furrows at the outbreak of an infestation would drive away the "destructive creatures", while applications of amurca and red earth, possibly sandarach (the red arsenic of Greeks), would keep vines free from beetles and ants (Columella, IV, 26). When mixed with soot, gnats could be driven away, and locusts were dispelled by using amurca containing extracts of cucumber or lupins, while caterpillars on cabbages were killed by an application of amurca and ox urine (Palladius, I, 122, 125, 135, 136). In addition to the above methods fumigation

---

[30]Palladius Rutilus Taurus Emilianus, Roman author who lived and wrote around the 4th century A.D. Very little is known about Palladius except that his books *On husbondrie* and *De re rustica* obtained some celebrity.

[31]For the citations to the works of Columella, Palladius, Pliny, Theophrastus, Varro, Virgil, and the "Geoponika" the Latin numeral refers to the book number and the arabic number to the chapter. In the case of the writings of Palladius and Virgil, the arabic numerals are to the stanzas and paragraphs, respectively. The works of Cato are not divided into books; thus the Roman numbers associated with these references are to the numbered sections.

[32]Virgil or Vergil (Publius Vergilius Maro, 70 B.C–19 B.C.), greatest of Roman poets; born in Andes dist., near Mantua, in Cisalpine Gaul. The poet's boyhood experience of life on his father's farm was an essential part of his education. In 41 B.C. Virgil went to Rome, where he became a part of the literary circle patronized by Maecenas and Augustus and where his *Eclogues*, or *Bucolics*, were completed in 37 B.C. In these poems he idealizes rural life in the manner of his Greek predecessor Theocritus. From the *Eclogues*, Virgil turned to rural poetry of a contrasting kind, realistic and didactic. In his *Georgics*, completed in 30 B.C., he seeks, as had the Greek Hesiod before him, to interpret the charm of real life and work on the farm. His perfect poetic expression gives him the first place among pastoral poets. For the rest of his life Virgil worked on the *Aeneid*, a national epic honoring Rome and foretelling prosperity to come.

procedures were carried out. The mixture of amurca, sulfur, and bitumen was heated in a copper vessel and the resulting gluey substance applied to the trunks and branches of vines for control of caterpillars (Cato, XCV; XVII, 47). Both Pliny the Elder (XVII, 47) and Palladius (I, 127) remark that the smoke from the boiling mixture of amurca, bitumen, and sulfur (Cato, XCV) was successful in preventing caterpillars from attacking vines. A remedy for blight (Pliny the Elder, XVIII, 45) was to sprinkle the infected plant (vines) with amurca.

Protection of store grain by amurca has often been recommended by various Roman agriculturists from the 2nd century B.C. until the 4th century A.D., indicating the usefulness of this treatment. Amurca, when incorporated into threshing floors, was also helpful in keeping ants away (Cato, XCI; Varro, I, 51) and, on being made into a paste with straw and applied to granary walls, appeared to be instrumental in keeping the grain free from weevils (Cato, XCII). In his well-known treatise *De agri cultura* (XCII), Cato stated that granary insects and mice can be prevented from damaging stored grain by applying amurca as follows: "mix amurca with a small amount of ground straw and clay; knead the mixture until it turns into a viscous paste. Smear this paste on the interior walls, floor, and ceiling of the granary and sprinkle aqueous amurca on the dried coatings. When the latter are dried up, deposit cool grain in the treated store. Granary pests will be incapable of damaging grain kept in such stores". Moreover, Varro, Pliny the Elder, and Palladius improved the above description by supplementing amurca with ground chalk and an insectistatic crushed foliage of coriander (*Coriandrum sativum*), fleabane (*Inula conyza*), or wormwood (*Arthemisia absinthium*) and recommended distributing those mixtures in the stored grain (Beavis J.C., 1988; Levinson H. and Levinson A., 1998).

It is likely the above blends of amurca to have acted by clogging the wall crevices and cracks, which could have otherwise served as hiding and oviposition niches for pest species, as well as by suppressing substances which are capable of repelling and suppressing pest populations. A revived interest in pest-averting procedures practiced in antiquity may be worthwhile in view of the destructive side-effects of certain pesticides on the environment as well as the alarming increase and spread of pesticide resistance in storage insect species (Levinson H. and Levinson A., 1998).

Amurca was also used as a means of protecting clothes from moths and as a preservative for dried fruits (Cato, XCVIII, XCVIX).

The role of amurca is difficult to assess though it does seem to have been a universal remedy against insects, weeds, and plant diseases. The composition of amurca is unfortunately difficult to deduce. Pliny the Elder mentioned (XV, 4) that amurca was a bitter, watery liquid and this bitterness is now known to be chiefly due to the easily hydrolysable glycoside oleuropein (Fernández-Diez M.J., 1971) whose structure has been elucidated (Inouye H. et al., 1970). It is not known whether this glycoside possesses any pesticidal properties. The amurca would also have contained traces of phytocidal, insecticidal, and fungicidal glyceride oils as well as oleic acid. Salt was sometimes added to the olives prior to pressing which may have resulted in additional phytotoxic properties (Columella, XII, 52; Palladius, XI, 16). Thus, it is impossible to state whether the ancient amurca was effective against some weeds, or the salt associated with it was the control agent. A further and complicating factor was that during the preparation of amurca, as described by Varro (I, 64), the liquid was boiled to about two-thirds of its original volume in a copper vessel. In this way not only would the amurca become contaminated with traces of copper, but also a number of extra products could be formed by hydrolytic processes. As copper salts are now known to be extremely effective against certain fungal diseases, it is possible that the amurca prepared in this manner contained fungicidally active amounts of the metal.

Capasso R. et al. (1994b) investigated the possible utilization of OMWW and its bioactive polyphenols in protection of olives against *Bactrocera oleae* (Insecta: Diptera: Tephritidae). Among the main polyphenols occurring in OMWW, catechol showed the most deterrent action on the oviposition of *B. oleae*; 4-methylcatechol was less active, whereas hydroxytyrosol and tyrosol were inactive. In contrast, synthetic *o*-quinone was found to be stimulant at 7.5×10^{-2} M. Two other synthetic derivatives of catechol, diacetylcatechol, and guaiacol (2-methoxyphenol), were also deterrent, suggesting these compounds undergo a biochemical transformation into catechol by means of the bacterial symbionts of *B. oleae*. OMWW and their phenolic extracts showed deterrence only when highly concentrated, while natural olive juice was strongly deterrent. Experiments carried out to evaluate the effect of olive juice and catechol on the fecundity of *B. oleae* showed that they strongly reduce this function.

Capasso R. et al. (1995) investigated also the possibility of using OMWW as a pesticide to protect the olive plant against the knot disease caused by *Pseudomonas syringae* pv. *savastanoi* and the tomato plant against the serious disease caused by *Corynebacterium michiganese* (Gram-positive). Among the main polyphenols present in OMWW, 4-methylcatechol proved to be the most toxic to *P. syringae* pv. *savastanoi* at 10^{-4} mol/l, and also demonstrated bactericidal activity, while on *C. michiganese* was only slightly active; catechol and hydroxytyrosol were less active on *P. syringae* pv. *savastanoi*, but inactive on *C. michiganese*; tyrosol was completely inactive on both bacteria — see Chapter 2: "Characterization of olive processing waste", section: "Antimicrobial activity of OMWW" and Chapter 3: "Environmental effects", section: "Effects on soil biological properties".

In the light of the obtained results, it is recommended that the use of raw OMWW should be avoided, since it causes leaf and fruit abscission (Fiume F. and Vita G., 1977) and is highly toxic on Hep2 human cells. Therefore, one can conclude that OMWW needs to be fractionated in order to separate and isolate the antibacterial catechol, 4-methylcatechol, and hydroxytyrosol. However, catechol and 4-methyl-catechol could not be used because they are toxic on Hep2 and phytotoxic (Bartolini S. et al., 1994). As a matter of fact, hydroxytyrosol is the only promising OMWW polyphenol, which could be used in agriculture on *P. syringae* pv. *savastanoi*, since it is not phytotoxic (Bartolini S. et al., 1994) and not toxic on Hep2 human cells.

Table 10.1. Liquid foliar fertilizer based on OMWW
(ES2139505, 2000)

Nitrogen	10.4%
Hydrogen	3.4%
Oxygen	14.0%
Sulfur	14.0%
Carbon	1.0%
Iron	0.2%
OMWW	(100-sum of above elements)%

ES2139505 (2000) describes a composition based on OMWW, to be used as a pest repellent, herbicide, and fertilizer for vegetative growth on tree trunks. Liquid foliar fertilizer acts as a pesticide and activator of sap circulation in ligneous plants, and contains nitrogen, hydrogen, oxygen, sulfur, carbon, and iron. Incorporating these elements or their compounds into OMWW produces mixtures with diverse effects. The new composition is shown in Table 10.1. Fertilizer-pesticide mixed with water up to 50% when applied in concentrated form to the trunks of ligneous plants causes falling out of the old bark, and exposes pests to inclement weather, in addition to repellent action of the product. The composition also inhibits vegetative growth on trunks and helps healing wounds in trunk, improving harvest results.

Boz Ö. et al. (2003) investigated the herbicidal effect of OMWW on some weed species (*Portulaca oleracea*, commonly known as purslane) in wheat, maize, and sunflower crops. In trials with maize and sunflower, OMWW was applied as an air-dried solid form at 3 and 4.5 kg/m^2. It provided an effectiveness level on *Portulaca oleracea* of 63–98%. In trials with wheat, OMWW was applied as solid and liquid forms, each at two different doses, namely 4.5 and 6 kg/m^2 (solid), and 5 and 10 l/m^2 (liquid). Solid OMWW provided a reduction in total weed coverage of 75 and 81% at doses of 4.5 and 6 kg/m^2, respectively. The weed coverage reduction by liquid OMWW was 39 and 62% with 5 and 10 l/m^2, respectively. Apart from 12–26% reduction of the number of germinating seeds, OMWW showed no toxic effects on maize and sunflower. Wheat was affected in the initial stages but no adverse effect was detected at harvest. It can be concluded that the herbicidal effect of OMWW may be considered as an alternative to chemical weed control in some important summer crops (maize and sunflower) and for most of the weeds in winter wheat.

The organic substrate produced from OMWW within the framework of the EU project: LIFE00 ENV/IT/000223 "TIRSAV" — see Chapter 9: "Combined and miscellaneous processes" — has shown to have a suppressant effect on gall-forming nematodes of the *Meloidogyne* genus, in addition to the numerous agronomic benefits that derive from the input of organic material into the soil. This effect is of particular interest in combating species of nematoid phytoparasites as an alternative to strategies that employ synthetic nematoid suppressants. Its use could, therefore, be considered in nematode-suppressant strategies having low environmental impact, aimed not so much at complete eradication of the parasite as at reducing population levels to non-detrimental levels.

Use in Animal Feeding and Human Consumption

Many authors have described the use of olive cake in animal feeding as air-dried (Zumbo A. et al., 2001), ensiled (Nefzaoui A., 1991; Hadjipanayiotou M. and Koumas A., 1996) and soda- or ammonia-treated feedstuff (Aguilera J.F. and Molina Alcaide E., 1986; Molina Alcaide E. and Anguilera J.F., 1991).

Olive cake is not attractive as an animal feed, in that it contains, on a dry matter basis, fiber (58%), crude protein (5.5%), lipids (3.5%), soluble carbohydrates (20%), and ash (13%) (Harb M., 1986). The nutritive value of olive cake is also very low: from 4.22 (untreated) to 6.46 (alkali-treated) MJ of metabolizable energy (ME)/kg dry matter (DM) (Molina Alcaide E. and Aguilera J.F., 1988; Nefzaoui A., 1991). Hadjipanayiotou M. and Koumas A. (1996) estimated that untreated olive cake worth 3.85 MJ ME/Kg DM. This nutritional value is close to that of straw. In particular, the low protein and energy values make it impossible to use dried olive cake directly. Assuming that feed contributes around 70% of the total cost of animal production, the conversion of even a proportion of the available olive cake to a feed suitable for chickens could prove attractive to egg/meat producers and olive growers alike (Haddadin M.S. et al., 1999).

As to the nutritional acceptability of OMWW, negative opinions exist which are due to the presence of antidigestive principles. It has been reported that the use of dried OMWW by ruminants induces diarrhea because of high concentration of potassium and phenolic compounds (Salvemini F., 1985). The excessive amounts of these compounds may prevent the recovery as animal feeding of by-products from the purification of this wastewater.

Martilotti F. (1983) developed a method, the so-called "Dalmolive" method, outlined in Fig. 10.4, which could prove useful in certain conditions. The method combines approximately: 50 kg OMWW, 20 kg partly destoned exhausted olive cake, and 12.6 kg of several agricultural residues and by-products which can produce 29 kg of feed in pellets whose chemical composition is given in Table 10.2. The "Dalmolive" method seems to have been of interest because solid-state fermentation of the mixture of OMWW with solid by-products of olive oil extraction and other agricultural wastes reduces the inhibition of phenolic compounds.

Bufano G. et al. (1982) conducted a feeding trial using 18-month-old ewes receiving hay and a mixture containing 0, 20, 40, or 60% OMWW paste (moisture content varying from 35 to 57%), straw or crushed olive branches and a 20% protein supplement. The results obtained were very unsatisfactory — see Table 10.3. Average daily live weight gains were low for all the lots during the entire experimental period and did not seem to vary with different rations. However, they showed that OMWW paste was accepted willingly and that rations containing up to 34% of the total (60% of the mixture) with low nutritive value fodders such as straw or olive

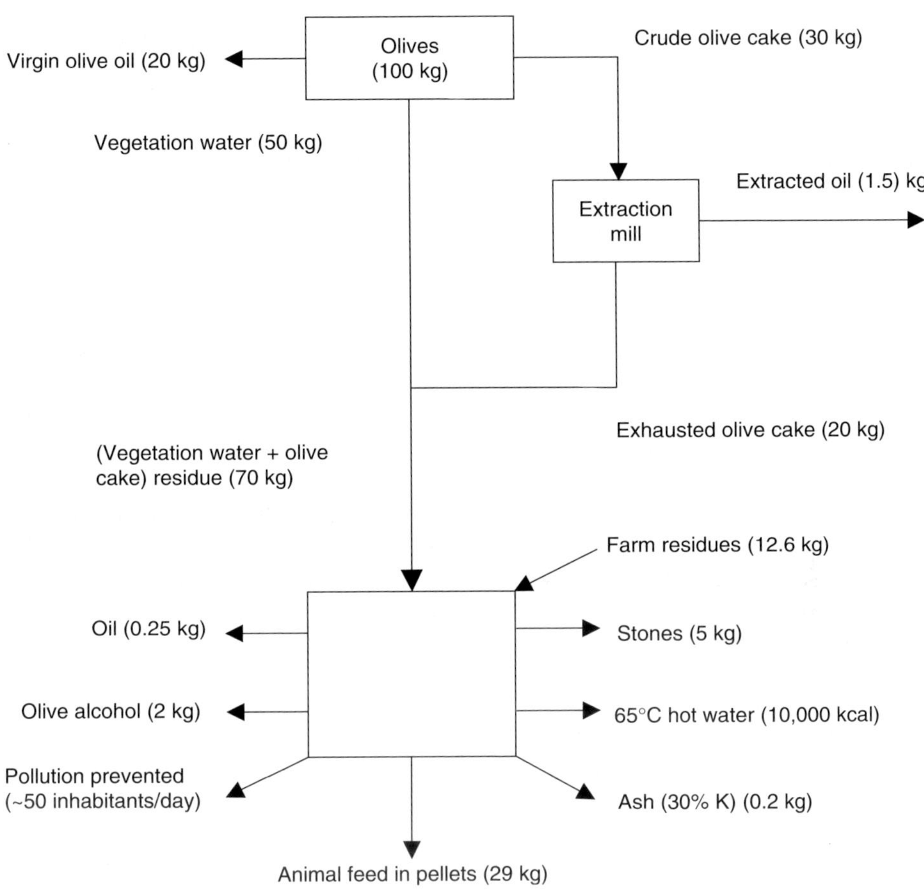

Fig. 10.4. The "Dalmolive" method (Martilotti F., 1983).

Table 10.2. Chemical composition of OMWW paste obtained by the "Dalmolive method" (compiled by Sansoucy R. et al., 1985)

	Italy[1]	Greece[2]
Dry matter (%)	85.3	93.6
Crude protein (%)	21.6	21.8
Ether extract (%)	4.0	3.3
Crude fiber (%)	13.1	4.7
Ash (%)	8.9	9.5
N-free extract (%)	52.5	60.7
Digestible (crude) protein (%)	17.2	–

(1) Agricultural chemistry laboratory of Milan.
(2) Zoiopoulos P.E., 1983.

Table 10.3. Feeding of 18-month-old ewes with a mixture containing (0–60%) OMWW paste, straw or olive branches, 20% protein complement and an additional supply of hay (Bufano G. et al., 1982)

% OMWW paste in the mixture	Mixture with straw				Mixture with olive branches			
	0	20	40	60	0	20	40	60
Initial weight, kg	29.2	32.4	32.5	30.9	31.3	32.9	32.9	33.1
Final weight, kg	32.4	33.3	35.5	34.9	32.3	34.6	35.3	36.3
Duration, days	90	90	90	90	90	90	90	90
Average daily gain, g/d*	35.8	10.4	34.5	45.5	11.0	33.3	26.6	34.7
Consumption: Mixture, g/d	792	992	1099	922	896	951	948	908
Consumption: Hay, g/d	641	602	687	700	610	664	629	676

*The average daily gains obtained show very high variation coefficients within the different lots.

branches, hay, and a protein supplement (about 11% of the total ration) would at least ensure maintenance of the sheep and even a slight live weight increase.

Paste for feeding animals has been prepared from concentrated, deoiled unfermented OMWW and dried, exhausted olive cake or whole olives, which have been destoned and crushed. Optional additives include meatmeal, fishmeal, soya flour, arachis flour, and mineral salts in proportions suitable to animal feeding. A typical composition of a 1:1 mixture of concentrated deoiled OMWW and destoned olive paste contains 8.5% lipids, 8.0% protein, 13.5% fiber, 12% ash, and 58% non-nitrogenous extract (PT64109, 1976).

OMWW contains about 8% sugars and minerals that are ideal substrate for yeast or other suitable fungi. This situation can be used for the production of a protein mass that contains carbohydrates, lipids, minerals, and vitamins, i.e. animal feed. The treatment results also in the reduction of the BOD_5 by 80%. Several tests have been undertaken for production of single cell protein (SCP) from OMWW and the use of lipophilic microorganisms, especially yeasts, has been extensively reviewed (Zoiopoulos P.E., 1983; Hamdi M., 1993a).

The *Saccharomyces* and *Candida* yeasts have been tested by several researchers because of their easy culture and high protein content. Table 10.4 compiled by Hamdi M. (1993a), gives the yeasts used in these works. The continuous culture was tested with dilution values of $0.2–0.3\,h^{-1}$ (Giulietti A. et al., 1984). The use of this type of SCP for animal feeding was limited by phenolic compounds fixed on the yeasts. In order to remove the phenolic compounds centrifuged OMWW was treated with 40% of H_2O_2 and NaOH solution and inoculated with *Saccharomyces cerevisiae* (Amat A.M. et al., 1986). Yeasts used in these processes degraded only sugars and lipids; polluting compounds such as pectins, tannins, and polyphenols were not removed.

The use of fungi with a high utilization spectrum, like *Aspergillus* sp. and *Geotricum candidum* gives a biomass of high digestibility with a crude protein content up to 30% potentially suitable for ruminants' alimentation (Vaccarino C. et al.,

Table 10.4. Yeasts for single cell production from OMWW; compiled by Hamdi M., 1993a

Strain	Biomass (g/l)	References
Torulopsis utilis (*Candida utilis*)	13	Fiestas Ros de Ursinos J.A., 1961b
Saccharomyces lipolitica	18–26	Ercoli E. and Ertola R., 1983
Saccharomyces and *Candida*	20–26*	Giulietti A.M., 1984
Saccharomyces cerevisiae	–	Amat A.M. et al., 1986
Saccharomyces cerevisiae	30	FR2620439, 1989

*Continuous culture.

1986). With media lacking additional nitrogen and sulfate, the growth of *Aspergillus niger* was limited (Hamdi M. et al., 1991b). On the basis of design experiments, the highest biomass and the greatest COD removal were obtained with a $COD:N:SO_4^-$ ratios averaging 100:3:1.5 (Hamdi M. et al., 1991a). A bubble-column (tower) fermentator is more adequate to the growth of *A. niger* than stirred tank fermentator for OMWW treatment (Hamdi M. et al., 1991b, 1992c).

Studies have also been made on a number of white rot fungi (*Coriolus versicolor*) and yeasts (*Yarrowia lipolytica* in combination with *Pseudomonas putida*, *Candida krusei*, *Saccharomyces chevalierie*, and *Saccharomyces rouxii*) for their potential to produce SCP from OMWW — see Chapter 8: "Biological processes", section: "Use of specific aerobic microorganisms".

In conclusion, the process of using OMWW as a mean for developing yeasts suitable for feeding shows the drawback of being stationary, as contrasted with other similar ones which work continuously. In addition, with this procedure a homogeneous effluent is obtained with a strong antibiotic activity that prevents or hinders its further biodegradation. Microbial protein has a high nucleic acid content, so levels need to be limited in the diets of monogastric animals. Finally, this process is not economically desirable because of its high consumption of energy.

Some components of OMWW are also useful as nutrients for growing microalgae. These components include sugars and salts, which comprise, respectively 1.6–4% and 1.8–2% of OMWW. The salts include potassium and sodium, phosphates, and carbonates. Microalgae comprise a vast group of photosynthetic, heterotrophic organisms which have an extraordinary potential for cultivation as energy crops. Microalgae converts sunlight, CO_2, and nutrients like nitrates and phosphates into proteins, lipids and carbohydrates, pigments, and specific bioactive substances. Sánchez-Villasclaras S. et al. (1995) examined the use of OMWW as a nutrient medium for producing biomass of microalgae. The influence of the aeration level and the composition of the culture medium were examined in relation to the concentration of OMWW and KNO_3 added in a batch culture of the microalgae. The organisms used were the Chlorophyta: *Chlorella pyrenoidosa* and *Scenedesmus obliquus*. The most suitable conditions for obtaining quality biomass with a composition balanced in proteins and lipids included an OMWW concentration of 10%, without the addition of nitrates and with an aeration level of 1 vvm (volume air/volume liquid/min). Once harvested microalgae could be dried and made into high-protein feed for livestock and fish, fertilizer and, high value chemicals. At the same time by reducing certain components in OMWW — essentially sugars and salts — the microalgae diminish the pollutant effect of OMWW.

Lastly, the direct use of OMWW as drinking water for livestock was tested at the Food, Oils and Fats Experimental Station in Milan (Sansoucy R. et al., 1985). This water was suggested to replace drinking water for hens and turkeys (Fedeli and Camurati F., 1981). In the case of turkeys, it was estimated that the cost per kilogram of meat produced was lower if OMWW were used and a marked decrease of mortality rates was also observed although the authors did not provide statistical data. However, the use of OMWW as drinking water for turkeys, and probably ducks, would have only a small impact on the quantity of OMWW used.

Despite all the provided evidence, the studies in the nutritive value of OMWW and possibilities for their use in animal rations are too scarce to be able to draw accurate conclusions.

As far as 2POMW is concerned, its protein content is a factor of major significance that should be taken under serious consideration, in case of protein recovery, as it links process yield to process cost. Recovery of proteins from 2POMW through cellulose ion exchange and ultrafiltration processes is an attractive alternative only if the amino acid profile is balanced and comparable to reference protein of FAO/WHO. Several samples of 2POMW were physico-chemically characterized and the evaluation of the results revealed that protein content was low in comparison to other wastes. Moreover, the protein recovery costs by cellulose ion exchange and ultrafiltration techniques are extremely high considering the amount of protein that could possibly be recovered (EU project: FAIR-CT96-1420 "IMPROLIVE").

ES2180423 (2003) proposes a compound feed for ruminants comprising extracted and dried 2POMW (100–700) g/kg, barley grains (870–210) g/kg and a mixture of minerals, which include ashes generated from the incineration of 2POMW (30–60) g/kg.

WO2004110171 (2004) describes a process for producing an olive powder from an aqueous olive paste derived from olive cake, 2POMW or whole olives which is suitable for human consumption. A typical process comprises the following steps: (i) pretreating the olive cake or 2POMW with a stabilizing additive or preservative (e.g. citric acid, NaCl, ethanol, sodium metabisulfite, etc.) to avoid microbial deterioration; (ii) blanching the stabilized olive cake or 2POMW at moderate heat to inactivate the polyphenol oxidase enzyme; (iii) adding to the blanched paste an esterase enzyme (preferably, a thermostable β-glucosidase) to hydrolyze the oleuropein; (iv) drying the olive paste to remove water and provide a particulate intermediate; (v) optionally dry comminuting the particulate intermediate having a water content of less than 20% by weight (preferably less than 5%), in a mill whilst the temperature of the material in the mill is maintained at a temperature less than 10°C to form a powder of which at least 99% by weight has a particle size less than 0.55 mm. A commercial process describing the production of olive powder from 2POMW is illustrated schematically in the flow diagram of Fig. 10.5.

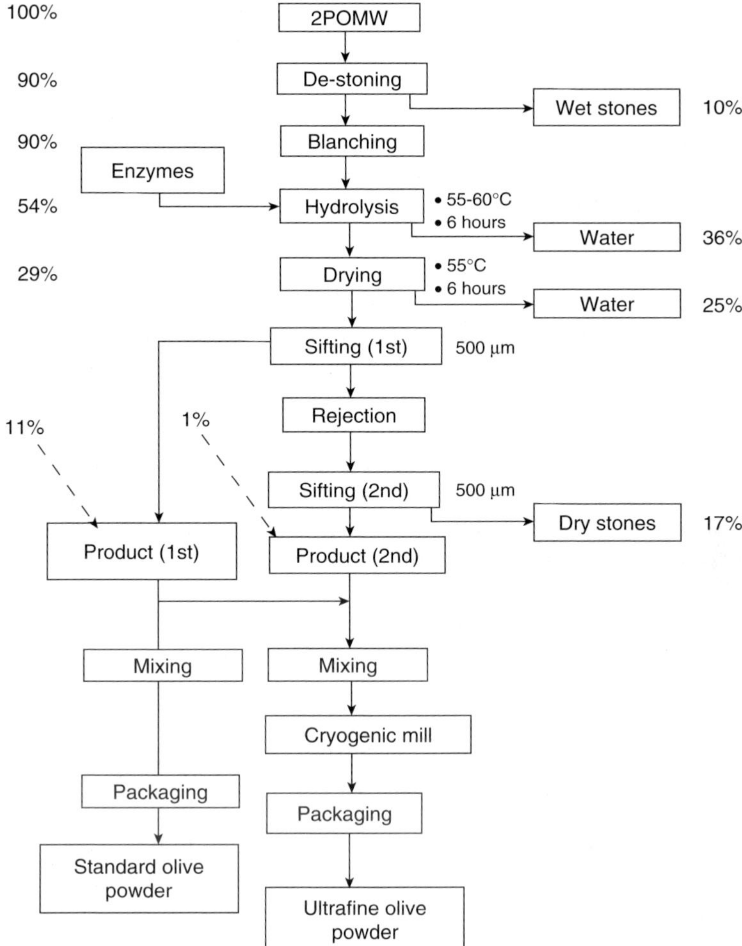

Fig. 10.5. Flow diagram of a process for producing olive powder (WO2004110171, 2004). In the flow diagram, the figures to the left hand side of the figure refer to the percentage by weight of the starting 2POMW (alperujo) remaining following the specified step. On the right of the flow diagram, figures are given for the percentage by weight removed during the respective step.

Recovery of Residual Oil

The (semi-) solid olive waste has a variable oil content, which depends on the extraction system — see Table 10.5. The residual oil (pomace- or orujo- or seed-oil) has been traditionally extracted with organic solvents — mainly hexane or benzene — after its moisture content has been reduced to about 5–8%. Because the solid waste is very humid, has a high enzyme content and the constituents have been broken

Table 10.5. Composition of (semi-) solid olive-mill waste

	Pressure	3-Phase system	2-Phase system
Solids (%)	64–74	46–57	42.6–27
Olive oil (%)	4.5–9	3–4	2.4–3
Water (%)	20–27	40–50	55–70

down, it forms the ideal place for hydrolysis and oxidation, unless it is quickly dried or extraction takes place immediately. For these reasons, residual oil is often high in oxyacids and in general has high acid levels, which make the refining process difficult. The quality of this oil is not acceptable for direct human consumption (Alvarado C.A., 1998) and is mainly used for making soap (Amirante P. et al., 1993).

Ortega Jurado A. and Ramos Ayerbe F. (1978a) developed a modification of the traditional extraction process of residual oil from olive cake. The olive cake contains about 40% of an easily separated fraction, consisting of stone fragments with hardly any oil. The proposed modification consists of a separation of these stone fragments followed by solvent extraction of the oil-rich pulp agglomerated by granulation or rolling to facilitate extraction. The process is claimed to increase the capacity and profitability of the extractors, produces more concentrated miscellas and reduces the quantity of solvent used and the damage to the plant. Owing to the difficulties presented by olive cake pulp in extraction, a granulating process to facilitate solvent extraction was also proposed by the same authors. Various types of granulating and compacting machines are described, together with an equation for calculating compacting pressure, the effect of granule size and compaction on the extraction process, and the advantages of adding bentonite to fatty pulp for granulation. The new process gave better results compared to the traditional extraction: 96% of the total oil was extracted from granulate compacted with 2% soda bentonite (Ortega Jurado A. and Ramos Ayerbe F., 1978b; Ramos Ayerbe F. and Ortega Jurado A., 1980a,b).

ES2006904 (1989) describes a process for extracting residual oil from olive cake. The process consists of: (i) separating the stones from the olive cake in flotation tanks; (ii) breaking up and thermo-heating the resulting pulp so that the oil globules can combine and separate from the solids and water; and (iii) phase separating according to the selected method.

ES2048667 (1994) describes a process for extracting residual oil from olive cake without using solvent. The olive cake is: (i) mixed with hot water (40–70°C), (ii) homogenized by beating, (iii) centrifuged in a decanter or horizontal centrifuge to obtain a liquid fraction and a secondary residue; (iv) filtering the liquid through a high precision sieve; (v) separating in a vertical centrifuge into an olive residue oil and a second liquid fraction; and (vi) heating the latter to 90°C and recycling it to stage (i).

Supercritical fluid extraction is a viable alternative process for extracting oil from olive cake. Lucas A. de et al. (2002) studied the influence of various operational variables on quality parameters of olive cake extracted with CO_2. Effects of pressure (P: 100–300 bar), temperature (T: 40–60°C), solvent flow (1–1.5 l/min), and particle size (D: 0.30–0.55 mm) on extraction yield, and 3 oil quality parameters, namely, acidity, peroxide value, and phosphorous content were studied. Response surface methodology based on statistical analysis of experimental data allowed mathematical expressions to be obtained that related the operational variables and parameters studied. Under optimum extraction conditions, in the experimental range analyzed ($P = 300$ bar, $T = 60°C$, $D = 0.30$ mm, and solvent flow $= 1.25$ l/min at standard conditions), oil yield was 80% (w/w) with respect to hexane extraction, whereas olive acidity, peroxide value, and phosphorous content were 14% (w/w), 8 meq/kg, and $2.3 \times 10^{-3}\%$ (w/w), respectively. Results were compared to those obtained by hexane Soxhlet extraction. Quality of the supercritical extract was superior, requiring only simple refining. This advantage may result in improved economics of the supercritical process in relation to the conventional extraction with hexane. In a later study, Lucas A. de et al. (2003) investigated the influence of the same operational variables, pressure (100–300 bar), temperature (40–60°C), solvent flow (1–1.5 l/min), and particle size (0.30–0.55 mm) on a different set of oil quality parameters, namely, tocopherol concentration, extinction coefficient at 232 and 270 nm, and saponification value. Results from these experiments were also used to design a 3-step sequential CO_2 extraction procedure to obtain a higher-quality extract. The optimal operational sequence consisted of a 1st extraction step at 75 bar for 1 h using 1% (v/v) ethanol modifier, followed by a 2nd extraction stage at 350 bar, for 2.5 h without ethanol and a 3rd step, also at 350 bar, for 2.5 h but using ethanol. These extraction conditions resulted in an intermediate fraction of oil with 64% yield and normal parameters according to EU food legislation, which is suitable for food use without any further refining. On the contrary, the oils obtained by hexane extraction and by conventional supercritical CO_2 extraction at optimal conditions would only be suitable for human consumption after further refining.

The introduction of the two-phase extraction system, in the early 1990s, had dramatic consequences for this sector of the olive industry, especially in Spain. The number of extracting facilities (seed-oil extraction plants) was decreased, while their average capacity increased (Alvarado C.A., 1998). The new waste (2POMW) has a lesser oil content of about 2.4–3% — which represents 12% of the total olive oil — but a moisture content of around 55–70%. The greater moisture, together with the sugars and fine solids that in the three-phase system were contained in OMWW give 2POMW a doughy consistency. 2POMW is dried and subjected to chemical extraction with hexane in order to produce an extra yield of oil. The high moisture content of 2POMW demands much more energy and the sugars present in it make it sticky and difficult to dry. The reuse of this olive by-product in the conventional seed-oil extraction plants for an additional extraction of the remaining oil contained therein, poses a difficult problem associated with its special characteristics — see Chapter 6: "Thermal processes", section: "Physico-thermal processes". In addition, the recently discovered problems concerning the detection of polycyclic aromatic hydrocarbons (PHAs) in the recovered oil as a result of drying 2POMW before oil chemical extraction has obliged manufactures to perform a further purification step, which greatly increases production costs (Alburquerque J.A., 2004).

ES2076899 (1995) describes a process to separate 2POMW into residual oil, stones, and pulp. The process consists of the following steps: (i) subjecting 2POMW to a separation, which is carried out by mechanical means, between the stones and the pulp; (ii) the pulp subsequently being subjected to the action of a decanter, in which separation takes place also by mechanical means between the oil and the oil-free residual pulp, which can be used as animal feed or fertilizer; (iii) the stones, which can in principle be impregnated with pulp, are subjected to the action of an air cleaner, which separates the pulp from the stones, the latter being used as fuel. In this way a residual oil of high quality is obtained, because it is additive-free and has suffered little thermal degradation.

ES2144359 (2000) describes a biotechnological process for recovering the oil retained in 2POMW, by applying an anaerobic process which hydrolyses and solubilizes selectively the carbohydrates (celluloses and hemicellulose), which are the main components of 2POMW, thus releasing the oil, which can be recovered by centrifugation. The process is based on the use of 2POMW as a culture medium for growing hydrolytic bacteria (*Bacillus stearothermophilus*) which, through the activity of their extracellular enzymes, convert the carbohydrates constituting the olive pulp into a low molecular weight compound, which is soluble in the vegetation water, and the oil thus released can be separated out from the aqueous phase and the solid phase by simple centrifugation in a three-phase decanter.

Recovery of Organic Compounds

Several techniques exist, which allow some potentially valuable organic compounds contained in the OMWW to be extracted. The current state-of-the-art uses specific solvents and ultrafiltration/reverse osmosis techniques, which require the application of sophisticated technologies, which in turn, require that complex chemical facilities are available.

Pectins

Pectins are natural hydrocolloids found in higher plants that are widely used as gelling agents, stabilizers, and emulsifiers in the food industry. Basically, they are complex polysaccharides containing 1,4-linked-α-D-galactosyluronic acid residues that are occasionally interrupted by $(1 \rightarrow 2)$-α-L-rhamnose residues carrying sugar side chains, typically galactose and arabinose. Commercial pectins are only available from two important sources: apple pomace and citrus peels. The availability of other pectic sources is always being searched and the possibility of using waste products as raw materials is an important economical aspect. The waste beet solids from sugar extraction and the sunflower heads residues obtained after the oil extraction were very promising sources as they contain 10–20% of pectic material. However, pectins obtained from those sources have poor gelling ability and those two raw materials have low commercial value.

The olive pulp cell walls are known to contain about one-third of arabinose-rich pectic polysaccharide — see Chapter 2: "Characterization of olive processing waste", section: "Organic compounds". The gelation potential of pectins obtained from 2POMW was investigated to see if new applications could be found for this low-value by-product from olive oil production (Cardoso S.M. et al., 2003). Pectic raw material was extracted from the alcohol insoluble residue of 2POMW. The purified olive pectic extract contained 48% galacturonic acid and 31% arabinose with a total sugar content of 72% and a degree of methylesterification of 43%. Compared with a commercial low methoxy citrus pectin, the olive pectic extract demonstrated higher critical galacturonic acid and calcium concentration for gelation to take place, and gels showed lower viscoelastic moduli at corresponding galacturonic acid and calcium concentration. From these results, it could be inferred that 2POMW can be a potential source of gelling pectic material, with useful properties for practical applications.

Antioxidants

Both the stone and the pulp of olive fruit are rich in simple and complex water-soluble phenolic compounds with potent antioxidant properties, which may have a protective action on human health (Owen R.W. et al., 2000). The majority of these compounds, depending on their partition coefficients (K_p), end up in OMWW and 2POMW during olive processing, for which reason these wastes may constitute a suitable source of phenolic antioxidants. The most abundant phenolic compounds found in OMWW and 2POMW are mainly hydroxytyrosol and tyrosol together with *p*-hydroxybenzoic, vanillic, caffeic, and ferulic acid in less quantity — see Chapter 2: "Characterization of olive processing waste", section: "Organic compounds", and Chapter 4: "The effect of olive-mill technology", section: "Olive oil production systems".

The phenols protect the oil from degrading, while much of the benefits of olive oil consumption in the Mediterranean diet have been attributed to the presence of these natural antioxidant compounds (Visioli F. and Galli C., 1995; Trichopoulou A. et al., 1995). Interest in natural antioxidants is increasing because of the growing body of evidence indicating the involvement of oxygen-derived free radicals in several pathologic processes, such as cancer and atherosclerosis (Manna C. et al., 1997; Petroni A. et al., 1994, 1995, 1997; Visioli F. et al., 1995b). Recent studies, which involve administration of the phenolic fraction of the olive vegetation water in rats exposed to oxidative stress from secondary smoke, show a dramatic reduction of stress and protective activity by polyphenols. These water-soluble phenols could, therefore, be employed in preservative chemistry and, possibly as prophylactic agents in the prevention of certain free radical-induced diseases such as skin damage produced by overexposure to sunrays and environmental stress. Their antioxidant effect has been proven as they block the oxidation process in the initiation phase, either oxidizing themselves by forming more stable natural peroxidic compounds, or acting on the already formed radicals to prevent the propagation phase.

The ability of the phenolic compounds to scavenge superoxide, already reported for hydroxytyrosol and oleuropein (Visioli F. et al., 1999; Leger C.L. et al., 2000), is suggestive of a potential use of these phenols in environments in which Fenton — see Chapter 7: "Physico-chemical processes", section: "Fenton reaction" — and Haber-Weiss[33] reactions take place, and in which the concomitant production of superoxide and nitric oxide would yield the powerful oxidant peroxynitrile. It is noteworthy that the established antioxidants vitamin E and BHT[34] do not scavenge superoxide, and, thus the phenols may add stability to products exposed to high O_2^- levels (Visioli F. et al., 1999). The antioxidant activities of phenols were also tested against hypochlorous acid (HOCl), which could be considered as a source of reactive chlorine species. The protection of hypochlorous acid-induced damage of catalase is of biological significance due to the well-known protein-damaging activity of HOCl, which is produced in biological systems, at the site of inflammation by activated neutrophils through the enzyme myeloperoxidase. Also, because foods often come into contact with chlorine-based bleaches, employed as disinfectants in food plants, the use of HOCl scavengers may provide additional protection against reactive chlorine species (Visioli F. et al., 1999).

The phenolic compounds have shown *in vitro* biological activity on the human metabolism. The most important is the inhibition of low-density lipoprotein (LDL) oxidation, which contributes to the progression of human atherosclerosis (Visioli F. et al., 1995b; Visioli F. and Galli C., 1995). Animal studies have shown that LDL from olive oil-fed rodents is significantly more resistant to oxidation than control samples (Scaccini C. et al., 1992). The phenolic compounds have also been found to prevent platelet aggregation (Petroni A. et al., 1995), counteract cytotoxicity induced by reactive oxygen species in various human cellular systems (Manna C. et al., 1997) and inhibit the formation of thromboxane B_2 (TxB_2) and leukotriene B_4 (LTB_4) (Petroni A. et al., 1997). Thromboxane B_2 and leukotriene B_4 are two important substances secreted by polymorphonuclear leukocytes involved in pathophysiological processes, related to chronic inflammation and vascular injury. The potent

---

[33]The Haber-Weiss cycle consists of the following two reactions:

$$H_2O_2 + OH \rightarrow H_2O + O_2^- + H^+ \quad \text{and} \quad H_2O_2 + O_2^- \rightarrow O_2 + OH^- + OH^\bullet$$

The second reaction achieved notoriety as a possible source of hydroxyl radicals. However, it has a negligible rate constant. It is believed that iron (III) complexes can catalyze this reaction: first Fe(III) is reduced by superoxide followed by oxidation by dihydrogenperoxide. See also Fenton reaction.

[34]Butylhydroxy toluene or butylated hydroxytoluene; commercial antioxidant.

inhibition of calcium ionophore[35]-stimulated production of LTB$_4$ and its metabolites by human neutrophils[36] suggests that the olive-derived phenols exert biological effects beyond their antioxidant capacities. The activity of several enzymes, including those involved in the production of eicosanoids[37], for example phospholipases and oxygenases, is modulated by the intracellular peroxide tone (Kohyama N. et al., 1997). Thus, by scavenging reactive oxygen species, the phenols could lower the activity of such enzymes and, in turn, decrease the production of pro-inflammatory factors, which are associated with colon and breast pathologies.

In addition, the olive-derived phenols have shown antibiotic activity with both antimicrobial (Juven B. and Henis Y., 1970) and antifungal properties (Mahmoud A.L., 1994). The polyphenols have been demonstrated to inhibit or delay the rate of growth of bacteria such as *Salmonella, Cholerae, Staphylococcus, Pseudomonas,* and *Influenza in vitro*. These data suggest a potential role for the polyphenol antioxidants in promoting intestinal and respiratory human and animal wellness, and as an anti-microbial food additive in pest management programs.

Thus, in light of the increasing amount of evidence showing the potential health benefits of olive-derived phenolic compounds, it would be desirable to have processes for extracting antioxidant components from olive-based starting materials.

The olive vegetation water is used to prepare a skin cosmetic causing no skin irritancy and having excellent rough skin preventive effect (JP2000319161, 2000). For the vegetation water, an aqueous solution fraction produced in the process of obtaining ordinary olive oil can be used directly, however, it is preferable that it is used in a refined state after removing lipid components, fibrous components, seed crusts, and the like as contaminants through filtration or centrifugal separation. This skin cosmetic may be formulated, as necessary, with various ingredients generally used in cosmetic compositions, and can be prepared into any formulation such as ointment, lotion, milky lotion, pack, cataplasm, granules, or base makeup. The formulation is stable and makes the skin glow.

Lesage-Meessen L. et al. (2001) tested olive oil residues for their composition in simple phenolic compounds as a function of the extraction system i.e. the three- and two-phase centrifugation systems. Phenolic compound extraction with ethyl acetate was efficient and allowed recovery of 28.8 and 42.2% of total phenols present in dry olive oil residues originating from three-phase and two-phase systems, respectively. The qualitative and quantitative HPLC analyses of the extracts showed that hydroxytyrosol and *p*-tyrosol were the most abundant phenolic compounds.

---

[35]A compound which can carry specific ions through membranes of cells or organocells.

[36]White blood cells. Neutrophils are the predominant cell type involved in acute inflammation.

[37]Eicosanoids are a family of compounds derived from polyunsaturated eicosanoic acids. The eicosanoids include prostaglandins, leukotreines, and the intermediate hydroperoxyeicosatetraenoic (HPETE) and hydroxyeicosatetrenic (HETE) acids. The prostaglandins and leukotrienes act as paracrine and autocrine regulators through a family of transmembrane receptors. They regulate many cell functions and play crucial roles in a variety of physiological and pathophysiological processes, including regulation of smooth muscle contractility and various immune and anti-inflammatory functions.

p-Coumaric, caffeic, ferulic, and vanillic acids were also present. The phenolic extract from the two-phase system had the highest concentration in hydroxytyrosol (1.16% w/w dry residue) and the strongest antioxidant activity. Olive oil residues were confirmed as a cheap source of large amounts of natural phenolic antioxidants. However, all these methods have the drawback of their high cost as regards the use of solvents and the generation of wastes which are difficult to eliminate.

Despite the need for olive-derived antioxidant compositions, the prior art, up to now, does not provide simple and effective processes for producing such compositions. EP811678 (1997) discloses a process for extracting antioxidants from olives, in which olives are crushed, vacuum dried, and pressed to obtain a lipid fraction and a cake. The cake is then extracted with a hot medium comprising a mixture of triglycerides or a C_2 to C_6 alkylene glycol at a pressure of at least 40 bar, to obtain an antioxidant-enriched extract. The extract contains hydrosoluble antioxidants, namely, hydroxytyrosol, tyrosol, phenolic acids, and oleuropein. This method requires the use of a pressure-piston apparatus for extraction, lyophilization equipment and supplies for freeze-drying, and other equipment and chemicals that result in a relatively complex, expensive process. Moreover, this process uses fresh green or ripe olives, which can be expensive.

OMWW is a potential rich source of antioxidant compounds, which have not been effectively exploited, due to the impracticality of extracting usable amounts of antioxidant compounds using conventional technology (Visioli F. et al., 1995a). It is estimated that the content of OMWW in phenolic compounds is 0.5–1.8% — see Chapter 2: "Characterization of olive processing waste", section: "Organic compounds". Additionally, other components, in minor quantities, such as flavonoids, anthocyanins, and tannins are of potential biological interest due to their antioxidant activities. During olive oil processing a large fraction of these phenols is lost in OMWW and poured into the Mediterranean environment. In addition, several inferior grades of olive oil now used in industrial (rather than culinary) applications, and therefore, relatively inexpensive compared to culinary grade olive oil, offer potentially rich sources of antioxidant compounds. To date, however, these potential sources of beneficial antioxidants have not been effectively exploited.

The aim of many recent studies has been to improve the knowledge on the biological activities of antioxidant phenols obtained from OMWW and to define procedures for their extraction and valorization as health food supplements and/or natural food antioxidants. In one of the earliest studies three technologies for economic recovery of by-products from OMWW were studied on laboratory and pilot plant scale: (i) solvent extraction of flavor and phenolic compounds (antioxidants); (ii) recovery of phenols by adsorbent resins; (iii) selective concentration by ultrafiltration and reverse phase osmosis. Diagrams and tables illustrate the recoveries and economic aspects and the antioxidant activity of the recovered phenols (Camurati F. et al., 1984). In a more recent study three different extraction procedures were employed to optimize the recovery of phenols from OMWW (EU project: FAIR-CT97-3039). In particular, solid–liquid extraction, liquid–liquid extraction, and adsorption techniques with the use of resins were tested. The first one is the most effective but is also very expensive and the adsorption of phenols with *ad hoc* resins was the technology of choice. It is estimated that from 1 l of OMWW a recovery of ~1 g of raw material containing ~130 mg of hydroxytyrosol can be obtained. Hydroxytyrosol the most active component of OMWW has been revealed to be the most interesting because of its remarkable pharmacological and antioxidant properties (Visioli F. et al., 2000). In particular, hydroxytyrosol is a natural antioxidant showing antimicrobial and phytotoxic activity and is useful as a food preservative, in agriculture for the protection of olives, for the prophylaxis of radical-induced human diseases, and in topical preparations having anti-aging and anti-inflammatory action. Hydroxytyrosol is also of particular interest because it is amphiphilic and, thus it acts at the oil–water interface and in systems where both oil and water phases are present, such as emulsions (Visioli F. et al., 1999).

There are less references on the use of the olive solid wastes, such as crude cake and 2POMW that remain after the extraction of untreated olive oil, as a substrate for the recovery of polyphenols.

ES2143939 (2000), which describes a steam explosion process for the production of mannitol from olive cake coming from a three-phase centrifugation system, discloses also the extraction of phenols from this waste — see section: "Alcohols". The results obtained show that hydroxytyrosol is extracted from the olive stone in soluble extract concentrations of up to 1% in dry weight of the stone, less quantity than that obtained in the pulp, where this polyphenol is mainly found. In the case of the stone wall, tyrosol is obtained in concentrations of up to 0.5% in dry weight and it is verified that the addition of acid to the material before the treatment appreciably increases the quantities of phenols detected in the soluble extract.

WO0145514 (2001) describes a method of extracting antioxidant compositions from olives and olive by-products. One of the embodiments includes the steps of passing OMWW through a solid matrix to trap antioxidant components in OMWW on the matrix, and washing the matrix with a polar organic solvent to remove the antioxidant composition in the polar organic solvent. Suitable solvents include polar alcohols, acetone, ethyl acetate, acetonitrile, dioxane, and mixtures thereof. The polar organic solvent can be partially removed to form a liquid concentrate, or preferably substantially and completely removed to produce a solid antioxidant composition. The antioxidant activity of the composition can be enhanced, either by acidifying OMWW, or by acidifying the solution of the antioxidant composition in the polar organic solvent, or by dissolving the solid antioxidant composition in an acid, and redrying the antioxidant composition. The extracted antioxidant composition has a total phenolic content of about 10–30% gallic acid equivalents by weight based on the dry weight of the composition and an antioxidant activity of about 0.4 to 2.0 ascorbic acid equivalents based on the dry weight of the composition. Specifically, the antioxidant composition comprises about 1–5% by weight hydroxytyrosol, about 0.4–1.5% by weight tyrosol and about 0.05–1% by weight oleuropein based on the dry weight of the composition.

The solid matrix can be any material having a stronger affinity for at least some of the antioxidant components than for the aqueous phase. The solid matrix is preferably composed of a plurality of small particles having a large surface area, such as chromatographic beads. Preferably, the solid matrix is a solid phase resin, and is disposed in a bed of a chromatographic column. A particularly preferred solid matrix material is a polymeric adsorbent material marketed under the trademark AMBERLITE® (Rohm and Haas GmbH). The AMBERLITE® material is a macroreticulated crosslinked copolymer having a plurality of microscopic channels resulting from the liquid expulsion of a precipitating agent during polymerization of a monomer mixture under suspension conditions. These resins are typically styrenic, acrylic, or phenolic-based. Among the AMBERLITE® resins the polystyrene-based resins are preferred and, especially the grades AMBERLITE® XAD-2, AMBERLITE® XAD-4, AMBERLITE® XAD-7, and AMBERLITE® XAD-16. Other preferred polystyrene resins are DUOLITE®, particularly DUOLITE® S-761 (Duolite Company) — see also Chapter 7: "Physico-chemical processes", section: "Adsorption".

In a specific example, a 250 ml sample of OMWW is first filtered, and then passed through an AMBERLITE® XAD-16HP column at 5 ml/min. After OMWW has been added to the column, another 135 ml (1 bed volume) of a fresh water wash is passed through the column. Next, methanol is passed through the column, and about 90 ml of the methanol effluent wash collected. The methanol is evaporated to give 0.35 g of solid (WO0145514, 2001).

The extracted antioxidant composition may be used to impart antioxidant activity to a product or enhance any activity already present. Thus, the product can be a food product subject to oxidation, or an oil, such as an edible oil or a cooking oil. Alternatively, the product can be a topical antioxidant composition, a preservative composition, a nutritional supplement or a cosmetic. In a particular example the antioxidant extract was dissolved in 50/50 (v/v) water/ethanol mixture to a concentration of 2.34 mg/ml; 25 μl of the solution was added to each of three cosmetic products: (1) Aveeno moisturizing lotion (Ryoelle Laboratories, Division of S.C. Johnson & Son, Inc. Racine, WI) 0.80; (2) Yves Rocher revitalizing cream for hands (soin beauté des mains, la Gacilly, France) 0.76; and (3) USANA hand and body lotion (USANA Inc., Salt lake City, UT) 0.98 g. Each composition was mixed well, so that no visible changes were observed. All compositions were allowed to stand for 30 minutes, then each starting cosmetic and antioxidant-enhanced mixture was assayed for antioxidant activity by the photochemiluminescence assay. The results displayed in Table 10.6 show that the antioxidant composition increases the antioxidant activity of cosmetic products by as much as nearly 300%.

A related process describes the isolation of antioxidants from OMWW by fluidized bed adsorption, especially using polymeric ion-exchanging adsorbents (e.g. AMBERLITE® XAD or LEWATIT® EP), followed by elution of the absorbed antioxidants and removal of the solvent. The process is free of energy requirement and problems associated with extraction processes (EP1310175, 2003).

WO2005003037 (2005), discloses in one of its embodiments, a process for the recovery of antioxidant compounds, present in OMWW by using the system of filters described earlier — see also Chapter 5: "Physical processes", section: "Filtration".

Table 10.6. Antioxidant enhancement of cosmetics (WO0145514, 2001)

Cosmetic product	Antioxidant activity (AsA eq.)*		% Increase
	Starting cosmetic	Enhanced cosmetic	
Aveeno	1.27	3.08	143
Yves Rocher	0.94	3.71	295
USANA	0.57	2.07	263

*The results are reported as ascorbic acid equivalents (AsA eq.), calculated from the lag time according to the Esterbauer reference (Esterbauer H. et al., 1989).

The process comprises the following steps: (i) retention of the phenolic compounds on the filters of resins (e.g. AMBERLITE® XAD-16 or XAD-4 or any compatible mixed bed, or cationic resin or PVPP) by passing OMWW through said system of filters; (ii) separating the filters of resins from the other filters; (iii) washing of the filters of resins by passing a solvent through said filters; (iv) regenerating the filters of resins by passing an organic solvent though said filters and, thereby, recovering said compounds.

WO02064537 (2002) describes a method for obtaining purified hydroxytyrosol from olive production by-products by means of a two-step chromatographic treatment. The olive production by-products include 2POMW, olive cake (three-phase) and stones if they are subjected to a steam explosion process. The olive by-product is introduced into a column of a non-activated ion exchange resin and eluting with water to give a solution containing at least 85% of the hydroxytyrosol present in the olive by-product, and having a hydroxytyrosol content of 60–70% (based on solids); introducing the solution into a second column of a XAD-type absorbent non-ionic resin and eluting with a mixture of methanol or ethanol and water (30–33%) to give a solution containing at least 75% of hydroxytyrosol present in the olive by-product and having a hydroxytyrosol purity of at least 95%. Hydroxymethyl-furfural is completely separated and a highly pure hydroxytyrosol is obtained. The resins used in the columns are relatively inexpensive, mechanically strong and easily regenerated.

Fernández-Bolaños J. et al. (2002) developed a hydrothermal process claiming to produce large quantities of highly purified hydroxytyrosol from 2POMW. The effect of hydrothermal processing of 2POMW on the solubilization of hydroxytyrosol was studied by assaying different saturated steam conditions. A high amount of hydroxytyrosol was solubilized and increased with increasing steaming temperature and time, reaching 1.4–1.7 g/100 g of dry 2POMW. The effect of acidic (H_2SO_4) and basic (NaOH) catalysts was also evaluated. Acid-catalyzed treatment was more effective at milder conditions, whereas the alkali-catalyzed conditions were not very suitable. Approximately, 4.5–5 kg of hydroxytyrosol could be obtained from 1000 kg of 2POMW with a moisture content of 70%. After a purification process, at least 3 kg of hydroxytyrosol, at 90–95% purity, could be obtained.

A work published by Ballesteros Perdices I. et al. (2002) for the production of ethanol from 2POMW discloses that, after a hydrothermal treatment performed to promote the subsequent transformation of the cellulose contained in the substrate to ethanol, phenolic compounds are obtained such as tyrosol and hydroxytyrosol. However, 2POMW has undergone a washing and drying process before being used with the aim of promoting the subsequent fermentation process to which it is subjected.

WO2004009206 (2004) discloses an improved process to extract phenolic compounds, mainly present in 2POMW, by the use of a hydrothermal treatment which uses hot water in liquid phase. 2POMW in its crude state, i.e. as it is received from the olive-mill, is treated in a closed autoclave-type reactor at a temperature of 180 to 240°C for an appropriate time period of 4 to 30 minutes, maintaining during this time the water in liquid phase by applying suitable pressure. Next, the reactor is cooled, the humid material is filtered, and its phenol concentration is determined by HPLC, GC-MS, or spectroscopy techniques. During the treatment of the substrate, an autohydrolysis of the hemicellulosic debris occurs in 2POMW that produces the liberation of acetyl groups and, in consequence a reduction in the pH of the liquid fraction, thus promoting the solubilization of the phenols of interest. Using this treatment, one can obtain an aqueous extract with a content of up to 1.9% w/w of hydroxytyrosol and 0.7% w/w tyrosol, from which the compounds of interest can be obtained by conventional extraction/purification techniques. The application of the hydrothermal treatment to the recovery of soluble phenols does not use solvents and/or acids, nor do abrupt depressurizations occur as in the steam explosion treatment. In addition, the use of crude 2POMW, whose elimination could become a problem, avoids the expense of water in the washing phase and the subsequent generation of a liquid residue, whose management may be problematic from an environmental point of view. Furthermore, it is advantageous from an energy-saving point of view as it avoids the drying phase, after that of washing, which is disclosed in the aforementioned work, while permitting an increase in the yield of the extracted phenolic compounds.

Supercritical fluid extraction of 2POMW, and supercritical fluid chromatographic separation of the extracts were performed to study the content of tocopherols, a group of compounds of interest for the food industry owing to their antioxidant activity (Ibáñez E. et al., 2000). The tocopherols, well known components of vitamin E, have been detected in olive by-products — see Chapter 2: "Characterization of olive processing waste", section: "Organic compounds". The developed method consists of supercritical CO_2 extraction at a pilot plant scale and subsequent fractionation by two successive depressurizations. Enrichment of α-, β-, and γ-tocopherol was achieved in a second separator when working at low densities in the first separator. Fractions obtained using high densities in the first separator contained major proportions of triglycerides, waxes, and sterols. Tocopherols from olive by-products were separated and quantified in an environmentally friendly way by using supercritical fluid chromatography with packed capillary columns coated with polyethylene glycol and pure CO_2. The studied olive by-products can be considered a natural source of antioxidants because a substantial concentration of tocopherols has been obtained in the extracts. The described method could lead to a scaling up of the supercritical fluid chromatography column for fractionation of the tocopherol analogs found in olive by-products.

Olive leaves were treated with supercritical CO_2 to analyze the possibility of obtaining tocopherol concentrates (Lucas A. de et al., 2002). Oil and tocopherol extraction rates were determined as a function of pressure (25–45 MPa), particle size (0.25–1.5 mm), solvent flow (0.5–1.5 l/min), and temperature (40–60°C). Two optimal extraction conditions were determined, considering the maximum recovery or concentration criterion. These conditions led to a highly valuable extract of 74.5 and 97.1% (w/w) tocopherol concentration, respectively.

Extracts from olive leaves and OMWW are used as antimicrobial agents in detergents, rinsing, and cleaning agents because they have an antimicrobial effect against aerobic and anaerobic germs as well as yeasts and moulds (WO03079794, 2003). They are also used as antidandruff agents because they have an antimicrobial effect against the yeast produced during the formation of dandruff (WO03080006, 2003).

A study investigated the possibility of recovering the flavonoid (anthocyanin) pigments from OMWW (Codounis M. et al., 1983). The effluent passed through an ultrafiltration unit and the permeate obtained was concentrated under vacuum from 5% refractometer solids (RS) to 65% RS and held at 0°C until use. The concentrate was subsequently diluted to 10, 20, 30, and 40% RS and passed through two types of resin. Results indicated that the use of alcohol acidified with 0.01% citric acid is most suitable for elution of adsorbed anthocyanins from the resins. The eluted anthocyanins can be concentrated or dried to yield products, which can be used as food colorants.

A useful component of the olive-mill wastes with potential biological interest due to its antioxidant activity is squalene. Squalene is a naturally occurring polyprenyl[38] compound primarily known for its key role as an intermediate in cholerestol synthesis. Squalene is not very susceptible to peroxidation and appears to function in the skin as a quencher of single oxygen, protecting human skin from lipid peroxidation due to exposure to UV and other sources of ionizing radiation. Squalene may also act as a "sink" for highly lipophilic xenobiotics. The primary therapeutic use of squalene is as an adjunctive therapy in a variety of cancers. It received its name because of its occurrence in shark liver oil (*Squalus* spp.), which contains large quantities and considered the richest source of squalene. However, it is widely distributed in nature, with reasonable amounts found in olive oil, palm oil, wheat-germ oil, amaranth oil, and performs critical biological functions. Squalene is found in high concentrations in olive-oil residues after the last production step (deodorization) and is regarded as a waste product of the refineries. Olive oil deodorization distillate contains squalene in a concentration range 10–30%. Squalene and squalane are high value, biological raw materials and are required

---

[38]Prenyl also known as isoprenoid or isoprene.

in large quantities in the health food and pharmaceutical industries. Existing attempts to obtain squalene from olive-oil residues by distillation methods have not resulted in producing squalene in economically viable quantities.

ES8602102 (1986) describes a process for the recovery of squalane and squalene from the by-products of the refining of olive oil by subjecting the by-products, optionally after hydrogenation, to successive crystallizations in organic solvents to remove the insoluble impurities, and distilling the filtrates arising from the crystallization steps. Hydrogenation is effected under pressure using an Ni catalyst (1–2% wt). Organic solvents with a boiling point of 40–150°C, selected from saturated, unsaturated, aromatic, or halogenated hydrocarbons are used for crystallization, which is effected continuously. Distillation is effected at pressures below 650 Pa.

ES2004269 (1988) describes a process for obtaining squalane by: (i) subjecting the by-products from the refining of olive oil to a saponification using a 5% excess of a strong alkali (e.g. soda, potash, etc.); (ii) extracting the non-saponifiable substances by means of distillation by dragging a steam; (iii) hydrogenating the distilled product (containing over 90% squalene) with a Raney nickel catalyst (0.1–0.2% wt.) at a temperatures of 180–250°C; (iv) purifying the hydrogenated product (crude squalane) by means of deodorization, deparaffination by crystallizing and filtering at low temperature.

An alternative process for the recovery of squalene uses supercritical carbon dioxide (CO_2) extraction. This process consists mainly of converting the free fatty acids and the methyl and ethyl esters normally occurring in this by-product into their corresponding triglycerides. The latter are then extracted with supercritical CO_2 to provide a highly enriched squalene fraction. The process has been carried out on a pilot plant scale with a column operating in the counter-current mode. By the use of this process, squalene can be recovered in high purity and yields of about 90% (Bondioli P. et al., 1993). An improved process was developed by the partners of EU project: FAIR2-CT95-1075. The new process comprises the steps: (a) saponifying olive oil residues; (b) drying the saponified material to a residual water content below 1%; (c) integrally extracting the resulting soap with CO_2 as the solvent; (d) purifying the extract by chromatography; and optionally (e) hydrogenating the squalene obtained to squalane. The apparatus for the production of both compounds includes one production installation for the integral and continuous performance of steps (c), (d), and (e) with supercritical CO_2. The results of this work have been patented (DE19934834, 2001). The process avoids the disadvantages of the processes known from ES2004269 (1988) and ES8602102 (1986). The former involves the use of high temperatures with accompanying squalene decomposition and isomerization as well as the use of Raney-nickel for the hydrogenation, which results in an allergic squalane product by virtue of the presence of nickel residues. The latter also involves the use of a nickel catalyst as well as solvents which cause environmental problems.

Another interesting compound for recovering from olive-mill wastes is a glucoside known as oleuropein — see Chapter 2: "Characterization of olive processing waste", section: "Organic compounds". A number of scientific studies have shown this compound to have certain antiviral, antifungal, antibacterial, antioxidant, and anti-inflammatory properties.

Currently, most of the oleuropein commercially available to consumers is derived from olive leaves. To date, the fruit of the olive plant, which is rich in oleuropein, has largely been ignored as a source of oleuropein due to certain problems associated with the production of olive oil. The oleuropein and its derivatives are water-soluble polyphenolic compounds, produced in olive pulp and are found in abundance in OMWW. Similarly, a number of monophenolic compounds, such as tyrosol and its derivatives, produced in olive stones, are also abundant in OMWW. Current technology does not permit the isolation of oleuropein and its derivatives from such highly polluting monophenolic compounds in OMWW except through time-consuming and expensive separation processes. For these reasons, OMWW is currently treated as waste and is discarded without realizing its content of oleuropein.

Recently, a process was disclosed for the recovery of oleuropein from the olive vegetation water (WO0004794, 2000; WO0218310, 2002). According to one aspect of the process, the stones are removed from the olives prior to pressing[39]. The destoned pulp is then pressed to obtain a liquid-phase mixture including olive oil, vegetation water and solid by-products. The vegetation water is separated from the rest of the liquid-phase mixture and collected.

It should be appreciated that the vegetation water produced in this manner is substantially free of compounds that are found primarily in olive pits, such as tyrosol and other highly polluting, monophenolic compounds. The vegetation water thus obtained may be used, for example, in a variety of ways not amenable to conventional vegetation water: (i) as a natural antibacterial, antiviral and/or fungicidal product for agricultural and/or pest control applications, (ii) as a raw material for the production of oleuropein and other antioxidants for a variety of medical purposes (e.g. holistic medicine), and (iii) as a therapeutic and/or an antioxidant beverage for a variety of health purposes. The vegetation water or a concentrate or isolate thereof can be administered orally or parenterally to human bodies.

The vegetation water or extract may be concentrated by distillation under vacuum. The concentrate may be dried by spray drying or oven drying under vacuum to obtain a powder containing oleuropein. It may be desirable to conduct such steps at a temperature no greater than about 88°C to avoid degradation of the glucoside. The oleuropein can then be purified, for example, by chromatographic separation procedures[40].

In conclusion, the results indicate that OMWW has a powerful antioxidant activity, and thus might be a cheap source of natural antioxidants. The water of the

---

[39]For purposes of commercial production, the apparatuses disclosed in US4452744 (1984), US4522119 (1985), and US4370274 (1983) are recommended. Additional devices that may be used are disclosed in IT1276576 (1997), IT1278025 (1996).

[40]Techniques suitable for concentrating and/or isolating oleuropein from aqueous and aqueous–alcoholic solutions are taught, for example in US5714150.

olives has 300 to 500 times higher levels of polyphenols than in extracted oil (Dr Roberto Crea in "Olive Oil Source"[41]). This is a particular advantage that OMWW, which is an undesirable by-product of the olive oil manufacturing process, can be productively used. Formulations of antioxidant polyphenols derived from OMWW are now being used in health foods such as the CreAgri's Olivenol polyphenol extract (WO0004794, 2000; WO0218310, 2000; "Olive Oil Source"). One tablet of Olivenol® is claimed to contain the equivalent amount of antioxidant polyphenols present in approximately 4 to 6 ounces of high quality extra virgin olive oil without the calories. The French company Societé distillerie also commercializes its polyphenol extracts, which in a case have been used by a client for antioxidant-enhancement of cosmetics.

Oleanolic acid and/or maslinic acid and physiologically acceptable salts thereof can be obtained from by-products generated in the olive-mill manufacturing processes such as strained lees, extraction residues, squeezed oil, extracted oil, degummed oil scum, deacidified oil scum, dark oil, waste decoloring agent, deodorized scum, exhausted cake, and OMWW. Processes for the industrial recovery of oleanolic and/or maslinic acid from olive by-products resulting from the milling and processing of olives, either proceeding from three-phase or two-phase extraction systems is described in: WO9804331 (1998) and WO0212159 (2002). These processes enable to obtain, by separation and with purities higher than 80%, of both acids with yields comprised between 0.2 and 1.5%, as a function of the product and prime material processed. Fundamentally, they comprise selective extractions and fractionation of resulting mixtures with the use of solvents, including liquefied gas extractions. The extracted oleanolic and/or maslinic acid can be converted into physiologically acceptable salts — for the purpose of making the product water-soluble — by treatment with a basic medium. The physiologically acceptable salts of oleanolic and/or maslinic acid can be subjected to concentration and/or fractionation–purification treatments to thus give highly purified physiologically acceptable salts of oleanolic and/or maslinic acid (WO0212159, 2002).

The oleanolic acid (3-β-hydroxy-28-carboxyoleanene) is a triterpenic acid (2-α, 3-β-dihydroxy-28-carboxyoleanene) and it has been found in almost a hundred plants. It has been attributed with a number of proven biological activities (abortifacient, anti-carcinogenic, antifertility, antihepatotoxic, anti-inflammatory, antisarcomic, cancer-preventive, cardiotonic, diuretic, hepatoprotective, and uterotonic). The maslinic acid, also known as cratzegolic acid, has been found in a dozen plants. It is known to have antihistaminic and anti-inflammatory activity although it has not been extensively studied because of its scarcity. The isolation of oleanolic and maslinic acids from waxes on the surface of the fruit of the *Olea europaea* has been described by means of methanol extraction from olives previously washed with chloroform (Binachi G. et al., 1994).

---

[41]http://www.oliveoilsource.com.

It has been found that the extracts of the olive leaves titrated in oleanolic acid show a strong inhibitory activity against the enzyme testosterone 5-α-reductase. Similarly, the oleanolic acid and the extracts of olive leaves show an antimicrobial activity against *Propionibacterium acnes* and *Acinetobacter calcoaceticus*. The use of this mixture, optionally associated with *Larrea divaricata* extract, which is titrated with nordihydroguaiaretic acid (NDGA)[42], can be used for the treatment of acne, hyperseborrhea, and skin with acneic tendencies (FR2830195, 2003).

Enzymes

OMWW has been proposed as a substrate for laccase production by white rot fungi (Kahraman S. and Yesilada O., 2001; Fenice M. et al., 2003). The possible use of OMWW as a growth medium for the production of extracellular laccase and manganese peroxidase (MnP) from the white rot fungus *Panus tigrinus* CBS 577.79 was studied using a properly formulated OMWW-based medium (2-fold diluted OMWW supplemented with 0.5% sucrose and 0.1% yeast extract) either in a stirred-tank or an air-lift reactor (Fenice M. et al., 2003). Solid-state fermentation was also performed in a rotary drum reactor using maize stalks moistened with the OMWW-based medium. Highest levels of laccase and manganese peroxidase activity were obtained in the stirred-tank reactor (4600 ± 98 U/l on day 13) and in the air-lift reactor (410 ± 22 on day 7). Based on total enzyme activities, solid state fermentation appears to be more suitable than liquid submerged fermentation but the latter exhibits better volumetric productivities.

The enzymatic product obtained by the biological treatment of OMWW has been used in the olive oil extraction process to improve olive oil yield and quality. A multiphase disposal process for OMWW was studied in relation to two main phases of the technological process: biological treatment of OMWW to obtain an enzymatic concentrate and direct recycling of this concentrate in the mechanical olive oil extraction process (Montedoro G.F., 1993). The disposal treatment for OMWW includes static settling, sterilization, fermentation, centrifugation (in which biomass is recovered and used as an animal feed), and ultrafiltration (resulting in a permeate that is discarded and an enzymatic retentate concentrate). The enzymatic retentate concentrate is recycled and used in the malaxation step in the mechanical olive oil extraction process. The fermentation step with the yeast *Cryptococcus albidus* var. *albidus* IMAT 4735 led to production of a pectinase (polygalacturonase) with an activity of 25 VU/ml[43] in the culture broth. After centrifugation, the broth was

---

[42]Substance found in abundance in the oleoresins of Larrea (Chaparral) and the Guaiacum genus (Lignum Vitae). It is strongly antioxidant to lipids and is antifungal, antimicrobial, and antibacterial. Synonyms: β,γ-dimethyl-α,δ-bis-(3,4-dihydroxyphenyl) butane; 4,4′-(2,3-dimethyltetramethylene) dipyrocatechol.

[43]Enzyme activity was expressed in viscometric units (VU). 1 VU was defined as the amount of enzyme that decreased the initial viscosity of the substrate solution by 50% in 1 min.

concentrated by ultrafiltration (4–5 times) and used in the extraction process. This operation resulted in an increase of 8–9% in olive oil yield and improved oil quality; turbidity, oxidation induction time, chlorophyll, and contents of aromatic compounds were generally improved.

Preliminary results have shown that a commercial strain of *P. ostreatus* can be used for setting up a solid state fermentation for the treatment of OMWW (Setti L. et al., 1998). This study was made using a laboratory-scale bioreactor in which some of the limiting parameters for the growth of both the mycelia and the fruiting bodies of *P. ostreatus* such as temperature, relative humidity, and light can be controlled. Edible white rot fungi were cultivated on expanded clay beads supplemented with OMWW and compared to those grown in liquid-state fermentation Folin-detectable phenols were removed by at least 80–90% from the OMWW within 2 days in a recycle solid state bioreactor more efficiently than in liquid-state fermentation. This detoxification process of OMWW can also be accompanied by a decrease in the COD of OMWW. This study demonstrated that high value products of industrial interest, such as edible biomass and enzymes, peroxidases, and phenol oxidases, could be produced in great quantities from agro-industrial wastewater by treatment with basidiomycetes in solidstate fermentation. These enzymes are studied for their capability of reducing the toxicity of many aromatic compounds. However, the laboratory results of the EU project: AIR3-CT94-1987 "BIOWARE" indicated that while *Pleurotus* grew well in batch culture on OMWW, this requires addition of specific nutrients the addition of which would entail costs and complications in a full-scale plant that could not be justified.

Production of Various Products

Alcohols

OMWW has a sugar content of about 1.6–5% (w/v) that can serve as a source for alcohol production (Fiestas Ros de Ursinos J.A., 1961a,b, 1967; Fernández-Bolaños J. et al., 1983). A possibility for the utilization of sugars is their transformation to ethanol and recovery of the alcohol by distillation (Martinengri G.B., 1963; Oliveira de J.S., 1974). Oliveira de (1974) studied the effect of the yeasts *Saccharomyces* wine 31 B2, *S. mollasses*, bread yeast, *Candida utilis* and the natural fauna on ethanol production from OMWW and, with the exception of some reports (Martinengri G.B., 1963), no essential differences in the amounts of alcohol produced 0.5–0.57% (w/v) were found. Some of early efforts extracted alcohol from OMWW by evaporation. The extracted alcohol was used in foods, fuels, cosmetics, etc. (PT69240, 1979; PT69785, 1979).

OMWW has, though, a toxic effect on yeasts and to counteract this effect some investigators have recommended a dilution to 2% sugar (Fiestas Ros de Ursinos J.A., 1967). Bambalov G. et al. (1989) confirmed that fresh OMWW was unfavorable to yeast growth. Eight culture-collection yeast strains of various species and five yeast strains isolated and identified by the authors were tested for both growth in OMWW and fermentation of the sugars in the same media. The culture-collection yeast strains did not grow in an effluent containing 2.86% sugar (w/v), 8 g/l phenolic substances, 4.58 g/l titratable acidity and pH 4.96, whereas the isolated strains of *Torulopsis* sp[44]. MK-1, *Saccharomyces norbensis* MC-1, *S. oleaceus* MC-2, and *S. oleaginous* grew well and fermented the sugars and produced alcohol in amounts of 1.63 to 1.38%, respectively. None of the yeasts grew in OMWW vacuum-concentrated to over 13–14% of dry matter. The strain of *Torulopsis* sp. MK-1 showed a higher stability.

One of the objectives of the EC project: AIR3-CT94-1987 "BIOWARE" was the recovery of ethanol by selecting yeast strains able to degrade the sugars present in OMWW, with a conversion efficiency near the theoretical values. It was concluded that the quantity of ethanol that might be produced using yeast strains to treat the OMWW is too low, due to the low levels of sugars and hence the production of ethanol was of no economical interest. All these studies confirmed that alcohol fermentation by yeasts is not an economical way of OMWW utilization, mainly due to the toxicity of the substrate and to a low alcohol concentration in the fermentation broth.

In another study solventogenic *Clostridium* spp. were used for butanol production from OMWW considering also the removal of COD (Wähner R.S. et al., 1988). As the OMWW did not support growth of the strains even if it was supplemented with nutrients, all the experiments were performed with diluted OMWW. A 50% dilution (COD value 100–120 g/l) was the lowest one, which allowed normal growth and solvent production. Butanol yields, ranging from 0.09 to 0.29 g per gram of sugar content and COD removals as high as 85% were achieved in small-scale experiments. The yield of butanol based on total sugar content was of the same order as the values reported for other agricultural wastes (Maddox I.S. and Murray A.E., 1983). The concentration of butanol achieved (2.8–8 g/l) and the COD reduction of 85% for OMWW with a high COD value (100–120 g/l) suggest that OMWW is an interesting raw material for butanol production. In addition, the acetone-butanol fermentation produces a significant reduction of COD value.

The recent development of a new two-phase centrifugation process for extracting olive oil in Spain has substantially reduced water consumption, thereby minimizing wastewater. However, a new high sugar content residue is still generated (2POMW). In a study the two fractions present in this residue (olive pulp and fragmented stones) were assayed as substrate for ethanol production by the simultaneous saccharification and fermentation (SSF) process (Ballesteros Perdices I. et al., 2001). Pretreatment of fragmented olive stones by sulfuric acid-catalyzed steam explosion was the most effective treatment for increasing enzymatic digestability. However, a pretreatment step was not necessary to bioconvert the olive pulp into ethanol.

---

[44]*Torulopsis* is considered by some authorities to be a synonym of *Candida*.

The olive pulp and fragmented olive stones were tested by the SSF process using a fed-batch procedure. By adding the pulp three times at 24 h intervals, 76% of the theoretical SSF yield was obtained. Experiments with fed-batch pretreated olive stones provided SSF yields significantly lower than those obtained at standard SSF procedure. The preferred SSF conditions to obtain ethanol from olive stones (61% of theoretical yield) were 10% substrate and addition of cellulases at 15 filter paper units/g of substrate.

ES2056745 (1994) discloses a process for obtaining mannitol[45] and derived products from OMWW, olive twigs, leaves, and stalks. The process comprises the steps of (i) partial or complete drying; (ii) extraction of mannitol by means of alcohols or hydroalcohols; (iii) the extracts obtained are defecated by the addition of basic lead acetate so as to eliminate the materials which accompany them; (iv) the extracts so purified, are concentrated and the mannitol is isolated by crystallization with alcohol (96°). The process is slightly different depending on the raw material which is processed, needed prior concentration, and even drying where the raw material is OMWW. A modified process for obtaining mannitol and its derivatives from 2POMW is illustrated in Fig. 10.6 (ES2060549, 1994).

ES2143939 (2000) discloses the use of a steam explosion process to extract mannitol from olive cake coming from a three-phase centrifugation system. With this process the olive cake is treated in a 21 steam explosion unit at temperatures around 200°C for time periods of 2–4 minutes, there then occurring an abrupt decompression and the subsequent unloading of the reactor. All the mannitol present in olive cake is separated out and recovered and by means of various simple purification stages (ultrafiltration, ion exchange, and fractionated crystallization) permissible in food industry, can achieve a yield with a high degree of purity.

Biosurfactants

As part of the effort to contribute to the recycling of OMWW, it was investigated whether OMWW could be used as raw material for biosurfactant production (Mercadè M.E. et al., 1993). Different biosurfactant-producing strains were assayed and several strains of *Pseudomonas* sp. were able to grow on OMWW as the sole carbon source and accumulate rhamnolipids. Samples of OMWW were diluted depending on their composition and it was only necessary to add $NaNO_3$ (0.25 g/l). Conversion yields of 0.058 g of rhamnolipid per gram of substrate (OMWW) were achieved and COD of OMWW was reduced approximately 50% in 72 h. An improved process was developed for rhamnolipid production from OMWW as feedstock in a stirred tank fermentor with *Pseudomonas aeruginosa* (Mercadé M.E. and Manresa M.A., 1994). *P. aeruginosa* JAMM (NCIB 400440) was selected for its capacity to decrease surface tension when grown on OMWW (Mercadè M.E., 1990 and Mercadè M.E. et al., 1993). The surfactant concentration increased during

---

[45]White crystalline, sweetish, water-soluble, carbohydrate alcohol, $HOCH_2(CHOH)_4CH_2OH$, occurring in three optically different forms.

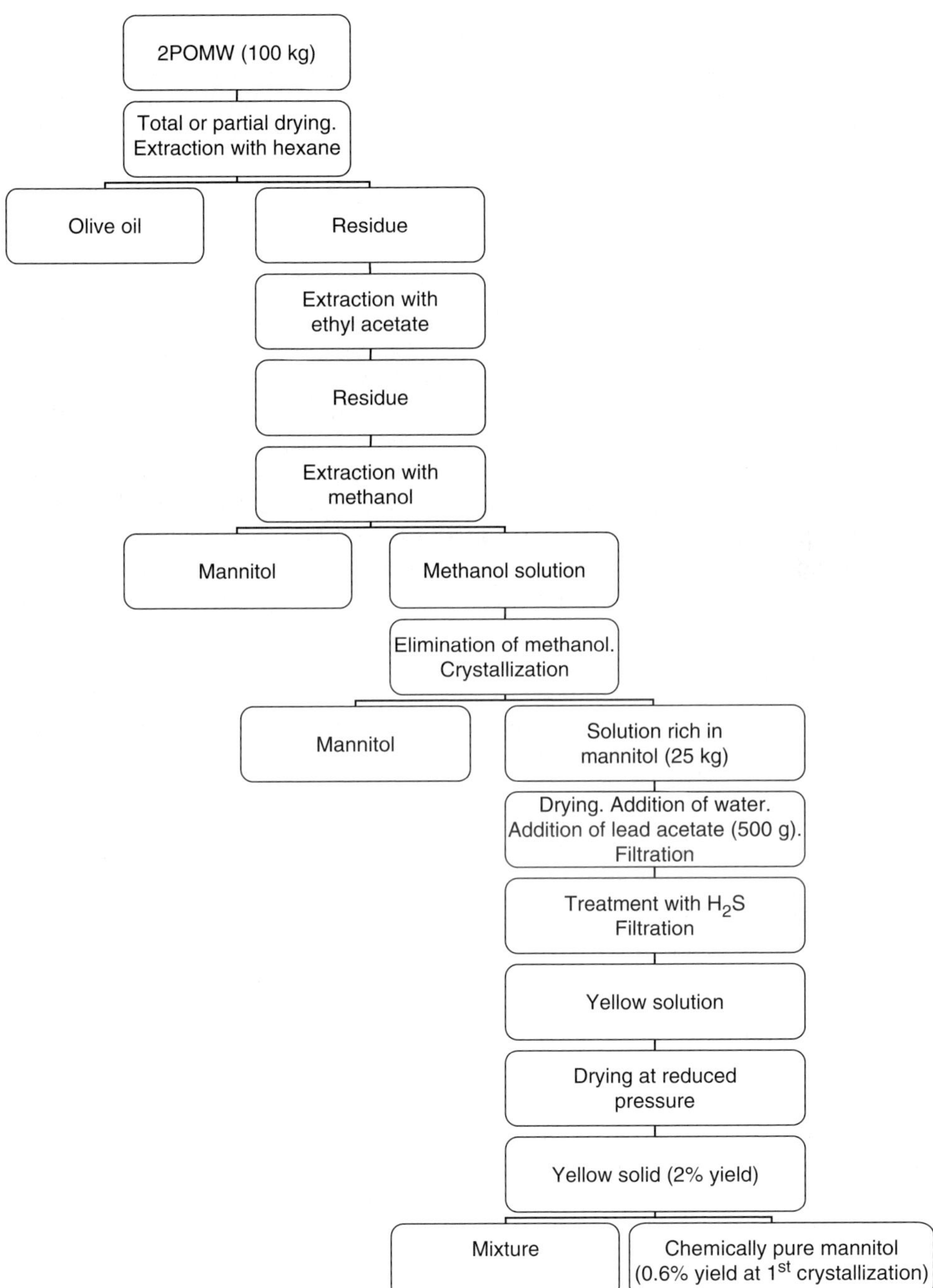

Fig. 10.6. Modified process for obtaining mannitol and its derivatives from 2POMW (ES2060549, 1994).

incubation, achieving a final value of 1.4 g/l. The conversion yield was 0.058, calculated on the basis of the COD (24 g/l) of the culture medium.

However, the yields for biosurfactant production from OMWW are low. From the economic standpoint, biological surfactants are not yet competitive with their synthetic counterparts.

Biopolymers

OMWW has been proposed as a low-cost substrate for xanthan production with the additional environmental benefit of this use (López López M.J., 1996; López López M.J. and Ramos-Cormenzana A., 1996, 1997; López López M.J. et al., 2001a,b). Xanthan gum, an extracellular heteropolysaccharide produced by the bacterium *Xanthomonas campestris* is the most commercially accepted microbial polysaccharide. Because of its special rheological properties, this biopolymer is widely used as thickener or viscosifier in food, cosmetics, pharmaceuticals, paper, paint, textiles, adhesives, and tertiary oil recovery. The idea of utilizing OMWW for xanthan production is supported on the basis of its high C/N ratio with a concentration of free sugars up to 4–5% — see Chapter 2: "Characterization of olive processing waste", optimal for polymer production. Moreover, OMWW includes organic acids and other compounds such as carbohydrates and phenolics that could serve as a carbon source for polymer production, and *X. campestris* is able to metabolize some phenolic substances. The use of OMWW as a substrate could reduce the cost of xanthan production which is the greatest factor limiting the use of xanthan in large-scale fermentation processes when compared with similar polymers from algae or plants.

Growth and xanthan production on dilute OMWW as a sole source of nutrients were obtained at OMWW concentrations below 60%, yielding a maximal xanthan production of 4.4 g/l at 30–40% OMWW concentration. Addition of nitrogen and/ or salts led to significantly increased xanthan yields with a maximum of 7.7 g/l. The nitrogen/salts supplements also allowed an increase in the optimal OMWW concentration. Inocula pregrown on OMWW can be used. Results suggest that an improved xanthan yield could be obtained with adequate balance between waste concentration and nitrogen or salt supplementation. This process could be improved by selecting the proper *X. campestris* strain. Differences among strains were found in the range of tolerance to OMWW concentration and xanthan amount obtained. The most valuable strain was *X. campestris* NRRL B-1459-S4L4II because of its capability for producing xanthan from 50–60% OMWW as the sole nutrient source (López López M.J. et al., 2001a).

OMWW has also been used for the production of homo- and copolymers of polyhydroxyalkanoates (PHAs) (Martínez-Toledo M.V. et al., 1995; González-López J. et al., 1996; Pozo C. et al., 2002). PHAs are reserve polyesters that are accumulated as intracellular granules in a variety of bacteria. These polymers are usually synthesized under unbalanced growth conditions, whereby depletion of an essential nutrient other than the carbon source promotes the formation of high-energy storage polymers. Of these polymers, poly-β-hydroxybutyrate (PHB) is the most common. PHB production by *Azotobacter chroococcum* strain H23 was reported in chemically defined medium and OMWW medium (Martínez-Toledo M.V. et al. (1995). Preliminary data (González-López J. et al., 1996) show that *A. chroococcum* strain H23 is also able to form PHA copolymers containing β-hydroxybutyrate and β-hydroxyvalerate (P[HB-co-HV]) in NH_4^+ — amended media, using either glucose (1%) or OMWW (15%) as sole carbon source. Large amounts of homopolymers (PHB) and copolymers (P[HB-co-HV]) were produced by *A. chroococcum* strain H23 without nutrient limitation when growing in culture media amended with a high concentration of OMWW as the primary carbon source (Pozo C. et al., 2002). P[HB-co-HV] copolymer was formed when valerate (pentanoate) was added as a precursor to the OMWW medium, but it was not formed with the addition of propionate as a precursor. *A. chroococcum* formed homo- and copolymers of PHA up to 80% of the cell dry weight, when grown on NH_4^+ — medium supplemented with 60% (v/v) OMWW, after 48 h of incubation at 100 rev min^{-1} and 30°C. The results show that OMWW supports the growth of strain H23 and also that this waste could be utilized as a carbon source. Production of PHAs by using OMWW looks promising, since the use of inexpensive feed-stocks for PHAs is essential if bioplastics are to become competitive products.

PHAs were also produced by growing the nutritionally versatile *Pseudomonas putida* on OMWW. The transformation with the plasmid pSK2665, harboring *Alcaligenes eutrophus* (*Ralstonia eutropha*) genes needed for synthesis of poly (3-hydroxybutyric acid), allow *P. putida* strain to grow in high concentration of OMWW accumulating biodegradable thermoplastic (Ribera R.G. et al., 2001).

The dark polymeric organic fraction rich in potassium recovered from OMWW and named polymerin and the potassium salified deglycosylated polymerin derivate (K-SD polymerin) could be recovered for possible use in agriculture as bioamendments, macro- and microelement biointegrators, and due to their similarity with humic acids, as a biofilter for toxic metals (Capasso R. et al., 2002a,b) — see Chapter 2: "Characterization of olive processing waste", section: "Organic compounds" and Chapter 3: "Environmental effects", section: "Effect on soil".

The EU project: QLK5-2000-00766 "BIOLIVE" aims to develop a technology capable of using exhausted olive cake (*orujillo*) for the fabrication of new biopolymers. The exhausted olive cake has a high lignin, cellulose, and hemicellulose content. It is possible that these natural biopolymers could be transformed by liquefaction in new monomer compounds for the fabrication of polyurethane and phenolic resins. Significant amount of biomass based phenol and polyol compounds can substitute traditional phenol and polyol from fossil origin.

Activated Carbons

Activated carbons have been prepared from olive stones, and solvent extracted olive pulp. Raw material is an abundant and cheap waste by-product of oil production, making these activated carbons economically feasible. The solid residues are carbonized at 850°C and activated physically either with CO_2 or steam at 800°C (Mameri M. et al., 2000a; Galiatsatou P. et al., 2001, 2002). Another possibility of using olive waste as raw material to produce activated carbons by chemical and physical activation methods has also been investigated (Moreno-Castilla C. et al., 2001). In the first case, KOH and H_3PO_4 were used as activating agents, and in the second case, CO_2 at 840°C for different periods of time. Results obtained indicate that the chemical activation of olive waste with KOH at 800°C, in an inert atmosphere, produced activated carbons with much lower ash content, higher nitrogen surface area and much better developed porosity than in the case of either its chemical activation with H_3PO_4 or its physical activation with CO_2 at 840°C. The activated carbons can have different uses, such as absorbents in liquid and gas phases, catalysts, and support for catalysts. The activated carbons were proved to be efficient adsorbents for the removal of phenols and COD decrease in OMWW. They are used for the treatment of contaminated water (El-Sheikh A.H. et al., 2004) and they showed also a high capacity to adsorb herbicides (2,4-dichlorophenoxy-acetic acid, 2,4-D; and 2-methyl, 4-chlorophenoxyacetic acid, MCPA) from water, with adsorption capacity values higher than those corresponding to a commercial activated carbon used from drinking water treatment.

So far only olive stones have been used to produce different activated carbons. Alternatively, OMWW was used to mix with the residual fly ash produced by coal combustion in thermoelectric power plants in order to increase the adsorptive capacity of fly ash, by means of a new way of aggregation (Rovatti M. et al., 1992). The product of the aggregation was submitted to a pyrolysis and activation process in order to obtain an adsorbent material. The pyrolysis produced an oily liquid fraction, with a good calorific value, a high hydrogen content gaseous fraction, and a carbonaceous matrix dry residue. This solid product had good mechanical strength, wet strength, high porosity, and high specific surface area and good adsorptive properties. Due to its good microstructural characteristics, the solid product was utilized in adsorption processes, on laboratory apparatus, in order to evaluate its adsorptive capacity for organic vapors. The results obtained with gas-solid adsorption experiments in fixed-bed bench scale plant, using toluene vapors as the adsorbed gas, verified the possibility of getting useful products at a high added value from waste materials of difficult disposal qualities.

Generation of Energy

Waste treatment technologies aimed at energy recovery may represent an interesting alternative for a sustainable disposal of residues from olive oil production, able to reduce the environmental impact and to generate electric energy for sale or satisfy the energy needs of olive-mills. However, these disposal systems are characterized by a rather high technological level requiring remarkable capital investments and qualified personnel; moreover, plant management is onerous and complex. Therefore, such solutions are well suited for centralized approaches, thanks to the economy of scale and the opportunity of pursuing high levels of efficiency and reliability (Caputo A.C. et al., 2003).

The residual biomass of olive processing with a potential energy use is classified into two groups. The first group is constituted by residual biomass produced during olive tree culture (pruning and harvest residues). The second group is constituted by residual biomass produced during the various stages of olive oil extraction. Depending on the extraction system the available energy from the by-products is different. For instance, exhausted olive cake and 2POMW are characterized by an average heating value of 19,000 and 14,000 kJ/kg, respectively. The by-products of both groups present, from an energy point of view, favorable aspects in their use, e.g. ensured annual production, relative concentration in a place, proper humidity conditions, low sulfur content, and other harmful emissions, and finally, high thermal value. Not using these resources originates environmental problems due to limited storage life, plague propagation, and forest fires (Jurado F. et al., 2003). Furthermore, energy produced by using biomass helps "Sustainable Development" and meeting targets in the agreement of Kyoto. However, an appropriate technology must be employed to avoid the production of pollutants and other problems, while maximizing process efficiency.

Solid olive-mill wastes have been traditionally used as fuel, both domestic and industrial. These wastes have been burnt to feed the boilers which provide the thermal energy for the evaporation/distillation of OMWW — see also Chapter 6: "Thermal processes", section: "Irreversible thermo-chemical processes". Olive cake has also been used commercially as a fuel for pottery kilns and as an alternative energy source by the brick industry (Nicoletti G., 1999).

There are three main thermo-chemical methods by which this renewable energy source can be utilized, namely gasification, briquetting and combustion (direct firing), or co-combustion (co-firing). Another type of gasification involves the generation of biogas (methane) by the anaerobic degradation of olive-mill wastes.

Gasification

Gasification — known also as pyrolitic distillation — is a thermo-chemical process that converts biomass into a combustible gas called producer gas (syngas). Producer gas contains carbon monoxide, hydrogen, water vapor, carbon dioxide, tar vapor, and ash particles. Gasification produces a low- or medium-Btu gas, depending on the employed process, which can be used in many combustion systems such as boilers, furnaces, and gas engines. Some technology issues regarding the fluctuation in the quality of the gas and change in the gas composition need to be resolved before the gas can be used in combustion systems (Dally B. and Mullinger P., 2002).

The gasification technology is in the development stage. The main drawback from such an approach is the high cost associated with initial setup and operation of these facilities. There are a few demonstration projects that use varied gasifier designs and plant configurations. However, pretreatment of biomass feedstock is generally the first step in gasification. Pretreatment involves drying, pulverizing, and screening.

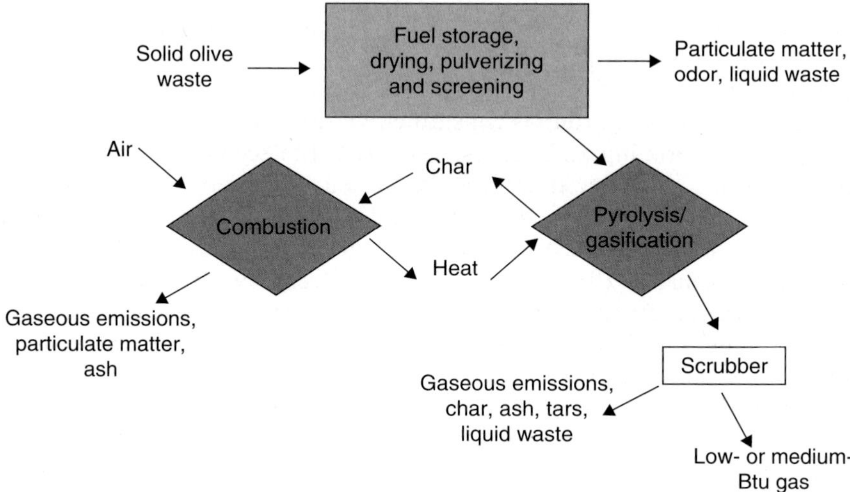

Fig. 10.7. Schematic representation of the biomass gasification process.

Optimal gasification requires dry fuels of uniform size, with moisture content no higher than 15–20%.

Biomass gasification is a two-stage process — see Fig. 10.7. In the first stage, called pyrolysis, heat vaporizes the volatile components of biomass in the absence of air at temperatures ranging of 450–600°C. Pyrolysis vapor consists of carbon monoxide, hydrogen, methane, carbon dioxide, volatile tars, and water. The residue, about 10–25% of the original fuel mass, is charcoal. The second stage of gasification is called char conversion. This occurs at temperatures of 700–1200°C. The charcoal residue from the pyrolysis stage reacts with oxygen, producing carbon monoxide. Gasification, pyrolysis, and degradation kinetics experiments on olive cake and agricultural residues (wood chips, wheat straws, grape residues, and rice husks) revealed that char from olive residues has the least nitrogen and sulfur content between all residues (Di Blasi C. et al., 1999a–c).

In the process of combustion, both stages of gasification occur. When the residue burns, the heat of combustion produces pyrolytic vapors. Some gasification of these vapors also occurs. In combustion, however, the pyrolytic vapors are immediately burned at temperatures in the range of 1500–2000°C. In contrast, the process of gasification is controlled, allowing the volatile gases to be extracted at a lower temperature before combustion.

Fluidized bed gasification is considered to be the most advanced method for thermo-chemical conversion of various biomass fuels (agroresidues, wood-energy crops, etc.) to energy offering economical and environmental benefits. Ash-related problems such as sintering, agglomeration, deposition, erosion, and corrosion, which are due to the low melting point of ash in the agroresidues, are the main obstacles for economical and viable application of this conversion method for energy exploitation of the specific residues. Among the different ash constituents, chlorine, followed by potassium appears to play the most important role regarding the reactivity of the ash fraction in biomass and its behavior during the gasification process. Leaching (washing) and fractionation pretreatment techniques have been tested on their ability to handle the ash-related problems caused during the gasification of olive residues. (Arvelakis S. et al., 2001a,b, 2002, 2003; García-Ibañez A. et al., 2004).

Arvelakis S. et al. (2003) studied the effect of leaching and fractionation pretreatment techniques on the gasification of olive cake. Gasification tests were performed in a laboratory scale fluidized bed gasifier at the temperature level of 800°C using silica sand as the bed inert material. Fractionation pretreatment shows to lead to a substantial increase of the agglomeration problems during the gasification process. The removal of the fine particles from the olive material during the fractionation procedure led to a substantial increase of the ash reactivity. As a result agglomeration in the case of tests with fractionated olive cake appeared in a significantly shorter time and decreased the operation time of the reactor to almost the half compared to the tests with the untreated olive cake. On the contrary, leaching pretreatment showed a highly positive effect as far as the ash thermal behavior of the olive cake is concerned. Leaching led to a significant expulsion of alkali metals and chlorine. The leaching of inorganic constituents from the olive material led to changes in inorganic composition and substantial improvements in ash thermal behavior under gasification conditions. As a result, the ash of the leached samples appears to have a very low tendency to cause agglomeration/deposition problems. Leaching proved to extend significantly the operation time of the gasifier (3 to 6 times) compared to the tests with the untreated and fractionated olive cake.

Leached and exhausted 2POMW has been tested in a 300 kWth atmospheric circulating fluidized bed (CFB) gasification facility, using air as a fluidization agent (García-Ibañez A. et al., 2004). The first tests have demonstrated that the CFB test rig operates adequately and makes it possible to carry out gasification experiments with exhausted 2POMW as a fuel. The lower heating value of the producer gas obtained is 3.8 MJ/Nm^3 at the lowest temperature (780°C). The carbon conversion in exhausted 2POMW gasification at 800°C was in the range of 81.0–86.9%. The increase in equivalence ratio did not improve carbon conversion significantly but increased the gas yield.

A new technique to perform the gasification of solid olive-mill wastes was developed within the framework of the EU project: FAIR CT96-1420 "IMPROL-IVE". The gasifier is a fluidized/moving system, a rather new concept of reactor because of the special configuration of the reactor zones. In the bottom part, the fluidized bed allows the required combustion, consisted of exothermic reactions, necessary to maintain the thermal balance inside the whole reactor. In the upper part, the moving bed zone does not allow the combustion process but only the endothermic gasification processes. This is due to the fact that the raising gas that reaches the moving bed contains a very low concentration of oxygen and has a high temperature (800–850°C). Therefore, only the gasification process can be performed in the moving bed. The olive waste used for the gasification was exhausted 2POMW of mean particle size 1.4 mm and olive stones of mean particle size 2.57 mm. The fluidized bed was filled with sand of mean particle size 0.21 mm, or in some runs, with dolomite with a mean particle size of 0.35 mm. The gasification is made in autothermal conditions, that is, a fraction of the solid waste (about 50%) is burned to maintain the high temperature required while the other 50% suffers gasification. The electrical heating is only used during the start-up and slightly during the operation. The low heating value of the flue gas is similar to other biomass gasification processes (4–6 MJ/Nm^3). The typical composition of the flue gas is: 7–10% H_2, 2.5–6% CH_4, 6–18% CO, 0.06–1.6% C_2H_4, and 64–84% of no combustible gases, mainly CO_2, N_2, and H_2O. The presence of sand and dolomite in the fluidized bed does not affect appreciably neither the tar production in the moving bed nor the flue gas composition.

In another study, olive cake samples were subjected to direct and catalytic pyrolysis to obtain hydrogen-rich gaseous products at desired temperatures (Demirbaş A., 2001; Caglar A. and Demirbaş A., 2004). The samples, both untreated and impregnated with catalyst, were pyrolyzed at 775, 850, 925, 975, and 1025 K temperatures. The total volume and the yield of gas from both pyrolysis were found to increase with increasing temperature. The largest hydrogen-rich gas yield obtained from olive cake, using about 17% $ZnCl_2$ as catalyst at about 1025 K temperature, is 70.6%. In general, in the pyrolysis of biomass, the yield of the hydrogen-rich gaseous product increases with $ZnCl_2$ catalyst, but the yield of pyrolytic gas decreases in spite of increasing the yield of charcoal and liquid products. The catalytic effect of K_2CO_3 was greater than that of Na_2CO_3 for the olive cake.

Gasification technologies under development would enable the solid olive-mill wastes to be used in gas turbines. A variety of relatively large-scale biomass gasification technologies are at various advanced stages of development. Three gasifier/gas cleanup designs are considered: (i) atmospheric-pressure air-blown fluidized bed gasification with wet scrubbing; (ii) pressurized air-blown fluidized bed gasification with hot-gas cleanup; (iii) atmospheric-pressure indirectly heated gasification with wet scrubbing. Jurado F. et al. (2002, 2003) developed a detailed model simulating the performance of a combined-cycle power plant based on biogas gasifier/gas turbine technologies. The modeled performance of the alternative gasifiers is given in Table 10.7. The feed stock in all cases is olive residues with 20% moisture content with the following composition (dry mass basis): 50.2% carbon, 5.4% hydrogen, 34.4% oxygen, 0.2% nitrogen, and 4% ash. Its higher heating value

Table 10.7. Modeled performance of alternative gasifiers

	Low-pressure indirect heat	Low-pressure air-blown	High-pressure air-blown
Carbon to gas*	70.1	96.9	97.4
HHV**, MJ/kg	18.1	6.47	5.48

*% Carbon in fuel divided by carbon into gasifier.
**HHV: Higher heating value.

(HHV) is 20–47 MJ/dry kg. The gasifier is capable of converting tons of olive residues into a gaseous fuel that is fed into a gas turbine. The tested gasifiers enable the use of advanced power systems that will nearly double the efficiency of today's industry. The gasifier heats the residues in a chamber filled with hot sand until the olive residues break into basic chemical components. The solids — sand and char — are separated from the gases, which then flow through a scrubber. The final result is a very clean-burning gas fuel suitable for direct use in modern power systems such as gas turbines.

Briquetting

Briquetting is a low-cost technique used to agglomerate a wide range of materials into fuel blocks to be transported and utilized as solid fuel. It has been widely used to upgrade coal dust and mine waste for use as barbecue fuel. Different biomass products have been considered for biobriquetting including, among other agricultural wastes, solid olive-mill wastes. In such an approach one needs to consider five main issues, namely, shatter index, compressive strength, water resistance, combustion characteristics, and emission of pollutants (Dally B. and Mullinger P., 2002). Solid olive residues have low compression strength and shattering index, even when milled down to 0.25 mm size particles, which requires the addition of a binding agent for them to become usable. Olive residues also have reasonable water resistance when compared to other biomass products. This strength decreases with the amount of moisture in the residue. This depends on the initial water content and the applied pressure (Yaman I. et al., 2000).

One way to improve the properties of briquettes from olive residues is to add paper waste, which contains fibrous material increasing in this way the shatter index substantially. In addition, the waste paper has similar combustion characteristics to those of olive residues and will have minimum effect on the burning rate (Dally B. and Mullinger P., 2002).

Emissions from the combustion of briquettes can vary substantially. Burning is usually undertaken in a relatively uncontrolled environment and can be very harmful to the environment. However, considering that the need for alternative fuels will increase in the near future, briquettes offer a substantially better alternative to coal (Dally B. and Mullinger P., 2002).

An early process describes the use of OMWW for making solid fuel (ES8404708, 1984). The process comprises grinding and crushing of a solid fraction made up of vegetable residue, the organic fraction of solid domestic or industrial residues and the sludge from the purification of OMWW. This is followed by incorporation of a liquid fraction consisting of OMWW. By mixing and rotating the induced fermentation breaks the vegetable fibers homogenizing and increasing the energy content of the product, which is then granulated, palletized, briquetted, or compacted.

Several environmentally acceptable smokeless briquettes have been prepared with different coals and olive stone as biomass. Blesa M.J. et al. (2001) and Moliner R. et al. (2004) studied the effect of the pyrolysis process on the physico-chemical and mechanical properties of these briquettes. Feedstocks for smokeless fuel briquettes manufacture were carbonized in order to reduce the volatile matter and the sulfur content of the coal. Coal was pyrolyzed at temperatures of 500–700°C and the temperature chosen to carry out pyrolysis was 600°C due to the lowest content of sulfur per thermie in the pyrolyzed material. In order to study the influence of the pyrolysis process on the properties of the briquettes, biomasses were pyrolyzed separately at 400 and 600°C and together with the coal at 600°C of temperature. The materials pyrolyzed at 600°C showed a lower content of volatile matter and a higher calorific value than the standard levels reported in the literature for materials to prepare smokeless briquettes. The briquettes were prepared by mixing the pyrolyzed materials with humates as binder and $Ca(OH)_2$ as sulfur sorbent. The briquetting process was followed by Fourier transform infrared spectroscopy (FTIR), scanning electron microscopy (SEM), CO_2 adsorption and the mechanical properties were tested evaluating their impact resistance, water resistance, and compression strength. The best briquettes with respect to the mechanical properties were those prepared with coal and biomasses co-pyrolyzed at 600°C although some of them fixed a higher percentage of sulfur during pyrolysis due to the metal content of the biomasses.

In two other papers Blesa M.J. et al. (2003a,b) studied the effect of the curing temperature and time on the physico-chemical and mechanical properties of these briquettes. Humates and molasses were used as binders which act with different roles, as a film or matrix depending on the curing. Additives, like H_3PO_4, were also tested. This acid was added to favor the polymerization of the binder. The effect of the curing temperature and time on the briquettes has been studied by FTIR, temperature programmed decomposition (TPD) followed on-line by mass spectrometry (MS) and optical microscopy (OM). FTIR and TPD have been used to determine the presence of different oxygenated functional groups and the reactions of polymerization. OM is used to find the morphology of the briquettes which influence their final properties. TPD experiments help to predict the final properties of the briquettes more clearly than FTIR. The aliphatic structures and methoxy groups as well as the hydrogen bonds decrease during the curing. On the other hand, the carboxylic groups tend to be formed due to the oxidation produced by the effect of curing temperature. In addition, the briquettes cured at 200°C for 2 h showed the highest mechanical strength. Moreover, the mechanical resistance of the studied briquettes is improved by the effect of the curing time. These curing conditions also produce waterproof briquettes due to the presence of carboxylic groups which contribute to the stabilization of the briquettes because of the formation of hydrogen bonds. The effect of the H_3PO_4 on briquettes prepared with molasses produce a stabilization of these materials.

Co-Combustion

The olive cake can be considered as an alternative fuel, which does not contain sulfur. The olive cake is quite dense ($0.5\,g/cm^3$ at moisture level of 5–6%) and has a calorific value of 12,500–21,000 kJ/kg. It is comparable with the calorific values of wood and soft coal, which are 17,000 and 23,000 kJ/kg, respectively. The sulfur content of olive cake is about 0.05–0.1% (Atimtay A. and Topal H., 2004). Efficient use of olive cake in energy production solves two problems in one step: clean energy production and acceptable disposal of olive-mill waste.

Research work on direct combustion of olive cake using a fluidized bed combustor showed that olive cake could not be burnt without other energy being supplied to the process (Kraisha Y.M. et al., 1998). In this study olive cake was burnt in a fluidized bed combustor 0.146 m in diameter and 1 m long. The waste was predried to minimize its water content by placing a known quantity on a tray and then drying it in an oven at 105°C for 1 h; 8 kW electrical heating coil was used to preheat the waste to self-ignition as it moved through the bed.

2POMW is also a potential fuel with high calorific value. Drying of 2POMW should always be a previous operation for combustion by reducing its moisture content (EU project: FAIR CT96-1420 "IMPROLIVE"). 2POMW has been used as a fuel for electricity production in a fluidized bed combustor, although operation difficulties with failures in the bed during the fluidization have imposed severe constraints for commercial operation. These difficulties are related to the high moisture content and alkaline content in the ash (Armesto L. et al., 2003).

Co-combustion (co-firing) of solid olive-mill wastes refers to the use of one or more additional fuels (e.g. wood or coal) simultaneously in the same combustion chamber of a power plant. Co-combustion of these olive residues with coal is generally viewed as the most cost-effective approach. Solid olive-mill wastes have similar density, heat release, and general burning characteristics as that of coal. Research has shown that blending olive residues with coal is an attractive approach where efficiency is maintained, emissions are reduced and it requires minimal modification to the feeding system.

In the literature, several studies have been reported on the co-combustion of olive cake with various combustible materials, such as coal, shale oil, and diesel oil, in a fluidized bed combustor for energy production. (Abu-Qudais M., 1996; Abu-Qudais M. and Okasha G., 1996; Alkhamis T.M. and Kablan M.M., 1999; Cliffe K.R. and Patumsawad S., 2001; Suksankraisorn K. et al., 2003). Abu-Qudais M. (1996) investigated the combustion characteristics of olive cake (with 6% moisture content) in a fluidized bed. He found that the combustion efficiency ranges from 83 to 95%, depending on the excess air supplied.

Topal H. et al. (2003) investigated the combustion characteristics of olive cake and lignite coal (Tunçbilek) by burning them separately in a CFB of 125 mm diameter and 1800 mm height. The results obtained for olive cake and lignite coal were compared with each other to show the effect of high volatile content of olive cake on combustion characteristics and emission behavior of the CFB. In a complementary study of the previous work by Atimtay A. and Topal H. (2004) various mixtures of the same olive cake and lignite coal — 25, 50, and 75% olive cake mixed with lignite — were co-combusted and the combustion characteristics of olive cake/lignite coal mixtures were investigated by using the same bed. The combustion experiments were carried out with various excess air ratios. When the excess air is increased, there is a sharp decrease in the CO and hydrocarbons (C_mH_n) concentrations. The optimum value seems to be about 50% excess air for the fuels used in this study. The combustion efficiency increases as the excess air ratio increases, indicating that volatiles released from the fuel burn more completely as the air amount per unit mass of fuel is increased. The results suggest that olive cake is a good fuel that can be mixed with lignite coal for cleaner energy production in small-scale plants by using CFB. The mixing ratio of olive cake to lignite coal is suggested to be below 50 wt% in order to be within the limits set for biomass by the EC Directive-2001/180/EC (200 mg/N m^3 for SO_2, 400 mg/N m^3 for CO, and 400 mg/N m^3 for NO_x, if the size of CFB > 50 MWh).

Cliffe K.R. and Patumsawad S. (2001) studied the feasibility of co-firing 2POMW and coal mixtures in a bubbling fluidized bed combustor designed for coal combustion. The olive residue used in this study had a moisture content of around 60%, which has the effect of reducing the overall combustion efficiency. 2POMW with up to 20% mass concentration was co-fired with coal in a fluidized bed combustor with a maximum drop of efficiency of 5%. 2POMW, mixed with coal, greater than 20% by mass caused the bed temperature to drop to a level such that combustion could not be sustained. A 10% olive cake concentration in the fuel mixture gave a lower CO emission (33% reduction) than 100% coal firing even though the bed temperature was dropped by 25°C. This behavior is attributed to the improved combustion in the freeboard region or just above the bed surface. A 20% olive cake mixture gave a higher CO emission than both 100% coal firing and 10% olive cake mixture, but the combustion efficiency was higher than the 10% olive cake mixture. This was because the freeboard temperatures were lower than when burning 10% concentration 2POMW due to the heat in flue gas being used to evaporate the water in the mixed fuel so leading to a lower reaction rate.

Suksankraisorn K. et al. (2003) tested the co-combustion of 2POMW with coal in a fluidized bed combustor of 0.15 m in diameter and 2.3 m height. Bed depths up to 0.3 m with 2 m in freeboard height were used. CO emissions are relatively insensitive to changes in the fraction of 2POMW in the fuel mixture. The SO_2 emissions were reduced as the amount of olive cake in the mixture increase as a result of fuel-sulfur dilution. A slight increase in NO and N_2O emissions was also observed.

Armesto L. et al. (2003) assessed also the feasibility of co-firing 2POMW and coal in a bubbling fluidized bed. Two different Spanish coals were selected for this study, lignite and anthracite. The combustion tests were carried out in the CIEMAT bubbling fluidized bed pilot plant. In order to study the effect of different parameters on the emissions and combustion efficiency, the tests were done using different operating conditions: furnace temperature, share of 2POMW in the mixtures and coal type. The pilot plant tests show that the combustion of 2POMW/lignite or anthracite mixtures in bubbling fluidized bed is one way to utilize this biomass residue in energy generation. The presence of 2POMW in the mixtures has not any significant effect on the combustion efficiency. SO_2 and NO_x emissions decrease when the amount of 2POMW in the mixtures increases, while N_2O emission increases. The SO_2 emissions correlate quite well with coal sulfur content. As could be expected, when the share of 2POMW in the mixtures increases the SO_2 emissions decrease. The amount of 2POMW in mixtures causes a slight decrease of the NO_x emissions because of higher volatile content. On the contrary, the N_2O emissions increase is attributed to the decrease of the flame temperature caused by the high moisture content of 2POMW.

It is evident from the above discussion that combustion of solid olive-mill wastes (olive cake or 2POMW) is an attractive option and energy companies are starting to exploit the potential of these residues as biomass fuel for energy production. Although the combustion characteristics of solid olive-mill wastes are not dissimilar to that of low to medium rank coals, where most of the mass is released in the temperature range 260–350°C and char burnout occurs at about 700°C, care would need to be taken in utilization in any power station boiler. This is because in the modern electricity plant availability is the overriding concern. Furthermore, owing to the ash content of the wastes they could only be utilized in a boiler designed to handle ash and equipped with dust collectors. Moreover, exhaust gases have to be treated, leading to additional costs (Dally B. and Mullinger P., 2002). Extensive research is under way to fully understand the burning characteristics of such fuels and the initial results suggest that, in the near term, the highest-efficiency, lower-cost, lower-risk technology is co-combustion with coal in industrial and utility boilers. By co-firing olive waste with coal, a continuous supply of waste would not be an issue, since the boiler plant would always have the primary fuel (coal) for 100% utilization. Co-firing also offers an approach to both waste management and energy production since the equipment needed to handle and burn away wastes is similar to that which is required for coal. Solid olive-mill wastes could replace some coal for energy production or capture a portion of the expanded production (Cliffe K.R. and Patumsawad S., 2001).

Fluidized bed combustion has been shown to be the technology of choice capable of burning practically any waste combination with low emissions. The significant advantages of fluidized bed combustors over conventional combustors include their compact furnace, simple design, effective burning of a wide variety of fuels, relatively uniform temperature, and the ability to reduce emission SO_2 and NO_x emissions (Cliffe K.R. and Patumsawad S., 2001).

Biogas Production

Anaerobic biogas production is an effective process for converting a broad variety of biomass to methane to substitute natural gas and medium calorific gases. For instance, biogas obtained by the anaerobic treatment of $1\,m^3$ of OMWW contains 60–80 kWh of energy. The process can be carried out in relatively inexpensive and simple reactor designs and operating procedures.

The biological disposal of OMWW by anaerobic degradation has been investigated by several researchers for the production of methane (Fiestas Ros de Ursinos J.A. et al., 1982; Aveni A., 1983, 1984; Boari G. et al., 1984; Rigoni-Stern S. et al.,

1988; Rozzi A. et al., 1989a; Dalis D. 1991; Martín-Martín A. et al., 1991; Georgacakis D. and Dalis D., 1993; Tekin A.R. and Dalgiç A.C., 2000). However, this technique has a low efficiency due to the toxicity of the waste and demands high investment and working cost, see also Chapter 8: "Biological processes", section: "Anaerobic processes".

Laboratory experiments demonstrated that biogas could be produced from dilute OMWW by using a suitable anaerobic filter (EC project: AIR3-CT94-1987 "BIOWARE"). Considerable problems arose when attempting to transfer the laboratory results to a semi-industrial scale, in particular the need for a digester larger than $200\,m^3$ capacity to attain an economic break-even point. As this effectively excludes small- and medium-scale mills, state support is considered essential if the technology involved is to be made generally available (Bonfanti P. and Lazzari M., 1999).

Biogas production from a slurry obtained by mixing finely ground exhausted olive cake in water was investigated using anaerobic digesters of 1 l working volume at 37°C (Tekin A.R. and Dalgiç A.C., 2000). A start-up culture was obtained from a local landfill area and it was adapted to the slurry within 10 days at this temperature. The biogas generation rates were determined by varying the total solids (TS) concentration in the slurry and the hydraulic retention time (HRT) during semi-continuous digestion. The maximum rate was found to be 0.70 l of biogas per liter of digester volume per day, corresponding to a HRT of 20 days and 10% TS with a yield of 0.08 l biogas per gram COD added to the digester. The methane content of the biogas was in the range of 75–80% for both batch and semi-continuous runs, the remainder being principally carbon dioxide. Considering optimum anaerobic biogas production, this corresponds to about $1 \times 10^7\,m^3$ (STP) of methane, which has an approximate energy value of $4.0 \times 10^8\,MJ$. Biogasification of this residue and utilization of its stabilized form as a fertilizer is considered an alternative to its utilization as a fuel.

Erŏglu E. et al. (2004) reported the production of photobiological hydrogen by using OMWW as a sole substrate source. The photobiological hydrogen generation is receiving considerable attention in solar energy-based biotechnological research as a potential source of renewable and pollution-free fuel. Among the various organisms capable of hydrogen production, photosynthetic bacteria are more promising due to their relatively higher conversion yields of organic substrates into hydrogen, ability to trap energy at a wide range of the light spectrum and versatility in sources of metabolic substrates that increases their potential to be used in association with waste treatment. *Rhodobacter sphaeroides* O.U.001 (DSM 5864) was used in this study and the hydrogen production experiments were performed in 400 ml jacketed glass-column photobioreactors. Hydrogen production studies on diluted-OWMW were investigated in the range of 20% (v/v) and 1% (v/v) OMWW containing media. Below 5% OMWW containing media, bacterial growth rate fitted well to the logistic model where hydrogen production was observed for the ones below 4% OMWW. A maximum hydrogen production potential (HPP) of $13.9\,l\,H_2/l$ OMWW was obtained at 2% OMWW. During the biological hydrogen production process, COD of the diluted wastewater decreased from 1100 to 720 mg/l; BOD_5 decreased from 475 to 200 mg/l, and the total recoverable phenol content (ortho- and meta-substitutions) decreased from 2.32 to 0.93 mg/l. In addition, valuable by-products such as carotenoid (40 mg/l OMWW) and polyhydroxybutyrate (PHB) (60 mg/l OMWW) were obtained. According to these results, OMWW was concluded to be a very promising substrate source for biohydrogen production process, with additional benefits of its utilization with regard to environmental and economical aspects.

Liquid Fuels

Demirbas A. et al. (2000) applied thermo-chemical conversion processes, including mainly pyrolysis and liquefaction to convert olive cake to liquid fuel. Alkali catalysts, such as NaOH, KOH, Na_2CO_3, K_2CO_2, and $Na_2C_2O_4$, were used in catalytic liquefaction experiments. The pyrolysis was applied within 600–850 K and the liquefaction experiments were performed within the range 425–625 K. The maximum yields of liquid products was 40.4% from pyrolysis and 85% from KOH catalytic (1/1, w/w) liquefaction runs at 575 K.

Dorado M.P. et al. (2004) investigated the possibility of making fuel out of used olive oil with an alkali-catalyzed transesterification process. Better results were obtained using KOH and methanol instead of NaOH and ethanol, which decreases transesterification rates. The presence of KOH and methanol above or below the optimum quantity decreases the ester yield because of the presence of soaps or unreacted glycerides, respectively. Settling at ambient temperature under 25°C increases the difficulty of ester and glycerol separation because of a conflict between glycerol solubility and low temperatures. This could be solved by increasing the settling temperature or the time for settling. In summary, the reaction was optimized at ambient temperature using 1.26% KOH, 12% methanol, 1 min of stirring, with 90 min of pour-off time, 11.38% distilled water by volume at 25°C to purify the ester, and drying over 0.5% Na_2SO_4. Losses of esters during the washing process were less than 4%. The ester yield of the reaction was 94%. The small presence of unreacted glycerides did not drop the engine performance. Fuel specifications were close to those of diesel fuel, thus indicating that methyl esters from used olive oil can be considered as a fuel candidate.

Miscellaneous

OMWW with or without pretreatment has been used to replace part of the water added at the various stages of olive oil extraction. (ES2010535 (1989) discloses a process where the hot water added to the milled olive beater, and the horizontal and vertical centrifuges is replaced by OMWW after it has been filtered to remove solid particles and preheated to the required temperature.

OMWW has been used traditionally to make soap. For instance, the so called *Marseille* soap was originally made in Gallipoli (Italy) from OMWW mixed with soda.

A process produces vitamin B_{12} from OMWW. Preparation of vitamin B_{12} from residual contaminants of the oil industry comprises cultivating *Propionibacterium shermanii* ATCC 13673 on predigested OMWW at pH 6.5–8.2 and 28–32°C and isolating the product (ES2122927, 1999).

OMWW has been tested as growth medium for the production of plant growth hormones from white rot fungi; *Funalia trogii* ATCC 200800 and *Trametes versicolor* ATCC 200801 have been tested for this purpose (Yurekli F. et al., 1999). Gibberellic acid, abscisic acid, indoleacetic acid, and cytokinin were determined in the culture media of these fungi. Both organisms produced enhanced levels of all three hormones in the presence of OMWW. The hormone indoleacetic acid was also experimentally produced from OMWW by *Arthobacter* spp. (Tomati U. et al., 1990).

A microbiological process for the biodegradation of aromatic compounds and synthesis of pigments, dyes, alkaloids, and polymers is effected using the recombinant strain *Escherichia coli* P-260 (access no. CET 4627) characterized by the constitutive expression of the enzyme 4-HPA (4-hydroxyphenylacetic acid) hydroxylase of *Klebsiella pneumoniae* (WO9804679, 1998). The use of this enzyme is of special interest due to its double catalytic activity: the hydroxylation of monohydroxylated compounds into others that are di-hydroxylated and the oxidation of the latter into quinones. The low specificity of the enzyme substrate allows a great variety of aromatic compounds to be transformed into di-hydroxylated compounds, which may be metabolized through the different biodegradation routes that exist in bacteria. The process leads to the decontamination of toxic or recalcitrant compounds in the environment of OMWW, such as phenolic compounds, among which 4-hydroxyphenylacetic acid (4-HPA), the enzyme's preferred substrate, constitutes one of the major ones (Balice V. and Cera O., 1984). Moreover, in the absence of concomitant catabolic routes, the oxidation of the di-hydroxylated intermediate into quinone leads to the formation of cyclical organic structures of possible industrial interest: pigments similar to melanin, as well as alkaloids and polymers.

The process is specifically used for:

- degrading benzene derivatives having a hydroxy group in the 3- or 4-position of the ring, especially 4-hydroxyphenylacetic acid, one of the compounds present in OMWW;
- degrading aromatic aminoacids (or their derivatives) with characteristics as in the benzene derivatives;
- degrading toxic aromatic compounds contaminating the environment, particularly the phenolic compounds of OMWW;
- synthesis of dihydroxy-aromatic compounds and quinones, especially the following compounds of industrial interest:
  i. pigments in general, particularly melanin-like pigments are useful as tints and colorants in cosmetics;

  ii. alkaloids;

 iii. substrate blocks for the synthesis of new penicillin antibiotic derivatives;

 iv. polymers of the "semi-ladder" or phenoplast type. The "semi-ladder" polymers are useful in the production of fibers, films, and adhesives.

FR2838451 (2003) discloses a process for removing organic pollutants from OMWW and converting them to a material useful as a wood glue, especially for chipboard manufacture. The process comprises the addition of a polyisocyanate to OMWW to form a polyurethane-containing precipitate and separating the precipitate by filtration.

A thermal lithosynthesis process is proposed for disposal of liquid and solid industrial wastes in the production of inert clay agglomerates (EP471132, 1992). The lithosynthesis process consists of a cold working step comprising: milling of clay into granules, mixing said clay granulates with liquid wastes such as OMWW, laminating and maturing said mixture to obtain a homogeneous paste; and a warm working step comprising: heating said paste by burners with conventional fuels to evaporate the water content and to obtain a granular material, heating said granular material by burners fed with conventional fuels and/or by exhausted and/or emulsioned oils to expand the inner portion of granules and to obtain a glassy outer layer thereof, cooling said expanded granules to attain clay agglomerates. The process does not need a special plant, but may be carried out in conventional kilns already employed for the production of the clay agglomerate. A plant treating a clay quantity of about 400–500 tons/day is possible to dispose about 40–50 tons/day of OMWW. The final granular product, which is chemically inert, may be advantageously used as a lightweight aggregate for concrete and cements.

One of the proposals of the EU project: ICA3-CT-1999-00011 "WAWAROMED" is to use the treated OMWW — see also Chapter 8: "Biological processes", section: "Phytoremediation (Wetlands)" — in the brick production. This will reduce the demand for fresh water of the brickyard, provided the treated wastewater fulfills the quality demand for water utilization in the brick production. But, because of the high temperatures employed in the process of brick backing the water quality can be "lower" than for other possible uses of the treated OMWW. The adhesive and stabilizing properties of OMWW have been used in Tunisia for stabilizing clay bricks and in the construction of roads and agricultural tracks (Friaâ A. et al., 1986a,b; Mensi R. and Kallel A., 1990).

Vegetation water or an aqueous solution part obtained by squeezing olive fruit or an extract of them using a water-insoluble organic solvent has been proposed as a deodorant particularly effective in deodorizing malodor components of tobacco odors and capable of being stably supplied (JP2003019192, 2003).

Powder produced from ground olive cake has been used as a drilling fluid additive in oil wells (US5801127, 1998). The olive powder is prepared by drying olive cake to remove residual water, and then grinding the cake to a particle size of less than 1500 μm, and preferably less than 200 μm. The olive powder can be added to oil- or water-based drilling fluids in an amount of 2–23 kg per 160 l (one 42-gallon barrel) of drilling fluid. The additive is environmentally acceptable and provides a cheap alternative way of disposing solid olive-mill waste. The drilling fluid reduces well cake permeability, seals off underground formations, and cavities to prevent fluid loss from the drilling fluid; furthermore, it reduces bit balling, drag, and torque and is biodegradable and non-toxic.

Part IV
Table Olives

Chapter 11

Table Olives

Introduction

The average world production of table olives for the harvesting years 1999/
2000–2002/2003 was 1,485,300 metric tons (IOOC, 2004)[46]. The world production of
table olives has been increasing and reached a record level of 1,773,500 metric tons in
the 2002/2003 season (nearly 50% increase based on the 1990/1991 season).
Production for the 2004/2005 season is estimated at 1,465,500 metric tons (IOOC,
2004)[47]. Most production occurs in countries around the Mediterranean basin and
in the Middle-East — see Fig. 11.1. The world's largest producer is the European
Union (EU) (43.9% of the total). The average production in EU for the harvesting
years 1999/2000–2002/2003 was 651,500 tons, with Spain accounting for 467,600
(71.8%), Greece 104,300 tons (16.0%) and Italy 66,500 (10.2%) tons of this figure.
Portugal and France produce 11,100 and 2000 tons of table olives, i.e. 1.7 and 0.3%
of the EU total, respectively — see Fig. 11.2. Apart from EU, other significant
producers of table olives are Turkey (10.5%), Egypt (10.6%), Morocco (5.6%),
Syria (8.2%), and the United States (6.6%).

The wide variety of table olives and the development of new presentations (whole,
pitted, sliced, stuffed) has made product diversification possible and has led to the
growth recorded in the table olives sector in the past few years. As in the case of olive
oil, the production of table olives fluctuates year by year as a result of the uncertainties of the weather and the process of alternate bearing of olive trees. Many of
the varieties used for table olives are also suitable for oil production and the output
of a particular grove in a given year may go partially for oil and partially for table
olives, depending on various factors including market prices, size, and quality of

---

[46]http://www.internationaloliveoil.org; last accessed March 2005.

[47]http://www.internationaloliveoil.org; last accessed March 2005. Table olive balances compiled at the
91st session of the IOOC, Madrid, Spain, June 2004.

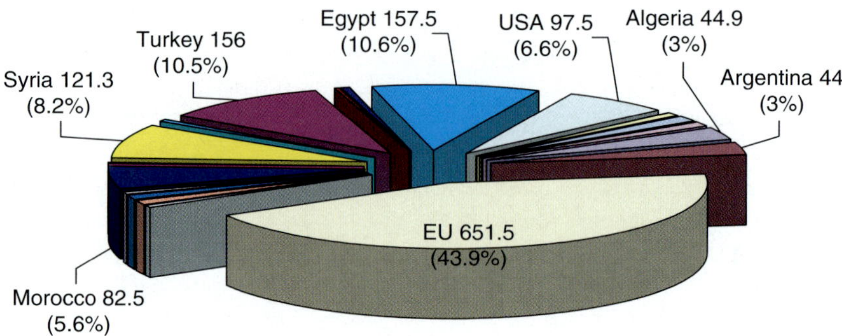

Fig. 11.1. Average world production of table olives (1000 tons) for the harvesting years 1999/2000–2002/2003. IOOC data, December 2004.

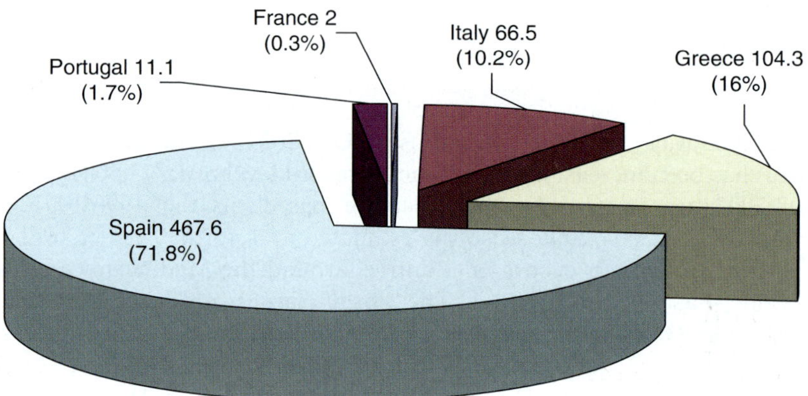

Fig. 11.2. Average production of table olives (1000 tons) in EU for the harvesting years 1999/2000–2002/2003. IOOC data, December 2004.

the olives. The differences in production practices are relatively minor; for example, table olive production usually involves more frequent and labor-intensive pruning and earlier harvesting. Olive groves which produce table olives cover a far smaller area than those producing olive oil. In Spain, less than 6% of the total area is devoted to table olive production, whereas the figure in Italy is less than 3%.

Table olives (also called edible or eating olives) are prepared from healthy, specifically cultivated, olive varieties picked at the right maturation stage and whose quality, after appropriate processing, corresponds to that of an edible well-preserved product. The most important industrial preparations of table olives are the Spanish (or Sevillian) for green olives, the Californian for oxidized black olives, and the Greek for naturally black olives. Half of the world production corresponds to Spanish-style table olives.

In the Spanish and Californian procedures, olives are treated with a diluted aqueous NaOH solution (lye) to hydrolyze their natural bitterness (oleuropein). After the treatment the olives are rinsed to remove the excess alkali, and the fruit is then left to ferment in brine (NaCl) for several months. In preparing Spanish-style green olives the appropriate olives are totally immersed in a lye solution until the lye penetrates about 1/8 inch into the flesh. In the Californian-style process, the dilute lye solution is administered in several repetitive steps to develop a dark oxidation ring at the surface of the olive, ending with a "pit lye" step wherein the alkaline solution is allowed to penetrate through the olive flesh to the pit of the olive. The total penetration results in a nearly neutral pH, e.g. between 7.0 and 8.0. The processing of naturally black olives, according to the Greek traditional method, includes only fermentation in brine. The preparation of green olives has a seasonal character, mainly occurring from September to November, while that of naturally black olives is spread more uniformly throughout the year.

In all stages of table olive processing large quantities of clean water are consumed for cleaning, debittering, washing, and fermentation and wastewater is produced. Kopsidas (1992) has reported a detailed description of the contents of different wastewaters generated in these stages for both green and black olives. Among these olive types, the preparation of green and black table olives can yield 3.9–7.5 and 0.19–1.9 m^3 of wastewater per ton of olives, respectively, depending on olive variety, maturity, and treatment process. Even though table olive production has undergone comprehensive modernization in the last decade, 1.2 l of fresh water per kg olives is still needed and the amount of wastewater generated remains unchanged. The table olive sector in EU is estimated to produce more than 750,000 m^3 of wastewater per year to be treated and disposed of. Currently, most of the wastewater generated in these processes is released untreated into the water environment. At best it is sent to evaporation ponds, where it stays for the rest of the year with subsequent bad and disagreeable odors, breeding of insects and risks of surface and ground contamination.

The wastewater that results from table olive processing is similar in nature to OMWW, although somewhat weaker in strength. Like OMWW, table olive processing wastewater has relatively high organic content and several polyphenolic and other toxic compounds. Both types of wastewater constitute a major environmental concern since they are characterized by high organic content and are generated in large quantities for a specific period of the year. Furthermore, they contain compounds such as polyphenols that are toxic to plants and/or soil microbes when discarded to soil receptors. Although OMWW has been extensively studied and a variety of applicable treatment methods have been proposed, the wastewater arising from table olive processing has received much less attention (Kyriacou A. et al., 2005).

A variety of methodologies, that have been used for the treatment of OMWW, have been proposed for the treatment of table olive processing wastewater, although not all are applicable to green olives, as the debittering stage leads to a wastewater with special characteristics (alkaline pH in combination with high phenol content) (Kyriacou A. et al., 2005).

The various processes for eliminating or recycling wastewater from table olive processing have been reviewed in an early paper by Garrido Fernández A. (1975), and in the book "Table Olives: Production and Processing" by the same author (Garrido Fernández A. et al., 1997). According to these reviews the most suitable processes are oxidation of organic material, purification by activated charcoal, physico-chemical processes (combined flocculation–filtration, reverse osmosis, hyperfiltration), and submerged combustion. A later paper studied the characteristics of the several procedures that might be used to regenerate green table olive fermentation brines with the aim of reusing them in other stages of the elaboration process and, especially for the final packing of the olives (Garrido Fernández A. et al., 1992a,b). Romero Barranco C. et al. (2001) reviewed the treatment and recycling techniques of spent brines and osmotic solutions, with emphasis to recently developed procedures for the management of spent brines resulting from Spanish-style green table olive production.

The table olive industry is under increasing pressure by the controlling authorities to find a disposal system that meets anti-pollution standards; at the same time such a solution has to be economically viable.

Characterization of Table Olive Processing Waste

The various types of wastewater generated during the various stages of table olive processing have very different characteristics. In addition, the values of the main parameters of each type of table olive wastewater vary widely depending on several factors such as olive variety and harvesting and processing conditions. Some representative values of a table olive wastewater, reported by Kopsidas G.C. (1992), are: pH, 3.6–13.2; suspended solids, 0.03–0.4 g/l; dissolved solids, 0.2–80.0 g/l; BOD_5, 0.1–6.6 g/l; COD, 0.3–18.2 g/l; NaCl, 0–80.0 g/l. A table olive wastewater contains also polyphenols, which hinder its biological treatment due to their biotoxic properties. In most cases it has a light yellow to brown colored appearance and is followed by odors.

The stages of debittering and subsequent washing produce the largest fraction of wastewater. Lye solutions and washingwaters present the highest COD and BOD_5 in some cases reaching values of up to 40 and 20 g/l, respectively. Some typical values of the main physico-chemical parameters of spent lye solutions are given in Table 11.1.

Brines from the green table olive processing constitute 80–85% of the total pollution load, but only 20% of the total liquid volume (Garrido Fernández A. et al., 1992b). In addition to salt (about 7 to 9%), brines contain organic substances such as lactic and other acids, proteinous materials, pigments, etc. The principal characteristics of brines are shown in Table 11.2.

The microbial succession in green olive fermentation is characterized by three distinct phases (Garrido Fernández A. et al., 1997). In the first phase, the microbial

Table 11.1. Typical values of the main physico-chemical characteristics of a lye solution (adapted from Beltrán F.J. et al., 1999)

Parameter	Value
pH	12.9
Total carbon (g/l)	8.1–8.8
A_{254}	5.79
A_{410}	2.94
BOD_5 (g/l)	2.5–3.3
COD (g/l)	22–25

A_{254} and A_{410} are absorbances at 254 and 410 nm, respectively, measured in 1 cm path length cell.

Table 11.2. Main physico-chemical characteristics of fermentation brines (Garrido Fernández A. et al., 1992b)

Parameter	Range
pH	3.60–4.30
NaCl (g/l)	60–90
Free acidity, as lactic acid (g/l)	5–15
Combined acidity (meq/l)	80–120
Color (A_{440}–A_{700})	0.20–0.60
Polyphenols, as tannic acid (g/l)	0.18–0.30
Suspended solids (g/l)	0.20–2.00
BOD_5 (g/l)	14–18
COD (g/l)	16–26

A_{440} and A_{700} are absorbances at 440 and 700 nm, respectively.

groups that dominate the process are Gram-negative bacteria, mostly belonging to *Enterobacteriaceae* family. The second phase is characterized by a progressive growth of lactic acid bacteria and yeasts and a gradual decrease of Gram-negative bacteria. During the third phase, an abundant growth of lactobacilli is observed, mainly from species of *Lactobacillus plantarum*, becoming the dominant microflora of the fermentation. On the contrary, in underbittered table olives, such microorganisms are almost inhibited. Inhibition of *L. plantarum* in the brines of these olives is attributed to the presence of phenolic substances and particularly oleuropein, an ester of elenolic acid and hydroxytyrosol. Although oleuropein undergoes a notable decrease during the course of fruit ripening, this compound is of considerable importance in process technologies of table olives due to its intense bitter taste and noteworthy bactericidal properties. This substance is enzymatically hydrolyzed by β-glucosidase, releasing glucose, and aglycones (the latter compounds are also bitter and have a distinctly inhibiting function toward *L. plantarum*). Oleuropein appears to be completely degraded by the NaOH treatment that is commonly used in the preparation of Spanish-style green table olives, yielding hydroxytyrosol and elenolic acid, with a consequent sweetening of the fruits (Marsilio V. et al., 1996). During the fermentation process phenols diffuse from the pulp into the brine. Also, even during the alkaline treatment, there is rapid diffusion of the pentacyclic oleanolic and maslinic acids salts into the brine (Bianchi G., 2003).

Hydroxytyrosol has been reported as the main phenolic compound in table olive processing wastewater (Brenes Balbuena M. et al., 1990; Castro Gomez-Millan A. de and Brenes Balbuena M., 2001). Other detected phenolic compounds include tyrosol, 3,4-dihydroxyphenylacetic acid, 4-hydroxybenzoic acid, catechol, vanillic acid, vanillin (Brenes Balbuena M. et al., 1990), caffeic acid, *p*-coumaric acid (Brenes Balbuena M. et al., 1990, 1995; Castro Gomez-Millan A. de and Brenes Balbuena M., 2001), elenolide and luteolin-7-glucoside (Brenes Balbuena M. et al., 1995). The major phenolic compounds identified in table olive processing wastewater are presented in Table 11.3.

Among the different wastewaters generated during the various stages of table olive processing, wastewater from the debittering and washing of green olives have the highest phenol content with concentrations of up to 6 g/l (García-García P. et al., 1989). Although the great diversity of phenolic compounds and organic acids

Table 11.3. Phenolic compounds detected in table olive processing wastewaters

Phenolic compounds	Synonyms	References
Benzoic acid		Kyriacou A. et al., 2005
Caffeic acid	3-(3,4-Dihydroxyphenyl)-2-Propenoic acid; 3,4-Dihydroxycinnamic acid	Brenes Balbuena M. et al., 1990, 1995; Castro Gomez-Millan A. de and Brenes Balbuena M., 2001
Catechol	Benzocatechol; *o*-Dihydroxybenzene; *o*-Benzenediol; 1,2-Benzenediol; catechin; 1,2-Dihydroxybenzene; *o*-Hydroxyphenol; 2-Hydroxyphenol; oxyphenic acid; phthalhydroquinone	Brenes Balbuena M. et al., 1990
trans-Cinnamic acid	3-Phenyl-2-propenoic acid; β-phenylacrylic acid	Kyriacou A. et al., 2005
p-Coumaric acid	3-(4-Hydroxyphenyl)-2-propenoic acid; *p*-Hydroxycinnamic acid; β-(4-Hydroxyphenyl)-acrylic acid	Brenes Balbuena M. et al., 1990, 1995; Castro Gomez-Millan A. de and Brenes Balbuena M., 2001; Kyriacou A. et al., 2005

Table 11.3. Phenolic compounds detected in table olive processing wastewaters — Cont'd

Phenolic compounds	Synonyms	References
Cyclohexanecarboxylic acid	Hexahydrobenzoic acid; cyclohexylcarboxylic acid	Kyriacou A. et al., 2005
3,4-Dihydroxyphenyla-cetic acid		Brenes Balbuena M. et al., 1990
Elenolide		Brenes Balbuena M. et al., 1995
Ferulic acid	3-(4-Hydroxy-3-methoxyphenyl)-2-propenoic acid; 4-hydroxy-3-methoxycinnamic acid; 3-methoxy-4-hydroxycinnamic acid; caffeic acid 3-methyl ether	Kyriacou A. et al., 2005
p-Hydroxybenzoic acid		Kyriacou A. et al., 2005
Hydroxytyrosol	2-(3,4-Dihydroxyphenyl)ethanol; 4-(2-hydroxyethyl)benzene-1,2-diol; 3,4-DHPEA	Brenes Balbuena M. et al., 1990; Sánchez A.H. et al., 1995; Castro Gomez-Millan A. de and Brenes Balbuena M., 2001; Kyriacou A. et al., 2005
Luteolin-7-glucoside	Cynaroside	Brenes Balbuena M. et al., 1995
2-Phenoxyethanol	Ethylene glycol monophenyl ether; phenylcellosolve	Kyriacou A. et al., 2005
Phenylacetic acid		Kyriacou A. et al., 2005
D-3-Phenylacetic acid		Kyriacou A. et al., 2005
Syringic acid	4-Hydroxy-3,5-dimethoxybenzoic acid; 3,5-dimethoxy-4-hydroxybenzoic acid; suringic acid	Kyriacou A. et al., 2005
3,4,5-Trimethoxybenzoic acid		Kyriacou A. et al., 2005
Tyrosol	*p*-Hydroxy phenyl ethanol; 4-Hydroxy phenyl ethanol; 1-Hydroxy-2-(4-hydroxyphenyl) ethane; *p*-HPEA	Brenes Balbuena M. et al., 1990; Kyriacou A. et al., 2005
Vanillic acid	4-Hydroxy-3-methoxybenzoic acid; 3-methoxy-4-hydroxybenzoic acid; protocatechuic acid 3-methyl ester; 4-hydroxy-meta-anisic acid; para-vanillic acid	Brenes Balbuena M. et al., 1990; Kyriacou A. et al., 2005
Vanillin	4-Hydroxy-m-anisaldehyde; 4-hydroxy-3-ethoxybenzaldehyde; zimco; vanilline; vanillaldehyde; vanillic aldehyde; lioxin	Brenes Balbuena M. et al., 1990; Kyriacou A. et al., 2005
Veratric acid	3,4-Dimethoxybenzoic acid; 3,4-dimethylprotocatechuic acid	Kyriacou A. et al., 2005

in table olive processing wastewater merits special consideration, as it affects its characteristics and treatability, the information found in the literature is very limited.

Effect of Table Olive Processing

Extensive research work has been carried out to diminish the volume of water used at the various stages of table olive processing (lye treatment, washing, and fermentation) with considerable success in reusing lye solutions (Garrido Fernández A. et al., 1977, 1979) and in the partial or total elimination of washing waters (González Cancho F. et al., 1984) without any negative influence on the quality of the final product. Important variations in the analyzed components were not found, except an increase in carotene content (Castro Ramos R. de, 1980). The reused volume of alkaline solutions generated by this procedure is discarded in shallow ponds, where they undergo solar evaporation (Garrido Fernández A. et al., 1992b). Although the reuse of lye solution presents certain economic advantages, and reduces waste elimination, it also brings about undesired contamination during the microbiological process (fermentation), which leads to a low-quality brine that is not reusable.

In an attempt to reduce the volume of wastewater from table olive processing, pilot plant trials were conducted, which replaced washing of Spanish-style green olives by neutralization of the lye with food-grade HCl (Castro Gomez-Millan A. de et al., 1983; González Cancho F. et al., 1983, 1984). Addition of HCl in one step produced, in some cases, a strong inhibition of lactic fermentation. This was less pronounced when HCl was added in two steps: (i) half of it in the fresh brine and (ii) the rest of it 48 h later, when the active fermentation is finished. The elimination of washing (but not of repeated lye use) resulted in higher levels of polyphenols, dissolved organic solids, COD and BOD_5 in the fermentation brine, and suppression of gram-negative bacilli. Reducing the lye concentration from 2% to <1.5% NaOH caused reduced growth of lactic bacteria and adverse color and flavor changes (González Cancho F. et al., 1983, 1984).

Papamichael-Balatsoura V.M. and Balatsouras G.D. (1988) used modified spent lye as cover brine of Conservolea olives subjected to fermentation. Green olives (Conservolea var.) were debittered in a 2.4% NaOH solution in the conventional manner and subjected to fermentation by covering with spent lye with added HCl to neutralize the alkalinity and sufficient salt to give a final concentration of 10% NaCl. Inoculation of olives with lactic acid starter was unnecessary. The course of fermentation was comparable to that of conventional processing of Spanish-style green olives. The end product possessed acceptable sensory qualities. A precipitate of dirty-green color and slightly higher bitterness of the product compared with the standard type of Spanish green olives, are the drawbacks of the technique. Advantages of the process are reduced labor costs, reduced pollution, elimination of brine acidification with lactic acid, and preservation of all olive flesh components.

ES2198215 (2004) and WO2005002364 (2005) disclose a process for the neutralization of olives, which have been subjected to alkaline treatment, involving the application of carbon dioxide in order to remove the lye (NaOH) present in the olive pulp. The carbon dioxide is applied under dry conditions in a pressurized atmosphere, and the olives are gently, mechanically agitated so that the carbon dioxide can contact the entire surface area of the olives. In this way, dry neutralization is performed, i.e. disposing with the normal need for water, thereby eliminating the standard discharges and considerably reducing the neutralization time. The amount of acid needed to neutralize the lye is reduced, and in the case of black olives, the neutralization and oxidation steps can be carried out simultaneously in the absence of water at the same time as is required for neutralization normally.

While the attempt to reduce the amount of wastewater generated during the debittering and subsequent washing stages of green table olive processing has been met with success, the brine volume cannot be reduced. The concentration of the brine depends on the variety of olive under treatment, and usually ranges from 3 to 5% initially. The concentration is gradually increased over a period of 3–4 weeks to a level of 7–9%, again depending on variety. In most cases, the olives are held in the brine for a period of 1–6 months. At the end of this treatment, the olives are removed for further processing, leaving the spent brine as a waste material. Before packing, the brines are discarded with other liquid residues, producing serious problems in the receiving streams because of their high concentration of salt, free acidity, and BOD_5. Their regeneration could eliminate about 80% of the total pollution and facilitate the recycling of the remaining liquid wastes. In this sense, experiments to reuse brines in different degrees of dilution for a few fermentative processes failed, probably due to the presence of metabolites from the first fermentation with inhibitory action against lactic bacteria. A more suitable solution might be to use these brines, after partial regeneration, for packing.

Brenes Balbuena M. et al. (1989a) studied the effect of reusing brines, regenerated by ultrafiltration or adsorption with activated carbon, on some quality attributes of packed green table olives, when the product is conserved either by its physico-chemical characteristics or by pasteurization. Results indicated that color of fruits is unaffected while texture showed a tendency to improve when regenerated brines are used. Polyphenol concentration and color of the solution followed equilibrium laws, although pasteurization produced a slight darkening in the brine. Organoleptic tests did not show significant differences when brines regenerated by both systems were reused for packing, except in the case of regeneration by activated carbon, followed by pasteurization. Research carried out to study the influence of some additives for preventing darkening indicated that ascorbic acid and sulfur dioxide are effective for olive fruits and the later is also useful for improving the solution color.

The reuse of fermentation brines regenerated by ultrafiltration through a 4000 Da cut-off membrane in canned anchovy-stuffed olives and olives packed in pouches has been shown to be possible (Rejano Navarro L. et al., 1995). The pasteurization of the product packed in plastic pouches gave rise to undesirable browning in olives and brines as a consequence of the oxidation of orthodiphenols, particularly hydroxytyrosol. In contrast, packing regenerated brines in aluminum pouches gave a commercially acceptable product. When olives were packed with regenerated brines with the combined acidity corrected, a slow increase in the pH of brines with time was observed. The cut-off of the ultrafiltration membrane was important in determining the acceptability of anchovy-stuffed olives packed with regenerated brines. Thus, olives packed with fermentation brines ultrafiltrated through membranes of 200,000 Da had a poor flavor; but, when membranes with a 4000 Da pore size were used, the olives were acceptable.

Panagou E.Z. and Katsaboxakis C.Z. (2006) studied the effect of reusing brine on the spontaneous fermentation of cv. Conservolea green olives in comparison to other initial brining conditions. The different treatments included: (a) brine acidification with 2‰ (v/v) lactic acid (control), (b) addition of 25‰ (v/v) 1N HCl, (c) substitution of the initial brine by 20% (v/v) with a brine from a previous fermentation. Brine reuse process was the most effective in minimizing the likelihood of spoilage since it decreased the survival period of enterobacteria (24 days), followed by the HCl treatment (28 days), and the control (35 days). However, after 35 days of fermentation, pH values reached a plateau above 4.8 in all treatments indicating that supplementary treatments were necessary to enhance lactic acid fermentation and attain acidity/pH levels that would improve the physico-chemical characteristics of the final product and ensure its safety. Addition of 1.5% (w/v) glucose in the HCl treated and brine reuse processes as well as 5‰ (v/v) lactic acid in the control was performed. All supplementary treatments were effective in reducing pH to a final value of 4.3–4.5. However, glucose increased the final concentration of lactic acid in brine reuse and HCl treated processes (73.4 and 67.8 mM, respectively) compared with the control that was lacking in acidity (44.7 mM), denoting a clear advantage of glucose over lactic acid as a supplement.

Environmental Effects

The disposal of the table olive wastewater has become a serious problem to the industry. Long term acceptance of table olive wastewaters has contaminated the treated wastewater percolation ponds, underlying soils, and groundwater under a small city's wastewater treatment plant (Martin C.J., 1992).

The lye solution is contaminated with impurities which impart a deep brown, almost black, color and an unpleasant odor to the lye. Because the BOD_5 of this waste is very high, it cannot be released into municipal sewage treatment systems without first being diluted with enormous quantities of water.

The corrosive nature of the salt in the brine and the fact that the waste contains both a non-biodegradable salt and organic solids, make also the disposal problem a particularly difficult one. If, for example, the spent brine is discharged into a site from which it can enter into wells or streams used for irrigation, the quality of water from such source will be decreased and in severe cases the water will actually cause deterioration of the soil to which is applied so that it will no longer produce satisfactory crops. In addition to the effect on soil, sodium and chloride ions have been shown to exert a specific toxicity on certain plants. Almonds may develop tip burn, and avocados a leaf scorch due to excessive sodium. Among the crops that are sensitive to chloride ion are peaches and other stone fruits, pecans, some citrus varieties, avocados, and some grapes. In addition, organic acids, such as lactic acid, which are present in the spent brines, are detrimental to the soil because they lower the pH of the soil considerably. In general, neutral soil, that is pH 7, is necessary for good crops (US3732911, 1973). Very limited information, if any, exists, on the potential of recycling this kind of wastewater in agriculture.

A field experiment was conducted to examine the effect of drip irrigation using table olive wastewater on physiological, nutritional and yield parameters of olive trees (Murillo J.M. et al., 2000). Two types of wastewater were used in the experiment, the first with specific absorption rate (SAR) and electrical conductivity (EC) values of 12–56 and 3.5–4.2 dS/m, respectively, and the second 73–90 and 4.3–6.0 dS/m. In general, this kind of wastewater has a highly variable composition and SAR values that are too high for agricultural purposes. Olive trees rapidly responded to wastewater application. Compared to the control (fresh water), the more saline wastewater caused important decreases in leaf water potential, stomatal conductance to H_2O and the photosynthesis rate after only 15 days of irrigation, the reduction being more pronounced after two months of irrigation. This treatment also caused a rapid, significant reduction in leaf nitrogen concentration, as compared with the nitrogen level in the trees before irrigation. Both types of wastewater significantly reduced olive yield, compared to that obtained in the control. These results indicate that this kind of wastewater is unsuitable for application to olive orchards under irrigation.

Physical Processes

Filtration through membranes of different pore size and materials is increasingly being used by the table olive processing industry to reuse, concentrate, or eliminate wastewater, in spite of the fact that early tests failed because of the substances dissolved in olive brines and the chemical composition of the available membranes (Rose W.W., 1982).

NL1005938C (1998) applies micro-, ultra-, and nanofiltration for the removal of the organic compounds from spent brine. At least some of the brine is regenerated and can be reused.

Brenes Balbuena M. et al. (1990) reported the successful application of ultra-filtration through polysulfone membranes to recover green table olive brines. Experiments were conducted to select optimum membrane pore size and working conditions, and to study effects on characteristics of brines and olives packed with regenerated brines. Good decoloration was achieved with polysulfone membranes of 1000 Da pore size. Experiments with pilot plant equipment provided with tubular membranes permitted establishing as best working conditions: 18 bars, room temperature (25°C), and no previous filtration; in this case the flux was mainly influenced by the initial polyphenol concentration of brine, whereas pH did not have any effect. Very low reduction of NaCl levels and slight decrease of lactic acid levels were noted. No effect on quality of olives packed with the treated brines was detected.

However, there are still some difficulties to be solved before the industry could use this modification. Some of them are related to the optimization of the ultrafiltration performance in order to obtain the highest possible flux. Others are connected to the preservation of the regenerated brines, because sometimes they can be used immediately after their treatment, but usually they must be stored depending on the packing needs of the industry. This storage creates some disadvantages: (i) a progressive darkening of the solution and (ii) the growth of yeasts and molds on the surface in contact with the air. To overcome those problems, a later study investigated the effect of some operating variables (temperature, operation mode, and previous coagulation with bentonite) on the ultrafiltration of these brines (Brenes Balbuena M. et al., 1990). Flux improved with temperature increase, and a temperature of 30–40°C is recommended, especially when flocculation with bentonite was carried out. Sodium bisulfite was the most effective agent in preventing darkening of regenerated brines during storage for several days, although ascorbic acid was also effective. HPLC polyphenol analyses of the original brine, permeate and concentrate confirmed that ultrafiltration removed a large proportion of polymerized polyphenols from the permeate. Simple phenols were not removed effectively, except for vanillic acid, caffeic acid, and 4-hydroxybenzoic acid which were substantially eliminated. Field tests showed that recycling of such brines after treatment with both 0.6% (w/v) of activated carbon and ultrafiltration through a polysulfone membrane of 1000 Da pore size can produce quantities of brine that, with minor composition adjustments, can be added to the solution employed for packing the final product. This can then, be used in other operations where brine is required (Garrido Fernández A. et al., 1992b).

In California, the very important ripe olive processing industry developed a complex installation including different types of filtration. In summary, the proposed system treats 264 m³ of process water per day and then recycles it back to the olive operation. Filtration begins by passing the waste through a 50 μm screen followed by ultrafiltration. The process is completed by reverse osmosis. The concentrate from ultrafiltration and reverse osmosis is dehydrated by evaporation followed by spray osmosis (Romero Barranco C. et al., 2001). The process was selected because of its lower running costs and energy consumption, compared to other alternatives (cf. evaporation) and of eliminating the use of evaporation ponds and the consequent threat of salt contamination of groundwater.

The objective of the UREKA project: E! 3508-AGROBRINES is the development of a new combined membrane-based method for treating table olive processing wastewaters. The membrane-based process combination, known as "integrated membrane treatment method", combines ultrafiltration, nanofiltration, and reverse osmosis membranes. Ultrafiltration membranes can remove TSS, bacteria, and medium to large molecular weight organics and oil effectively based on sieving. The regeneration and reuse of spent brine, through correctly selected ultrafiltration membranes with a molecular weight between 20 and 150 Da, should result in a reduced total wastewater volume characterized by a lower concentration of pollutants. Nanofiltration and reverse osmosis thin film composite membranes are capable of separating and rejecting low molecular weight organics, divalent, and monovalent salts. The mechanisms of separation for nanofiltration and reverse osmosis membrane elements deal with diffusion and electrochemical interaction phenomena. The treatment will be capable of water recovery (as high as 80%) of the initial wastewater volume and purifying spent brine for reuse. The quality of permeate would depend on the type of membranes to be used in a final purification stage (nanofiltration or reverse osmosis membranes).

Thermal Processes

In the table olive sector, the most commonly used system for eliminating spent lye solutions and brines has been evaporation in shallow ponds. The water is allowed to evaporate from these ponds, leaving a sludge of lye and/or salt and organic contaminants. The wastewater must be piped or trucked away to uninhabited areas where pollution consequences hopefully are minimized. This disposal operation involves considerable expense both for handling these wastes and for land needed to house the ponds. Ponds built by the industry in California had a surface area of hundreds of hectares. Leakage was prevented first by a layer of clay and secondly by plastic films. Following decades of use, the groundwater was found to be contaminated with salt and use of the ponds was forbidden (Martin C.J., 1992). However, in Spain and other Mediterranean countries the ponds are still in use (Romero Barranco C. et al., 2001).

Evaporation can be improved by various systems. Some of them produce turbulence at the wastewater surface, others produce sprays, etc. In all cases, care must be taken not to cause damage to vegetation around the ponds. In Spain, a device consisting of a panel of porous material that increases evaporation has been commercialized recently. The equipment is suspended above the liquid surface and the solution is poured through at a rate high enough to avoid accumulation of solids in the holes. Evaporation is improved by orienting the device across the prevailing wind, or by forced air supplied by a fan. However, problems frequently arise as the concentration becomes greater and the viscosity of the solution increases. These systems have, therefore, had only limited acceptance by the industry (Romero Barranco C. et al., 2001).

The spent brine because of its acidic pH, its content of salt, and its content of proteinous materials cannot be effectively concentrated by conventional thermal processes — those involving contact with a hot metallic surface such as steam coils or jackets. If such procedure is attempted, the surfaces become corroded and fouled with hard scaly deposits so that heat transfer is impeded and the equipment impaired. US3732911 (1973) discloses a process for reconditioning spent brine, which obviates the problems outlined above, comprising a series of steps: (a) concentrating the spent brine by submerged combustion evaporation wherein hot gaseous products of combustion directly contact and flow through said brine whereby to cause boiling thereof at a temperature below the normal boiling of the brine, and whereby to effectuate the concentration with a minimum of scaling and corrosive effect; (b) continuing said concentration while the temperature of the brine is raised to about 93°C, and there is produced a slurry of salt crystals suspended in liquid with a solids content of about 60% and containing, in addition to the salt, about 6% combustible organic material; (c) incinerating the slurry at a temperature of about 650°C to burn or at least carbonize organic components; (d) dissolving the incinerated residue in water; (e) filtering the resulting solution to remove water-insoluble impurities; (f) neutralizing the filtered solution with hydrochloric acid to yield a regenerated brine which is suitable for reuse in the processing of olives. An advantage of the process is that the evaporation of water is achieved at a temperature below the normal boiling point of the brine. This situation is explained as follows: As the hot products of combustion flow through the brine, heat is transferred from the rising gas bubbles to the liquid through the bubble interface. The partial pressure of the water vapor in the rising bubble is less than the atmosphere so that boiling takes place at a temperature considerably below 100°C. For a saturated NaCl solution, the observed boiling point is about 93°C (in contrast to a normal boiling point of 108.5°C), corresponding to a depressed vapor pressure of about 443 mm Hg, and a heat of evaporation of approximately 978 BTU/lb. Since there are no heat transfer surfaces there are no problems with fouling or corrosion, so that equipment cost can be lowered. The disadvantage of the process is its relatively low thermal efficiency. The vapors produced during the concentration (a) step have a very disagreeable odor, due to the presence of lactic acid and other volatile organic substances, and it is preferred to scrub the vapors before discharging them. Since the solution prepared from the recovered salt contains some alkaline impurities, it is usually necessary to neutralize the solution by adding a small proportion of an acid such as hydrochloric acid (Durkee E.L. et al., 1973). Since cost of the submerged combustion process is dependent on the volume of water to be removed, the more concentrated brine is more economically treated.

Steam explosion can be used to increase the accessibility and separate the main components of olive stones obtained from pitted table olive and olive oil processing. Pretreatment with high-pressure steam followed by explosion at atmospheric pressure produced autohydrolysis of the hemicellulose and a good substrate for enzymatic hydrolysis by cellulases. The hemicellulose is partially hydrolyzed to water-soluble oligomers or to individual sugars. Fernández-Bolaños J. et al. (1998, 1999) studied the effectiveness of steam-explosion on olive stones. The water-soluble phenolic fraction generated during steam-explosion and the exploded solid residues were determined to be a function of the severity of the steam treatment. An effective recovery of hydroxytyrosol (Fernández-Bolaños J. et al., 1998), a lignin with high quality and a cellulose with different properties (degree of polymerization, crystallinity, and other properties) (Fernández-Bolaños J. et al., 1999) were obtained. In a following study Fernández-Bolaños J. et al. (2001) reported on the application of this process to olive stones under various conditions of temperature and time (200–236°C for 2–4 min), with or without previous acid impregnation (0.1 H_2SO_4 w/w). This study showed that steam explosion pretreatment converted hemicellulose into soluble carbohydrates and that susceptibility of olive stones to hydrolysis with cellulase was improved, with respect to material without steam explosion (ball-milled material), although little increase in the extent of saccharification occurred when the alkali-soluble lignin was removed. Only when the substrate was post-treated with sodium chlorite was the enzymatic hydrolysis improved, the water-insoluble residue being almost completely hydrolyzed in 8 h of incubation. However, further studies and a large scale experimentation will be necessary in order to evaluate the economic suitability of the whole process.

Physico-Chemical Processes

Flocculation/Adsorption

Several physico-chemical processes have been applied for the decontamination of table olive processing wastewater. According to them, a reduction, which might reach eventually up to 40% of the original pollution, is obtainable by correcting the pH to 11.0 units with calcium oxide. Decantation of the flocks is accelerated notably by the addition of an anionic polyelectrolyte (Prästol^{TM} 2540 TR) in a proportion of 0.3 ppm, fact that may reduce appreciably the size of the installations. The final color of the solutions may additionally be improved by the treatment with 200–300 ppm of chlorine or ozone due to the destruction of most colored residual polyphenol compounds (García-García P. et al., 1990).

Mercer W.A. et al. (1971) were able to recondition spent olive storage and processing brines with pilot-scale activated carbon treatment. Reconditioned brines of lower salt content were reused to store olives with no detectable effect on product quality. On the average, activated carbon treatment removed 37–65% of COD in the olive brines and 21–30% of the suspended solids.

US3975270 (1976) discloses a process for the purification of spent lye comprising the steps of: (i) adding in succession (at intervals of 15–30 min) lime, charcoal, and calcium carbonate; (ii) allowing the mixture to settle for 30–60 min to form a usable olive processing liquor and a sludge containing contaminants; (iii) separating usable processing liquor from the sludge for reuse and, optionally (iv) filtering the recovered liquor through sand, charcoal, and a diatomaceous earth aid in a ratio of 2:1:2, respectively. The process reduces waste disposal problems, as only a small quantity of sludge is produced, instead of large volumes of obnoxious and caustic liquid waste, and large amounts of alkali are reused instead of going to waste.

Brenes Balbuena M. and Garrido Fernández A. (1988a) studied the possible regeneration of fermentation brines from Spanish green table olives by decolorating earths and activated carbon. According to the results, the first products offer inconveniences for the odor, flavor, and high proportions that are required; the second adsorbents produce more favorable results diminishing the color in a high enough proportion as to permit reusing such brines in the final packing.

Chemical Oxidation Processes

Chemical oxidation using ozone or advanced oxidation technologies (AOPs) based on the generation of hydroxyl radicals (i.e. from the combination of ozone and hydrogen peroxide or UV radiation) is another possible way to reduce COD and polyphenol content and, in addition, improve the biodegradability of table olive processing wastewater (Benítez F.J. et al., 2001, 2003). Use of ozone is recommended for the pretreatment of spent lye solutions since it shows a high reactivity toward phenolic compounds (found in this type of wastewater) reducing, at the same time, the alkalinity of the media for further biological processing. An ozone dose of 45 mg/l (flow rate 20 l/h) for a period of 35 min has been found to achieve the following goals: decrease pH, decrease phenolic content, and increase biodegradability. A similar picture is presented when hydroxyl radicals generated from AOPs are the main oxidizing agents (Beltrán F.J. et al., 1999).

AOP technologies used for the oxidation of table olive processing wastewater include Fenton's reagent (Benítez F.J. et al., 2003; Rivas F.J. et al., 2003b), a combination of UV radiation and hydrogen peroxide (Benítez F.J. et al., 2003) as well as photo-Fenton (Benítez F.J. et al., 2001).

Beltrán F.J. et al. (1999) studied the chemical oxidation of laboratory prepared lye solutions using ozone alone and combined with H_2O_2 or UV radiation. COD reduction of 80 or 90% can be reached with ozone doses of 3–4 g in the presence of 10^{-3} M initial hydrogen peroxide concentration (2.4 g at the conditions investigated) or 254 nm UV radiation while moderate reductions of total carbon can be achieved (between 40 and 60%). Both the aromatic content and color nearly completely disappear with less than 0.5 g of ozone applied. Along the reaction the initial pH of wastewater diminishes from 12 to 7.5. Biodegradability of wastewater, measured as BOD_5/COD, increases from 0.16 (untreated wastewater) to nearly 0.7 or 0.8 when using ozone combined with hydrogen peroxide (2.95 g O_3 fed during 2 h and 2.4 g H_2O_2 initially charged) or UV radiation (3.98 g O_3 fed during 2 h and incident radiation: 2.65×10^{-6} Einstein s^{-1} as the oxidizing systems, respectively. However, when H_2O_2 is applied at high concentration (10^{-2} M), it is necessary to eliminate its excess at the end of oxidation in order to avoid the lack of biological activity during a possible and subsequent biological step.

In Fenton's reaction, the ferrous and/or ferric cation decomposes catalytically hydrogen peroxide to generate powerful oxidizing agents, capable of degrading a number of organic and inorganic substances. Fenton's oxidation is a complicated system that involves a large number of reactions, including redox reactions, complexation precipitation equilibria, etc. Also, there is an uncertainty in the nature of oxidizing species generated during the process (formation of radical species or generation of aquo- or organo-complexes of high valence iron) — see Chapter 7: "Physico-chemical processes", section: "Fenton reaction". A model of Fenton's reagent processing of brines from the table olive industry is described by Rivas F.J. et al. (2003b). Reagent concentration exerted a positive influence on COD removal. H_2O_2 uptake showed values in the range 0.3–1.6 mol of COD eliminated per mol of H_2O_2 consumed depending on operating conditions. The optimum working pH was found to be in the interval 2.0–3.5. Reaction temperature increased the COD degradation rate, although similar COD conversion values were obtained after 5 h of treatment regardless of the value of this parameter. An analysis of the biodegradability of this type of effluent demonstrated the beneficial effect of the chemical preoxidation. According to the experimental results, it is suggested that there is an inhibitory effect of the wastewater due to its COD content and nature rather than attributable to the presence of high amounts of sodium chloride. Biodegradation efficiency increased as temperature was raised up to 30°C. A further increase of this parameter up to 40°C resulted in the death of the microorganisms.

Benítez F.J. et al. (2003) compared several chemical oxidation processes for the purification of storage brines from the preservation of table olives. Ozone alone produced COD removals in the range 14–23%, and a higher average removal of 73% of the aromatic compounds. The additional presence of H_2O_2 and UV radiation increased these values to 39% for COD and 86% for aromatics. However, UV radiation alone only gave a removal of 9% for COD and 27% for aromatics, and the additional presence of 0.5 M H_2O_2 led to 13% for COD and 38% for aromatics, respectively. The Fenton's reagent oxidation achieved a COD removal of 24% for the higher concentrations of Fe^{2+} and H_2O_2 The most effective process was the combination $O_3/UV/H_2O_2$ with total removals of 65 and 92% for the COD and aromatics, respectively. The aerobic treatment of these effluents gave a 66% removal regardless of the initial biomass concentration used, and a rate constant of 0.19 per day was obtained for the process by using the Contois model. Finally, the aerobic treatment of the wastewater, previously ozonated alone, and ozonated with UV radiation, gave increases in the COD removal and a final rate constant of 0.44 per day. The enhancements were due to the chemical oxidations, these procedures being suitable technologies as pretreatments to subsequent biological processes for the purification of these residues.

Wet oxidation (in autoclave) has been used for the decontamination of spent lye solutions (García-García P. et al., 1989; Rivas F.J. et al., 2000b). When the best conditions were used, the phenol concentration is reduced to 1% of the original level if there is in the liquid an excess of NaOH which is not neutralized by the CO_2 and other acids produced during the degradation process. However, the oxidation by this technology is not strong enough because the organic carbon concentration remains very high, 75% of the initial value.

Biological Processes

Biological processes that have been used for the treatment of OMWW may be also applicable to table olive processing wastewater (Beltrán F.J. et al., 2001; Aggelis G. et al., 2001; Assas N. et al., 2001). Only limited experience is available on the application of these processes to the different types of wastewater generated during the various stages of table olive processing but one can assume they will not work properly because these solutions are rich in polyphenol complexes that are resistant to microbial degradation. In the anaerobic treatment of ripe olive processing wastewater, inhibition of gas production was observed when the COD of the influent was above 1 g/l, but no effect of salt was apparent (Borja-Padilla R. et al., 1993a). However, this means that they will need to be diluted initially as the normal concentration is about 4.5–6.0 g/l. In the case of green table olive packaging wastewater, aerobic treatment removed only 80–85% of the total BOD_5, even when the required amount of oxygen was supplied. Neither the salt content (in the range of 1–3% w/v), nor the pollutant load (in the range 1.0–5.0 g/l) influenced the yield, although in the latter case the load remaining became progressively further out with the authorized limits (Brenes Balbuena M. et al., 2000).

Depending on the concentration of pollutants, one may consider anaerobic (COD > 3–5 g/l) or aerobic (for lower concentrations) treatment. Some polymeric carbohydrates, proteins, etc. could also be degraded during the acidic phase of anaerobic treatment. When no inhibitory substances are present, it may be possible to obtain 100% degradation, for example in the case of sugars, organic acids, etc. Biological treatment of table olive processing wastewater requires a prolonged acclimatization period for the biomass, even if a pretreatment stage precedes (Beltrán F.J. et al., 2000, 2001). A significant advantage for the use of a selected species is the immediate initiation of the process, which may be very important in the case of seasonal production of large quantities of wastewater. Such a species is *Aspergillus niger* which has been used successfully for the reduction of the organic and phenolic load of spent lye solutions and washing waters (Kotsou M. et al., 2004b). The *Aspergillus* genus has been also used successfully for the bioremediation of OMWW and satisfactory removal efficiencies have already been reported — see Chapter 8: "Biological processes", section: "Use of specific aerobic microorganisms".

An activated biofilter system (Neptune Micro-Floc), which is a combination of trickling filtration and activated sludge, has been proposed for the cleaning up of table olive processing wastewaters in Madera, California (Cortinovis D., 1975). The plant was first operated in a conventional trickling filter mode, with all final clarifier sludge being wasted to the primary tanks. In a later stage the activated biofilter system process was initiated by returning all final clarifier sludge to the influent of the single filter in operation at that time. A second filter began operation when additional wastes were added to the system. BOD_5 removal by the system was as high as 90%.

Brenes Balbuena M. et al. (2000) investigated the feasibility of using an activated sludge process for bioremediation of green table olive wastewater. Using activated sludge seed obtained from the aeration tank of a municipal wastewater activated sludge treatment factory, a reduction in COD of 75–85% was observed, due mainly to removal of organic acids and ethanol. However, only a small proportion of polyphenols, particularly polymerized polyphenols, was removed with the remainder being responsible for both the color and residual COD of effluents. The Grau model was used to describe the relationship between influent and effluent's COD; kinetic coefficient was 9.8/day. Increasing the hydraulic retention time from 0.369 to 0.512 days and the temperature from 10 to 32°C increased removal efficiency with a COD in the effluent of 200–300 mg/dm^3 being achieved routinely. The process worked well at NaCl concentration up to 20 g/dm^3 although sludge volume index increased to more than 200 cm^3/g at a NaCl concentration of 30 g/dm^3.

Aggelis G. et al. (2002) tested eight white rot fungi (*Abortiporus biennis, Dichomitus squalens, Inonotus hispidus, Irpex lacteus, Lentinus tigrinus, Panellus stipticus, Pleurotus ostreatus,* and *Trametes hirsuta*) for the detoxification and decolorization of spent lye solutions. The white rot fungi were grown in lye for 1 month and the reduction in total phenolics, the decolorization activity and the related enzyme activities were compared. Phenolics were efficiently reduced by *P. ostreatus* (52%) and *A. biennis* (55%), followed by *P. stipticus* (42%), and *D. squalens* (36%), but only *P. ostreatus* had high decolorization efficiency (49%). Laccase activity was the highest in all of the fungi, followed by manganese-independent peroxidase (MnIP). Substantial manganese peroxidase (MnP) activity was observed only in lye treated with *P. ostreatus* and *A. biennis*, whereas lignin peroxidase (LiP) and veratryl alcohol oxidase activities were not detected. Early measurements of laccase activity were highly correlated ($r^2 = 0.91$) with the final reduction of total phenolics and could serve as an early indicator of the potential of white rot fungi to efficiently reduce the amount of total phenolics in lye. The presence of MnP was, however, required to achieve efficient decolorization. Phytotoxicity of lye treated with a selected *P. ostreatus* strain did not decline despite large reductions of the phenolic content (76%). Similarly, in lye treated with purified laccase from *Polyporus pensitius*, a reduction in total phenolics which exceeded 50% was achieved; however, it was not accompanied by a decline in phytotoxicity. These results are probably related to the formation of phenoxy radicals and quinonoids, which repolymerize in the absence of veratryl alcohol oxidase but do not lead to polymer precipitation in the treated lye.

Aggelis G. et al. (2001) evaluated the performance of separate single step aerobic, anaerobic, and combined aerobic–anaerobic biotreatments of lye solutions for removal of toxic polyphenolic compounds. The aerobic treatment was performed in 11 aerobic reactors, using activated sludge from a municipal wastewater treatment plant, at 25°C. Anaerobic treatment involved the use of an initial anaerobic culture (pH 7.22, dissolved COD 1.13 g/l) obtained from dairy, olive-mill, and piggery-fed anaerobic digesters, to ensure the presence of a wide variety of bacterial species. Separate aerobic and anaerobic treatment resulted in maximum efficiency of organic matter removal of 71.6–75.9 and 49%, respectively. Anaerobic digestion led to removal of approximately 12% of the polyphenols present, whereas aerobic treatment had little effect on polyphenol levels. In addition, aerobic treatment required pH correction of the influent and resulted in high biomass production. However, the combined process (aerobic treatment of the anaerobic effluent) was more successful, resulting in an overall decrease in levels of phenolic compounds and COD of 28 and 83.5%, respectively, with no requirement for pH correction of either influent, and reduced levels of sludge production.

Combined and Miscellaneous Processes

In order to facilitate the degradation of toxic or non-biodegradable organic substances in table olive processing wastewater, many researchers have proposed integrated treatments comprising combination of various processes.

Field tests showed that recycling of spent brines after treatment with both 0.6% (w/v) of activated carbon and ultrafiltration through a polysulfone membrane of 1000 Da pore size can produce quantities of brine that, with minor composition adjustments, can be added to the solution employed for packing the final product. This can then be used in other operations where brine is required. A possible design of the regeneration process is also proposed (Garrido Fernández A. et al., 1992b).

An efficient way to eliminate the turbidity from spent brines is to treat them with activated carbon and then to apply either flocculation or filtration. The flocculation method uses a solution of bentonite to flocculate the suspension, prepare a precoat of a 50% mixture of Diatom 185-Speedplus and filter the clarified liquid using 0.3–1.5 g/l of a 50% mixture of Diaflow 2-Diatom 185 diatomaceous earth as filter aids. The filtration method applies tangential filtration to the decolored solution using a ceramic filtration element of 0.2 μm. Discussion of the efficacy and economics of the process suggests that both methods are promising. Industrial application of either method will depend on equipment and maintenance costs (Brenes Balbuena M. et al., 1988c).

Brenes Balbuena M. et al. (2004) used a combination of fermentation and evaporation processes for treatment of washing water from Spanish-style green olive processing. The washing water was fermented on a pilot plant scale (500 dm^3) and evaporated under vacuum. Simultaneous or sequential inoculation in alkaline conditions with *Enterococcus casseliflavus* cc45 and *Lactobacillus pentosus* LP99 did not prevent malodorous spoilage. By contrast, appropriate fermentation was achieved when the washing water was acidified with HCl, to pH 5, followed by inoculation with *L. pentosus*, or acidified up to pH 3.4 with no subsequent inoculation. The type of fermentation influenced, to a large extent, the COD of the distillate streams obtained when the washing water were evaporated under vacuum. The high amount of ethanol formed in the washing water, fermented only by yeasts, gave rise to distillates containing *circa* 30% of the COD originally present in the washing water, whereas this figure was 10% for washing water fermented by lactobacilli. This work also disclosed the presence of large amounts of valuable substances, such as lactic acid and hydroxytyrosol in the concentrates.

Ozonation of green table olive processing wastewater after dilution (1:25) with synthetic urban wastewater has been presented as a suitable treatment to render this effluent more biodegradable (Rivas F.J. et al., 2000a). The combination of acidic and basic cycles during the ozonation process leads to a notable improvement of the water quality if compared to results obtained for simple ozonation. A significant enhancement of the biodegradability properties is achieved after the chemical pre-oxidation step. Variations in pH may be accomplished not only by addition of mineral acids or bases, but by mixing wastewater having extreme pH values. Thus, the joint treatment of green table olive processing wastewater and OMWW may be an alternative route to change the pH by simultaneously saving the chemicals. Consequently, aerobic biodegradation experiments carried out by using non-acclimatized microorganisms from a municipal wastewater plant have been demonstrated to further reduce the remaining COD content. A kinetic model involving the integrated chemical–biological process acceptably predicts the COD depletion profile with time. Dissolved ozone in the bulk of the liquid is also well simulated by the proposed mechanism; however, it fails when computing the amount of non-reacted ozone at the reactor outlet. Aerobic biodegradation experiments were relatively well fitted by means of the Monod equation.

Beltrán-Heredia A.J. et al. (2000b,d) investigated the oxidative degradation of the organic matter present in the washing waters from the black table olive industry by an aerobic process and by the combination of two successive steps: ozonation pretreatment followed by aerobic degradation. In the single aerobic process, the average removal of COD and polyphenols was 75 and 50%, respectively. A kinetic study was performed using the Contois model, which applied to the experimental data, provided the following specific kinetic parameters: $4.81 \times 10^{-2}\,\mathrm{h}^{-1}$ for the kinetic substrate removal rate constant, $0.279\,\mathrm{g\ VSS\ g\ COD}^{-1}$ for the cellular yield coefficient and $1.92 \times 10^{-2}\,\mathrm{h}^{-1}$ for the kinetic constant for endogenous metabolism. Using a combination of ozonation followed by aerobic pretreatment, removal of COD and total phenols was 82 and 76%, respectively, higher than the values obtained in the single aerobic process. The kinetic parameters of the following aerobic degradation stage are also evaluated, being $5.42 \times 10^{-2}\,\mathrm{h}^{-1}$ for the kinetic substrate removal rate constant, $0.280\,\mathrm{g\ VSS\ g\ COD}^{-1}$ for the cellular yield coefficient and $9.1 \times 10^{-3}\,\mathrm{h}^{-1}$ for the kinetic constant for the endogenous metabolism.

GR1004115 (2003) and Kyriacou A. et al. (2005) disclose a combined aerobic and electro-chemical process for the treatment of green table olive processing wastewater. The first step of the process consists of aerobic treatment of the wastewater, using a selected strain of *Aspergillus niger*, for bulk COD removal and the biodegradation of phenols and most of the organic acids. Electrolysis in the presence of hydrogen peroxide constitutes the second step, to achieve oxidation of the remaining recalcitrant organic compounds. Results are reported for the different stages of the scale-up procedure, from laboratory shake-flask cultures to pilot plant of $4\,m^3$/day capacity — see Fig. 11.3. In the aerobic treatment step, COD removal efficiency varied between 66–86% while the concentration of selected phenols was reduced by 65%. The efficiency of the electro-chemical step depended on pH and the concentration of H_2O_2 used. In laboratory experiments using 500 ml electrolysis cell, 96% removal efficiency was achieved for both COD and measured phenols with 2.5% H_2O_2. As H_2O_2 represented a large fraction of the electro-chemical treatment costs, a lower concentration of 1.6% was used in the pilot plant, achieving a 75% COD reduction. Treatment efficiency was improved with coagulation with 0.4% $Ca(OH)_2$ paste resulting to an overall COD removal of 98% (COD $<$ 500 mg/l).

Kotsou M. et al. (2004b) used combined aerobic treatment and chemical oxidation with Fenton's reagent for the processing of green table olive processing wastewater. The aerobic treatment was performed in a bubble column bioreactor using an *A. niger* strain. After 2 days of aerobic treatment, COD was reduced by 70%, while the total and simple phenolic compounds were decreased by 41 and 85%, respectively. Fenton's reagent was used as a secondary chemical treatment step for the oxidation of the recalcitrant organic compounds or metabolites of those that could not be oxidized biologically. Control was achieved by adjusting the H_2O_2 concentration, relative to the COD load after the aerobic stage, while iron concentration determined the time required for the completion of the reaction. After chemical oxidation, COD removal was highly enhanced by coagulation with CaO. In contrast, coagulation with CaO of the wastewater after the aerobic treatment was not very effective (29% COD removal). This observation offers a confirmation to the position that Fenton's oxidation is particularly suited as a pretreatment for the subsequent flocculation, improving removal efficiencies.

Rivas F.J. et al. (2001e) treated table olive processing wastewater by means of an integrated wet air oxidation-aerobic biodegradation process. COD conversion in the range of 30–60% (6 h of treatment) has been achieved by using relatively mild conditions (443–483 K and 3.0–7.0 MPa of total pressure using air). Use of homogeneous catalysts [copper (II)] or radical promoters (H_2O_2) resulted in a higher efficiency of the process (roughly between 16 and 33% COD removal improvement depending on operating conditions). Biodegradability tests, conducted after the oxidation pretreatment, showed a negative effect with the addition of copper to the reaction media. Promoted experiments resulted in the highest values of biodegradability (measured as the ratio, BOD_5/COD).

The integrated process of WO03000601 (2003) — described already in Chapter 9: "Miscellaneous processes" — can also be used for the treatment of spent lye solutions and/or waste oxidation water used for debittering and blackening olives (agitation or aeration).

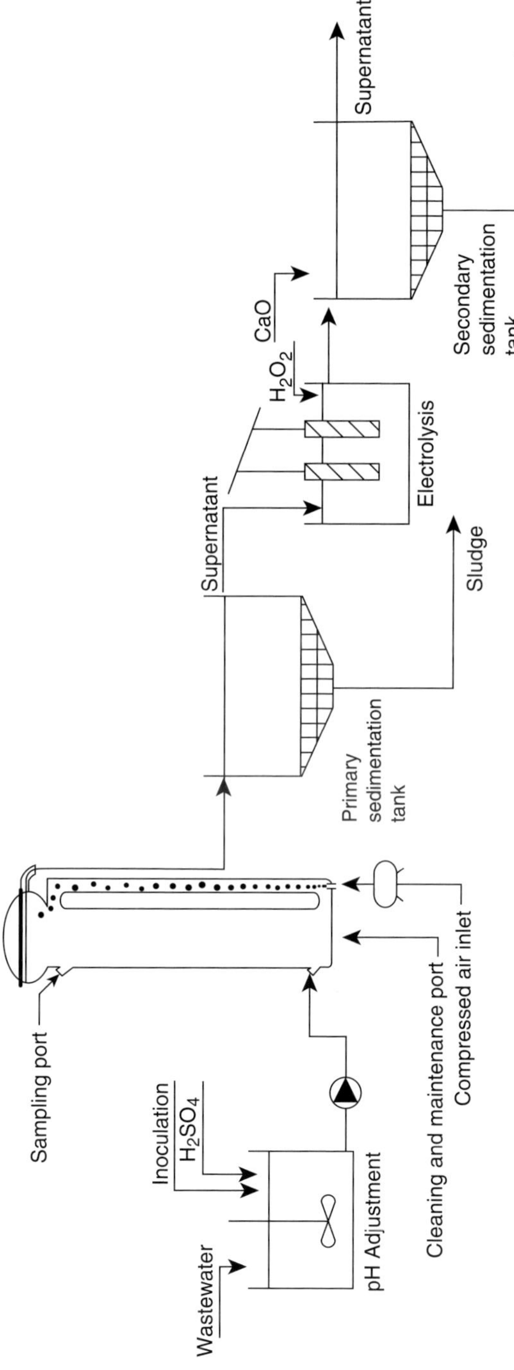

Fig. 11.3. Schematic diagram of the pilot plant at the premises of the Agricultural Cooperation of Peta, Greece (Kyriacou A. et al., 2005).

Uses

There are several patents to obtain antioxidants from the solid and liquid residues generated during the olive oil extraction process and from olive leaves — see Chapter 10: "Uses", section: Antioxidants". However, extraction of antioxidants from the wastewaters originated during table olive processing could be a better way since hydroxytyrosol is present in these liquids in a very high concentration (4–6 g/l). Brenes Balbuena M. and Castro Gomez-Millan A. de (1998) proposed a method to obtain hydroxytyrosol from the washing waters of the Spanish-style green olive processing. According to this method 30 l of washing water were placed in eight PVC cylindrical vessels and stored at ambient temperature for one year. HCl was added into four vessels to adjust the initial pH of the washing water (10.6) up to 4 and 3. HCl was employed in another two vessels to lower the pH up to 3 and sodium metabisulfite was also added. All the containers were sealed and anaerobic conditions were maintained during storage. Throughout the year of preservation a fermentation process occurred. A high population of lactic acid bacteria was detected in treatments with no initial correction of pH and those with the pH corrected up to 4. Yeasts grew in all the vessels although in a higher number in those treatments with pH initially corrected. As a consequence of the micro-organisms growth, lactic acid was formed in treatments without initial pH correction and pH corrected to 4, and acetic acid as well as ethanol were detected in all the assays. It must be stressed that no unpleasant odors or spoilage symptoms were detected. The concentration of phenolic compounds slightly diminished during the fermentation process as happens during fermentation of Spanish–style green olives (Brenes Balbuena M. et al., 1995). Elenolic acid glucoside, the other part of the oleuropein molecule, hydrolyzed in elenolic acid and glucose (Brenes Balbuena M. and Castro Gomez-Millan A. de, 1998), being influenced by the pH and temperature. Eventually, the elenolic acid disappeared due to the acidic conditions. After a year of storage of the washing waters, these solutions had a very high concentration of hydroxytyrosol and low level of other phenolic compounds or parts of the oleuropein molecule.

The Institute of Food Oils of CSIC (Spain) developed an improved version of the above method to obtain antioxidants and antimicrobial compounds from the wastewaters (lye, washing water, brines) generated during the various stages of table olive processing (ES2186467, 2003). The first step of the method consists in storing/ fermenting the wastewater in acid conditions for more than five months in order to maintain constant the concentration of hydroxytyrosol and decrease the amount of elenolic acid glucoside. Subsequently, the wastewater is ultrafiltrated through membranes having a pore size of between 1000 and 10,000 Da using tubular equipment. The permeate is passing through a non-ionic resin column (AMBERLITE® XAD-series) in order to adsorb the phenolics. Regeneration of the column is then made with ethanol. The solvent is evaporated and the aqueous residue is treated with activated carbon to eliminate color and undesirable odor. The liquid extract is frozen at temperatures below 0°C and lyophilized. Thus, a solid is obtained with a concentration in hydroxytyrosol higher than 60–80%. An extract rich in the antimicrobial elenolic acid glucoside can also be obtained.

US3975270 (1976) discloses a process for recovering usable olive processing liquor from a black olive processing waste lye solution comprising the steps of: (a) successively adding lime, charcoal, and calcium carbonate to said solution under stirring at intervals of about from 15 to 30 minutes; (b) allowing the mixture to settle to form a recycled olive processing liquor and a sludge containing contaminants; (c) separating the recycled olive processing liquor from the sludge containing contaminants. The required amounts of lime, charcoal, and calcium carbonate are 1.0, 1.0, and 2.5 part(s) per 900 parts of waste lye solution, respectively. The invention claims to have several economic advantages. The savings realized in purchasing lye, in piping or trucking the used processing liquor, and in purchasing real estate will more than offset the expense of purchasing lime, charcoal, and calcium carbonate. Furthermore, no additional equipment is necessary. Another advantage of the invention is that pollution from the lye waste is substantially reduced. Only the settled contaminants (sludge) must be disposed of and, since the volume is small, may be used as fertilizer because it contains mainly organic contaminants, the lye having been removed and recycled.

Part V
Economic and Legislative Overview

Chapter 12

Economic Evaluation

There is a lack of information related to the limits of applicability and economical viability of the various methods for the treatment and disposal of olive processing wastes. Most of the economical data regarding capital investment, running, and maintenance costs of the various management schemes for treating olive-mill wastes are not comparable or unreliable. The majority of the installations are mainly laboratory or pilot plants and therefore, the possible costs have to be derived by extrapolation. In addition, the available data refer to different periods, labor costs, production capacities, water availability, processing systems, pretreatment, etc. Quite often the selection of a method is taken under the increasing pressure of the controlling authorities without proper economical analysis and study of the environmental impacts. As a result an economically viable solution could be rejected because of social and environmental concerns.

One of the first economic evaluations of the costs and the energy consumptions of the various processes used for the OMWW treatment was carried by Boari G. et al. (1984) — see Table 9.1. Incineration and concentration by distillation were reliable but quite expensive and energy consuming. Aerobic processes were not advisable because of (i) high consumption of nutrients (to reach a ratio BOD_5:N:P = 100:5.1 from BOD_5:N:P = 100:1:0.5; (ii) very high production of secondary sludge which has to be disposed of; (iii) high capital cost. Anaerobic processes were advisable because of their well known advantages related to energy and chemicals saving and to low sludge production.

A recent economical analysis of the various treatment schemes has been carried out by Azbar N. et al. (2004), on the basis of the results obtained within the framework of the EU project: FAIR CT96-1420 "IMPROLIVE". The costs of the various treatment schemes were calculated by considering a three-phase olive-mill generating $5000 \, m^3$ OMWW per year during a 100-day campaign with 10 years of useful life for the treatment units, assuming an 1:5 oil to OMWW weight ratio. All treatment units discharge treated water with COD of about $4 \, g/l$ — see Table 12.1.

Table 12.1. Costs due to various treatment schemes for a three-phase olive-mill generating 5000 m^3 OMWW with 10 years of useful life for the treatment units, assuming an oil to OMWW weight ratio of 1:5 (Azbar N. et al., 2004.)

Treatment scheme	Investment cost, €	Operating cost, €/m^3 OMWW	Total cost, €/m^3 OMWW	Calculated cost, €/ton olive
Forced mechanical evaporation+lagooning	180,700	6.82	10.43	52.1
Physico-chemical+biological+ultrafiltration	150,600	8.68	11.69	58.4
Biological (both solid wastes and wastewater treatment)	180,700	6.21	9.82	49.1
Physico-chemical+reverse osmosis	138,600	5.27	8.04	40.2
Physico-chemical+ultrafiltration	216,900	Not known	–	–
Biological (anaerobic+aerobic) reverse osmosis	180,700	Not known	–	–
Vacuum evaporation	96,400	3.69	5.62	28.1
Forced natural evaporation	42,200	0.47	1.31	1.5
Improved natural evaporation	30,100	0.05	0.65	3.2
Mechanical biological pretreatment (biogas production)+ sludge management (aerobic stabilization+solar drying)	500,000–850,000	3.5–5.5	13.5–22.5	67.5–112.5
Commercial evaporator*	850,000	1	3.95	19.7

*The commercial evaporator has a capacity of 288 m^3/day and cost of power is calculated by using the data in the company catalogue and cost of electricity in Turkey in 2000.

Costs to be incurred on the olive oil product are in the order of 0.15 to 0.32 €-cents/kg of oil for natural evaporation and 1.9–11.2 €-cents/kg of oil for the remaining more sophisticated treatment options. The cheapest scheme among the second category is evaporation–distillation of OMWW; it is noted, however, that this alternative is for a 28,800 and not for a 5000 m^3 per year facility. It is anticipated by the author that smaller evaporator units of the same size as the other alternatives would create higher unit costs of treatment. In any case, even the highly sophisticated alternative of mechanical/biological treatment followed by an anaerobic process with biogas production and full sludge management, the so called AquatecOLIVIA technology (DE19829673, 2000), would require 6–11 €-cents/kg of oil under the circumstances described in this discussion — see Chapter 8: "Biological processes", section: "Anaerobic processes". Unit cost of treatment is given as 3–6 €-cents/l of oil[48]. The difference is possibly due to different oil to OMWW ratios in calculation, which may vary largely in practice.

From Table 12.1 it can be seen that natural evaporation both in forced or improved forms is the cheapest solution. If the costs of treatment with natural evaporation alternatives are compared with traditional domestic wastewater treatment costs varying between 0.25 and 0.50 €-cents/m^3, they are rather close to this range. A cost correction based on the corresponding COD loads of the domestic wastewater and OMWW yields that traditional domestic wastewater treatment cost is more expensive on the unit COD load basis, as COD is 400 mg/l for domestic wastewater and 80,000 mg/l for OMWW. In other words, domestic wastewater is 200 times more diluted than OMWW, but is only 2–40 times less costly in treatment per unit volume (Azbar N. et al., 2004).

Caputo A.C. et al. (2003) carried out a technical and economical analysis of different thermal disposal plant solutions based on gasification and combustion technologies. Energy recovery sections based on combined gas–steam cycle, steam or gas turbine cycle, and internal combustion engine, have been considered. The profitability of combining OMWW and olive cake treatment has been investigated in order to assess its feasibility. While adding OMWW to olive cake is counter-productive from the energy conversion point of view, the organizational and economic benefit coming from the avoidance of a separate OMWW disposal may prove to be profitable in a future legislative scenario when stricter limitation on OMWW disposal will force oil producers to bear disposal costs. In this context, a critical parameter affecting the energy, economic and environmental performances, and thus, the feasibility of combined disposal approaches, is the degree of mixing of OMWW and olive cake. This is defined as the dilution factor K, computed as the ratio of OMWW to olive cake flow rate. In fact, while from an environmental point of view, a combined disposal of the whole OMWW flow rate — equivalent to the largest value for K is desirable, a rise in K value is associated to a diminution of energy conversion efficiency, a reduction of revenues from sale of produced energy,

---

[48]http://www.aquatec-engineering.com/engl-src/products.htm.

as well as to an increase of purchased and operating equipment costs. On the other hand, if treatment of OMWW is compulsory (future legislative scenario), revenues from avoided disposal costs increase with K.

As reference for the analysis, a specific industrial basin constituted by 194 olive-mills processing about 500×10^5 kg of olives per year with a production of olive cake and OMWW around 186×10^5 and 248×10^5 kg, respectively, has been considered. Moreover, an operation period of six months (November–March) on three shifts (24 h/day) has been hypothesized. Considering that a density of 1233 kg/m^3 for olive cake and 1017 kg/m^3 for OMWW has been assumed, the resulting mass flow (M_{TOT}) is 4300 and 5740 kg/h for olive cake and OMWW, respectively. Accordingly, the dilution factor (K) ranges from 0 to 1.6, when the entire OMWW flow rate of 5.6 m^3/h is treated in combination with olive cake. Results are compared by using economic performance measures such as net present value (NPV), pay back time (PBT), and profitability index (PI), taking into account capital investments, operating costs, revenues from energy production, and avoided disposal costs. A sensitivity and risk analysis is also performed in order to assess the economic profitability of the proposed solutions.

In particular, centralized waste-to-energy plant solutions based on combustion or gasification systems have been considered, so that benefits from economy of scale can be obtained considering the complexity and capital investment nature of such plants, which are not suitable to small sized decentralized applications.

Table 12.2 shows the values of technical and economic performance indices computed for the five analyzed disposal solutions in the examined scenario. The highest NPV is shown by the centralized-combined gasification plant with combined gas–steam turbine cycle, which is also characterized by the largest power output, indicating a higher conversion efficiency (n). However, such plant solution is

Table 12.2. Technical and economic performance of analyzed plants (6 months operation) (Caputo A.C. et al., 2003)

Disposal solution	n	P (MW)	TCI (M€)	NPV (M€)	PI	IRR	PBT (months)
Gasification+combined gas–steam cycle	0.52	3.8	7.74	16.45	3.13	40.1	34
Gasification+gas turbine cycle	0.32	2.8	5.77	16.22	3.80	49.1	28
Gasification+internal combustion engine	0.33	2.8	5.41	16.25	4.00	52.8	28
Fluidized bed combustor+heat recovery steam generator	0.28	1.3	5.36	13.17	3.15	41.1	35
Steam generator+steam turbine	0.30	1.7	5.04	11.16	3.17	39.2	36

n: energy conversion efficiency;
P: net electric power output (MW);
TCI: total capital investment (€);
PI: profitability index;
IRR: internal rate of return (%year);
PBT: pay back time (months).

penalized by a higher capital investment, while the other less capital intensive solutions show better profitability as indicated by the higher values of internal rate of return, profitability index and pay back time even if revenues from produced energy are lower.

Therefore, even if from a thermal efficiency standpoint, mixing OMWW with olive cake is a penalizing solution, it proved to be economically profitable when savings from avoided external disposal costs are accounted for. In the considered scenario, where residues from 196 mills are treated, NPV values in the 11–16 M€ range result with capital investments in the order of 5–7 M€. In any case, satisfactory values of the economic evaluation indices are obtained proving the cost-effectiveness of the proposed centralized-combined disposal approach with energy recovery.

Chapter 13

Legislative Aspects and Environmental Policies

EU Policies

Regarding market measures, olive oil is part of the reform of the Common Agricultural Policy (CAP) initiated in 2003. All the sectors covered by the reform are subject to compulsory cross-compliance measures. Beneficiaries of direct payments are obliged to comply with agricultural and environmental conditions and statutory management requirements[49].

Regarding the specific problem of olive oil waste, it is noted that the CAP olive oil reform does not provide for specific measures on olive processing wastes. Producer Member States have national dispositions on olive oil waste in conformity with Directive 75/442/EEC (Article 3) on waste (OJ L 194, 25.07.1975, p. 39) as amended by Council Directive 91/156/EEC of 18 March 1991 and Council Directive 96/61/EC (Article 3(c)) on integrated pollution prevention and control (IPPC). The directives concern both large industrial installations and disposal and incineration of waste and landfills. The general principles are:

- prevention of waste,
- recovery of waste (firstly as material, secondly as energy),
- safe disposal.

The problem of olive processing wastes is aggravated by the lack of a common policy among the olive oil producing countries. Every country has its own legislation/regulations that often vary greatly among them with a consequent non-uniform application of generally accepted guidelines — see Chapter 1: "Introduction", section: "Current practices for olive processing waste management". For this

---

[49]http://europa.eu.int/comm/agriculture/markets/olive/index_en.htm.

reason there is a need to establish an international normative that will impose a unified strategy.

The Case of Italy

Italy is the only olive oil producing country with a special legislation for the disposal and/or recycling of olive processing wastes. The various aspects of treatment and disposal of olive-mill wastes are regulated by the following law and legislative decrees:

a) Law no 574 of 11/11/1996 containing "regulations pertaining to the agronomic use of vegetation waters and olive-mill effluents".
b) Legislative decree no. 22 of 05/02/1997, implementing Community directives pertaining to waste, hazardous waste, packaging, and packaging waste, which represents the "framework law" for pollution from waste.
c) Legislative decree no. 152 of 11/05/1999 implementing Community directives pertaining to waters, which represents the "framework law" regarding pollution prevention.

Law 574/1996 is exclusively dedicated to olive-mill effluents; legislative decree 22/1997, art. 8, par. 1, lett. e) excludes such effluents from its field of application but re-includes liquid wastes.

Carro F. (2004), in the framework of the EU project: LIFE00 ENV/IT/000223 "TIRSAV", carried out a critical analysis of the various legislative aspects at issue in Italy, the main points of which are presented below.

Many doubts remain as to which legislative regime is applicable to olive-mill effluents. Opinions vary as to the relative applicability of "*ad hoc*" law 574/1996, legislation pertaining to wastes (legislative decree 22/1997 and ministerial decree of 05/02/1998), or legislation pertaining to wastewaters (legislative decree 152/1999). In other words, some consider that dumping is exempt from the requirements for authorization, while others maintain permission must be requested, either as per legislative decree 22/1997 or, alternatively, as per legislative decree 152/1999.

The different interpretations[50] surrounding the issue prompted intervention by the Ministry for the Environment which in a ministerial note of 14/07/2003 declared that, vegetation waters and crude olive cake used in agriculture as per law 574/1996 are excluded from the field of application of regulations pertaining to waste, therefore, forms specifying composition do not need to be filled for the transportation of such materials. The conditions for the controlled spreading of wastes arising from the processing of olives, in accordance with law 574/1996, are listed in Table 8.5. If one or more of these conditions is not met, particularly that concerning the effective and objective use of the materials for agronomic purposes,

---

[50]Court of Cassation, 03/10/2003, no. 37562. Court of Cassation, 12/07/2002, no. 26614.

or if methods of distribution do not comply with legal prescriptions, then the waste shall be considered as "special waste" thereby coming under the general regulations for waste.

According to some experts, a solution to the correct interpretation of olive-mill waste lies in making a distinction between liquid waste and wastewaters[51]: Legislative decree 22/1997 removes "direct wastewaters" from its field of application. These, then fall within the scope of application of legislative decree 152/1999. This states that:

1) "ordinary liquid wastes" are entirely subject to legislative decree 22/1997, from the production stage to final disposal in an appropriate treatment plant;
2) "wastewaters" (direct wastewaters) are entirely subject to the prescriptions of legislative decree 152/1999, from production and preventive purification, until they are finally disposed of through direct channeling into a "receiving body";
3) "liquid waste composed of effluents" (formerly known as indirect waste, now abolished) are entirely subject to the prescriptions of legislative decree 22/1997, from production and temporary stage (at the site of production), to collection, transportation, and final disposal.

According to this interpretation a purification plant intended exclusively for the treatment of effluents from the production cycle on a site must be removed from the field of application of regulations pertaining to waste as per art. 8, par. 1, lett. e of legislative decree 22/1997 and must be re-included within the sphere of legislative decree 152/1999.

At this point the question arises as to whether law 574/1996 is applicable exclusively to either legislative decree 22/1997 or legislative decree 152/1999, or to both. Bearing in mind the definition of "dumping"[52], in the absence of a duct leading directly from the olive-mill to the land, it would appear that the vegetation waters from the mill are to be considered as liquid waste (hence, regulated in accordance with legislative decree 22/1997 and not wastewater). A distinction must be made between dumping and the possible subsequent use for agronomic purposes of part or all of the constituents of the dumped material[53]. It is, therefore worth underlining that permission for fertirrigation and permission for dumping are two different things[54].

The indications set out by law 574/1996 pertain exclusively to the agronomic use of OMWW and not to the management of effluents from the processing of olives. If no such agronomic use is made, the derogation clause does not come into effect

---

[51]In accordance with art. 2, par. 2, lett. bb), legislative decree 152/1999, "dumping" refers to: "any direct discharge via a duct or other means of conveyance of liquid effluents into surface waters, the soil, the subsoil, or a sewerage system, regardless of their pollutant nature, even when having undergone prior purification treatment".

[52]see note 2.

[53]Court of Cassation 17/01/2000, no. 425.

[54]Court of Cassation, 26/10/1999, no. 12174.

and the general rules and regulations are applied i.e. legislative decree 22/1997 if the effluent is discharged into a tank, and legislative decree 152/1999, if they are discharged by direct channeling into a receiving body.

Proposals

The future legislator should address the following points (Tomati U., 2002):

- Information about the state of the olive-mill industry. There is a lack of reliable data concerning the quantity of olive-mill waste produced and where it is produced.
- Environmental awareness. Pedomorphological, hydrological, and climatic situation; agricultural practices. For example, the possibility of land spread depends on the suitability of soils to accept the waste and the nutritional need of the cultures. The applicability of technologies in the environmental situation of each country should also be taken into consideration.
- Dissemination of information relating to the disposal and recycling techniques of olive-mill waste in an efficient and economic way.
- Harmonization of diverse national regulations. Comparison of the legislations/ regulations of the various olive oil producing countries and elaboration of common rules about the management of olive-mill wastes.
- Reutilization of olive-mill waste.
- Environmentally compatible disposal of waste.
- Economic evaluation.
- Environmental policy for protection of resources. Proposal of various measures to meet the demands of environmental policy (sustainable management of resources according to the precautionary principle). Besides prohibition and environmental levies, subsidies are also possible.

In conclusion, the olive oil industry could be faced with a new crisis, similar to the one in the nineties, if an economically viable and environmentally friendly solution is not found for the safe disposal or recycling of these wastes.

Supplement

Appendix 1

Major EU Research Projects

Contract no.	Project Acronym	Title
AIR3-CT94-1987	BIOWARE	Development of a Biological Integrated Process for Purifying Olive Oil Wastewater Recovering Energy and Producing Alcohol
AIR3-CT93-1355		Improvement of olive oil quality by biotechnological means
COLL-CT-2003-500467	INASOOP	Integrated Approach to Sustainable Olive Oil and Table Olives Production
CRAFT CR178091/BRE21281	CRAFT	Industrial system for the treating and use of the residues generated in the olive oil production
E! 3508-AGROBRINES	AGROBRINES	Innovative Method For Highly Polluted Wastewater Generated By Pickled Table Olives and Vegetables
ENDEMO C EE./00337/87	ENDEMO C	New olive mill to produce only first quality oil and total recovery of subproducts for industrial use
EVK1-CT2002-30018	MOWOM	Development of a new mobile waste water treatment process for SME olive mills
EVK1-CT2001-30011	UDOR	Technology for treatment and recycling of the water used to wash olives
EV5V-CT93-0249		Integrated chemical and biological treatment of industrial waters
EVK1-CT2001-30011		Technology for treatment and recycling of the water used to wash olives

(continued)

Contract no.	Project Acronym	Title
EVK1-CT-2002-30028	SOLARDIST	Development of a solar distillation waste-water treatment plant for olive oil mills
EVWA-CT92-0006		Bioremediation of olive-mill wastewater for use as fertilizer
FAIR2-CT95-1075		Ultrahydrophytosqualene: New Processes for the Generation of Squalene by Supercritical Fluid Extraction from Waste of Olive Oil Production and Hydrogenation of Squalene
FAIR-CT96-1420	IMPROLIVE	Improvements of Treatments and Validation of the Liquid-Solid Waste from the Two-Phase Olive Oil Extraction
FAIR-CT97-3039		Natural Antioxidants from Olive Oil Processing Wastewater
FAIR-CT97-3620	HUSK	Composting of husks produced by two phase centrifugation olive milling plants
FAIR5-CT97-3807	LAGAR	Water recovery from olive-mill wastewaters after photocatalytic detoxification and disinfection
FAIR-CT98-9584		Recycling and use of waste from olive oil production
2001 GR 16 0 PP 209	NAIAS	North Aegean Innovative Actions and Support; Action 7.6 "Innovative Olive-Mill Waste Management Systems"
ICA3-CT-1999-00010	MEDUSA-WATER	Mediterranean usage of biotechnological treated effluent water
ICA3-CT-1999-00011	WAWAROMED	Wastewater recycling of olive oil mills in Mediterranean countries — Demonstration and sustainable reuse of residuals
ICA3-CT-1999-00014	WAM-ME	Water resources management under drought conditions. Criteria and tools for conjunctive use of conventional and marginal waters in Mediterranean regions
LIFE00 ENV/IT/000223	TIRSAV	Innovative technologies for recycling olive residue and vegetation water
LIFE99 ENV/D/000424	OLIVIA	Innovative demonstration facility for the treatment of wastewater from olive oil presses (OMW) with material and energetic utilization of residues
QLK5-2000-00766	BIOLIVE	Development of industrial solutions for the recycling and valorization of the olive oil fabrication residues for biopolymers and fine chemicals

Appendix 2

Databases

Name	Description	URL
BIOSIS	Biological Abstracts (The Scientific World, Inc.)	http://www.biosis.org
CAS	Chemical Abstracts (STN)	http://www.cas.org
COMPEDEX	Computerized Engineering Index and El Engineering Meetings (Engineering Information Inc.)	http://www.ei.org/eicorp
EMBASE	Biomedical and pharmacological abstracts (ELSEVIER PUBLICATIONS)	http://www.embase.com
esp@cenet search site	Full-text patent database (European Patent Office)	http://ep.espacenet.com
FSTA	Food Science Technology Abstracts (IFIS)	http://www.ifis.org
PAJ search site of the JPO's IPDL (Intellectual Property Digital Library)	Computer Translation of Japanese full-text patents (Japanese Patent Office)	http://www.jpo.go.jp
"SCIENCE DIRECT" search site	Full-text collection and abstracts database (ELSEVIER SCIENCE B.V.)	http://www.sciencedirect.com
USPTO search site	Full-Text US patents (United States Patent and Trademark Office)	http://www.uspto.gov
WPI	World Patent Index (DERWENT)	http://www.derwent.com

Appendix 3

International Organizations

European Union	http://europa.eu.int
European Union Network for the Implementation and Enforcement of Environmental Law (IMPEL)	http://europa.eu.int/comm/ environment/impel
FAO (Food and Agriculture Organization of the United Nations)	http://www.fao.org
FEDOLIVA (Federation of the Olive Oil Industries of the EU)	http://europe.eu.int/civil_society/conecs
(IOOC) International Olive Oil Council	http://www.internationaloliveoil.org

Appendix 4

National Associations/Institutions/ Research Laboratories

AFIDOL (L'Association Française Interprofessionnelle de l'Olive)	http://www.afidol.org
Alfanet	http://www.alfanet.it/oliodioliva
AOOA (Australian Olive Oil Association)	http://www.aooa.com.au
ASOLIVA (The Olive Oil Exporters Association of Spain)	http://www.asoliva.com
ASSITOL (Associazione Italiana dell' industria Olearia)	http://www.federalimentare.it/ docassitol.html
California Olive-Oil Council	http://www.oliveoilsource.com
Casa do Azeite	http://www.casadoazeite.pt
EDA (Waste Management Laboratory, Department of Environmental Studies, University of the Aegean, Greece)	http://www.aegean.gr/environment/eda/ OliveNet
FEDEROLIO (Federazione Nazionale del Commercio)	http://www.frantoionline.it/profess/ ass&con/federolio.htm
Foundation for the Promotion and Development of olive	http://www.oliva.net
INASOOP (Integrated Approach to Sustainable Olive Oil and Table Olives Production)	http://www.inasoop.info
Institute for Olive Cultivation Research	http://www.iro.pg.cnr.it
NAGREF (Institute of Olive Tree and Subtropical Plants of Chania)	http://www.nagref-cha.gr
NAOOA (North American Olive Oil Association)	http://www.afius.org/public/naooa/ index.htm
ONH (Tunisian Olive Oil Office)	http://www.onh.com.tn
Sevitel (Greek Association of Industries and Processors of Olive Oil)	http://www.oliveoil.gr

(*continued*)

TDC-OLIVE (European network of Technology Dissemination Centres)	http://www.tdcolive.net
S.S. TARIS ZEYTIN VE ZEYTINYAGI TARIM SATIS KOOPERATI FLERI BIRLIGI (Taris Olive and Olive Oil Agricultural Sales Cooperatives Union)	http://www.tariszeytin.com.tr
UNAPROL (Unione Nazionale tra le Associazioni di Produttori di Olive)	http://www.agriline.it

Bibliography

Literature

Abu-Qudais, M. (1996). Fluidized bed combustion for energy production from olive cake. *Energy*, *21* (3), 173–178.

Abu-Qudais, M., & Okasha, G. (1996). Diesel fuel and olive-cake slurry: atomization and combustion performance. *Appl. Energy*, *54* (4), 315–326.

Abu-Zreig, M., & Al-Widyan, M. (2002). Influence of olive mills solid waste on soil hydraulic properties. *Commun. Soil Sci. Plant Anal.*, *33* (3–4), 505–517.

Aggelis, G., Gavala, H.N., & Lyberatos, G. (2001). Combined and separate aerobic and anaerobic biotreatment of green olive debittering wastewater. *J. Agric. Eng. Res.*, *80* (3), 283–292.

Aggelis, G., Ehaliotis, C., Nerud, F., Stoychev, I., Lyberatos, G., & Zervakis, G. (2002). Evaluation of white rot fungi for detoxification and decolorization of effluents from the green olive debittering process. *Appl. Microbiol. Biotechnol.*, *59* (2–3), 353–360.

Aggelis, G., Iconomou, D., Christou, M., Bokas, D., Kotzailias, S., Christou, G., Tsagou, V., & Papanikolaou, S. (2003). Phenolic removal in a model olive oil mill wastewater using *Pleurotus ostreatus* in bioreactor cultures and biological evaluation of the process. *Water Res.*, *37* (16), 3897–3904.

Aguilera, J.F. (1987). "Improvement of olive cake and grape by-products for animal nutrition". *Degradation of Lignocellulosic in Ruminants and in Industrial Processes* (ed. Meer van der, J.M. et al.), Elsevier Applied Science, London, UK. ISBN 1-85166-165-4.

Aguilera, J.F., & Molina Alcalde, E. (1986). Valorisation nutritive d'un grignon d'olive traité et à la soude. *Ann. Zootech*, *35* (3), 205–218 (in French).

Aguilera, J.F., García, M.A., & Molina Alcalde, E. (1992). The performance of ewes offered concentrates containing olive by-products in late pregnancy and lactation. *Anim. Prod.*, *55*, 219–226.

Aktas, E.S., Imre, S., & Ersoy, L. (2001). Characterization and lime treatment of olive mill wastewater. *Water Res.*, *35* (9), 2336–2340.

Al-Malah, K., Azzam, M.O.J., & Abu-Lail, N.I. (2000). Olive mills effluent (OME) wastewater post-treatment using activated clay. *Separ. Purif. Technol.*, *20* (2), 225–234.

Alba-Mendoza, J., Ruiz-Gómez, A., & Hidalgo-Casado, F. (1990). Technological evolution of the different processes for olive oil extraction. In: Edible Fats and Oils Processing: Basic Principles and Modern Practices. *Am. Oil Chem. Soc.*, 341–347 (ed. Erickson, D.R.), Champaign, Illinois, USA.

Albarrán, A., Celis, R., Hermosín, M.C., López-Piñeiro, A., & Cornejo, J. (2004). Behaviour of simazine in soil amended with the final residue of the olive-oil extraction process. *Chemosphere, 54* (6), 717–724.

Albarrán, A., Celis, R., Hermosín, M.C., López-Piñeiro, A., Ortega-Calvo, J. J., & Cornejo, J. (2003). Effect of solid olive-mill waste addition to soil on sorption degradation and leaching of the herbicide. *Soil Use Manage., 19* (2), 150–156.

Albi Romero, M.A., & Fiestas Ros de Ursinos, J.A. (1960). Estudio del alpechín para su aprovechamiento industrial. Ensayos efectuados para su posible utilización como fertilizante. *Grasas y Aceites, 11,* 123–124 (in Spanish).

Alburquerque, J.A., Gonzálvez, J., García, D., & Cegarra, J. (2004). Agrochemical characterization of "alperujo", a solid by-product of the two-phase centrifugation method for olive oil extraction. *Bioresource Technol., 91* (2), 195–200.

Alianiello, F. (1997). Biochemical effects of olive mill wastewater on soil. *Proc. Int. Symp. on the olive wastes,* Kalamata, Greece, Nov. 1997.

Alianiello, F., Trinchera, S., Dell'Orco, S., Pinzari, F., & Benedetti, A. (1998). Somministrazione al terreno delle acque di vegetazione da frantoio: efffetti sulla sonstanza organica e sull'attività microbiologica del suolo. *Agricoltura Ricerca, 173,* 64–74 (in Italian).

Alkhamis, T.M., & Kablan, M.M. (1999). Process for producing carbonaceous matter from tars, oil shale and live cake. *Energy, 24* (10), 873–881.

Alkhamis, T.M., & Kablan, M.M. (1999). Olive cake as an energy source and catalyst for oil shale production of energy and its impact on the environment. *Energy Convers. Manage., 40* (17), 1863–1870.

Alliotta, G., Fiorentino, A., Oliva, A., & Tesmussi, F. (2002). Olive oil mill wastewater: isolation of polyphenols and their phytotoxicity in vitro. *Allelopahthy J., 9,* 9–17.

Allouche, N., Fki, I., & Sayadi, S. (2004). Toward a high yield recovery of antioxidants and purified hydroxytyrosol from olive mill wastewaters. *J. Agric. Food. Chem., 52* (2), 267–273.

Altieri, R., Cultrera, N., & Fontanazza, G. (2001). Impiego del substrato IRO-CNR da reflui oleari quale terreno di crescita per *Pleurotus eryngii* (D.C. ex. Fr.) Quel. J. var. Mara, per *Fragaria vesca,* L. var. Regina delle Valli e per barbatelle di *Olea europaea,* L. cv. Fs-17. *Atti del Convegno Proposte Innovative ed Applicazioni Industriali di Biotechnologie Agrarie ed Alimnetari,* Giulianova Lido, 25–26 Oct. 2001, 68–74 (in Italian).

Altieri, R., Fontanazza, G., & Cultrera, N. (2002). Utilizzazione di una tecnologia innovative per il trattamento dei sottoprodotti dell'oliveto e del frantoio a fini agronomici ed energetici. *Atti del Convegno CNR-ISAFoM,* Portici (NA), 23–24 Sep. 2002 (in Italian).

Alvarado, C.A. (1998). Environmental Performance in the Olive Oil Industry: Comparing Technologies using Environmental Performance Indicators. *M.Sc. Thesis in Environmental Management and Policy,* Lund University.

Amat, A.M., Arques, A., & Miranda, M.A. (1999). p-Coumaric acid photodegradation with solar light using a 2,4,6-triphenylpyrylium salt as photosensitizer: A comparison with other oxidation methods. *Appl. Catal. B: Environ., 23* (2–3), 205–214.

Amat, A.M., Di San Filippo, P., Caredda, E., & Schivo, M. (1987). Biomasse utili da acque di vegetazione dell'industria olearia. *Riv. Merceol., 26,* 255–264 (in Italian).

Amat, A.M., Di San Filippo, P., Rinaldi, A., Sanjust, E., Satt, G., & Viola, A. (1986). Acque di vegetazione dell' industria olearia: Materia prima o rifuto inquinante? "Vegetable material in water in the olive oil industry: raw material or polluting waste?" *Riv. Merceol., 25* (3), 183–199 (in Italian).

American Public Health Association (1989). St*andard Methods for the Examination of Water and Wastewater*, 17th edn., APHA, Washington, D.C., USA.

Amhajji, A., El-Jalil, M.H., Faid, M., Vasel, J. L., & El-Yachioui, M. (2000). Polyphenol removal from olive mill waste waters by selected moulded strains. *Grasas y Acietes, 51* (6), 400–404.

Amiot, M.J., Fleuriet, A., & Macheix, J.J. (1986). Importance and evolution of phenolic compounds in olive during growth and maturation. *J. Agric. Food Chem., 34*, 823–826.

Amiot, M.J., Fleuriet, A., & Macheix, J.J. (1989). Accumulation of oleuropein derivatives during olive maturation. *Phytochemistry, 28*, 67–69.

Amirante, P. (1998). Il compostaggio dei sottoprodotti dell'estrazione olearia: aspetti legislativi, tecnologici e risultati sperimentali. *Rifiuti Solidi*, anno XII, *1*, Jan.–Feb. 1998 (in Italian).

Amirante, P., & Di Renzo, G.C. (1991). Tecnologías e instalaciones para la depuración de alpechines por medio de procesos físicos de concentración y ósmosis. *Reunión Internacional Sobre: Tratamiento de Alpechines*. DGIEA. Casejeria de Agricultura y Pesca del la Junta de Andalucía, Seville, Spain, 1991, 49–61 (in Spanish).

Amirante, P., & Mongelli, G.L. (1982). Prove sperimentali di trattamento delle acque reflue di un oleificio con un impianto di incenerimento. "Experimental tests on wastewater treatment in an oil factory with an incinerator plant". *Riv. Ital. Sostanze Grasse, 59* (LIX) (6), 295–300 (in Italian).

Amirante, P., & Montel, G.L. (1998). Tecnologie ed impianti per il compostaggio intra ed interaziendali. *Proc. Symp. "Il compostaggio in ambito agricolo, promozione ed impiego di compost in agricoltura biologica"*, Istituto Agronomico Mediterraneo, Valenzano (BA) 12 June (in Italian).

Amirante, P., & Pipitone, F. (2002). Utilizzazione dei sottoprodotti della filiera olivicolo-olearia. "Re-use of the by-products of olive growing and olive oil production". *Olivae, 93*, 27–33 (in Italian).

Amirante, P., Di Renzo, G.C., Di Giovacchino, L., Bianchi, B., & Catalano, L. (1993). Evolution technologique des installations d'extraction d'huile d'olive. "Technological evolution of the installation of extracting olive-oil". *Olivae, 48*, 43–53 (in French).

Amirante, P. et al. (1991). Il compostaggio di concentrati di reflui oleari. Prove sperimentali in un impianto pilota statico ad asse verticale. *Riciclo delle Biomasse Rifiuto e di Scarto e Fertilizzazione Organica del Suolo* (a cura di, N. Senesi, & T.M. Milano), Patron Editore, Bologna, 231–234 (in Italian).

Ammar, E., & Ben Rouina, B. (1999). Potencial horticultural utilization of olive oil processing waste water. *Acta Horticulturae, 474*, 741–744.

Ammar, E., Mekki, H., Ueno, S., & Ben Zina, M. (2003). Traitement de l'effluent des huiléries d'olive et son integration dans des briques de construction -Optimisation des propriétés physico-mecaniques par analyse statistique. "Treatment of olive oil mill waste water and its incorporation in building bricks - Optimization of their physico-mechanical properties by statistical analysis of principal components". *Ann. Chim. Sci. Mater., 29* (4), 33–46 (in French).

Amro, B., Aburjai, T., & Al-Khalil, S. (2002). Antioxidative and radical scavenging effects of olive cake extract. *Fitoterapia, 73* (6), 456–461.

Andreoni, N., Dellomonaco, G., Spandre, R., & Fiorentini, R. (1996). Olive-mill wastewater spreading and groundwater pollution. *Agricoltura Mediterranea. Int. J. Agric. Sci., 126* (2), 200–2005.

Andreoni, V., Ferrari, A., Ranaldi, G., & Sorlini, C. (1986). The influence of some phenolic acids present in oil mill wastes on microbic groups for methanogenesis. *Proc. Int. Symp. on Olive By-Products Valorization*, FAO, UNDP (Food and Agriculture Organization of the United Nations), Seville, Spain, 5–7 Mar. 1986, 11–12.

Andreoni, V., Bonfanti, P., Daffonchio, D., Sorlini, C., & Villa, M. (1993). Anaerobic digestion of olive mill effluents: Microbiological and processing aspects. *J. Environ. Sci. Health, Part A*, *A28* (9), 2041–2059.

Andreozzi, R., Caprio, V., D'Amore, M.G., & Insola, A. (1995). p-Coumaric acid abatement by ozone in aqueous solution. *Wat. Res.*, *29* (1), 1–6.

Andreozzi, R., Longo, G., Majone, M., & Modesti, G. (1998). Integrated treatment of olive oil mill effluents (OME): Study of ozonation coupled with anaerobic digestion. *Water. Res.*, *32* (8), 2357–2364.

Andres, M., Cuenca, J.M., & Pardo, J.E. (2001). Application of environmental management to the olive oil industry. I. The polluting power of vegetable water. *Alimentaria*, *28* (326), 85–88.

Angelakis, A.N., Marecos do Monte, M.H.F., Bontoux, L., & Asano, T. (1999). The status of wastewater reuse in the Mediterranean basin: Need for guidelines. *Wat. Res.*, *33* (10), 2201–2217.

Angelidaki, I., & Ahring, B.K. (1996). Codigestion of olive oil mill wastewaters together with manure, household waste or sewage sludge. *Biomass Energy Environ., Proc. 9th, Eur. Bioenergy Conf.*, *1*, 290–295 (ed. Chartier, P.), Elsevier, Oxford, UK.

Angelidaki, I., & Ahring, B.K. (1997a). Codigestion of olive oil mill wastewaters with manure, household waste or sewage sludge. *Biodegradation*, *8* (4), 221–226.

Angelidaki, I., & Ahring, B.K. (1997b). Codigestion of olive oil mill wastewaters together with manure, household waste or sewage sludge. *Global Environ. Biotechnol., Proc. 3rd Int. Symp. Int. Soc. Environ. Biotechnol.*, 1997, 173–180 (ed. Wise, D.L.), Kluwer, Dordrecht, The Netherlands.

Angelidaki, I., Ellegaard, L., & Ahring, B.K. (1997). Modelling anaerobic codigestion of manure with olive oil mill effluent. *Proc. 8th IAWQ Int. Conf. on Anaerobic Digestion*, Sendai, Japan, 25–29 May, 1997. *Water Sci. Technol.*, *36* (6–7), 263–270.

Angelidaki, I., Petersen, S.P., & Ahring, B.K. (1990). Effects of lipids on thermophilic anaerobic digestion and reduction of lipid inhibition upon addition of bentonite. *Appl. Microbiol. Biothechnol.*, *33*, 469–472.

Angelidaki, I., Ahring, B.K., Deng, H., & Schmidt, J.E. (2002). Anaerobic digestion of olive oil mill effluents together with swine manure in UASB reactors. *Water Sci. Technol.*, *45*, (10), 213–218.

Angerosa, F., & Di Giovacchino, L. (1996). Natural antioxidants of virgin olive oil obtained by two and tri-phase centrifugal decanters. *Grasas y Aceites*, *47* (4), 247–254.

Annesini, M.C., & Gironi, F. (1991). Olive oil mill effluent. Aging effects on evaporation behaviour. *Wat. Res.*, *25* (9), 1157–1160.

Annesini, M.C., Giona, A.R., Gironi, F., & Pochetti, F. (1983). Treatment of olive oil wastes by distillation. *Effl. Wat. Treat. J.*, *23* (6), 245–248.

Annesini, M.C., Giona, A.R., Gironi, F. & Pochetti, F. (1984). Distillation process for vegetable water treatment. *Effl. Water Treat. J.*, *24* (12), 450–456.

ANPA (1999). Primo rapporto sui rifiuti speciali (in Italian).

Antolin, G., Lara, A., & Peran, J.R. (1997). Depuración y aprovechamiento integral de las aguas residuales de la industria del aceite de oliva (alpechín). "Depuration and integral use of the residual water from olive mills (alpechín)". *Inf. Tecnol.*, *8* (1), 95–102 (in Spanish).

Antolovich, M., Prenzler, P.D., Robards, K., & Ryan, D. (2000). Sample preparation in the determination of phenolic compounds in fruits. *Analyst*, *125* (1), 989–1009.

Antolovich, M., Prenzler, P.D., Patsalides, E., McDonald, S., & Robards, K. (2002). Methods for testing antioxidant activity. *Analyst*, *127* (1), 183–198.

Antonacci, R., Brunetti, A., Rozzi, A., & Santori, M. (1981) Trattamento anaerobico di acque di vegetazione di frantoio. I: Risultati preliminary. "Anaerobic treatment of olive oil mill effluent. I: Preliminary results". *Ingegneria Sanitaria*, *6*, 257–262 (in Italian).

Aragón, J.M. (1999a). Tratamiento del Alperujo para su Mejor Aprovechamiento. *Jornada Fairflow Europe*. Instituto de la grasa y sus derivados (C.S.I.C.), Seville, Spain.

Aragón, J.M., María, C., and Palancar, M.C. (2000). Results of the Projects IMPROLIVE & IMPROLIVE-2000.

Aragón, J.M. (1999b). Proyecto Improlive: Tratamiento del Alpeorujo para Mejorar su Aprovechamiento. *Alimentación, Equipos y Tecnología*, *18* (3), 117–123 (in Spanish).

Aragón, J.M., Palancar, M.C., Torrecilla, J.S., & Serrano, M. (1997). Drying alpeorujo in fluidized-moving bed. *Proc. Int. Symp. on the olive wastes*, Kalamata, Greece, 5–8 Nov. 1997.

Aramendia, M.A., Boráu, V., García, I., Jiménez, C., Lafont, F., Marinas, J.M., & Urbano, F.J. (1996). Qualitative and quantitative analices of phenolic compounds by high performance liquid chromatography and detection with atmospheric pressure chemical ionization mass spectrometry. *Rapid Commun. Mass Spectrom.*, *10*, 1585–1590.

Arienzo, M., & Capasso, R. (2000). Analysis of metal cations and inorganic anions in olive oil mill waste waters by atomic absorption spectroscopy and ion chromatography. Detection of metals bound mainly to the organic polymeric fraction. *J. Agric. Food Chem.*, *48* (4), 1405–1410.

Arienzo, M., Martino, A. de, Capasso, R., Di Maro, A., & Parente, A. (2003). Analysis of carbohydrates and amino acids in vegetable waste waters by ion chromatography. *Phytochemical Analysis*, *14* (2), 74–82.

Arjona, R., García, A., & Ollero Castro, P. de (1999). The drying of alpeorujo, a waste product of the olive oil mill industry. *J. Food Eng.*, *41* (3), 229–234.

Arjona, R., Ollero Castro, P. de, & Vidal, B. (2005). Automation of an olive waste industrial dryer. *J. Food Eng.*, *68* (2), 239–247.

Armesto, L., Bahillo, A., Cabanillas, A., Veijonen, K., Otero, J., Plumed, A., & Salvador, L. (2003). Co-combustion of coal and olive oil industry residues in fluidized bed. *Fuel*, (http://www.fuelfirst.com) *82*, 993–1000.

Arpino, A., & Carola, C. (1978). Lo smaltimento delle acque di vegetazione provenienti dagli impianti di estrazione dell'olio di oliva. Nota II: L'incenerimento delle acque di vegetazione. "Disposal of vegetable water from olive oil extraction plants. II: Incineration of vegetable water". *Riv. Ital. Sostanze Grasse*, *55* (LV) (1), 24–28 (in Italian).

Arvelakis, S., Gehrmann, H., Beckmann, M., & Koukios, E.G. (2001a). Prospects for large scale fluidized bed gasification of olive-oil residue: Evaluation of leaching as a pretreatment. *5th Int. Biomass Conf. of the Americas; Session 29: Biomass Properties and Preparation*, Orlando, Florida, USA, 17–21 Sep. 2001.

Arvelakis, S., Gehrmann, H., Beckmann, M., & Koukios, E.G. (2002). Effect of leaching on the ash behavior of olive residue during fluidized bed gasification. *Biomass & Bioenergy*, *22* (1), 55–69.

Arvelakis, S., Vourliotis, P., Kakaras, P., & Koukios, E.G. (2001b). Effect of leaching on the ash behavior of wheat straw and olive residue during fluidized bed combustion. *Biomass & Bioenergy, 20* (6), 459–470.

Arvelakis, S., Gehrmann, H., Beckmann, M., & Koukios, E.G. (2003). Agglomeration problems during fluidized bed gasification of olive-oil residue: Evaluation of fractionation and leaching as pre-treatments. *Fuel, 82,* 1261–1270.

Assas, N., Marouani, L., & Hamdi, M. (2000). Scale down and optimization of olive mill wastewaters decolorization by *Geotrichum candidum*. *Bioprocess Eng., 22* (6), 503–507.

Assas, N., Ayed, L., Marouani, L., & Hamdi, M. (2001). Decolorization of fresh and stored-black olive mill wastewaters by *Geotrichum candidum*. *Process Biochem., 38* (3), 361–365.

Atanassova, D., Kefalas, P., & Psillakis, E. (2005a). Measuring the antioxidant activity of olive oil mill wastewater using chemiluminescence. *Environ. Int., 31* (2), 275–280.

Atanassova, D., Kefalas, P., Petrakis, C., Mantzavinos, D., Kalogerakis, N., & Psillakis, E. (2005b). Sonochemical reduction of the antioxidant activity of olive mill wastewater *Environ. Int.; Recent Advances in Bioremediation, 31* (2), 281–287.

Atimtay, A., & Topal, H. (2004). Co-combustion of olive cake with lignite coal in a circulating fluidized bed. *Fuel, 83* (7–8), 859–867.

Attom Mousa, F., & Al-Sharif Munjed, M. (1998). Soil stabilization with burned olive. *Appl. Clay Sci., 13* (3), 219–230.

Aveni, A. (1983). Biogas recovery from olive oil production waste by anaerobic digestion. *R and D program –recycling of urban and industrial waste Broni (Pavia)*. Instituto Ricerche, Breda, Italy, 17–19 May, 1983.

Aveni, A. (1984). Biogas recovery from olive oil mill waste water by anaerobic digestion. *Anaerobic Digestion and Carbohydrate Hydrolysis of Waste* (eds. Ferrero, G.L., Ferranti, M.P., & Naveau, H.), Commission of the European Communities, Luxembourg. Elsevier Applied Science Publishers, New York, USA, 489–491.

Aveni, A. (1985). *Water Res., 19* (5), 667–669.

Aveni, A. (1988). Use of membrane technology for processing wastewater from olive-oil plants with recovery of useful components. *Proc. Symp. Future Industrial Processes of Membrane Processes* (eds. Cecille, L., & Toussaint, J.C.), Commission of the European Communities, Brussels, Belgium, 179–189, Elsevier Applied Science.

Azbar, N., Bayram, A., Filibeli, A., Muezzinoglu, A., Sengul, F., & Ozer, A. (2004). A review of waste management options in olive oil production. *Critical reviews in Environ. Sci. Technol., 34* (3), 209–247.

Bacaoui, A., Dahbi, A., Yaacoubi, A., Bennouna, C., Maldonado-Hodar, F.J., & Rivera-Utrilla, J. (2002). Experimental design to optimize preparation of activated carbons for use in water treatment. *Environ. Sci. Technol., 36* (17), 3844–3849.

Baccioni, L. (1981). Recycling of water and combustion of sediments: a solution for purification of water in oil mills. *Riv. Ital. Sostanze Grasse, 58 (LVIII)* (1), 34–37.

BADIS (Bureau d' Audit et de Développement Industriel et Social) (1994). Traitement des effluents d'huilerie d'olive. 1è Partie: Le secteur de l'huile d'olive en Tunisié, Tunis, Tunisia; 30 pages plus Annexes (in French).

Balice, V., & Cera, O. (1984). Acidic phenolic fraction of the olive vegetation water determined by a gas chromatographic method. *Grasas y Aceites, 35* (5), 178–180.

Balice, V., Carrieri, C., & Cera, O. (1990). Caratteristiche analitiche delle acque de vegetazione. *Riv. Ital. Sostanze Grasse, 67* (LXVII) (1), 9–16 (in Italian).

Balice, V., Carrieri, C., Cera, O., & Di Fazio, A. (1986). Natural biodegradation in olive mill effluents stored in open basins. *Proc. Int. Symp. on Olive By-Products Valorization*, Seville, Spain, 4–7 Mar. 1986.

Balice, V., Carrieri, C., Cera, O., & Rindone, B. (1988). The fate of tannin-like compounds from olive mill effluents in biological treatments. *Proc. 6th Symp. on Anaerobic Digestion of Wastewater*, Bologne, Italy, 1988, 275–279 (eds. Hall, E.R. & Hobson, P.N.), Pergamon, Oxford, UK.

Balice, V., Carrieri, C., Liberti, L., Passino, R., & Santori, M. (1985a). Two-stage anaerobic-aerobic biological treatment of olive oil wastewater combined with municipal sewage. *Stud. Environ. Sci., 29* (*Chem. Prot. Environ.*, 1985), 517–527.

Balice, V., Carrieri, C., Liberti, L., Passino, R., & Santori, M. (1985b). Trattamento biologico abbinato anaerobico-aerobico delle acque di vegetazione di frantoio. " Combined biological treatment of olive-mill vegetation water". *Ingegneria sanitaria, 2,* 69–76 (in Italian).

Balis, C. (1983). Agricultural use of the olive oil mill wastewaters, biomethanization and co-composting. *Technical report.* Ministry of Agriculture and Agricultural University of Athens (in Greek).

Balis, C. (1994). Enrichment of olive oil mill wastes through microbiological processing. *Proc. 7th Mediterranean Conf. on Organic Wastes Recycling in Soils*, Vieste, Italy, 22–25 Sep., 1994, *1*, 99–115, Ordine Nazionale dei Biologi, (eds. Landi, E. & Dumonet, S.).

Balis, C., Chatzipavlidis, I., & Flouri, F. (1996). Olive mill waste as a substrate for nitrogen fixation. *Proc. Olive Oil Processes and By Products Recycling*, Granada, Spain, 10–13 Sep., 1995. *Int. Biodeterior. Biodegrad.*, 1996, *38* (3–4), 169–178.

Balis, C., Flouri, F., Servis, D., & Tjerakis, C. (1987). Study and measures for the prevention of pollution from the olive oil wastewaters. *Report to the Greek Ministry of Agriculture*, 1987.

Ballesteros Perdices, I., Oliva Domínguez, J.M., Saez, F., & Ballesteros Perdices, M. (2001). Ethanol production from lignocellulosic by-products of olive oil extraction. *Appl. Biochem. Biotechnol., 91–93*, 237–252.

Ballesteros Perdices, I., Oliva Domínguez, J.M., Negro Alvarez, M.J., Manzanares, P., & Ballesteros Perdices, M. (2002). Ethanol production from olive oil extraction residue pretreated with hot water. *Appl. Biochem. Biotechnol., 98–100*, 717–732.

Bambalov, G., Israilides, C.J., & Tanchev, S. (1989). Alcohol fermentation in olive oil extraction effluents. *Biol. Wastes, 27* (1), 71–75.

Bartolini, S., Capasso, R., Evidente, A., Giorgelli, F., & Vitagliano, C. (1994). Effect of olive oil mill waste waters and their polyphenols on leaf and fruit abscission. *Acta Horticulturae, 356*, 292–296.

Basca, R.R., & Kiwi, J. (1998). Effect of rutile phase on the photocatalytic properties of nanocrystalline titanina during the degradation of p-coumaric acid. *Appl. Catal. B. Environ., 16* (1), 19–29.

Basile, P., Florio, G., & Fragiacomo, P. (1998). Aspetti Energetici ed Ambientali del Processo di Produzione dell'Olio di Oliva. *Inquinamento, 7* (July/August) (in Italian).

Bassus Cassianus (6th–7th century, A.D.). Γεωπονικά *"Geoponika or Agricultural Pursuits".* Translated by the Rev. T. Owen, London; printed for the author by, W. Spilsbury, London, 1805–1806. "Geoponika", a collection of agricultural literature, was compiled by Cassianus Bassus, at the end of the 6th or beginning of the 7th century. This compilation was revised about 950 A.D. by an unknown writer.

Beaufoy, G. (2000). The environmental impact of olive oil production in the European Union: Practical options for improving the environmental impact. *Report produced by the European Forum on Nature Conservation and Pastoralism and the Asociatión para el Análysis y Reforma de la Politica Agro-rural*. Brussels, Environment Directorate-General, European Commission, 2000. pp. 74.

Beavis, J.C. (1988). *Insects and other invertebrates in classical Antiquity*. Exeter: Exeter University Publication.

Beccari, M., & Majone, M. (1997). Comment reply. *Water Res.*, *31* (12), 3168.

Beccari, M., Majone, M., & Torrisi, L. (1998). Two reactor system with partial phase separation for anaerobic treatment of olive oil mill effluents. *Proc. 19th Biennial Conf. of the Int. Association on Water Quality*. Part 4 (of 9), 21–26 June 1998, Vancouver, Canada. *Water Sci. Technol.*, *38* (4–5), 53–60.

Beccari, M., Bonemazzi, F., Majone, M., & Riccardi, C. (1996). Interaction between acidogenesis and methanogenesis in the anaerobic treatment of olive oil mill effluents. *Wat. Res.*, *30* (1), 183–189.

Beccari, M., Carucci, G., Majone, M., & Torrisi, L. (1999a). Role of lipids and phenolic compounds in the anaerobic treatment of olive oil mill effluents. *Environ. Technol.*, *20* (1) 105–110.

Beccari, M., Majone, M., Carucci, G., Lanz, A.M., & Petrangeli, P.M. (2002). Removal of molecular weight fractions of COD and phenolic compounds in an integrated treatment of olive oil mill effluent. *Biodegradation*, *13* (6), 401–410.

Beccari, M., Marjone, M., Petrangeli, N., Papini, M., & Torrisi, L. (2000). Enhancement of anaerobic treatability of olive oil mill effluents by addition of $Ca(OH)_2$ and bentonite without intermediate solid/liquid separation. *Proc. 1st World Congress of the Inter. Water Association*, Paris, 3–7 July 2000. *Water Sci. Technol.*, 2001, *43* (1), 275–281.

Beccari, M., Majone, M., Riccardi, C., Savarese, F., & Torrisi, L. (1999b). Integrated treatment of olive oil mill effluents: Effect of chemical and physical pretreatment on anaerobic treatability. *Proc. 4th International Symp. on Waste Management Problems in Agro-Industries*, Istanbul, Turkey, 23–25 Sep. 1998. *Water Sci. Technol.*, *40* (1), 347–355.

Becker, P., Köster, D., Popov, M.N., Markossian, S., Antranikian, G., & Märkl, H. (1999). The biodegradation of olive oil and the treatment of lipid rich wool scouring wastewater under aerobic thermophilic conditions. *Water Res.*, *33* (3), 653 –660.

Bejarano, M., & Madrid, L. (1992a). Solubilization of heavy metals from a river sediment by a residue from olive oil industry. *Environ. Technol.*, *13* (10), 979–985.

Bejarano, M., & Madrid, L. (1992b). Effect of alpechín on heavy metal solubilization. *Suelo y Planta*, *2* (4), 541–549.

Bejarano, M., & Madrid, L. (1996a). Solubilization of heavy metals from a river sediment by an olive mill effluent at different pH values. *Environ. Technol.*, *17* (4), 427–432.

Bejarano, M., & Madrid, L. (1996b). Release of heavy metals from a river sediment by a synthetic polymer and an agricultural residue: Variation with time of contact. *Toxicol. Environ. Chem.*, *55*, 95–102.

Bejarano, M., & Madrid, L. (1996c). Solubilization of Zn, Cu, Mn and Fe from a river sediment by olive mill wastewater: Influence of Cu or Zn. *Toxicol. Environ. Chem.*, *55* (1–4), 83–93.

Bejarano, M., & Madrid, L. (1996d). Complexation parameters of heavy metals by olive mill wastewaters determined by a cationic exchange resin. *J. Environ. Sci. Health, Part B*, *31* (5), 1085–1101.

Bejarano, M., Mota, A.M., Goncalves, M.L.S., & Madrid, L. (1994). Complexation of Pb(ii) and Cu(ii) with a residue from the olive-oil industry and a synthetic-polymer by DPASV. *Sci. Total Environ.*, *158* (1–3), 9–19.

Bellido, E. (1986). Un nuevo proceso e instalación para la del alpechín y/o aguas residuales de las industrias de obtención de aceite de oliva. *P. de, I.*, 533–670 (in Spanish).

Bellido, E. (1987). Consideración en torno a la contaminación por alpechín. Diseño de una instalación para realizar un proceso industrial de depuración y prímenos resultados obtenidos a escala gemí-industrial. *Encuentro Euroamericano sobre Innov. Tecnológica, Medio Ambiente y Desarrollo*. Seville (in Spanish).

Bellido, E. (1989a). Un nuevo concepto de la depuración de las aguas residuales de las almazaras. *Reunión Internacional sobre Innovación Tecnológica*. Madrid (in Spanish).

Bellido, E. (1989b). Contaminación por alpechín. Un nuevo tratamiento para disminuir su impacto en diferentes ecosistemas. *Ph.D. Thesis*. University of Córdoba (in Spanish).

Beltrán, F.J., García-Araya, J.F., Navarrete, V., & Gimeno, O. (2001). Treatment of wastewater from table olive industries: Quantum yield of photolytic processes. *Bull. Environ. Contam. Toxicol.*, *67* (2), 195–201.

Beltrán, F.J., García-Araya, J.F., Frades, J., Alvarez, P., & Gimeno, O. (1999). Effects of single and combined ozonation with hydrogen peroxide or UV radiation on the chemical degradation and biodegradability of debittering table olive industrial wastewaters. *Water Res.*, *33* (3), 723–732.

Beltrán-Heredia, A.J., Torregrosa, A.J., Domínguez-Vargas, J.R., & García-Rodríguez, J. (2000a). Treatment of black-olive wastewaters by ozonation and aerobic biological degradation. *Water Res.*, *34* (14), 3515–3522.

Beltrán-Heredia, A.J., Torregrosa, A.J., Domínguez-Vargas, J.R., & García-Rodríguez, J. (2000b). Aerobic biological treatment of black table olive washing wastewaters: Effect of an ozonation stage. *Process Biochem.*, *35* (10), 1183–1190.

Beltrán-Heredia, A.J., Torregrosa, A.J., Domínguez-Vargas, J.R., & Peres, J.A. (2001a). Oxidation of p-hydroxybenzoic acid by UV radiation and by TiO_2/UV radiation: Comparison and modeling of reaction kinetic. *J. Hazard. Mater.*, *83* (3), 255–264.

Beltrán-Heredia, A.J., Torregrosa, A.J., Domínguez-Vargas, J.R., & Peres, J.A. (2001b). Kinetics of the reaction between ozone and phenolic acids present in agro-industrial wastewaters. *Water Res.*, *35* (4), 1077–1085.

Beltrán-Heredia, A.J., Torregrosa, A.J., García-Rodríguez, J., & Domínguez-Vargas, J.R. (2000c). Ozone treatment of olive mill wastewater. *Grasas y Aceites*, *51* (5), 301–306.

Beltrán-Heredia, A.J., Torregrosa, A.J., Domínguez-Vargas, J.R., & García-Rodríguez, J. (2000d). Ozonation of black-table-olive industrial wastewaters: Effect of an aerobic biological pretreatment. *J. Chem. Technol. Biotechnol.*, *75* (7), 561–568.

Beltrán-Heredia, A.J., Torregrosa, A.J., García-Araya, J.F., .Domínguez-Vargas, J.R., & Tierno, J.C. (2001c). Degradation of olive mill wastewater by the combination of Fenton's reagent and ozonation with an aerobic biological treatment. *Water Sci. Technol.*, *44* (5), 103–108.

Beltrán-Heredia, A.J., Torregrosa, A.J., García-Rodríguez, J., Ramos-Viseas del Pilar, M., & Domínguez-Vargas, J.R. (2000e). Aerobic biological treatment of olive mill wastewater previously treated by an ozonation stage. *Grasas y Aceites*, *51* (5), 332–339.

Benítez, F.J., Acero, J.L., & Leal Ana, I. (2003). Purification of storage brines from the preservation of table olives. *J. Hazard. Mater.*, *96* (2–3), 155–169.

Benítez, F.J., Beltrán Heredia, A.J., & Acero, J.L. (1993a). Protocatechuic acid ozonation in aqueous solutions. *Water Res.*, *27* (10), 1519–1525.

Benítez, F.J., Beltrán-Heredia, A.J., & Acero, J.L. (1993b). Protocatechuic acid degradation by chlorine dioxide. *Afinidad, 50* (443), 28–32.

Benítez, F.J., Beltrán Heredia, A.J., & Acero, J.L. (1996a). Oxidation of vanillic acid as a model of polyphenolic compounds in olive oil wastewaters. III. Combined UV radiation hydrogen peroxide oxidation, *Toxicol. Environ. Chem.*, *56* (1–4), 199–210.

Benítez, F.J., Acero, J.L., González, T., & García, J. (2001). Organic matter removal from wastewaters of the black olive industry by chemical and biological procedures. *Process Biochem.*, *37* (3), 257–265.

Benítez, F.J., Acero, J.L., González, T., & García, J. (2002a). Application of ozone and advanced oxidation processes to the treatment of lye-wastewaters from the table olives industry. *Ozone Sci. Eng.*, *24* (2), 105–116.

Benítez, F.J., Acero, J.L., González, T., & García, J. (2002b). The use of ozone, ozone plus UV radiation, and aerobic microorganisms in the purification of some agro-industrial wastewaters. *J. Environ. Sci. Health Part A Toxic Hazard. Subst. Environ. Eng.*, *37* (7), 1307–1325.

Benítez, F.J., Beltrán-Heredia, A.J., Acero, J.L., & Cercas, V. (1996b). Chemical and biological pretreatments of olive mill wastewaters. 212th *American Chemical Society National Meeting*; Orlando, Florida, USA, 25–29 Aug. 1996. *Abstr. Pap. Am. Chem. S.*, *212* (1–2), 103-Envr. Part 1.

Benítez, F.J., Beltrán Heredia, A.J., Acero, J.L., & González, T. (1994a). Oxidation of vanillic acid as a model of polyphenolic compound present in olive oil wastewaters. I. Ozonation process. *Toxicol. Environ. Chem.*, *46* (1–2), 37–47.

Benítez, F.J., Beltrán Heredia, A.J., Acero, J.L., & González, T. (1995). Oxidation of vanillic acid as a model of polyphenolic compound present in olive oil wastewaters. II. Photochemical oxidation and combined ozone-UV oxidation. *Toxicol. Environ. Chem.*, *47* (3/4), 141–153.

Benítez, F.J., Beltrán Heredia, A.J., Acero, J.L., & González, T. (1996c). Degradation of protocatechuic acid by two advanced oxidation processes: Ozone/UV radiation and H_2O_2/ UV radiation. *Water Res.*, *30* (7), 1597–1604.

Benítez, F.J., Beltrán-Heredia, A.J., Acero, J.L., & Pinilla, M.L. (1997a). Ozonation kinetics of phenolic acids present in wastewaters from olive oil mills. *Ind. Eng. Chem. Res.*, *36* (3), 638–644.

Benítez, F.J., Beltrán-Heredia, A.J., Acero, J.L., & Pinilla, M.L. (1997b). Simultaneous photodegradation and ozonation plus UV radiation of phenolic acids major pollutants in agro industrial wastewaters. *J. Chem. Technol. Biotechnol.*, *70* (3), 253–260.

Benítez, F.J., Beltrán Heredia, A.J., González, T., & Acero, J.L. (1994b). Photochemical oxidation of protocatechuic acid. *Water Res.*, *28*, 2095–2100.

Benítez, F.J., Beltrán Heredia, A.J., González, T., & Real, F. (1998). Kinetics of the elimination of vanillin by UV radiation catalyzed with hydrogen peroxide. *Fresen. Environ. Bull.*, *7* (11/12), 726–733.

Benítez, F.J., Beltrán-Heredia, A.J., Peres, J.A., & Domínguez-Vargas, J. R (2000). Kinetics of p-hydroxybenzoic acid photodecomposition and ozonation in a batch reactor. *J. Hazard Mater.*, *73* (2), 161–178.

Benítez, F.J., Beltrán Heredia, A.J., Torregrosa, A.J., & Acero, J.L. (1996d). Chemical and biological degradation of olive-mill wastewaters. *Rev. R. Acad. Cienc. Exactas, Fis. Nat. 90* (3), 205–209.

Benítez, F.J., Beltrán Heredia, A.J., Torregrosa, A.J., & Acero, J.L. (1997c). Improvement of the anaerobic biodegradation of olive mill wastewaters by prior ozonation pretreatment. *Bioprocess Eng.*, *17* (3), 169–175.

Benítez, F.J., Beltrán-Heredia, A.J., Torregrosa, J., & Acero, J.L. (1997d). Treatment of wastewaters from olive oil mills by UV radiation and by combined ozone UV radiation. *Toxicol. Environ. Chem. 61* (1–4), 173–185.

Benítez, F.J., Beltrán-Heredia, A.J., Torregrosa, A.J., & Acero, J.L. (1999a). Treatment of olive mill wastewaters by ozonation, aerobic degradation and the combination of both treatments. *J. Chem. Technol. Biotechnol.*, *74* (7), 639–646.

Benítez, F.J., Beltrán-Heredia, A.J., Torregrosa, A.J., Acero, J.L., & Cercas, V. (1997e). Aerobic degradation of olive mill wastewaters. *Appl. Microbiol. Biotechnol.*, *47* (2), 185–188.

Bernal, M.P., Paredes, C., Sánchez Monedero, M.A., & Cegarra, J. (1998). Maturity and stability parameters of composts prepared with a wide range of organic wastes. *Bioresourc. Technol.*, *63* (1), 91–99.

Benítez, F.J., Beltrán-Heredia, A.J., Torregrosa, A.J., et al. (1999b). Aerobic treatment of black olive wastewater and the effect of an ozonation stage. *Bioprocess Eng.*, *20* (4), 355–361.

Bertin, L., Majone, M., Di Gioia, D., & Fava, F. (2001). An aerobic fixed-phase biofilm reactor system for the degradation of the low-molecular weight aromatic compounds occurring in the effluents of anaerobic digestors treating olive mill wastewaters. *Biotechnology*, *87* (2), 161–177.

Bertoldi, M. de, Vallini, G., Pera, A., & Zucconi, F. (1982). Comparison of three windrow compost systems. *BioCycle*, *23* (2), 45–50.

Bertoldi, M. de, Filippi, C., & Picci, G. (1986). Olive residue composting and land utilization. *Proc. Int. Symp. on Olive By-Products Valorization*, FAO, UNDP (Food and Agriculture Organization of the United Nations), Seville, Spain, 5–7 Mar. 1986.

Bianco, A., Mazzei, R.A., Melchioni, C., Romeo, G., Scarpati, M.L., Soriero, A., & Uccella, N. (1998). Microcomponents of olive oil III. Glucosides of 2-(3,4-dihydroxy-phenyl)ethanol. *Food Chem.*, *63* (4), 461–464.

Bianco, A., Buiarelli, F., Cartoni, G., Coccioli, F., Jasionowska, R., & Margherita, P. (2003). Analysis by liquid chromatography-tandem mass spectrometry of biophenolic compounds in olives and vegetation waters, Part I. *J. Separation Sci.*, *26* (5), 409–416.

Bianchi, G. (2003). Lipids and phenols in table olives. *European J. Lipid Sci. Techn.*, *15* (5), 229–242.

Binachi, G., Pozzi, N., & Vlahov, G. (1994). *Phytochemistry*, *37*, 205–207.

Bing, U., Cini, E., Cioni, A., & Laurendi, V. (1994). Smaltimento-recupero delle sanse di oliva provenienti da un "due fasi" mediante distribuzione in campo. *L'Informatore Agrario*, *47*, 75–78 (in Italian).

Bini, C., Gentili, L., Bini, L.M., et al. (1998). Effects of wine wastewaters on soils and olive trees in Tuscany (Italy). *Fresen. Environ. Bull.*, *7* (9A–10A), 756–763, Sp. Iss.

Blánquez, P., Caminal, G., Sarra, M., Vicent, M.T., & Gabarrell, X. (2002). Olive oil mill waste waters decoloration and detoxification in a bioreactor by the white rot fungus *Phanerochaete flavido-alba. Biotechnol. Prog.*, *18* (3), 660–662.

Blesa, M.J., Miranda, J. L., Izquierdo, M.T., & Moliner, R. (2003a). Curing temperature effect on mechanical strength of smokeless fuel briquettes prepared with molasses. *Fuel*, *82* (8), 943–947.

Blesa, M.J., Miranda, J.L., Izquierdo, M.T., & Moliner, R. (2003b). Curing time effect on mechanical strength of smokeless fuel briquettes. *Fuel Process. Technol.*, *80* (2), 155–167.

Blesa, M.J., Fierro, V., Miranda, J.L., Moliner, R., & Palacios, J. M. (2001). Effect of the pyrolysis process on the physicochemical and mechanical properties of smokeless fuel briquettes. *Fuel Process. Technol.*, *74* (1), 1–17.

Boari, G., & Mancini, I.M. (1990). Combined treatments of urban and olive mill effluents in Apulia, Italy. *Proc. IAWPRC Symp. on Waste Management Problems in Agro Industries*; Istanbul, Turkey, 25–27 Sep. 1989. *Water Sci. Technol.*, *22* (9), 235– 240.

Boari, G., Carrieri, C., & Santori, M. (1980). Depurazione con processi a membrana di acque di vegetazione: diversamente pretrattate. "Purification by membrane processes of pretreated wastewater from olive oil manufacture". *Olie, Grassi, Deriv.*, *16* (1), 2–5 (in Italian).

Boari, G., Mancini, I.M., & Trulli, E. (1993). Anaerobic digestion of OMW pretreated and stored in municipal solid waste sanitary landfills. *Proc. 2nd IAWQ Int. Symp. on Waste Management Problems in Agro-Industries*; Istanbul, Turkey, 23–25 Sep. 1992. *Water Sci. Technol.*, *28* (2), 27–34.

Boari, G., Brunetti, A., Passino, R., & Rozzi, A. (1984). Anaerobic digestion of olive oil mill wastewaters. *Agric. Wastes*, *10* (3), 161–175.

Bonari, E., & Ceccarini, L. (1991). Spargimento delle acque di vegetazione dei frantoi sul terreno agrario. *L'Informatore Agrario*, *13*, 49–57 (in Italian).

Bonari, E., & Ceccarini, L. (1993). Sugli effetti dello spargimento delle acque di vegetazione su terreno agrario: risultati di una ricerca sperimentale. *Genio Rurale*, *5*, 60–67 (in Italian).

Bonari, E., Macchia, M., & Ceccarini, L. (1993). The wastewaters from olive oil extraction: Their influence on the germinative characteristics of some cultivated and weed species. *Agric. Med.*, *123*, 273–280.

Bonari, E., Ceccarini, L., Silvestri, N., Tonini, M., & Sabbatini, T. (2001). Spargimento delle acque di vegetazione dei frantoi oleari sul terreno agrario. *L'Informatore Agrario*, suppl.1, *50*, 8–12 (in Italian).

Bonari, E., Silvestri, N., Tonini, M., Sabbatini, T., Giannini, C., & Ceccarini, L. (2001). Spargimento delle acque di vegetazione dei frantoi oleari sul terreno agrario. *L'Informatore Agrario*, suppl.1, *50*, 19–21 (in Italian).

Bondioli, P., Lanzani, A., & Fedeli, E. (1991). Digestione anaerobica degli effluenti liquidi. Nota 1: particolarità, prospettive e campi di impiego dell'industria delle sostanze grasse. "Anaerobic digestion of fluid effluents. Note 1: Characteristics and application prospects in oils and fats industry". *Riv. Ital. Sostanze Grasse*, *68* (LXVIII), 1–4 (in Italian).

Bondioli, P., Lanzan, A., Fedeli, E., Sala, J.M., & Gerali, G. (1992). Valutazione della possibilita di pretrattare le acque di vegetazione dei frantoi oleari con ozono. "Evaluation of the possibility of pretreating olive oil mill waste water by means of ozone". *Riv. Ital. Sostanze Grasse*, *69* (LXIX) (10), 487–497 (in Italian).

Bondioli, P., Mariani, C., Lanzani, A., Fedeli, E., & Muller, A. (1993). Squalene recovery from oil deodorizer distillates. *JAOCS, J. Am. Chem. Soc.*, *70* (8), 763–766.

Bonfanti, P., & Lazzari, M. (1998). Analisi delle prestazioni tecnico-economiche di impianti per il trattamento anaerobico delle acque di vegetazione di oleifici a tre fasi. *La Rivista di Ingegneria Agraria*, 4/98 (in Italian).

Bonfanti, P., & Lazzari, M. (1999). Analysis of the technical-economic performances of three-phase OMWW anaerobic treatment plants. *La Rivista di Ingegneria Agraria*, *30* (1), 12–18 (in Italian).

Bonfanti, P., Lazzari, M., & Zolli, S. (1997). Risultati ottenuti in un biennio di prove sperimentali su di un impianto per il trattamento biologico delle acque di vegetazione di frantoio ai fini del recupero energetico. "Results obtained in a biannual experimental test planned for the biological treatment of olive mill wastewater" (project BIOWARE). *Atti V Convegno Nazionale ingegneria agraria,* Ancona, 11–12 Sep. 1997, *l.1,* 101–110 (in Italian).

Bonfanti, P., Fregonese, A., Lazzari, M., & Zolli, S. (1996). A pilot plant for the biological treatment of olive oil waste water. *Ag. Eng. Congress '96,* report, N.96-E-036,1–8, Madrid.

Borja-Padilla, R. (1994). Influence of substrate concentration on the macro-energetic parameters of anaerobic digestion of black-olive wastewater. *Biotechnol. Letters, 16* (3), 321–326.

Borja-Padilla, R., & González, A.E. (1994). Comparison of anaerobic filter and anaerobic contact process for olive mill wastewater previously fermented with *Geotrichum candidum. Process Biochem., 29* (2), 139–144.

Borja-Padilla, R., Alba-Mendoza, J., & Banks, C.J. (1996a). Anaerobic digestion of wash waters derived from the purification of virgin olive oil using a hybrid reactor combining a filter and a sludge blanket. *Proc. Biochem., 31* (3), 219–224.

Borja-Padilla, R., Alba-Mendoza, J., & Banks, C.J. (1995a). Activated sludge treatment of wash waters derived from the purification of virgin olive oil in a new manufacturing process. *J. Chem. Technol Biotechnol., 64* (1), 25–30.

Borja-Padilla, R., Alba-Mendoza, J., & Banks, C.J. (1997). Impact of the main phenolic compounds of olive mill wastewater (OMW) on the kinetics of acetoclastic methanogenesis. *Proc. Biochem., 32* (2), 121–133.

Borja-Padilla, R., Alba-Mendoza, J., & Durán Barrantes, M.M. (1992a). Effect of additives used in olive oil processing on the kinetics of anaerobic digestion of olive mill wastewater. *Grasas y Aceites, 43* (4), 204–211.

Borja-Padilla, R., Alba-Mendoza, J., & González Berecca, A. (1992b). Kinetic study of the anaerobic digestion of olive mill wastewater, obtained from oil extraction using Olivex, and previously biotreated with *Geotrichum candium. Grasas y Aceites, 43* (4), 219–225.

Borja-Padilla, R., Banks, C.J., & Alba-Mendoza, J. (1995b). A simplified method for determination of kinetic parameters to describe the aerobic biodegradation of two important phenolic constituents of olive mill wastewater by a heterogeneous microbial culture. *J. Environ. Sci. Health Part A, 30* (3), 607–626.

Borja-Padilla, R., Durán Barrantes, M.M., & Luque González, M. (1992c). Aerobic treatment of effluents from biomethanation of olive mill wastewater. *Grasas y Aceites, 43* (1), 20–25.

Borja-Padilla, R., Fernández, A.G., & Durán Barrantes, M.M. (1992d). Kinetic-study of anaerobic-digestion process of wastewaters from ripe olive processing. *Grasas y Aceites, 43* (6), 317–328.

Borja-Padilla, R., Martín-Martín, A., & Durán Barrantes, M.M. (1991a). Kinetic study of biomethanation of condensation water from the thermal concentration of olive oil mill wastewater. *Grasas y Aceites, 42* (6), 437–443.

Borja-Padilla, R., Martín-Martín, A., & Durán Barrantes, M.M. (1992e). Kinetic study of the biomethanation of olive mill wastewater previously subjected to aerobic treatment with *Geotrichum candidum. Grasas y Aceites, 43* (2), 82–86.

Borja-Padilla, R., Martín-Martín, A., & Fiestas Ros de Ursinos, J.A. (1990a). Estudio cinético de la depuración anaerobia del alpechín en presencia de diversos soportes para immobilization de los microorganismos responsables del proceso. "Kinetic study of the anaerobic purification of olive mill wastewater by using various supports for the immobilization of the microorganisms". *Grasas y Aceites*, *41* (4–5), 347–356 (in Spanish).

Borja-Padilla, R., Martín-Martín, A., & Fiestas Ros de Ursinos, J.A. (1991b). Substrate concentration effects on the kinetics of olive mill wastewater biomethanation in fluidized bed reactors with immobilized microorganisms. *Grasas y Aceites*, *42* (5), 363–370.

Borja Padilla, R., Martín-Martín, A., & Garrido Fernández, A. (1994). Kinetics of black olive wastewater treatments by the activated sludge system. *Biochem.*, *29* (7), 587–593.

Borja-Padilla, R., Alba-Mendoza, J., Martín-Martín, A., & Mancha, A. (1998a). Effect of organic loading rate on anaerobic digestion process of wastewaters from the washing of olives prior to the oil production process in fluidized bed reactor. *Grasas y Aceites*, *49* (1), 43–50.

Borja-Padilla, R., Alba-Mendoza, J., Martín-Martín, A. and Mancha, A. (1999). Kinetic study of anaerobic digestion of wastewaters from the washing of olives prior to the oil production process in a completely mixed reactor with immobilised microorganisms. *Grasas y Aceites*, *50* (2), 87–93.

Borja-Padilla, R., Banks, C.J., Maestro-Durán, R., & Alba-Mendoza, J. (1996b). The effects of the most important phenolic constituents of olive mill wastewater on batch anaerobic methanogenesis. *Environ. Technol.*, *17* (2), 167–174.

Borja-Padilla, R., Martín-Martín, A., Durán Barrantes, M.M., & Maestro-Durán, R. (1992f). Kinetic study of anaerobic digestion of olive mill wastewater in the mesophilic and thermophilic temperature ranges. *Grasas y Aceites*, *43* (6), 341–346.

Borja-Padilla, R., Martín-Martín, A., Fiestas Ros de Ursinos, J.A., & Maestro-Durán, R. (1990b). Efecto de inhibicion en el proceso de biometanization del alpechín en biorreactores con microorganismos inmovilizados en diversos tipos de soportes. "Effect of the inhibition of olive mill wastewater biomethanation in bioreactors with microorganisms immobilized on various types of support". *Grasas y Aceites*, *41*, 397–403 (in Spanish).

Borja-Padilla, R., Martín-Martín, A., Gómez, L.F., & Ramos-Cormenzana, A. (1993a). Anaerobic digestion of olive mill wastewater pretreated with *Azotobacter chroococcum*. *Resou. Conserv. and Recycl.*, *9* (3), 201–211.

Borja-Padilla, R., Carrido, S.E., Martínez, L., Ramos-Cormenzana, A., & Martín-Martín, A. (1993b). Kinetic study of anaerobic digestion of olive mill waste water previously fermented with *Aspergillus terreus*. *Process Biochem.*, *28* (6), 397–404.

Borja-Padilla, R., Martín-Martín, A., Alonso, V., García, I., & Banks, C.J. (1995c). Influence of different aerobic pretreatments on the kinetics of anaerobic digestion of olive mill wastewater. *Water Res.*, *29* (2), 489–495.

Borja-Padilla, R., Martín-Martín, A., Fiestas Ros de Ursinos, J.A., Olias, J.M., & Durán Barrantes, M.M. (1992g). Use of sepiolite and bentonite in degradation of phenolic compounds in olive oil-extraction wastewater. *Inquinamento*, *34* (3), 114–117.

Borja-Padilla, R., Martín-Martín, A., Maestro-Durán, R., Alba-Mendoza, J., & Fiestas Ros de Ursinos, J.A. (1991c). Cinética del proceso de depuración anaerobia de alpechín previamente tratado por vía aerobia. "Kinetics of the anaerobic purification of aerobically prebio-treated olive mill wastewater". *Grasas y Aceites*, *42* (3), 194–201 (in Spanish).

Borja-Padilla, R., Martín-Martín, A., Maestro-Durán, R., Alba-Mendoza, J., & Fiestas Ros de Ursinos, J.A. (1992h). Enhancement of the anaerobic digestion of olive mill wastewater by the removal of phenolic inhibitors. *Process Biochem.*, *27* (4), 231–237.

Borja-Padilla, R., Martín-Martín, A., Sánchez, E., Rincón, B., & Raposo, F. (2005). Kinetic modeling of the hydrolysis, acidogenic and methanogenic steps in the anaerobic digestion of two-phase olive pomace (TROP). *Process Biochem.*, *40* (5), 1841–1847.

Borja-Padilla, R., Rincón, B., Raposo, F., Alba-Mendoza, J., & Martín-Martín, A. (2003). A study of anaerobic digestibility of two-phases olive mill solid waste (OMSW) at mesophilic temperatures. *Process Biochem.*, *38* (5), 733–742.

Borja-Padilla, R., Alba-Mendoza, J., Mancha, A., Martín-Martín, A., Alonso, V., & Sánchez, E. (1998b). Comparative effect of different aerobic pretreatments on the kinetics and macroenergetic parameters of anaerobic digestion of olive mill wastewater in continuous mode. *Bioprocess Eng.*, *18* (2), 127–134.

Borja-Padilla, R., Alba-Mendoza, J., Garrido-Hoyos, S.E., Martínez, L., García-Pareja, M.P., Incerti, C., & Ramos Cormenzana, A. (1995d). Comparative study of anaerobic digestion of olive mill wastewater (OMW) and OMW previously fermented with *Aspergillus terreus*. *Bioprocess Eng.*, *13* (6), 317–322.

Borja-Padilla, R., Alba-Mendoza, J., Garrido-Hoyos, S.E., Martínez, L., García-Pareja, M.P., Monteoliva-Sánchez, M., & Ramos-Cormenzana, A. (1995e). Effect of aerobic pretreatment with *Aspergillus terreus* on the anaerobic digestion of olive mill wastewater. *Biotechnol. Appl. Biochem.*, *22* (2), 233–246.

Borsani, R., & Ferrando, B. (1996). Ultrafiltration plant for olive vegetation waters by polymeric membrane batteries. *Desalination*, *108*, 281–286.

Boudouropoulos, I.D. (2000). Potential and perspectives for application of environmental management system (EMS) and ISO 14000 to food industries. *Food Reviews Int.*, *16* (2), 177–237.

Burali, A., & Boeri, G.C. (2003). *IMPEL Olive Oil Project Report - CMA & NOA*. Number Report 2003/3, Project Manager: Méndez Miguel, Rome, Nov. 2003; number of pages, report: 33 and annexes: 61. European Union Network for the Implementation and Enforcement of Environmental Law (IMPEL) (http://europa.eu.int/comm/environment/impel).

Bouranis, D.L., Vlyssides, A.G., & Drossopoulos, J.B. (1995). Some characteristics of a new organic soil conditioner from the co-composting of olive oil processing wastewater and solid residue. *Commun. Soil Sci. Plant Anal.*, *26* (15–16), 2461–2472.

Box, J.D. (1983). Investigation of the Folin-Ciocalteau phenol reagent for the determination of polyphenolic substances in natural waters. *Water Res.*, *17*, 511–525.

Boz, Ö., Doğan, M.N., & Albay, F. (2003). Olive processing wastes for weed control. *Weed Res.*, *43*, 439–443.

Bradley, R.M., & Baruchello, L. (1980). Primary wastes in the olive oil industry. *Effl. Wat. Treat. J.*, *20* (4), 176–177.

Brenes Balbuena, M., & Castro Gomez-Millan, A. de (1998). Washing waters of the Spanish-style green olive processing as a source of phenolic antioxidants. *2nd Int. Electronic Conf. On Synthetic Organic Chemistry (ECSOC-2)* (http://www.mdpi.org/ecsoc/), September 1–30, 1998.

Brenes Balbuena, M., & Garrido Fernández, A. (1988a). Regeneración de salmueras de aceitunas verdes estilo sevillano con carbón activo y tierras decolorantes. "Regeneration of fermentation brines from Sevillian green table olives by decolorating earths and activated carbon". *Grasas y Aceites*, *39* (2), 96–101 (in Spanish).

Brenes Balbuena, M., García-García, P., & Garrido Fernández, A. (1988b). Regeneration of Spanish style green table olive brines by ultrafiltration. *J. Food Sci.*, *53* (6), 1733–1736.

Brenes Balbuena, M., García-García, P., & Garrido Fernández, A. (1989a). Influence of reusing regenerated brines on some characteristics of the packing green olives. *Grasas y Aceites*, *40* (3), 182–189.

Brenes Balbuena, M., Montano, A., & Garrido Fernández, A. (1990). Ultrafiltration of green table olive: influence of operating parameters and effect on polyphenol composition. *J. Food. Sci.*, *55* (1), 214–217.

Brenes Balbuena, M., Romero Barranco, C., & Castro Gomez-Millan, A. de (2004). Combined fermentation and evaporation processes for treatment of wash waters from Spanish-style green olive processing. *J. Chem. Technol. Biotechnol.*, *79* (3), 253–259.

Brenes Balbuena, M., Sánchez Roldán, F., & Garrido Fernández, A. (1988c). Regeneration of brines from Spanish green table lives by coagulation-filtration. *Grasas y Aceites*, *39* (4/5), 264–271.

Brenes Balbuena, M., Vicente Fernández, A. de, García-García, P., & Garrido Fernández, A. (1989b). Characteristics of the waste waters from the elaboration of table olives. *Grasas y Aceites*, *40* (4–5), 287–290.

Brenes Balbuena, M., García-García, P., Romero Barranco, C., & Garrido Fernández, A. (2000). Treatment of green table olive waste waters by an activated-sludge process. *J. Chem. Technol. Biotechnol.*, *75* (6), 459–463.

Brenes Balbuena, M., Rejano Navarro, L., García-García, P., Sánchez, A.H., & Garrido Fernández, A. (1995). Biochemical changes in phenolic compounds during Spanish-style green olive processing. *J. Agric. Food Chem.*, *43*, 2702–2706.

Briccoli-Bati, C., & Lombardo, N. (1990). Effect of olive waste irrigation on young olive plants. *Acta Horiculture* (Olive Growing), *286*, 489–491.

Briccoli-Bati, C., Marsilio, V., & Di Giovacchino, L. (1990). Ulteriori osservazioni sull'effetto dello spandimento di acque di vegetazione su terreno agrario. *Quaderno di Scienza e Tecnología* (ed. NIA Ricerche), *1*, 90–93 (in Italian).

Brunetti, A. (1991). Anaerobic treatment of wastewater from olive oil extraction and from the agro-food industry: Startup procedures for UASB-type reactors. *Ist. Ric. Acque*, *94*, 10.1–10.9.

Brunetti, A., Rozzi, A., Antonacci, R., Labellarte, G., & Longobardi, C. (1983). Trattamento anaerobico di acque di vegetazione di frantoio. "Anaerobic treatment of olive mill wastewater". *Ingegneria Sanitaria*, *31* (4), 18–24 (in Italian).

Bufano, G., Cianci, D., Montemurro, O., Palermo, D., & Tasca, M.L. (1982). Prove di razionamento degli ovini con paste di vegetazione dei frantoi oelari. *Scienza e tecnica Agraria*, (1–2), 1–10 (in Italian).

Buldini, P.L., Mevoli, A., & Quirini, A. (2000). On line microdialysis-ion chromatography determination of inorganic anions in olive oil mill wastewater. *J. Chromatogr.*, *882* (1–2), 321–328.

Busacca, A. (1981). Smaltimento ed impiego di acque di vegetazione dalla spremitura delle olive. "Utilization and treatment of vegetation waters from olive oil extraction plants and their possible utilization". *AES*, *1*, 23–25 (in Italian).

Cabrera, F. (1996). The problem of the olive oil wastes in Spain: Treatment or recycling? *Proc. 7th Mediterranean Conf. on Organic Wastes Recycling in Soils*, Vieste, Italy, *1*, 1117–1125, Ordine Nazionale dei Biologi (eds. Landi, E., & Dumonet, S.).

Cabrera, F., López, R., Murillo, J.M., & Brenas, M.A. (1993). Olive vegetation water residues composted with other agricultural by-products as organic fertilizer. *10th World Fertilizer Congress of CIEC: Efficient Fertilization, Manuring and Irrigation for Improving Crop Yield, Food Quality and Renewable Resources*, CIEC, Nicosia, Cyprus, 79–96, 1993 (eds. Welte, F., & Szablocs, I.).

Cabrera, F., Toca, C.G., Díaz-Barrientos, E., & Arambarri, P. (1983). Influencia de los alpechines en la calidad del agua del río Guadiamar. *Actas Congr. Nacl. Quím. Tecnol. Agua, 3*, 535–543 (in Spanish).

Cabrera, F., Toca, C.G., Díaz-Barrientos, E., & Arambarri, P. (1984). Acid minewater and agricultural pollution in a river skirting the Doñana National Park (Guadiamar river, South West Spain). *Water Res., 18*, 1469–1482.

Cabrera, F., Soldevilla, M., Cordón, R., & Arambarri, P. (1987). Heavy metal pollution in the Guadiamar river and the Guadalquivir estuary (south west Spain). *3rd Int. Congress on Environ. Poll. and its Impact on Life in the Mediterranean Region, 16 Istanbul, 463–468 (eds. Herman, M., Kotzias, D., & Parlar, H.).*

Cabrera, F., López, R., Martínez-Bordiú, A., Dupuy de Lome, E., & Murillo, J.M. (1995). Land treatment of olive oil mill waste water. *Proc. Olive Oil Processes and By-Products Recycling*, Granada, Spain, 10–13 Sep. 1995. *Int. Biodeterior. and Biodegrad.*, 1996, *38* (3–4), 215–225.

Caglar, A., & Demirbaş, A. (2002). Hydrogen rich gas mixture from olive husk via pyrolysis. *Energy Convers. Manage., 43* (1), 109–117.

Calvet, C. et al. (1985). Composting of olive marc. *The Use of Composts as Horticultural Substrates* (ed. Verdonck, O.), Gent, Belgium, 255–259.

Camurati, F., & Fedeli, E. (1982). Attività antiossidante di estratti fenolici delle acque de vegetazione delle olivee. *Riv. Ital. Sostanze Grasse, 59* (LIX), 623–626 (in Italian).

Camurati, F., Lanzani, A., Arpino, A., Ruffo, C., & Fedeli, E. (1984). Le acque di vegetazione dalla lavorazione delle olive: Tecnologie ed economie di recupero di sottoprodotti. "Vegetation water from olive processing: Technologies and economical aspects in the recovery of by-products". *Riv. Ital. Sostanze Grasse, 61* (LXI) (4), 283–292 (in Italian).

Canepa, P., Marignetti, N., & Gagliardi, A. (1987). Trattamento delle acque di vegetazione mediante processi a membrana "Treatment of olive oil manufacturing wastewater by membrane processes". *Inquinamento, 5*, 73–77 (in Italian).

Canepa, P., Marignetti, N., Rognoni, U., & Calgari, S. (1988a). Considerazioni tecnico economiche per la realizzazione di un impianto di trattamento delle acque di vegetazione delle olive con un processo integrato a membrana. *Acqua-Aria, 10*, 1253–1257 (in Italian).

Canepa, P., Marignetti, N., Rognoni U., & Calgari, S. (1988b). Olive mills wastewater treatment by combined membrane processes. *Water Res., 22* (12), 1491–1494.

Capasso, R. (1997). The chemistry, biotechnology and ecotoxicology of the polyphenols naturally occurring in vegetable wastes. *Curr. Top. Phytochem., Res. Trends, 1*, 145–156.

Capasso, R. (1999). A review on the electron ionization and fast atom bombardment mass spectrometry of polyphenols naturally occurring in olive wastes and some of their synthetic derivatives. *Phtytochemical Analysis, 10* (6), 299–306.

Capasso, R., & Martino, A. ⁻de (1998). Recovering a polymeric polyphenolglycosilated pigment from olive oil mill wastewaters. *Proc. 2nd Int. Electronic Conf. on Synthetic Organic Chemistry (ECSOC-2)*, 1–30 Sep. 1998.

Capasso, R., Cutignano, A., & Evidente, A. (1997). Recovering hydroxytyrosol from, olive oil mill waste waters. *Abstracts of the 9th Int. Symp. on Environmental Pollution and its Impact on Life in the Mediterranean Region*, Sorrento, Italy, 4–9 Oct. 1997, 232.

Capasso, R., Evidente, A., & Visca, C. (1994a). Production of hydroxytyrosol from olive oil vegetation waters. *Agrochimica, 37* (XXXVII), 165–171.

Capasso, R., Evidente, A., & Scognamiglio, F. (1992a). A simple thin layer chromatographic method to detect the main polyphenols occurring in olive oil vegetation waters, *Phytochemical Analysis 3*, 270–275.

Capasso, R., Martino, A. de, & Arienzo, M. (2002a). Recovery and characterization of the metal polymeric organic fraction (polymerin) from olive oil mill wastewater. *J. Agric. Food Chem.*, *50* (10), 2846–2855.

Capasso, R., Martino, A. de, & Cristinzio, G. (2002b). Production, characterization, and effects on tomato of humic acid-like polymerin metal derivatives from olive oil mill waste waters. *J. Agric. Food Chem.*, *50* (14), 4018–4024.

Capasso, R., Cristinzio, G., Evidente, A., & Scognamiglio, F. (1992b). Isolation spectroscopy and selective phytotoxic effects of polyphenols from vegetable waste waters. *Phytochemistry*, *31* (12), 4125–4128.

Capasso, R., Evidente, A., Avolio, S., & Solla, F. (1999). A highly convenient synthesis of hydroxytyrosol and its recovery from agricultural waste waters. *J. Agr. Food Chem.*, *47* (4), 1745–1748.

Capasso, R., Evidente, A., Tremblay, E., & Sala, A. (1994b). Direct and mediated effects on *Bactrocera oleae* (Gmelin) (Diptera: Tephritidae) of natural polyphenols and some of related synthetic compounds: Structure-activity relationships. *J. Chem. Ecol.*, *20* (5), 1189–1199.

Capasso, R., Evidente, A., Schivo, L., Orru, G., Marcialis, M.A., & Cristinzio, G. (1995). Antibacterial polyphenols from olive oil mill waste waters. *J. Appl. Bacter.*, *79*, 393–398.

Capasso, R., Cristinzio, G., Martino, A. de, Arienzo, M., Lattanzio, V., & Cicco, N. (2000). Recovering the polymeric fraction from olive oil mill waste waters: Chemical characterization and potential biotechnological applications. *11th Int. Biotechnology Symp.*, Berlin, 3–8 Sep. 2000, 203–205.

Cardoso, S.M., Silva, A.M.S., & Coimbra, M.A. (2002). Structural characterisation of the olive pomace pectic polysaccharide arabinan side chains. *Carbohydrate Res.*, *337*, 917–924.

Cardoso, S.M., Coimbra, M.A., & Lopes-da- Silva, J.A. (2003). Calcium-mediated gelation of an olive pomace pectic extract. *Carbohydr. Polym.*, *52*, 125–133.

Cardoso, S.M., Guyot, S., Marnet, N., Lopes-da-Silva, J.A., Renard, C. MGC, & Coimbra, M.A. (2005). Characterisation of phenolic extracts from olive pulp and olive pomace by electrospray mass spectrometry. *J. Sci. Food Agric.*, *85* (1), 21–32(12).

Carlini, M. (1992). *CEEP/Ambiente, 31*.

Caro, N. de, & Ligori, C.N. (1959). Activita antibiotica di un estratto desunto delle acque di vegetazione delle olive. *Rend. Inst. Sup. Sanit.*, *22*, 223–243 (in Italian).

Carola, C., Arpino, A., & Lanzani, A. (1975). Lo smaltimento delle acque di vegetazione provenienti dagli impianti di estrazione dell'olio dalle olive e studio della loro possibile utilizzazione. Nota preliminare. "Disposal of vegetation waters coming from olive oil extraction plants and their possible utilization. Preliminary note". *Riv. Ital. Sostanze Grasse, 52 (LII)* (10), 335–340 (in Italian).

Carrieri, C. (1978). Ultrafiltration of vegetation waters from olive oil extraction plants; preliminary experiences. *Olii grassi derivati, 14*, 29–31.

Carrieri, C. (1991). Anaerobic-aerobic treatment of wastewater from olive oil extraction mixed with urban sludge in fully mixed reactors. *Ist. Ric. Acque, 94*, 11.1.–11.9.

Carrieri, C., Balice, V., & Rozzi, A. (1988). Comparison of three anaerobic treatment processes on olive oil mill effluents. *2nd Int. Conf. on Environ. Protection*, S. Angelo d'Ischia, Italy, Oct. 1988, 37–44.

Carrieri, C., Di Pinto, A.C., & Rozzi, A. (1992). Anaerobic co-digestion of sewage sludge and concentrated soluble wastewaters. *Proc. 2nd IAWQ Int. Symp. on Waste Management Problems in Agro Industries*, Istanbul, Turkey, 23–25 Sep. 1992. *Water Sci. Technol.*, 1992, *28* (2), 187–197.

Carrieri, C., Balice, V., Rozzi, A., & Santori, M. (1986). Anaerobic treatment of olive mill effluents mixed with sewage sludge. Preliminary results. *Proc. Int. Symp. on Olive By-Products Valorization*, Seville, Spain, 4–7 Mar. 1986.

Caporali, F., Anelli, G., Paolini, R., Campiglia, E., Benedetti, G., & Contini, M. (1996). Crop application of treated waste waters from olive oil extraction. *Ag. Med.*, *126*, 388–395.

Caputo, A.C., Scacchia, F., & Pelagagge, P.M. (2003). Disposal of by-products in olive oil industry: Waste-to-energy solutions. *Appl. Therm.*, *23* (2), 197–214.

Carro, F. (2004). Regulations concerning waste and vegetation water. *EU project: LIFE00 ENV/IT/000223 "TIRSAV" (Innovative technologies for recycling olive residue and vegetation water). Final Report.*

Casa, R., D'Annibale, A., Pieruccetti, F., Stazi, S.R., Sermanni, G. G., & Lo Cascio, B. (2003). Reduction of the phenolic components in olive mill wastewater by an enzymatic treatment and its impact on durum wheat (*Triticum durum* Desf.) germinability. *Chemosphere*, *50* (8), 959–966.

Castro Gomez-Millan, A. de, Durán Quintana, M.C., García-García, M.C., Garrido Fernández, A., González Cancho, F., Rejano Navarro, L., Sánchez Roldán, F., & Sánchez Tebar, J.C. (1983). Processing fruits of Gordal olive variety, without washing and with reusing lye solution, as Spanish style green olives. *Grasas y Aceites*, *34* (3), 162–167.

Castro Gomez-Millan, A. de, & Brenes Balbuena, M. (2001). Fermentation of washing waters of Spanish–style green olive processing. *Process Biochem.*, *36* (8–9), 797–802.

Castro Ramos, R. de, Nosti Vega, M., & Vázquez Ladron, R. (1980). Composition and nutritive value of some Spanish varieties of table olives. IV. Effects of reusing cooking lye and wash water. *Grasas y Aceites*, *31* (2), 91–95.

Catalano, L., & Felice, M. de (1989). Utilizzazione delle acque reflue come fertilizzante. *Atti. Sem. Int. Su Tecnologie e Impianti per il trattamento dei reflui dei frantoi oleari*, Lecce, Italy, 16–17 Nov. 1989 (in Italian).

Catalano, L., Gomes, T., Felice, M. de, & Leonardis, A. de (1985). Smaltimento delle acque di vegetazione dei frantoi oleari. Quali alternative alla depurazione? *Inquinamento*, *27* (2), 87–90 (in Italian).

Cato Marcus Porcius (234–149 B.C.) *De agri cultura, ad fidem Florentini codicis deperditi.* Bibliotheca Scriptorum Graecorum et Romanorum Teubneriana, Lipsiae, 1962 (ed. Mazzarino, A.).

Cayuela, M.L., Bernal, M.P., & Roig, A. (2004). Composting olive mill waste and sheep manure for orchard use. *Compost Sci. Util.*, *12* (2), 130–136.

Cegarra, J., Amor, J.B., González-López, J., & Bernal, J. (1999). Characteristics of a new solid olive-mill by-product ("alperujo") and its suitability for composting. *Proc. Int. Composting Symp. (ICS'99)*, *1*, 124–140. 2000, Halifax, Canada (eds. Warman, P.R., & Taylor, B.R.). ISBN 0-09685651-0-7.

Cegarra, J., Paredes, C., Roig, A., Bernal, M.P., & García, D. (1996a). Use of olive mill waste-water compost for crop production. *Proc. Olive Oil Processes and By Products Recycling*; Granada, Spain, 10–13 Sep. 1995. *Int. Biodeterior. Biodegrad.*, 1996, *38* (3–4), 193–203.

Cegarra, J., Paredes, C., Roig, A., & Bernal, M.P.(1996b). Composting of fresh olive-mill wastewater added to plant residues. *The Science of Composting*, *2*, 1100–1104, 1996 (eds. Bertoldi, M. de, Sequi, P., Lemmes, B., & Papi, T.), Blackie Academic & Professional, Glasgow, UK. ISBN 0-7514-0383-0.

Cegarra, J., Roig, A., Paredes, C., & Sánchez-Monedero, M.A. (1994). The influence of fresh olive-mill wastewater added to solid organic wastes for composting. *Proc. Mediterranean Conf. on Organic Wastes Recycling in Soils.*, *1*, 191–197, Vieste, Italy, 22–25 Sep. 1994, Ordine Nazionale dei Biologi (eds. Landi, E., & Dumonet, S.).

Cegarra, J., Roig, A., Navarro, A.F., Bernal, M.P., Abad, M., Climent, M.D., & Aragón, P. (1993). Características, compostaje y uso agrícola de residuos sólidos urbanos. *Jornadas de Recogidas Selectivas en Origen y Reciclaje*, Cordoba, Spain (in Spanish).

Ceccon, L., Saccu, B., Procida, G., & Cardinali, S. (2001). Liquid chromatographic determination of simple phenolic compounds in waste waters from olive oil production plants. *J. AOAC Int.*, *84* (6), 1739–1744.

Cereti, C.F., Rossini, F., Federici, F., Quaratino, D., Vassilev, N., & Fenice, M. (2004). Reuse of microbially treated olive mill wastewater as fertiliser for wheat (*Triticum durum* Desf.). *Bioresource Technol.*, *91* (2), 135–140.

Cert, A., Alba-Mendoza, J., Leon-Camacho, M., Moreda, W., & Pérez-Camino, M.C. (1996). Effects of talc addition and operating mode on the quality and oxidative stability of virgin oil obtained by centrifugation. *J. Agric. Food Chem.*, *44* (12), 3930–3934.

Chakchouk, M., Hamdi, M., Foussard, J.N., & Debellefontaine, H. (1994). Complete treatment of olive mill wastewaters by a wet air oxidation process coupled with a biological step. *Environ. Technol.*, *15* (4), 323–332.

Chamkha, M., Labat, M., Patel Bharat, K.C., & García, J.L. (2001). Isolation of a cinnamic acid-metabolizing *Clostridium glycolicum* strain from oil mill wastewaters and emendation of the species description. *Int. J. Systematic and Evolutionary Microbiol.*, *51* (6), 2049–2054.

Chartzoulakis, K. (2002). *Workshop on innovative olive mill waste management systems*, Mytilene, Greece, 1 Nov. 2002 (Co-ordinator Halvadakis, C.P.).

Chatzipavlidis, I., Antonakou, M., Demou, D., Flouri, F., & Balis, C. (1996). Bio-fertilization of olive oil mills liquid wastes. The pilot plant in Messinia, Greece. *Proc. Olive Oil Processes and By-Products Recycling*; Granada, Spain, 10–13 Sep. 1995. *Int. Biodeterior. Biodegrad.*, 1996, *38* (3–4), 183–187.

Ching-Shyung Hwu, Lier van, J.B., & Lettinga, G. (1996). Comment. *Wat. Res.*, *30* (9), 2229.

Ching-Shyung Hwu, & Lettinga, G. (1996). Acute toxicity of oleate to acetate-utilizing methanogens in mesophilic, and thermophilic anaerobic sludges. *Enzym. Microb. Technol.*, *21* (4), 297–301.

Chtourou, M., Ammar, E., Nasri, M., & Medhioub, K. (2004). Isolation of a yeast, *Trichosporon cutaneum*, able to use low molecular weight phenolic compounds: application to olive mill waste water treatment. *J. Chem. Technol. Biotechnol.*, *79* (8), 869–878.

Ciafardini, G., Zullo, B.A., & Cioccia, G. (1998). Effetti delle acque reflue dei frantoi su due batteri azotofissatori. *L'Informatore Agrario*, *15*, 35–37 (in Italian).

Cichelli, A., & Solinas, M. (1984). I composti fenolici delle olive e dell'óloi di oliva. "Phenolic compounds of olive and olive oil". *Riv. Merceol.*, *22*, 55–69.

Cicolani, B., Seghetti, L., D'Alfonso, S., & Di Giovacchino, L. (1992). Vegetation water spreading on land and wheat cultivation: Effects on invertebrate communities. *Proc. Int. Conf. on Treatment and Reutilization of Farm Effluents and Sludges*, Lecce, Italy, 10–12 December, 1992.

Cicolani, B., Seghetti, L., D'Alfonso, S., & Di Giovacchino, L. (1992). Spargimento delle acque di vegetazione dei frantoi oleari su terreno coltivato a grano: effetti della pedofauna. *L'Informatore Agrario, 34*, 69–75.

Civantos, L. (1981a). Aprovechamiento de ramones y leña en el olivar. *Agricultura, 55*, 1801–1181 (in Spanish).

Civantos, L. (1981b). Utilisation de broyeurs mobiles en vue de la valorisation de bois de taille de l'olivier. *Séminaire International sur la Valorisation de Sous-Produits de l'Olivier; "Proc. Int. Symp. on the Reuse of Olive Tree By-Products"* PNUD/FAO/COI; Monastir, Tunisia, Dec. 1981, 81–84 (in French).

Civantos, L. (1999). Comparación entre sistemas de extracción. *Obtención del aceite de oliva virgen.*, (ed. Civantos, L.), 189–201, Madrid: Editorial Agricola Española. (in Spanish).

Cliffe, K.R., & Patumsawad, S. (2001). Co-combustion of waste from olive oil production with coal in a fluidized bed. *Waste Manage., 21* (1), 49–53.

Codounis, M., Katsaboxakis, K., & Papanicolaou, D. (1983). *Progress in the extraction and purification of anthocyanin pigments from the effluents of olive oil extracting plants.* European Federation of Chemical Engineering; Food Working Party; Food Eng. Symp., 567–572, 1983, Athens, Greece.

Colucci, R., Di Bari, V., Ventrella, D., Marrone, G., & Mastrorilli, M. (2002). *Advances in Geoecology, 35*, 91–100.

Columella Lucius Junius Moderatus (1st century, A.D.) *De re rustica "On Agriculture"* (11 books) and *De Arboribus "On Trees"*. Translated by, H.B. Ash, E.S. Forster, & E. Heffner, Loeb Classical Library, Wm. Heinemann Ltd., London, 1968.

Cortinovis, D. (1975). ABF is cleaning up –economically. *Water Wastes Eng., 12* (6), 22–25.

Cossu, R., Blakey, N., & Cannas, P. (1993). Influence of co-disposal of municipal solid waste and olive vegetation water on the anaerobic digestion of a sanitary landfill. *Water Sci. Technol., 27* (2), 261– 271.

Cox, L., Celis, R., Hermosín, M.C., Becker, A., & Cornejo, J. (1997). Porosity and herbicide leaching in soils amended with olive-mill wastewater. *Agric., Ecosyst. Environ., 65* (2), 151–162.

Cox, L., Becker, A., Celis, R., López, R., Hermosín, M.C., & Cornejo, J. (1996). Movement of clopyralid in a soil amended with olive oil mill wastewater as related to soil porosity. *Fresen. Environ. Bull., 3–4*, 167–171.

Croce, F., Poulsom, S., & Hendricks, D.W. (1993). Combined treatment of olive mill effluent and municipal wastewater in a small tourist community. *Proc. IAWQ Int. Specialized Conf. on Pretreatment of Industrial Wastewaters*, Athens, Greece, 13–15 Oct. 1993. *Water Sci. Technol.*, 1994, *29* (9), 105–110.

Cummings, S.P., & Russell, N.J. (1996). Osmoregulatory responses of bacteria isolated from fresh or composted, olive-mill waste-waters. *World, J. Microb. Biot., 12* (1), 61–67.

Curi, K., Velioğlu, S.G., & Diyanmandoğlu, V. (1980). Treatment of olive oil production wastes. Treatment and Disposal of Liquid and Solid Industrial Wastes: *Proc. 3rd Turkish German Environmental Engineering Symp.*, 189–205 Istanbul, Turkey, July 1979 (ed. Curi, K.), Pergamon Press, Oxford, UK.

Curi, K., Velioğlu, S.G., & Sur, M.H. (1985). Anaerobic treatment of olive oil wastewater. *Proc. 1st Int. Symp. on Environmental Technology for Developing Countries. In Appropriate Waste Management for Developing Countries*, 291–310, Istabul, Turkey, 1985, (ed. Curi, K.) Plenum Press, New York, USA.

D'Addabbo, T., Sasanelli, N., Lamberti, F., & Carella, A. (2000a). Control of root-knot nematodes by olive and grape pomace soil amendements. *Proc. 5th Int. Symp. on Chemical and Non-Chemical Soil and Substrate Disinfestation*; Turin, Italy, 11–15 Sep. 2000, 53–57. *Acta Hortic.*, 2000, *532*, 53–58.

D'Addabbo, T., Fontanazza, G., Lamberti, F., Sasanelli, N., & Patumi, M. (1997). The suppressive effect of soil amendments with olive residues on *Meloidogyne incognita*. *Nematol. Medit.*, *25*, 195–198.

D'Addabbo, T., Sasanelli, N., Lamberti, F., Greco, P., & Carella, A. (2000b). Effect of olive and grape pomace in the control of root-knot nematodes. *Ann. Int. Res. Conf. On Methyl Bromide Alternatives and Emissions Reductions*; Orlando, Florida, USA, 6–9 Nov. 2000, 1–3.

D'Annibale, A., Crestini, C., Vinciguerra, V., & Sermanni, G.G. (1998). The biodegradation of recalcitrant effluents from an olive mill by a white rot fungus. *J. Biotechol.*, *61* (3), 209–218.

D'Annibale, A., Stazi, S.R., Vinciguerra, V., & Sermanni, G.G. (2000). Oxirane-immobilized *Lentinula edodes* laccase: Stability and phenolics removal efficiency in olive mill wastewater. *J. Biotecnol.*, *77*, 265–273.

Dalis, D. (1991). Optimisation of the C/N ratio during the anaerobic digestion of olive-oil waste waters. *Proc. 13th National Congress of the Hellenic Biological Sciences Union*, Heraklion, Crete, Greece, 1991 (in Greek).

Dalis, D., Hartmann, L., & Anagnostidis, K. (1991). Anaerobic treatment of olive oil waste waters. *R&D program –recycling of urban and industrial waste (RUW-056)*, Brussels, Belgium.

Dalis, D., Anagnostidis, K., López, A., Letsiou, I., & Hartmann, L. (1996). Anaerobic digestion of total raw olive oil wastewater in a two stage pilot plant (Up flow and fixed bed bioreactors). *Bioresource Technol.*, *57* (3), 237–243.

Dally, B., & Mullinger, P. (2002). Utilization of Olive Husks for Energy Generation: A Feasability Study. *Final Report – SENRAC Grant 9/00*, South Australina State Energy Research Advisory Committee, 17 pages.

Davies, L.C., Novais, J.M., & Martins-Dias, S. (2004). Detoxification of olive mill wastewater using superabsorbent polymers. *Environ. Technol.*, *25* (1), 89–100.

Delgado-Pertiñez, M. (1994). Valoración nutritive de la hoja de olivo. Efecto de su origen y de los tratamientos de manipulación. *Ph. D. Thesis*. University of Cordoba, Spain (in Spanish).

Delgado-Pertiñez, M., Gómez-Cabrera, A., Garrido Varo, A., & Guerrero Ginel, J.E. (1997). Characterization of olive leaf from cleaning oil presses: Moisture level, soil contamination and nutritive value. *Archivos de Zootecnia*, *46*, 173, 85–88.

Della Greca, M., Previtera, L., Temussi, F., & Zarrelli, A. (2004). Low-molecular-weight components of olive oil mill waste-waters. *Phytochem. Anal.*, *15* (3), 184–188.

Della Greca, M., Monaco, P., Pinto, G., Pollio, A., Previtera, L., & Temussi, F. (2000). Phenolic components of olive mill waste-water. *Nat. Prodc. Ltt.*, *14*, 429–434.

Della Greca, M., Monaco, P., Pinto, G., Pollio, A., Previtera, L., & Temussi, F. (2001). Phytotoxicity of low-molecular-weight phenols from olive mill waste waters. *Bull. Environ. Toxicol.*, *67* (3), 352–359.

Della Monica, M. (1979). Il trattamento delle acque provenienti dalla lavorazione delle olive. *Notizario Agricolo Regionale. Regione Puglia*, 24–25 Mar. (in Italian).

Della Monica, M., Potenz, D., Righetti, E., & Volpicella, M. (1978). Effetto inquinante delle acque reflue della lavorazione delle olive su terreno agrario. Nota 1. Evoluzione del pH, dei composti azotati e dei fosfato. "Pollution effect of waste water from olive manufacturing on agricultural soil. Note 1. Evolution of pH, nitrogen and phosphorus compounds". *Inquinamento, 20* (10), 81–84 (in Italian).

Della Monica, M., Potenz, D., Righetti, E., & Volpicella, M. (1979). Effetto inquinante delle acque reflue della lavorazione delle olive sul terreno agrario. Nota 2. Evoluzione dei lipidi, dei polifenoli, e delle sostanze organiche in generale. "Pollution effect of waste water from olive manufacturing on agricultural soil. Note 2. Evolution of lipids, polyphenols and organic substances in general". *Inquinamento, 21* (1), 27–30 (in Italian).

Della Monica, M., Agostiano, A., Potenz, D., Righetti, E., & Volpicella, M. (1980). Degradation treatment of waste water from olive processing. *Water Air Soil Pollut., 13* (2), 251–256.

Demichelli, M., & Bontoux, L. (1996). Survey on current activity on the valorization of by-products from the olive oil industry. *Eur. Commission, Joint Res. Ctr.,* 1997, 4–24 (Report EUR 16466).

Demirbaş, A. (2001). Yields of hydrogen-rich gaseous products via pyrolysis from selected biomass samples. *Fuel, 80* (13), 1885–1891.

Demirbaş, A., Caglar, A., Akdeniz, F., & Gullu, D. (2000). Conversion of olive husk to liquid fuel by pyrolysis and catalytic liquefaction. *Energy Sources, 22* (7), 631–639.

Di Blasi, C., Buonanno, F., & Branca, C. (1999a). Reactivities of some biomass chars in air. *Carbon, 37* (8), 1227–1238.

Di Blasi, C., Signorelli, G., Di Russo, C., & Rea, G. (1999b). Product distribution from pyrolysis of wood and agricultural residues. *Ind. Eng. Chem. Res., 38* (6), 2216–2224.

Di Blasi, C., Signorelli, G., & Portoricco, G. (1999c). Countercurrent fixed-bed gasification of biomass at laboratory scale. *Ind. Eng. Chem Res., 38* (7), 2571–2581.

Di Chio, D., Potenz, D., & Righetti, E. (1999). Degradation of organic substances in the soil: Proposal for a mathematical model. *Bioresource Technol., 67* (3), 267–278.

Di Giacomo, G. (1990). Research Report–ERSA, Apr. 1990.

Di Giacomo, G., Brandani, V., & Del Re, G. (1991). Evaporation of olive oil mill vegetation waters. *Proc. 12th Int. Symp. on Desalination and Water Re use;* Malta, 15–18 Apr. 1991. *Inst. Chem. Eng. Symp. Ser., 1* (125), 249–259; pub. by Inst. of Chemical Engineers, Rugby, England. *Desalination,* 1991, *81* (1–3), 249–259.

Di Giacomo, G., Bonfitto, E., Brunetti, N., Del Re, G., & Jacoboni, S. (1989). Pyrolysis of exhausted olive oil husks coupled with two-stages thermal decomposition of aqueous olive oil mills effluents. *Pyrolysis and Gasification* (eds. Ferrero, G.L., Maniatis, K., Buenkes, A., & Bridgwater, A.V.), Elsevier, New York, 586–590.

Di Gioia, D., Bertin, L., Fava, F., & Marchetti, L. (2001a). Biodegradation of hydroxylated and methoxylated benzoic, phenylacetic and phenylpropenoic acids present in olive mill wastewaters by two bacterial strains. *Res. Microbiol., 152* (1), 83–93.

Di Gioia, D., Barberio, C., Spagnesi, S., Marchetti, L., & Fava, F. (2002). Characterization of four olive mill wastewater indigenous bacterial strains capable of aerobically degrading hydroxylated and methoxylated monocyclic aromatic compounds. *Archives Microbiol., 178* (3), 208–217.

Di Gioia, D., Bertin, L., Fava, F., & Marchetti, L. (2001b). Biodegradation of synthetic and naturally occurring mixtures of mono-cyclic aromatic compounds present in olive mill wastewaters by two aerobic bacteria. *Appl. Microbiol. Biotechnol., 55* (5), 619–626.

Di Giovacchino, L. (1985). Sulle caratteristische delle acque di vegetazione delle olive. Nota, I. "On the characteristics of oil mills effluents. Note I". *Riv. Ital. Sostanze Grasse, 62* (LXII), 411–418 (in Italian).

Di Giovacchino, L. (1994). Resultados de la extracción del aceite de las aceitunas com un nuevo decantador de dos fases. *Olivae, 50*, 42–44 (in Spanish).

Di Giovacchino, L., & Seghetti, L. (1990). Lo smaltimento delle acque di vegetazione delle olive su terreno agrario destinato alla coltivazione di grano e mais. *L'Informatore Agrario, 45*, 58–62 (in Italian).

Di Giovacchino, L., Mascolo, A., & Seghetti, L. (1988). Sulle caratteristiche delle acque di vegetazione delle olive. Nota II. "On the characteristics of oil mills effluents. Note II". *Riv. Ital. Sostanze Grasse, 65* (LXV), 481–488 (in Italian).

Di Giovacchino, L., Sestili, S., & Di Vincenzo, D. (2002). Influence of olive processing on virgin olive quality. *Eur. J. Lipid Sci. Technol., 104*, 587–601.

Di Giovacchino, L., Solinas, M., & Mascolo, A. (1976). Evaluation of pollution of the rivers Vomano, Saline and Foro, in Abruzzo, Italy, by effluent from olive oil mills. *Annali dell Istituto Sperimentale per la Elaiotecnica, 6* (in Italian).

Di Giovacchino, L., Solinas, M., & Miccoli, M. (1994). Effect of extraction systems on the quality of virgin olive oil. *J. Am. Oil Chem. Soc. (JAOCS), 71* (11), 1189–1194.

Di Giovacchino, L., Basti, C., Costantini, N., & Surricchio, G. (2000). Olive vegetable water spreading and soil fertilization. *Int. Management and Irrigation of Olive Orchards*, Limassol, Cyprus, Apr. 2000, 68–72.

Di Giovacchino, L., Mucciarella, M.R., Costantini, N., & Surricchio, G. (2002). Double oil extraction from olive paste and olive pomace by centrifugal decanter at 2-phases, type integral. *Riv. Ital. Sostanze Grasse, 79* (LXXIX) (10), 351–355 (in Italian).

Di Giovacchino, L., Basti, C., Costantini, N., Ferrante, M.L., & Surricchio, G. (2001). Effects of olive vegetable water spreading on soil cultivated with maize and grapevine. *Agr. Med., 131*, 33–41.

Di Giovacchino, L., Basti, C., Costantini, N., Ferrante, M.L., & Angelis, A. de (1996). Risultati di esperienze pluriennali di spargimento di acque di vegetazione delle olive sul terreno agrario. *Atti Convegno "L'utilizzo dei residui dei frantoi oleari"*, Viterbo, 12 Apr. 1996, 35–42 (in Italian).

Di Giovacchino, L., Costantini, N., Serraiocco, A., Surricchio, G., & Basti, C. (2001). Natural antioxidants and volalite compounds of virgin olive oils obtained by two or three-phases centrifugal decanters. *Eur. J. Lipid Sci. Technol., 103* (5), 279–285.

Di Giovacchino, L., Basti, C., Costantini, N., Surricchio, G., Ferrante, M.L., & Lombardi, D. (2002). Effects of spreading olive vegetable water on soil cultivated with maize and grapevine. *Olivae, 91*, 37–43.

Dias Albino, A., Bezerra Rui, M., & Pereira Nazare, A. (2004). Activity and elution profile of laccase during biological decolorization and dephenolization of olive mill wastewater. *Bioresource Technol., 92* (1), 7–13.

Diodorus Siculus, (1st century, B.C.) Library of History. Volume **IV,** books 9-12.40, translated by, C.H. Oldfather, Loeb Classical Library, Wm. Heinemann Ltd., London, 1971. Series No. 375 / 474 pages. ISBN 0-674-99413-2.

Dorado, M.P., Ballesteros, E., Mittelbach, M., & López Aparicio, F.J. (2004). Kinetic parameters affecting the alkali-catalyzed transesterification process of used olive oil. *Energy & Fuels, 18* (5), 1457–1462.

Drachmann, A.G. (1932). Ancient oil mills and presses. Copenhagen.

Duarte, E.A., & Neto, I. (1996). Evaporation phenomenon as a waste mamangement technology. *Water Sci. Technol., 33* (8), 53–61.

Dubois, M., Gilles, K.A., Hamilton, J.K., Rebers, P.A., & Smith, F. (1956). Colorimetric method for determination of sugars and related substances. *Anal. Chem.*, *28*, 350–356.

Durán Barrantes, M.M. (1990). Depuración anaerobia del alpechín. Influencia de los compuestos fenólicos sobre et desarrollo del proceso de biometanizatión. Memoria Grado de Masteren Ciencias e Ingenieria de Alimentos. Unniversidad Politécnica de Valencia.

Durkee, E.L., Lowe, E., Baker, K.A., & Burgess, J.W. (1973). Field tests of salt recovery system for spent pickle brine. *J. Food Sci.*, *38* (3), 507–511.

East Cretan Section Technical Chamber of Greece (1980). *Environmental pollution from the liquid wastes of olive oil mills.* Report, Iraklion, Greece (in Greek).

Ehaliotis, C., Papadopoulou, K., Kotsou, M., Mari, I., & Balis, C. (1999). Adaptation and population dynamics of *Azotobacter vinelandii* during aerobic biological treatment of olive-mill wastewater. *FEMS Microbiol. Ecol.*, *30* (4), 301–311.

El-Asli, A., Boles, E., Hollenberg, C.P., & Errami, M. (2002). Conversion of xylose to ethanol by a novel phenol-tolerant strain of Enterobacteriaceae isolated from olive mill waste. *Biotechnol. Letters*, *24* (13), 1101–1105.

El-Sheikh, A.H., Newman, A.P., Al-Daffaee, H.K., Phull, S., & Cresswell, N. (2004). Characterization of activated carbon prepared from a single cultivar of Jordanian olive stones by chemical and physicochemical techniques. *J. Anal. Appl. Pyrolysis.*, *71* (1), 151–164.

Ercoli, E. (1984). Producción deproteínas unicelulares a partir de un residuo de la industria del aceite de oliva. *Ph.D. Thesis*, Universidad Nacional de San Luis, Argentina (in Spanish).

Ercoli, E., & Ertola, R. (1983). SCP production from olive black water. *Biotechnol. Letters*, *7*, 457–462.

Erguder, T.H., Guven, E., & Demirer, G.N. (2000). Anaerobic treatment of olive mill wastes in batch reactors. *Process Biochem.*, *36* (3), 243–248.

Erŏglu, E., Gündüz, U., Yücel, M., Türker, L., & Erŏglu, I. (2004). Photobiological hydrogen production by using olive mill wastewater as a sole substrate source. *Int. J. Hydrogen Energy*, *29* (2), 163–171.

Escolano Bueno, A. (1975). Tests on removal of waste liquid from olive oil extraction (alpechín) by disposal in ponds or lagoons for percolation and evaporation. *Grasas y Aceites*, *26* (6), 387–396.

Estaùn, V., & Calvet, C. (1985). Chemical determination of fatty acids, organic acids and phenols during olive marc composting processes. *The Use of Composts as Horticultural Substrates* (ed. O. Verdonck), Gent, Belgium, 263–270.

Esterbauer, H., Striegel, G., Puhl, H., & Rotheneder, M. (1989). Continuous monitoring of in *vitro* oxidation of human low density lipoprotein. *Free Radic. Res. Commun.*, *6* (1), 67–75.

Ettayebi, K., Errachidi, F., Jamai, L., Tahri-Jouti, M.A., Sendide, K., & Ettayebi, M. (2003). Biodegradation of polyphenols with immobilized *Candida tropicalis*, under metabolic induction. *Microbiol. Lett.*, *223* (2), 215–219.

European Commission – Directorate-General for Environment (2001). *Survey of wastes spread on land* –Final Report. Study contract B4-3040/99/110194/MAR/E3. ISBN 92-894-1732-3.

Fadil, K., Chahlaoui, A., Ouahbi, A., Zaid, A., & Borja-Padilla, R. (2003). Aerobic biodegradation and detoxification of wastewaters from the olive mill industry. *Int. Biodeterior. and Biodegrad.*, *51* (1), 37–41.

Favi, E., Fossi, F., & Giovannelli, P. (1990). Lo spandimento delle AA.VV.: indicazioni agronomiche e caratteristiche pedologiche. *Genio Rurale*, *5*, 78–79 (in Italian).

Fedeli, E., & Camurati, F. (1981). *Valorarisation des margines et des grignons épuisés par récupération de quelques composants.* Séminaire International sur la valorisation des sous-produis de lólivier. "Proc. Int. Symp. on the Reuse of Olive Tree By-Products". PNUD/FAO/COI; Monastir, Tunisia, Dec., 1981 (in French).

Felice, B. de, & Catalana, L. (1988). Smaltimento delle acque di vegetazione sui terreni agari. *Atti della rotonda su acque reflue dei frantoi oleari*, Spoleto, Italy (in Italian)

Felice, B. de, Pontecorvo, G., & Carfagna, M. (1997). Degradation of waste waters from olive oil mills by *Yarrowia lipolytica* ATCC 20255 and *Pseudomonas putida*. *Acta Biotechnol.*, *17* (3), 231–239.

Felizon, B., Fernández-Bolaños, J., Heredia, A., & Guillén Bejarano, R. (2000). Steam-explosion pretreatment of olive cake. *J. Am. Oil Chem. Soc.* (JAOCS), *77* (1), 15–22.

Fenice, M., Sermanni, G.G., Federici, F., & D'Annibale, A. (2003). Submerged and solid-state production of laccase and Mn-peroxidase by Panus tigrinus on olive mill wastewater-based media. *J. Biotechnol.*, *100* (91), 77–85.

Ferrieres, B. (2004). L'élimination des margines par bassin d'évaporation à la Coopérative Oeicole de Sommières. *Le Nouvel Olivier*, *41*, September/October 2004, 11 (in French).

Fernández-Bolaños, J., Felizon, B., Heredia, A., Guillén Bejarano, R., & Jiménez, A. (1999). Characterization of the lignin obtained by alkaline delignification and of the cellulose residue from steam-exploded olive stones. *Bioresource Technol.*, *68* (2), 121–132.

Fernández-Bolaños, J., Fernández-Diez, M, J., Rivas Moreno,, Gill Serrano, A., & Pérez, T. (1983). A sucares y polioles en aceitunas verdes III. *Grasas y Aceites*, *34*, 168–171 (in Spanish).

Fernández-Bolaños, J., Felizon, B., Brenes, M., Guillén Bejarano, R., & Heredia, A. (1998). Hydroxytyrosol and tyrosol as the main compounds found in the phenolic fraction of steam-exploded olive stones. *J. Am. Oil. Chem. Soc. (JAOCS)*, *75*, 1643–1649.

Fernández-Bolaños, J., Rodríguez, G., Rodríguez, R., Heredia, A., Guillén Bejarano, R., & Jiménez, A. (2002). Production in large quantities of highly purified hydroxytyrosol from liquid-solid waste of two-phase olive oil processing (alperujo). *J. Agric. Food Chem.*, *50* (23), 6804–6811.

Fernández-Bolaños, J., Felizon, B., Heredia, A., Rodríguez, R., Guillén Bejarano, R., & Jiménez, A. (2001). Steam-exposion of olive stones: hemicelluloe solubilization and enhancement of enzymatic hydrolysis of cellulose. *Bioresource Technol.*, *79* (1), 53–61.

Fernández-Diez, M.J. (1971). The Biochemistry of Fruits and their Products (ed. Hulme, A.C.), Vol. 2, Academic Press, London, 1971, Chapter 7, p. 255.

Fiestas Ros de Ursinos, J.A. (1953). Estudio del alpechín para su aprovechamiento industrial, j. concentracion delos azicares y damas substancias quellare en emulcion y disolucion por tratamiento con oxido de calcio. *Grasas y Aceites*, *4* (2), 63–67 (in Spanish).

Fiestas Ros de Ursinos, J.A. (1958). Alpechines. *Grasas y Aceites*, *9* (3), 126–135 (in Spanish).

Fiestas Ros de Ursinos, J.A. (1961a). Estudio del alpechín para su aprovechamiento industrial. V. Cinética del desarrollo de la levadura *Torulopsis utilis* en el alpechín. *Grasas y Aceites*, *12* (2), 57–66 (in Spanish).

Fiestas Ros de Ursinos, J.A. (1961b). Estudio del alpechín para su aprovechamiento industrial. VI. Amioacidos presentes en la levadura *Candida utilis*. *Grasas y Aceites*, *12*, 161–165 (in Spanish).

Fiestas Ros de Ursinos, J.A. (1966). Estudio del alpechín para su aprovechamiento industrial. VII. Instalacion comercia para la obtencion de la leva duras-pienso. *Grasas y Aceites, 17,* 41–47 (in Spanish).

Fiestas Ros de Ursinos, J.A. (1967). Estudio del alpechín para su aprovechamiento industrial. "Study of the vegetation water of olives for its industrial use". *Grasas y Aceites, 18* (2), 45–47 (in Spanish).

Fiestas Ros de Ursinos, J.A. (1977). Depuración de las aguas residuales en las industrias de aceitunas y aceites de oliva. *Grasas y Aceites, 28* (2), 113–121 (in Spanish).

Fiestas Ros de Ursinos, J.A. (1981a). Anaerobic fermentation of wastewater with high organic load. *Ingenieria Quimica, 47,* 85–91 (in Spanish).

Fiestas Ros de Ursinos, J.A. (1981b). Différentes utilisations des margines: Rechersches en cours, resultats obtenus et applications. *Séminaire Int. sur la valorisation de sous-produits de l' olivier*; "Several uses of olive oil effluents. Current research, obtained results and applications. Proc. Int. Symp. on the Reuse of Olive Tree By-Products". FAO-PNUD/COI; Tunisia, Dec. 1981, 93–110 (in French).

Fiestas Ros de Ursinos, J.A. (1981c). The anaerobic digestion of wastewaters from olive oil extraction. *2nd Int. Symp. on Anaerobic Digestion*, Travemünde, FRG, Sep. 1981, 6–11.

Fiestas Ros de Ursinos, J.A. (1986a). Current status of research and technology concerning the problems posed by vegetation water. *Proc. Int. Symp. on Olive By-Products Valorization*, FAO, UNDP (Food and Agriculture Organization of the United Nations), Seville, Spain, 5–7 March, 1986, 11–15.

Fiestas Ros de Ursinos, J.A. (1986b). Possibilities of using olive mill wastewater (alpechín) as a fertilizer. *Proc. Int. Symp. on Olive By-Products Valorization*, FAO, UNDP (Food and Agriculture Organization of the United Nations), Seville, Spain, 5–7 Mar. 1986, 321–330.

Fiestas Ros de Ursinos, J.A. (1991). Reuse and complete treatment of vegetable water: current situation and prospects in Spain. *Proc. Int. Conf. in Olive Oil Processing Wastewater Methods*, Hania, Crete, Greece, 1991.

Fiestas Ros de Ursinos, J.A. (1992). Plan de puesta en marcha de plantas experimentales de depuración y eliminacion de alpechines en las cuencas de los rios Guadalquivir y Guadalete-Evaluacion de la experiencia. Ministerio de Obras Publicas y Transportes-Confederacion hidrografica del Guadalquivir, Seville, Spain, July 1992 (in Spanish).

Fiestas Ros de Ursinos, J.A., & Borja-Padilla, R. (1990). Aprovechamiento y depuración integral del alpechín. Instituto de la grasa y sus derivados (C.S.I.C.), Seville, Spain (in Spanish).

Fiestas Ros de Ursinos, J.A., & Borja-Padilla, R. (1991). Tratamientos de alpechines mediante procesos biologicos. *Int. Symp. on Treatment of OMW*, Cordoba, Spain, 31 May–1 June 1990 (in Spanish).

Fiestas Ros de Ursinos, J.A., & Borja-Padilla, R. (1992). Use and treatment of olive mill wastewater: current situation and prospects in Spain. *Grasas y Aceites, 43* (2), 101–106 (in Spanish).

Fiestas Ros de Ursinos, J.A., & Borja-Padilla, R. (1996). Biomethanization. Proc. Olive Oil Processes and By-Products Recycling; Granada, Spain, 10–13 Sep. 1995. *Int. Biodeterior. Biodegrad.*, 1996, *38* (3–4), 145–153 (in Spanish).

Fiestas Ros de Ursinos, J.A., Martín-Martín, A., & Borja-Padilla, R. (1990). Influence of immobilization supports on the kinetic constants of anaerobic purification of olive mill wastewater. *Biol. Wastes, 33* (2), 131–142.

Fiestas Ros de Ursinos, J.A., García, A. J., Leon-Cabello, R., & Maestro-Durán, R. (1984). Energy production in the olive oil industry from biomethanisation of its effluents. *Food Ind. Environ.*, *9*, 163–172.

Fiestas Ros de Ursinos, J.A., Navarro Gamero, R., Leon-Cabello, R., García-Buendia, A.J., & Mastro Juan de Jauregui, J.M. (1982). Depuración anaerobia del alpechín como fuente de energia. "Anaerobic depollution of olive oil effluents as source of energy". *Grasas y Aceites*, *33* (5), 265–270 (in Spanish).

Figueira, F. (2003). *IMPEL Olive Oil Project Report - CMA & NOA*. Number Report 2003/3, Project Manager: Méndez Miguel, Rome, Nov. 2003; number of pages, report: 33 and annexes: 61. European Union Network for the Implementation and Enforcement of Environmental Law (IMPEL) (http://europa.eu.int/comm/environment/impel).

Filidei, S., Masciandaro, G., & Ceccanti, B. (2003). Anaerobic digestion of olive mill effluents: Evaluation of wastewater organic load and phytotoxicity reduction. *Water Air Soil Pollut.*, *145* (1–4), 79–94.

Filippi, C., Bedini, S., Levi-Minzi, R., Cardelli, R., & Saviozzi, A. (2002). Co-composting of olive oil mill by-products: Chemical and microbiological evaluations. *Compost Sci. Util.*, *10* (1), 63–71.

Fiorelli, F., Pasetti, L., & Galli, E. (1995). Fertility promoting metabolites produced by *Azotobacter vinelandii* grown on olive mill wastewaters. *Proc. Olive Oil Processes and By-Products Recycling*; Granada, Spain, 10–13 Sep. 1995. *Int. Biodeterior. Biodegrad.*, 1996, *38* (3–4), 165–167.

Fiorentino, F., Gentili, A., Isidori, M., Monaco, P., Nardelli, A., Parrilla, A., & Temussi, F. (2003). Environmental effects caused by olive mill wastewaters. *J. Agr. Food Chem.*, *51* (4), 1005–1009.

Fiume, F., & Vita, G. (1977). L'impiego delle acque di vegetazione del frutto di olivo per il controllo del Daccus oleae Gmel. *Bollettino Laboratorio di Entomologia Agraria 'F. Silvestri' (Portici)*, *34*, 25–36 (in Italian).

Flouri, F., Sotirchos, D., Ioannidou, S., & Balis, C. (1996). Decolorization of olive oil mill liquid wastes by chemical and biological means. *Proc. Olive Oil Processes and By Products Recycling*; Granada, Spain, Sep. 10–13, 1995. *Int. Biodeterior. Biodegrad.*, 1996, *38* (3–4), 189–192.

Flouri, F., Chatzipavlidis, I., Balis, C., Servis, D., & Tjerakis, C. (1990). Effect of olive oil mills wastes on soil fertility. *Int. Symp. on Treatment of OMWW*, Cordoba, Spain, May 31–June 1, 1990, 85–101.

Fodale, A.S., Mule, R., & Briccoli-Bati, B. (1999). The antifungal activity of olive oil wastewater on isolates of *Verticillium dahliae* Kleb. in vitro. *Proc. 3rd Int. Symp. on Olive Growing*, Chania, Crete, Greece, 22–26 Sep. 1997 (eds. I.T. Metzidakis & D.G. Voyiatzis (eds). *Acta Hortic.*, 2, (474), 753–756.

Folin, O., & Ciocalteau, V. (1927). On tyrosine and tryptofan determinations in protein. *J. Biol. Chem.*, *73*, 627–650.

Fontanazza, G., Regis Milano, S., Patumi, M., & Altieri, R. (1993). Un sistema innovativo per il trattamento dei reflui di frantoio oleario. *Olivo & Olio*, 7/8, 26–31 (in Italian).

Fountoulakis, M.S., Dokianakis, S. N., Kornaros, M.E., Aggelis, G., & Lyberatos, G. (2002). Removal of phenolics in olive mill wastewaters using the white-rot fungus *Pleurotus ostreatus*. *Water Res.*, *36* (19), 4735–4744.

Friaâ, A., Mensi, R., & Kallel, A. (1986a). Application of vegetation water in civil engineering –Soil treatment with vegetation water. *Int. Symp. on Olive By-Products Valorization*, Seville, Spain, 5–7 Mar. 1986, 263–273.

Friaâ A., Mensi, R., & Kallel, A. (1986b). Briques de terre stabilisées à la margine –Matériaux de construction pour un habitat économique. *Joint Symp. on the Use of Vegetable Plants and their Fibres as Building Materials*, Baghdad, Iraq, 7–9 Oct. 1986, E95–E102 (in French).

Frankel, R. (1984). The history of the processing of wine and oil in the Galilee in the period of the Bible, the Misbna and the Talmud, Tel Avi (in Hebrew).

Franzione, G. (1986). *Proc. Int. Symp. on Olive By-Products Valorization*, FAO, UNDP (Food and Agriculture Organization of the United Nations), Seville, Spain, 5–7 Mar. 1986, 285–298.

Galiatsatou, P., Metaxas, M., & Kasselouri-Rigopoulou, V. (2001). Mesoporous activated carbon from agricultrural by-products. *Microchimica Acta*, *136*, (3/4) 147–152.

Galiatsatou P., Metaxas, M., Arapoglou, D., & Kasselouri-Rigopoulou, V. (2002). Treatment of olive mill waste water with activated carbons from agricultural by-products. *Waste Manage.*, *22* (7), 803–812.

Gallardo-Lara, F., Azcon, M., & Polo, A. (2000). Phytoavailability and extractability of potassiun, magnesium and manganese in calcareous soil amended with olive oil wastewater. *Environ. Sci. Health, B.*, *35* (5), 623–643.

Galli, E., Pasetti, L., Fiorelli, F., & Tomati, U. (1997). Olive mill wastewater composting: Microbiological aspects. *Waste Manage. Res.*, *15* (3), 323–330.

Galli, E., Pasetti, L., Volterra, E., & Tomati, U. (1994). Compost from olive processing industry wastewaters. *Proc. 7th Mediterranean Conf. on Organic Wastes Recycling in Soils*, Vieste, Italy, *1*, 185–190, Ordine Nazionale dei Biologi (eds. Landi, E., & Dumonet, S.).

Galli, E., Tomati, U., Grappelli, A., & Buffone, R. (1988). Recycle of olive oil waste for *Pleurotus* mycelium production in submerged culure. *Agrochemica*, *32*, 451–456.

Galoppini, C., Andrich, G., & Fiorentini, R. (1994). Trasporto e biodegradabilità dei reflui di frantoio nel terreno agrario. *Agrochimica*, *38* (XXXVIII), 97–107 (in Italian).

Gamel, T.H., & Kiritsakis, A. (1999). Effect of methanol extracts of rosemary and olive vegetable water on the stability of olive oil and sunflower oil. *Grasas y Aceites*, *50* (5), 345–350.

García, A., Brenes Balbuena, M., Martínez, F., Alba-Mendoza, J., García, P., & Garrido Fernández, A. (2001a). High-performance liquid chromatography evaluation of phenols in virgin olive oil during extraction at laboratory and industrial scale. *J. Am. Oil Chem. Soc. (JAOCS)*, *78* (6), 625–629.

García, A., Brenes Balbuena M., Moyano, M.J., Alba-Mendoza, J., García-García, P., & Garrido Fernández, A. (2001b). Improvement of phenolic content in virgin oils by using enzymes during malaxation. *J. Food Eng.*, *48* (3), 189–194.

García-Barrionuevo, A., Moreno, E., Quevedo-Sarmiento, J., González-López, J., & Ramos-Cormenzana, A. (1993). Effect of wastewater from olive oil mill on nitrogenase activity and growth of *Azotobacter chroococcum. Environ. Toxicol. Chem.*, *12* (2), 225–230.

García-Barrionuevo, A., Moreno, E., Quevedo-Sarmiento, J., González-López, J., & Ramos-Cormenzana, A. (1992). Effect of wastewater from olive oil mill (alpechín) on *Azotobacter chroococcum* nitrogen fixation in soil. *Soil Biol. Biochem.*, *24* (3), 281–283.

García-García, I., Jiménez-Pena, P.R., Bonilla-Venceslada, J.L., Martín-Martín, A., Martín-Santos, M.A., & Ramos-Gomez, E. (2000). Removal of phenol compounds from olive mill wastewater using *Phanerochaete chrysosporium*, *Aspergillus niger*, *Aspergillus terreus* and *Geotrichum candidum*. *Process Biochem.*, *35* (8), 751–758.

García-García, P., & Garrido Fernández, A. (1984). Partial purification by precipitation of the lye solutions and washing water from Spanish style green olive industries. *Grasas y Aceites*, *35* (5), 295–299.

García-García, P., Brenes Balbuena, M., Vicente Fernández, A. de, & Garrido Fernández, A. (1990). Physicochemical depuration of the waste waters from the green table olive packing industries. *Grasas y Aceites*, *41* (3), 264–269.

García-García, P., Garrido Fernández, A., Chakman, A., Lemonier, J.P., Overend, R.P., & Chornet, E. (1989). Purification of polyphenolic-rich wastewaters-Application of the wet oxidation to the aqueous effluents from the olive industries. *Grasas y Aceites*, *40* (4–5), 291–295.

García-Gomez A., Bernal, M.P., & Roig, A. (2002). Growth of ornamental plants in two composts prepared from agroindustrial wastes. *Bioresource Technol.*, *83* (2), 81–87.

García-Gomez, A., Roig, A., & Bernal, M.P. (2003). Composting of the solid fraction of olive mill wastewater with olive leaves: Organic matter degradation and biological activity. *Bioresource Techn.*, *86* (1), 59–64.

García-Ibañez, A., Cabanillas, A., & Sánchez, J.M. (2004). Gasification of leached orujillo (olive oil waste) in a pilot plant circulating fluidized bed reactor. Preliminary results. *Biomass & Bioenergy*, *27* (2), 183–194.

García-Ortíz, R.A. (1998). Extraction technology and the by-product problems. *Revista de Ciencias Agrarias*, *21* (1–4), 377–384 (in Spanish).

García-Ortíz, R.A., Giráldez Cervera, J.V., González Fernández, P., & Ordóñez Fernández, R. (1993). El riego con alpechín. Una alternative al lagunaje. *Agricultura*, *730*, 426–431 (in Spanish).

García-Ortíz, R.A., Beltrán, G., González Fernández, P., Ordóñez Fernández, R., & Giráldez Cervera, J.V. (1999). Vegetation water (alpechín) application effects on soils and plants. *Acta Horticulturae*, *474*, 749–752.

García-Pareja, M.P., Ramos-Cormenzana, A., Gómez-Palma, L.P., Martínez-Nieto, L., & Garrido-Hoyos, S.E. (1990). Estudio comparado de la biotransformación microbiana de ligninas y alpechines. *I Congreso Internitional de Qufmica de la ANQUE*. Puerto de la Cruz, Tenerife, Spain (in Spanish).

Garrido-Hoyos, S.E., Martínez-Nieto, L., Camacho Rubio, F., & Ramos-Cormenzana, A. (2002). Kinetics of aerobic treatment of olive-mill wastewater (OMW) with *Aspergillus terreus*. *Process Biochemistry*, *37* (10), 1169–1176.

García-Rodríguez, A. (1990). Eliminatión y aprovechamiento agricola del alpechín, In *Reunión Iternacional Sobre: Trataminento de aplechines, DGIEA*. Casajeria de Agricultura y Pesca de la Junta de Andalucia, Seville, Spain, 105-124 (in Spanish).

Gariboldi, P., Jommi, G., & Verotta, L. (1986). Secoiridoids from *Olea europaea*. *Phytochemistry*, *25*, 865–869.

Garrido Fernández, A. (1975). Tratamiento de las aguas residuales de la industria del alderezo. Metodo para su eliminacion o reacondicionamiento para su posterior empleo. "Treatment of waste waters from the table olive". *Grasas y Aceites*, *26* (4), 237–244 (in Spanish).

Garrido Fernández, A. (1978). Natural black olives in brine. V. Regeneration trials with fermentation brines. *Grasas y Aceites*, *29* (2), 111–118 (in Spanish).

Garrido Fernández, A. (1983). Study of waste waters from processing of black olives and their repeated use. *Grasas y Aceites, 34* (5), 317–322.

Garrido Fernández, A. (1984). Study of wastewaters from ripe olive processing. II. Effects of re-using lye or aeration solutions on colour, texture and canning brines. *Grasas y Aceites, 36* (3), 165–171.

Garrido Fernández, A. (1992). Treatment of brines for the fermented vegetables industry. *Alimentacion Equipos y Tecnologia, 11* (4), 128–130.

Garrido Fernández, A., Brenes Balbuena M., & García-García, P. (1992a). Treatment for green table olive fermentation brines. *Grasas y Aceites, 43* (5), 291–298.

Garrido Fernández, A., Fernández-Diez, M.J., & Adams, M. (1997). Table Olives: Production and Processing. Plenum (US), pp. 495; ISBN: 0412718103.

Garrido Fernández, A., García-García, P., & Brenes Balbuena, M. (1992b). The recycling of table olive brine using ultrafiltration and activated carbon adsorption. *J. Food Eng., 17* (4), 291–305.

Garrido Fernández, A., Cordon Casanueva, J.L., Rejano Navarro, L., González Cancho, F., & Sánchez Roldán, F. (1979). Production of Spanish style green olives with re-use of lye and omission of washing. *Grasas y Aceites, 30* (4), 227–234.

Garrido Fernández, A., González Pellisso, F., González Cancho, F., Sánchez Roldán, F., Rejano Navarro L., Cordon Casanueva, J.L., & Fernández Diez, M.J. (1977). Changes in processing and packaging of green table olives in relation to elimation and reuse of wastes. *Grasas y Aceites, 28* (4), 267–285.

Gasparrini, R. (1999). Treatment of olive oil processing residues. *Oils and Fats Int., 15* (1), 32–33.

Gavala, H.N., Skiadas, I.V., & Lyberatos, G. (1998). On the performance of a centralized digestion facility receiving seasonal agroindustrial wastewaters. *Proc. 4th Int. Symp. on Waste Management Problems in Agro-Industries*, Istanbul, Turkey, 23–25 Sep. 1998. *Water Sci. Technol.* 1999, *40* (1), 339–346.

Gavala, H.N., Skiadas, I.V., Bozinis, N.A., & Lyberatos, G. (1996). Anaerobic codigestion of agricultural industries' wastewaters. *Proc. 18th Biennial Conf. of the Int. Association on Water Quality*. Part 7; Singapore, 23–28 June 1996. *Water Sci. Technol., 34* (11), Part 7, 67–75.

G.E.O.C.G. (1994). Management of liquid waste from oil mills. *Proc. G.E.O.C.G. 3rd Int. Conf. on Management of Liquid Waste from Oil Mills*. Sitia, Crete, Greece, 16–17 June 1994.

General Electric Company (1997–2003). *Nanofiltration -109 Olive Flume Wastewater*. Technical papers (http://www.gewater.com/library/tp/834_Nanofiltration_.jsp).

Georgacakis, D., & Christopoulou, N. (2002). Olive oil mill wastewaters treatment and disposal. A case application study at Samos island. *Research report on the results of a full scale demonstration installation at Marathokampos of Samos island*. Laboratory of Agricultural Structures, Agricultural University of Athens (in Greek).

Georgacakis, D., & Dalis, D. (1993). Controlled anaerobic digestion of settled olive oil wastewater. *Bioresource Technol., 46* (3), 221–226.

Georgacakis, D., Andreadi, E., & Christopoulou, N. (2002). Exploitation of cost efficient biogas production and utilisation from Greek pig farms and olive oil mill wastes. *Research program report*. Laboratory of Agricultural Structures, Agricultural University of Athens (in Greek).

Georgacakis, D., Konstas, S., & Dalis, D. (1995). Study on energy recovery from olive oil mill wastes. *ITE ELKEPA*, 21 (in Greek).

Georgacakis, D., Kyritsis, S., Manios, B., & Vlyssides, A.G. (1986). Economic optimization of energy production from olive oil wastewater, Peza Heraklion (Crete). *Proc. 2nd Int. Conf. Energy and Agriculture*, Sirmione, Breccia, Italy, 13–16 Oct. 1986.

Georgacakis, D., Tziha, F., Zogzas, M., Nikolarou, Xr., & Tsabdaris, A. (1994). Management of olive oil mill wastewaters prior to their biological treatment. *Research program report*. Laboratory of Agricultural Structures. Agricultural University of Athens and Ministry of Public Health, Athens (in Greek).

Gernjak, W., Maldonado, M.I., Malato, S., Cáceres, J., Krutzler, A., Glaser, A., & Bauer, R. (2004). Pilot-plant treatment of olive mill wastewater (OMW) by solar TiO_2 photocatalysis & solar photo-Fenton. *Sol. Energy*, *77* (5), 567–572.

Gharaibeh, S.H., Moore, S.V., & Buck, A. (1998a). Effluent treatment of industrial wastewater using processed solid residue of olive mill products and commercial activated carbon. *J. Chem. Technol. Biotechnol.*, 71 (4), 291–298.

Gharaibeh, S.H., Abu-El-Sha'r Wa'ill, Y., & Al Kofahi, M.M. (1998b). Removal of selected heavy metals from aqueous solutions using processed solid residue of olive mil products. *Water Res.*, *32* (2), 498–502.

Gharsallah, N. (1993). Production of single cell protein from olive mill wastewater by yeasts. *Environ. Technol.*, *14* (4), 391–395.

Gharsallah, N. (1994). Influence of dilution and phase separation on the anaerobic digestion of olive mill wastewaters. *Bioprocess Eng.*, *10* (1), 29–34.

Gharsallah, N., Labat, M., Aloui, F., & Sayadi, S. (1998). The effect of *Phanerochaete chrysosporium* pretreatment of olive mill waste waters on anaerobic digestion. *Proc. 4th Int. Symp. of the Int. Soc. for Environ. Biotechnol.*, Belfast, UK, 20–25 June 1998. *Resour. Conserv. and Recycl.*, *27* (1–2), 187–192.

Gianfreda L., Sannino, F., Rao, M.A., & Bollag, J.-M. (2003). Oxidative transformation of phenols in aqueous mixtures. *Water Res.*, *37* (13), 3205–3215.

Giannes, A., Diamadopoulos, E., & Ninolakis, M. (2003). Electrochemical treatment of olive oil mill wastewater using a Ti/Ta/Pt/Ir electrode. *Proc. 3rd Int. Conf. on Oxidation Technologies for Water and Wastewater Treatment* (eds. Vogelpohl A., Geiïen, S.U., Kragert B., & Sievers M.), Goslar, Germany, 18–22 May 2003, 147–152.

Giannoutsou, E., Lambraki, M., & Karagouni, A.D. (1997a). Chemical and microbial characterization of olive oil waste by the two phase extraction technique. *Proc. 11th Forum for Applied Biotechnology*, Gent, Belgium, 25–26 Sep. 1997. *Meded. Fac. Landbouwkd. Toegepaste Biol. Wet.* (Univ. Gent), *62* (4b), 1905–1908.

Giannoutsou, E., Lambraki, M., & Karagouni, A.D. (1997b). Microbial treatment of olive mill effluents. *Proc. Symp. "Olive's Wastes"*, Kalamata, Greece, 5–8 Nov. 1997.

Giannoutsou, E., Meintanis, C., & Karagouni, A.D. (2004). Identification of yeast strains isolated from a two-phase decanter system olive oil waste and investigation of their ability for its fermentation. *Bioresour. Technol.*, *93* (3), 301–306.

Gil, M., Haidour, A., & Ramos, J.L. (2000). Degradation of o-methoxybenzoate by two-member consortium made of a gram-positive *Arthrobacter* strain and a gram-negative *Pantotea* strain. *Biodegradation*, *11* (1), 49–53.

Giorgio, L., Andreazza, C., & Rotunno, G. (1981). Esperienze sul funzionamento di un impianto di depurazione per acque di scarico civili e di oleifici. *Ingegneria Sanitaria, 5*, 296–303 (in Italian).

Girolami, V., Vianello, A., Strapazzon, A., Ragazzi, E., & Veronese, G. (1981). Ovipositional deterrents in *Dacus oleae*. *Entomologia Experimentalis et Applicat, 29* (2), 177–188.

Giulietti, A.M., Ercoli E., & Ertola, R. (1984). Purification and utilization of olive black water and distillery slops by microbial treatment. *Acta Cient. Venez*, *35* (1), 63–66.

Gómez-Cabrera, A., Garrido Fernández, A., Guerrero Ginel, J.E., & Ortiz, V. (1992). Nutritive value of the olive leaf: effects of cultivar, season of harvesting and system of drying. *J. Agric. Sci.*, *119*, 205–210.

Gómez-Cabrera, A., Parellada, J., Garrido Varo, A., & Ocaña, F. (1982). Utilización de ramón de olivo en alimentatión animal. II. Valor alimenticio. *Avances en Atimentación y Mejora Animal*, *22* (11), 75–77 (in Spanish).

González, A.E., Rodríguez, T.M., Sánchez, J.J.C., Espinosa, F.A.J., & Rosa de la, B.F.J. (2000). Assessment of metals in sediments in a tributary of Guadalquivir river (Spain). Heavy metal partitioning and relation between the water and sediment system. *Water Air Soil Pollut.*, *121* (1–4) 11–29.

González, M.D., Moreno, E., Quevedo-Sarmiento, J., & Ramos-Cormenzana, A. (1990). Studies on antibacterial activity of waste waters from olive oil mills (alpechín): Inhibitory activity of phenols, & fatty acids. *Chemosphere*, *20* (3–4), 423–432.

González Cancho, F., Rejano Navarro, L., Duran Quintana, M.C. Sánchez Roldán, F., García-García, P., Castro Gomez-Millan, A. de, & Garrido Fernández, A. (1983). Influence of HCl addition on the fermentation of Spanish green olives. *Grasas y Aceites*, *34* (6), 375–379.

González Cancho, F., Rejano Navarro, L., Durán Quintana, M.C., Sánchez Roldán, F., Castro Gomez-Millan, A. de, García-García, P., & Garrido Fernández, A. (1984). Preparation of Spanish green olives without washing. Solution to the problems caused by HCl additions, and effects of treatment with weak lye solutions. *Grasas y Aceites*, *35* (3), 155–159.

González-López, J., Bellido E., & Benítez, C. (1994). Reduction of total polyphenols in olive mill wastewater by physico chemical purification. *J. Environ. Sci. Health Part A*, *29* (5), 851–865.

González López, J., Pozo, C., Martínez Toledo, M.V., Rodelas, B., & Salmeron, V. (1995). Production of polyhydroxyalkanoates by *Azotobacter chroococcum* H23 in wastewater from olive oil mills (alpechín). *Proc. Olive Oil Processes and By Products Recycling*; Granada, Spain, 10–13 Sep. 1995. *Int. Biodeterior. Biodegrad.*, 1996, *38* (3–4), 271–276.

González-Vila, F.J., Verdejo, T., & Martín, F. (1992). Characterization of wastes from olive and sugar beet processing industries and effects of their application upon the organic fraction agricultural soils. *Int. J. Environ. Anal. Chem.*, *46* (1–3), 213–222.

González-Vila, F.J., Verdejo, T., Del Rio, J.C., & Martín, F. (1995). Accumulation of hydrophobic compounds in the soil lipidic and humic fractions as result of a long-term land treatment with olive oil mill effluents (alpechin). *Chemosphere*, *31* (7), 3681–3686.

Greco G. Jr., Toscanoa, G., Cioffi, M., Gianfreda L., & Sannino F. (1999). Dephenolisation of olive mill wastewaters by olive husk. *Water. Res.*, *33* (13), 3046–3050.

Haddadin, M.S., Abdulrahim, S.M., Al-Khawaldeh, G.Y., & Robinson, R.K. (1999). Solid state fermentation of waste pomace from olive processing. *J. Chem. Technol. Biotechnol.*, *74* (7), 613–618.

Hadjipanayiotou, C. (2003). *IMPEL Olive Oil Project Report - CMA & NOA*. Number Report 2003/3, Project Manager: Méndez Miguel, Rome, Nov. 2003; number of pages, report: 33 and annexes: 61. European Union Network for the Implementation and Enforcement of Environmental Law (IMPEL) (http://europa.eu.int/comm/environment/impel).

Hadjipanayiotou, M., & Koumas, A. (1996). Performance of sheep and goats on olive cake selage. *Technical Bulletin No. 176*. Agricultural Research Institute, Nicosia, Cyprus, pp. 10.

Hadjisavvas, S. (1992). Olive oil processing in Cyprus (From the bronze age to the Byzantine period). *Studies in Mediterranean Archaeology*, **XCIX**, Nicosia 1992, Paul Åströms Förlag. ISBN 91-7081-033-8.

Halet, F., Belhocine, D., Grib, H., Lounici, H., Pauss, A., & Mameri, N. (1997). Treament of oil mill washing water by ultrafiltration. *3rd Int. Symp.on Environmental Biotechnology*, Oostende, Belgium, 21–24 Apr. 1997; Part II, 369–372.

Hamdi, M. (1987). Digestion anaérobie des margines par filtres anéaérobies. Diplõme d'ingénieur, University of Provence, Marseille (in French).

Hamdi, M. (1991a). Effects of agitation and pretreatment on the batch anaerobic digestion of olive mill wastewater. *Bioresource Technol.*, *36* (2), 173–178.

Hamdi, M. (1991b). Nouvelle conception d'un procédé de dépollution biologique des margines, effluents liquides de l'extraction de l'huile d'olive. "A new conception of biological process for olive mill wastewaters treatment". Biologie cellulaire et Microbiologie, *Thèse doctorat*, Université de Provence, France, 1991, 168 pages (in French).

Hamdi, M. (1992). Toxicity and biodegradability of olive mill wastewaters in batch anaerobic digestion. *Appl. Biochem. Biotechnol.*, *37* (2), 155–163.

Hamdi, M. (1993a). Future prospects and constraints of olive oil mill wastewaters and treatment. A review. *Bioprocess Eng. 8* (5–6), 209–214.

Hamdi, M. (1993b). Thermoacidic precipitation of darkly colored polyphenols of olive mill wastewaters. *Environ. Technol.*, *14* (5), 495–500.

Hamdi, M. (1996). Anaerobic digestion of olive mill wastewaters. *Process Biochem.* (Oxford), *31* (2), 105–110.

Hamdi, M., & Ellouz, R. (1992a). Bubble column fermentation of olive mill wastewater by *Aspergillus niger*. *J. Chem. Technol. Biotechnol.*, *54* (4), 331–335.

Hamdi, M., & Ellouz, R. (1992b). Use of *Aspergillus niger* to improve filtration of olive mill wastewaters. *J. Chem. Technol. Biotechnol.*, *53* (2), 195–200.

Hamdi, M., & Ellouz, R. (1993). Treatment of detoxified olive mill wastewaters by anaerobic filter and aerobic fluidized bed process. *Environ. Technol.*, *14* (2), 183–188.

Hamdi, M., & García, J.L. (1991). Comparison between anaerobic filter and anaerobic contact process for fermented olive mill wastewater. *Bioresource Technol.*, *38* (1), 23–30.

Hamdi, M., & García, J.L. (1993). Anaerobic digestion of olive mill wastewaters after detoxification by prior culture of *Aspergillus niger*. *Process Biochem.*, *28* (3), 155–159.

Hamdi, M., Bou Hamed, H., & Ellouz, R. (1991a). Optimization of the fermentation of olive mill waste-waters by *Aspergillus niger*. *Appl. Microbiol. Biotechnol.*, *36* (2), 285–288.

Hamdi, M., Brauman, A., & García, J.L. (1992a). Effect of an anaerobic bacterial consortium isolated from termites on the degradation of olive mill wastewater. *Appl. Microbiol. Biotechnol.*, *37* (3), 408–410.

Hamdi, M., Festino, C., & Aubart, C. (1992b). Anaerobic digestion of olive mill wastewaters in fully mixed reactors and in fixed film reactors. *Process Biochem.*, *27* (1), 37–42.

Hamdi, M., García, J.L., & Ellouz, R. (1992c). Integrated biological process for olive oil mill wastewater treatment. *Bioprocess Eng.*, *8* (1–2), 79–84.

Hamdi, M., Khadir, A., & García, J.L. (1991b). The use of *Aspergillus niger* for the bioconversion of olive mill wastewaters. *Appl. Microbiol. Biotechnol.*, *34* (6), 828–831.

Hamman, O. Ben, Rubia, T. de la., & Martínez, J. (1999). Decolorization of olive mill wastewaters by *Phangerochaete flavido-alba*. *Environ. Toxicol. Chem.*, *18* (11), 2410–2415.

Hanaki, K., Matsuo, T., & Nagasse, M. (1981). Mechanism of inhibition caused by long chain fatty acids in anaerobic digestion process. *Biotechnol. and Bioeng.*, *23*, 1591–1610.

Harb, M. (1986). Using the olive pomace for fattening the Awassi lamb. *Dirasat.*, *13*, 37–53.

Hardisson, C., Sala, J.M., & Stainer, R.Y. (1969). Pathways for the oxidation of aromatic compounds by *Azotobacter*. *J. Genet. Microbiol.*, *59*, 1–11.

Hytiris, N., Kapelllakis, I.E., La Roij de, R., & Tsagarakis, K.P. (2004). The potential use of olive mill sludge in solidification process. *Resour. Conserv. and Recycl.*, *40* (2), 129–139.

Ibáñez, E., Palacios, J., Señoráns, F.J., Santa-María, G., Tabera, J., & Reglero, G. (2000). Isolation and separation of tocopherols from olive by-products with supercritical fluids. *J. Am. Oil Chem. Soc. (JAOCS)*, *77* (2), 187–190.

Iconomou, D., Diamantitis, G., Zanganas, P., Theochari, I., Israilides, K., & Kouloumbis, P. (2000). Reduction of phenolic concentration in olive-oil mill wastewater by biotechnological means. *Proc. Int. Conf. on Production and Restoration of the Environment V*, (eds. Tsihrintzis, V. A., Korfiatis, G.P., Katsifarakis, K.L., & Demetracopoulos, A.C.) Thassos, Greece, 2000. Publisher Bouris, Thessaloniki, Greece, vol. I, 569–572.

Inan, H., Dimolgo, A., Şimşek, H., & Karpuzcu, M. (2004). Olive oil mill wastewater treatment by means of electro-coagulation. *Sep. Purif. Technol.*, *36* (1), 23–31.

Inigo Leal, B. (1991). New technological trends in the production and preservation of table olives. *Alimentaria*, *28* (228), 61–65.

Iniotakis, N., Michaelides, P.G., & Koumakis, M. (1989). Καθαρισμός κατσίγαρου με ταυτόχρονη αξιοποίηση χρήσιμων υλικων. "Purification of OMW with simultaneous recovery of useful materials". *Proc. Meeting ΓΕΩUEE*, 81–88, Heraklion, Greece, 1989 (in Greek).

Iniotakis, N., Michaelides, P.G., Diamantis, G.M., Israilides, C.J., & Papanikolaou, S. (1991). Katsigaros: a profitable management of liquid wastes. *2nd Conf. Environmental Science and Technology*, Molyvos, Mytiline, Greece, 1991.

Inouye, H., Yoshida, T., Tobita, S., Tanaka, K., & Nishioka, T. (1970). *Tetrahedron Lett.*, 2459–2464.

International Olive oil Council (1996). *World Olive Encyclopedia*. Chapter 7, 271–290 (eds. Plaza & Janés), Barcelona, Spain.

Israilides, C.J. (1990). Present situation and prospective on the management of olive oil waste effluent in Greece. *Proc. Int. Meeting on the Treatment of Olive Oil Vegetable Water*, Cordoba, Spain, 1–9 June 1990.

Israilides, C.J., Vlyssides, A.G., Douligeris, A.G., & Waite, T.D. (1998). Use of ionizing radiation for the treatment of olive mill wastes. *Proc. Int. Meeting on Restoration and Protection of the Environment IV*, Halkidiki, Greece, 1–4 July 1998.

Israilides, C.J., Vlyssides, A.G., Mourafeti, V.N., & Karvouni, G. (1997). Olive oil wastewater treatment with the use of an electrolysis system. *Bioresource Technol.*, *61* (2), 163–170.

Janer del Valle, L. (1980). Contaminación de las aguas por el alpechín y posibles soluciones al problema. "Water pollution by waste waters from olive oil plants and possible solutions of the problem". *Grasas y Aceites*, *31* (4), 273–279 (in Spanish).

Jaouani, A., Jaspers, C., Penninckx, M., Sayadi, S., & Vanthournhout, M. (2000). Proposal of a treatment flowchart for the olive oil mill wastewaters. *Meded. Fac. Landbouwkd. Toegepaste Biol. Wet.* (Univ. Gent), *65* (3A), 111–113.

Jaouani, A., Sayadi, S., Vanthournhout, M., & Penninckx, M. (2003). Potent fungi for decolourisation of olive oil mill wastewaters. *Enzym. Microb. Technol.*, *33* (6), 802–809.

Jelmini, M., Sanna, M., & Pelosi, N. (1976). Indagine sulle acque di rifuto degli stabilimenti di produzione olearia in provincia di Roma: Possibilitá di depurazione. "Research on the waste water from oil manufacturing plant in Rome country". *Ind. Aliment.*, *15* (11), 123–131 (in Italian).

Jemmett, M.T., Carrieri, C., Leo, P. de, & Kaspiotis, G. D. (1983). Esperienze di osmosi inversas ed ultrafiltrazione di axque di vegetazione delle olive con moduli a membrane piane. "Experiences with reverse osmosis, & ultrafiltration of olive processing wastewaters using a flat membrane module". *Riv. Soc. Ital. Sci. Aliment.*, año *12* (1), 37–46 (in Italian).

Jones, C.E., & Russell, N.J. (1997a). Identification of the bacterial flora in wastes from olive oil production. *Proc. UK Extremoiphile Network*, Sep. 1997, Bath, England.

Jones, C.E., & Russell, N.J. (1997b). Primary identification of the bacterial flora present in wastes from olive oil production. *Proc. Symp. "Olive's Wastes"*, Kalamata, Greece, 5–8 Nov. 1997.

Jones, C.E., Murphy, P.J., & Russell, N.J. (2000). Diversity and osmoregulatory responses of bacteria isolated from two-phase olive oil extraction waste products. *World J. Microbiol. Biotechnol. 16*, 555–561.

Jones, N., Duarte, E.A., & Silva, M.J. (1998). Use of two-phase olive oil bagasse as organic amendment –environmental prevention strategy. *Proc. 1st Int. Conf. on Environmental Engineering and Management*; Barcelona, Spain. *WITT Trans. Ecol. and Environ.-Environ-Eng.-Environ. Eng. Manage.*, *29*, 241–249 (ed. Power, H.).

Junta de Andalucía: Consejería de Aricultura y Pesca (2002). Estimación de aceite de oliva en Andalucía para la campaña 2001/02, pp, 1–10.

Jurado, F., Cano, A., & Carpio, J. (2003). Modelling of combined cycle power plants using biomass. *Renew. Energy*, *28* (5), 743–753.

Jurado, F., Ortega, M., & Cano, A. (2002). Robust control for a gas turbine in biomass-based electric power plants. *Energy Sources*, *24* (7), 591–599.

Juven, B., & Henis, Y. (1970). Studies on the antimicrobial activity of olive-derived phenolic compounds. *J. Appl. Bacteriol.*, *33*, 721–732.

Kahraman, S., & Yesilada, O. (1999). Effect of spent cotton stalks on color removal and chemical oxygen demand lowering in olive oil mill wastewater by white rot fungi. *Folia Microbiologica*, *44* (6), 673–676.

Kahraman, S., & Yesilada, O. (2001). Industrial and agricultural wastes as substrates for laccase production by white rot fungi. *Folia Microbiologica*, *46* (2), 133–136.

Kalmiş E., & Sargin, S. (2004). Cultivation of two *Pleurotus* species on wheat straw substrates containing olive mill waste water. *Int. Biodeterior. Biodegrad.*, *53* (1), 43–47.

Kapetanios, G.E., Loizidou, M., & Valkanas, G. (1993). Compost production form Greek domestic refuse. *Biores. Technol.*, *44*, 13–16.

Karapinar, M., & Worgan, J.T. (1983). Bioprotein production from the waste products of olive oil extraction. *J. Chem. Technol. Biotechnol.*, *33*B (3), 185–188.

Kasirga, E. (1988). Treatment of olive oil industry wastewaters by anaerobic stabilization method and development of kinetic model. *Unpublished Ph.D. Thesis.* Dokuz Eylul University, Graduate School of Natural and Applied Sciences, Izmir, Turkey (in Turkish).

Kestioglu, K., Yonar, T., & Azbar, N. (2005). Feasibility of physico-chemical treatment and Advanced Oxidation Processes (AOPs) as a means of pretreatment of olive mill effluent (OME). *Process Biochem.*, *40* (7), 2409–2416.

Khabbaz, M.S., Vossoughi, M., & Shakeri, M. (2004). Performance of an anaerobic baffled reactor for olive mill oil wastewater treatment. *Proc. 39th Central Canadian Symp. on Water Quality Research.*, Burlington, Ontario, Canada, 9–10 Feb. 2004.

Khoufi, S., Aouissaoui, H., Penninckx M., & Sayadi, S. (2003). Application of electro-Fenton oxidation for the detoxification of olive mill wastewater phenolic compounds. *Proc. 3rd Int. Conf. on Oxidation Technologies for Water and Wastewater Treatment*, Goslar, Germany, 18–22 May 2003. *Water Sci. Technol.*, *49* (4), 97–102, 2004, (eds. Vogelpohl, A., Geißen, S.U., Kragert, B., & Sievers, M.).

Kissi, M., Mountadar, M., Assobhei, O., Gargiulo, E., Palmieri, G., Giardina, P., & Sannia, G. (2001). Roles of two white rot basidiomycete fungi in decolorisation and detoxification of olive mill wastewater. *Appl. Microbiol. and Biotechnol.*, *57* (1–2), 221–226.

Knupp, G., Rücker, G., Ramos Cormenzana, A., Garrido-Hoyos, S.E., Neugebauer, M., & Ossenkop, T. (1996). Problems of identifying phenolic compounds during the microbial degradation of olive mill wastewater. *Proc. Olive Oil Processes and By Products Recycling*, Granada, Spain, 10–13 Sep. 1995. *Int. Biodeterior. Biodegrad.* 1996, *38* (3–4), 277–282.

Komilis, D. P., Karatzas, E., & Halvadakis, C.P. (2005). The effect of olive mill wastewater on seed germination after various pretreatment techniques. *J. Environ. Management. 74* (4), 339–348.

Kopsidas, G.C. (1990). Treatment of table olive industry wastewater in Greece. *Symp. of the Commission of the European Communities on treatment and use of sewage sludge and liquid agricultural wastes*, Athens, Oct. 1990, 293–297.

Kopsidas, G.C. (1992). Wastewater from the preparation of table olives. *Water Res.*, *26* (5), 629–631.

Kopsidas, G.C. (1994). Wastewater from the table olive industry. *Water Res.*, *28* (1), 201–205.

Kopsidas, G.C. (1995). Multi-objective optimization of table olive preparation systems. *Eur. J. Oper. Res.*, *85* (2), 383–398.

Koster, I. (1987). Abatement of long chain fatty acid inhibition of methanogenesis by calcium adition *Biological Wastes*, *22* (4), 295–301.

Kotsou, M., Mari, I., Lasaridi, K., Chatzipavlidis, I., Balis, C., & Kyriacou, A. (2004a). The effect of olive oil mill wastewater (OMW) on soil microbial communities and suppressiveness against *Rhizoctonia solani* . *Appl. Soil Ecol.*, *26* (2), 113–121.

Kotsou, M., Kyriacou, A.A., Lasaridi, K., & Pilidis, G. (2004b). Integrated aerobic biological treatment and chemical oxidation with Fenton's reagent for the processing of green table olive wastewater. *Process Biochm.*, *39* (11), 1653–1660.

Koussemon, M., Combet-Blanc, Y., Patel Bharat, K.C., Cayol, J.L., Thomas, P., García, J.L., & Ollivier, B. (2001). *Propionibacterium microaerophilum* sp. nov., a microaerophilic bacterium isolated from olive millwaste. *Int. J. Systematic and Evolutionary Microbiol.*, *51* (4), 1373–1382.

Koutsaftakis, A., Kotsifaki, F., & Stefanoudaki, E. (1999). Effect of extraction system, stage of ripeness and kneading temperature on the sterol composition of virgin olive oils. *J. Am. Oil Chem. Soc. (JAOCS)*, *76* (12), 1477–1481.

Kohyama, N., Nagata, T., Fujimoto, S., & Sekiya, K.(1997). Inhibition of arachidonate lipoxygenase activities by 2-(3,4-dihydroxyphenyl)ethanol, a phenolic compound from olives. *Biosci. Biotechnol Biochem. 61*, 347–350.

Kraisha, Y.M., Hamdan, M.A., & Qalalweh, H.S. (1998). Direct combustion of olive cake using fluidized bed combustor. *Energy Sources, 21*, 319.

Kyriacou, A., Lasaridi, K.E., Kotsou, M., Balis, C., & Pilidis, G. (2005). Combined bioremediation and advanced oxidation of green table olive processing wastewater. *Process Biochem.*, *40* (3–4), 1401–1408.

Labat, M., Sayadi, S., Gargouri, A., Zorgani, F., Jaoua, M., Zekri, S., & Ellouz, R. (1996). Procédé aérobie-anaérobie pour le traitement biologique des résidus liquides de l'industri oleicole. *Journées industrielles sur la digestion anaérobie*, Narbone, France, June 1996, 153–158 (in French).

Lafont, F., Aramendia, M.A., García, I., Boráu, V., Jiménez, C., Marinas, J.M., & Urbano, F.J. (1999). Analyses of phenolic compounds by capillary electrophoresis electrospray mass spectrometry. *Rapid Commun. Mass Spectrom.*, *13* (7), 56–567.

Lagoudianaki, E., Manios, T., Geniatakis, M., Frantzeskaki, N., & Manios, V. (2003). Odor control in evaporation ponds treating olive wastewater through the use of $Ca(OH)_2$. *J. Environ. Sci. Health Part A: Toxic Hazard. Subst. Environ. Eng.*, *38* (11), 2537–2547.

Lallai, A., & Mura, G. (1996). Low load anaerobic filters for the treatment of olive oil mill effluent. *Ing. Ambientale*, *25* (9), 485–491.

Lallai, A., Mura, G., Palmas, S., Polcaro, A.M., & Baraccani, L. (2003). Degradation of para-hydroxybenzoic acid by means of mixed microbial cultures. *Environ. Sci. and Pollution Res. Int.*, *10* (4), 221–224.

Lanzani, A., & Fedeli E. (1988). Composizione ed utilizzazione delle acque di vegetazione. "Composition and utilization of olive-mill wastewaters". *Tavola Rotonda: Lo smaltimento delle acque reflue del frantoi* (ed. Accademia Nazionale dell' Olivo), Spoleto, Italy: Arti grafiche Panetto and Petrelli, 15–28, 29 Apr. 1988 (in Italian).

Lanzani, A., Bondioli P., Fedeli E., Ponzetti, A., & Pieralisi, G. (1988). Un processo per ilo smaltimento tegrale delle acque di vegetazione con contemporanea valorizzazione delle sanse nella lavorazione delle olive. "A technology for the elimination of waste waters and contemporaneous enrichment of olive husks after processing". *Riv. Ital. Sostanze Grasse*, *65* (LXV), 117 (in Italian).

Lanzani, A., Ruffo, C., Cozzoli O., Arpino, A., & Fedeli, E. (1983). Investigation on industrial waste waters. I: Relation between production cycles, pollution and energy savings in the fats and oils industry. *Riv. Ital. Sostanze Grasse*, *60* (LX) (5), 281–288.

Lasaridi, K.E., Papadimitriou, E.K., & Balis, C. (1996). Development and demonstration of a thermogradient respirometer. *Compost Sci. Util.*, *4* (3), 53–61.

Le Tutour, B., & Guedon, D. (1992). Antioxidative activities of *Olea europaea* leaves, & related phenolic compounds. *Phytochem.*, *31* (4), 1173–1178.

Le Verge, S. (2004). La fertilization à partir des résidues de la trituration des olives. *Le Nouvel Olivier (OCL)*, *42*, Nov./Dec. 2004, 5–21.

Le Verge, S. (2005). La fertilization à partir des résidues de la trituration des olives. *Le Nouvel Olivier (OCL)*, *43*, Jan./Feb. 2005, 9–13.

Le Verge, S., & Bories, A. (2004). Les basins d'évaporation naturelle des margines. *Le Nouvel Olivier (OCL)*, *41*, Sept./Oct. 2004, 5–10.

Leandri, A., Pomp V., Pucci, C., & Spanedda, A.F. (1993). Residues on olives, oil and processing waste waters of pesticides used for the control of *Dacus oleae* (Gmelin) (Diptera: Tephritidae). *Anzeiger für Schädlingskunde Pflanzenschutz Umweltschutz*, *66* (3), 48–51.

Leger, C.L., Kadiri-Hassani, M., & Descomps, B. (2000). Decreased superoxide anion production in cultured human promonocyte cells (THP-1) due to polyphenol mixtures from olive oil processing wastewaters. *J. Agr. Food Chem.*, *48* (10), 5061–5067.

Leon-Cabello, R., & Fiestas Ros de Ursinos, J.A. (1981). Evaluación de alpechíns en balsas de evaporación. *Proc. Congreso sobre biotechnologias de bajo costo para la depuración de aguas residuales*, 18–20 November 1981, Madrid, FAO (in Spanish).

Leoni, C., Grischott, F., & Francini, G. (1980). Treatment of specific effluents from the food industry particularly high in organic matter. *Industria Conserve, 56* (3), 173–177.

Lesage-Meessen, L., Navarro, D., Maunier, S., Sigoillot, J.C., Lorquin, J., Delattre, M., Simon, J.L., Asther, M., & Labat, M. (2001). Simple phenolic content in olive oil residues as a function of extraction systems. *Food Chem., 75* (4), 501–507.

Levi-Minzi, R., Riffaldi, R., Saviozzi, A., & Cardelli, R. (1995). Decomposition in soil of anaerobically digested olive mill sludge. *J. Environ Sci. Health Part A, 30* (7), 1411–1422.

Levi-Minzi, R., Saviozzi, A., Riffaldi, R., & Falzo, L. (1992) L'épandage au champ des margines: Effets sur les propriétés du sol. *Olivae, 40*, 20–25 (in French).

Levinson, H., & Levinson, A. (1998). Control of stored food pests in the ancient Orient and classical antiquity. *J. Appl. Entomology, 122* (4), 137–144.

Liberti, L. (1988). I problemi della depurazione delle acque reflue da frantoi oleari. *Agric. ed Innovaz.* (5–6), 86–91 (in Italian).

Limiroli, R., Consonni, R., Ranalli, A., Bianchi, G., & Zetta, L. (1996). ^{1}H-NMR study of phenolics in the vegetation water of three cultivars of Olea europaea: similarities and differences. *J. Agric. Food Chem., 44* (8), 2040–2048.

Lolos, G., Skordilis, A., & Parissakis, G. (1994). Polluting characteristics and lime precipitation of olive mill wastewater. *J. Environ. Sci. Health Part A: Environmental Science and Engineering, 29* (7), 1349–1356.

Longhi, P., Vodopivec, B., & Fiori, G. (2001). Electrochemical treatment of olive oil mill wastewater. *Annali di Chimica, 91* (3–4), 169–174.

López, C.J. (1993). Evaluación de la experiencia de las plantas prototipo de depuración de alpechines en la cuenca del río Guadalquivir. *Proc. IX Congeso Nacional de Quimica*, Seville, Spain, *2*, 295–317 (in Spanish).

López, R., Martínez-Bordiú A., Dupuy de Lome, E., Cabrera, F., & Sánchez, M.C. (1996). Soil properties after application of olive oil wastewater. *Fresen. Environ. Bull.,* *5*, 49–54.

López-Aparicio, F.J., García-Granados López de Hierro, A., & Rodríguez, A.M. (1977). Estudio del contenido en ácidos carboxilicos del OMWW de la aceituna y evolutión de los mismos. *Grasas y Aceites, 28*, 393–401 (in Spanish).

López López, M.J. (1996). Producción de biopolimeros a partir del alpechín: obtención de xantano. "Production of biopolymers from alpechín: Xanthan production". *Ph.D. Thesis*, University of Granada, Spain (in Spanish).

López López, M.J., & Ramos Cormenzana, A. (1995). Xanthan production from olive mill wastewaters. *Proc. Olive Oil Processes and By Products Recycling*; Granada, Spain, Sep 10–13, 1995. *Int. Biodeterior. Biodegrad.* 1996, *38* (3–4), 263–270.

López López, M.J., & Ramos Cormenzana, A. (1997). Producción de xantano a partir del alpechín. *Int. Biodeterior. Biodegrad., 39* (1), 85 (in Spanish).

López López, M.J., Moreno, J., & Ramos Cormenzana, A. (2001a). *Xanthomonas campestris* strain selection for xanthan production from olive mill wastewaters. *Wat. Res., 35* (7), 1828–1830.

López López, M.J., Moreno, J., & Ramos Cormenzana, A. (2001b). The effect of olive mil wastewaters variability on xanthan production. *J. Appl. Microbiol., 90* (5), 829–835.

Lo Scalzo, R., & Scarpati, M.L. (1993). A new secoiridoid from olive wastewaters. *J. Nat. Prod.*, *56* (4), 621–623.

Lucas, A. de, Rincón, J., & Gracia, I. (2002). Influence of operating variables on yield and quality parameters of olive husk oil extracted with supercritical carbon dioxide. *J. Am. Oil. Chem. Soc. (JAOCS)*, *79* (3), 237–243.

Lucas, A . de, Rincón, J., & Gracia, I. (2003). Influence of operation variables on quality parameters of olive- husk oil extracted with CO_2. Three-step sequential extraction. *J. Am. Oil Chem. Soc. (JAOCS)*, *80* (2), 181–188.

Lucas, A. de, Martínez de la Ossa, E., Rincón, J., Blanco, M.A., & Gracia, I. (2002). Supercritical fluid extraction of tocopherol concentrates from olive tree leaves. *J. Supercritical Fluids*, *22* (3), 221–228.

Maddox, I.S., & Murray, A.E. (1983). Production of n-butanol by fermentation of wood hydrolysate. *Biotechnol. Letters*, *5*, 175–178.

Madejon, E., Galli, E., & Tomati, U. (1998). Bioremediation of olive mill pomaces for agricultural purposes. *Fresen. Environ. Bull.*, *7*, 873–879.

Madrid L., & Díaz-Barrientos, E. (1994a). Nature of the action of olive mill waste-water on the mobility of heavy-metals added to a soil. *Fresen. Environ. Bull.*, *3* (4), 226–231.

Madrid, L., & Díaz-Barrientos, E. (1994b). Retention of heavy-metals by soils in the presence of a residue from the olive-oil industry. *Eur. J. Soil Sci.*, *45* (1) 71–77.

Madrid, L., & Díaz-Barrientos, E. (1996). Nature of the action of a compost from olive mill wastewater on Cu sorption by soils. *Toxicol. Environ. Chem.*, *54*, 93–98.

Madrid, L., & Díaz-Barrientos, E. (1997). Modelling the mobilizing effect of olive mill wastewater on heavy metals adsorbed by a soil. *Dev. Plant. Soil. Sci*, *71* (Modern Agriclulture and the Environment), 443–447.

Madrid, L., & Díaz-Barrientos, E. (1998a). Mobility of cadmium added to a soil treated with agricultural residues. *Fresen. Environ. Bull.*, *7* (12A), 849–858 Sp. Iss.

Madrid, L., & Díaz-Barrientos, E. (1998b). Release of metals from homogeneous soil columns by wastewater from an agricultural industry. *Environ. Pollut.*, *101* (1), 43–48.

Madrid, L., López Nunez, R., & Díaz-Barrientos, E. (1993). Efecto de la adición de alpechín sobre las caracteristicas del cambio iónico K/(Ca+Mg) en un suelo montmorillonítico. *Proc. IX Congreso nacional de Quimica*, Seville, Spain, *1*, 27–34 (in Spanish).

Maestro-Durán, R. (1989). Relation between the composition and ripening of the olive and quality of the oil. *Acta Horiculturae*, *286*, *Proc. Int. Symp. on olive growing* (ed. Ralo, L.), University of Cordoba, Spain.

Maestro-Durán, R., Borja-Padilla, R., Martín-Martín, A., Fiestas de Ursinos, J.A., & Alba-Mendoza, J. (1991). Biodegradación de los compuestos fenólicos presentes en al alpechín. "Biodegradation of phenolic compounds in olive oil mill wastewater". *Grasas y Aceites*, *42* (4), 271–276 (in Spanish).

Mahmoud, A.L. (1994). Antifungal acion and antiaflatoxigenic properties of some essential oil constituents. *Lett. Appl. Microbiol.*, *19*, 110–113.

Mameri, N., Aioueche, F., Belhocine, D., Grib, H., Lounici, H., Piron, D. L., & Yahiat, Y. (2000a). Preparation of activated carbon from olive solid residue. *J. Chem. Technol. Biotechnol.*, *75* (7), 625–631.

Mameri, N., Halet, F., Drouiche, M., Grib, H., Lounici, H., Pauss, A., Piron, D.L., & Belhocine, D. (2000b). Treatment of olive mill washing water by ultrafiltration. *Canadian, J. of Chem. Eng.*, *78* (3), 590–595.

Mancini, I.M., & Boari, G. (1991). Experiments conducted in Puglia on combined biological treatment of municipal sewage, and wastewater from olive oil extraction. *Ist. Ric. Acque, 94*, 9.1–9.11.

Manna, C., Galletti, P., Cucciolla, V., Moltedo O., Leone, A., & Zappia, V. (1997). The protective effect of the olive oil polyphenol (3,4-dihydroxyphenyl) ethanol counteracts reactive oxygen metabolite-induced cytotoxicity in caco-2 cells. *J. Nutr., 127*, 286–292.

Mantzavinos, D., & Kalogerakis, N. (2005). Treatment of olive mill effluents: Part, I. Organic matter degradation by chemical and biological processes-an overview. *Environ. Int., 31* (2), 289–295.

Mantzavinos, D., Hellenbrand, R., Livingston, A.G., & Metcalfe, I.S. (1997a). Kinetics of wet oxidation of p-coumaric acid over a $CuO.ZnO-Al_2O_3$ catalyst. *Chem. Eng. Res. Des. Part A. Trans. Inst. Chem. Eng., 75* (A1), 87–91.

Mantzavinos, D., Hellenbrand, R., Metcalfe, I.S., & Livingston, A.G. (1996a). Partial wet oxidation of p-coumaric acid: Oxidation intermediates, reaction pathways and implications for wastewater treatment. *Water. Res., 30* (12), 2969–2976.

Mantzavinos, D., Lauer, E., Hellenbrand, R., Livingston, A.G., & Metcalfe, I.S. (1997b). Wet oxidation as a pretreatment method for wastewaters contaminated by bioresistant organics. *Wat. Sci. Technol, 36* (2–3), 109–116.

Mantzavinos, D., Hellenbrand, R., Livingston, A.G., et al. (1996b). Catalytic wet oxidation of p-coumaric acid: Partial oxidation intermediates, reaction pathways and catalyst leaching. *Appl. Catal. B-Environ., 7* (3–4), 379–396.

Mantzavinos, D., Sahibzade, M., Livingston, A.G., et al. (1999). Wastewater treatment: Wet air oxidation as a precursor to biological treatment. *Catal. Today, 53* (1), 93–106.

Marco, C. de, Simone, C. de, D'Ambrosio, M., & Owczarek, M. (1998). Factors affecting the induction of toxic and genotoxic effects in Vicia faba seedlings by olive mill wastewaters. *1st Intern. Workshop on Environmental quality and Environmental Engineering in the Middle East Region*, Konia, Turkey, 5–7 Oct. 1998.

Mari, I., Ehaliotis, C., Kotsou, M., Balis, C., & Georgacakis, D. (2003). Respiration profiles in monitoring the composting of by-products from the olive agro-industry. *Bioresource Technol., 87* (3), 331–336.

Marignetti, N., Canepa, P., & Gagliardi, A. (1986). Retention of polyphenols on porous polymers. *Tec. Chimiche, 6*, 40–43.

Marinos, E. (1991). Lagooning concentration of olive oil processing wastewaters. *Proc. Int. Conf. on Olive Oil Processing Wastewater Treatment Methods*, Hania, Crete, Greece, 1991.

Marques, I.P. (2001). Anaerobic digestion treatment of olive mill wastewater for effluent reuse in irrigation. *Desalination, 137* (1–3), 233–239.

Marques, I.P., Teixeira, A., Rodrigues, L., Martins Dias, S., & Novais, J.M. (1997). Anaerobic co-treatment of olive mill and piggery effluents. *Environ. Technol., 18* (3), 265–274.

Marques, I.P., Teixeira, A., Rodrigues, L., Martins Dias, S., & Novais, J.M. (1998). Anaerobic treatment of olive mill wastewater with digested piggery effluent. *Water Environ. Res., 70* (5), 1056–1061.

Marques, P.A.S.S., Rosa, M.F., Mendes, F., Collares Pereira, M., Blanco, J., & Malato, S. (1996). Wastewater detoxification of organic and inorganic toxic compounds with solar collectors. *Desalination, 108*, 213–220.

Marsilio, V., Lanza, B., & Pozzi, N. (1996). Progress in table olive debittering: Degradation in vitro of oleuropein and its derivatives by *Lactobacillus plantarum*. *J. Am. Oil Chem. Soc. (JAOCS)*, *73* (5), 1996.

Marsilio, V., Di Giovacchino, L., Solinas, M., Lombardo, N., & Briccoli-Bati, C. (1989). First observations on the disposal effects on olive oil mills vegetation waters on cultivated soil. *Proc. Int. Symp. Olive Growing*, Córdoba, Spain, 26–29 September 1989 (eds. Ralio, L., Caballero, J.M., & Fernández-Escobar, R.). *Acta Horticulturae*, 1990, *286*, 493–496.

Martilotti, F. (1983). Use of olive by-products in animal feeding in Italy. *Animal production and Health Division*, FAO, Rome, 1983.

Martín-Martín, A., Borja-Padilla, R., & Banks, C.J. (1994). Kinetic model for substrate utilization and methane production during the anaerobic digestion of olive mill wastewater and condensation wastewater. *J. Chem. Technol. Biotechnol.*, *60* (1), 7–16.

Martín-Martín, A., Borja-Padilla, R., & Chica, A. (1993). Kinetic study of an anaerobic fluidized bed system used for the purification of fermented olive mill wastewater. *J. Chem. Technol. Biotechnol.*, *56* (2), 155–162.

Martín-Martín, A., Borja-Padilla, R., García, I., & Fiestas Ros de Ursinos, J.A. (1991). Kinetics of methane production from olive mill wastewater. *Process Biochem.*, *26* (2), 101–107.

Martín-Martín, A., Borja-Padilla, R., Maestro-Durán, R., Alba-Mendoza, J., & Fiestas Ros de Ursinos, J.A. (1990). Influencia de la concentración de polifenoles sobre la cinética del proceso de depuración anaerobia del alpechín. *I Congreso Internacional de Qufmica de la ANQUE*. Puerto de la Cruz, Tenerife, Spain (in Spanish).

Martin, C.J. (1992). Remediation of groundwater contaminated by olive processing waste. *Desalination*, *88* (1–3), 253–263.

Martín-Olmedo, P., Cabrera, F., López, R., & Murillo, J.M. (1996). Residual effect of composted olive oil sludge on plant growth. *Atti VII Congresso Internazionale "L'approccio integrato della moderna biologia: uomo, territorio, ambiente". Mediterranean Conf. on Organic Wastes Recycling in Soils*, Ordina Nazionale del Biologi (ed.), 1199–1207.

Martín-Olmedo, P., López, R., Cabrera, F., et al. (1995). Nitrogen mineralization in soils amended with organic by-products of olive oil and sugar-beet processing industries. *Fresen. Environ. Bull.*, *4* (1) 59–64.

Martinengri, G.B. (1963). Technologia Chimica Industriale degli Oli, Grassi e Derivat. 3rd Voepli-Milano (in Italian).

Martínez, J., Pérez J., Moreno, E., & Ramos-Cormenzana, A. (1986). Incidencia del efecto antimicrobiano del alpechín en su posible aprovechamiento. *Grasas y Aceites*, *37* (4), 215–223 (in Spanish).

Martínez-Nieto, L., & Garrido-Hoyos, S.E. (1994). El alpechín, un problema medioambiental en vias de solución. *Química e Industria*, *41*, 17–27 (in Spanish).

Martínez-Nieto, L., Ramos-Cormenzana, A., García-Pareja, M.P., & Garrido-Hoyos, S.E. (1992). Biodegradación de compuestos fenólicos del alpechín con *Aspergillus terreus*. *Grasa y Aceites*, *43*, 75–81 (in Spanish).

Martínez-Nieto, L., Garrido-Hoyos, S.E., Camacho Rubio, F., García-Pareja, M.P., & Ramos-Cormenzana, A. (1993). Biological purification of waste products from olive extraction. *Bioresourc. Technol.*, *43* (3), 215–219.

Martínez-Toledo, M.V., González-López, J., Rodelas, B., Pozo, C., & Salmeron, V. (1995). Production of poly-β-hydroxybutyrate by *Azotobacter chroococcum* H23 in chemically-defined medium and alpechin medium. *J. Appl. Bacteriol.*, *78* (4), 413–418.

Martirani, L., Giardina, P., Marzullo, L., & Sannia, G. (1996). Reduction of phenol content and toxicity in olive oil mill waste waters with the lignolytic fungus *Pleurotus ostreatus*. *Water. Res., 30* (8), 1914–1918.

Mascolo, A., Cucurachi, A., Di Giovacchino L., & Ranalli, A. (1990). Disposal of waste water from olive oil production, by means of activated sludge plants for treatment of urban sewage. *Annali dell'Istituto Sperimentale per la Elaiotecnica, 9*-1981-1983, pub. 1990 (in Italian).

Massari, S., & Russo, G. (1999). Olive oil waste waters: Problems concerning their reuse and discharge with particular interest in the area of Lecce and Brindisi. *Ann. Cim., 89* (7–8), 515–522.

Massignan, L., Leo, P. de, & Carrieri, C. (1985). Depurazione mediante osmosi inversa di acque de oleifici. "Purification of wastewater from olive oil factories using reverse osmosis" *Riv. Soc. Ital. Sci. Aliment.*, anno *14* (6), 421–428 (in Italian).

Massignan, L., Leo, P. de, Traversi, D., & Aveni, A. (1988). Use of membrane technology for processing wastewater from olive-oil plants with recovery of useful components. *Proc. Symp. Future Industrial Processes of Membrane Processes* (eds. Cecille, L., & Toussaint, J.C.), Commission of the European Communities, Brussels, Belgium, 179–189, Elsevier Applied Science.

Maymone, B., Battaglini, A., & Tiberio, M. (1961a). Richerche sul valore nutritivo della sansa di olive. *Alimentazione Animale, 5* (4), 219–250 (in Italian).

Maymone, B., Battaglini, A., & Tiberio, M. (1961b). Richerche sul valore nutritivo della sansa d'olive. *Ann., Ist. Sper. Zootec., 2*, 219–231 (in Italian).

Mazzanti, U. (1998). Ricerca sul sistema di trattamento della SAEM per reflui di frantoi oleari. *Convegno nazionale "Verdeoliva"-proposte per la depurazione dei reflui da frantoi oleari*, Rome, 20 Mar. 1988 (in Italian).

Mazzotti, M. (2004). Enlightered mills. Mechanizing olive oil production in Mediterranean Europe. *Technology and Culture, 45* (2), 277–304.

Mechichi, T., Labat, M., Woo, T.H.S., Thomas, P., García, J.L., & Patel Bharat, K.C. (1998). *Eubacterium aggregans* sp. *nov.*, a new homoacetogenic bacterium from olive mill wastewater treatment digester. *Anaerobe, 4* (6), 283–291.

Medici, F., Merli, C., & Spagnoli, E. (1985). Anaerobic digestion of olive oil mill wastewaters: A new process. Anaerobic digestion and carbohydrates hydrolysis of waste. *Proc. 4th Int. Symp. on Anaerobic Digestion*, 385–398 (eds. Ferrero, M.P., Naveau H.), Elsevier Publ. London, 1985. Publisher Shanghai Bioenergy Eng. Corp., China.

Mekki, H., Ammar, E., Anderson, M., & Ben Zina, M. (2003). Recyclage des déchets de la trituration des olives dans les briques de construction. "The recyling of olive oil mill by-products in bricks". *Ann. Chim. Sci. Mater., 28* (1), 109–127 (in French).

Mellouli, H.J. (1996). Modification des charactéristiques physiques d'un sable limoneux par les effluents (les margnies) des moulins à huile d' olive: incidente sur l'evaporation. *Ph. D. Thesis*, Universtity Gent, Belgium, 225 pages (in French).

Mellouli, H.J., Hartmann, R., Gabriels, D., & Cornelis, W.M. (1998). The use of olive mill effluents ("margines") as soil conditioner mulch to reduce evaporation losses. *Soil and Tillage Res., 49*, 85–91.

Mellouli, H.J., Wescmacl van, B., Pocscn, J., & Hartmann, R. (2000). Evaporation losses from bare soils as influenced by cultivation techniques in semi-arid regions. *Agric. Water Manage., 42* (3), 355–369.

Mendia, L., & Procino, L. (1964). Studio sul trattamento delle acque di rifuto dei frantoi oleari. *Proc. ANDIS Conf.*, Bologna, Italy (in Italian).

Mendia, L., Carbone, P., D'Antonio, G., & Mendia, M. (1985). Treatment of olive oil wastewaters. *Pollut. of the Mediterr. Sea, Proc. IAWPRC Int. Reg. Conf.*, Split. Yugoslavia, 2–5 Oct. 1985. *Water Sci. Technol.*, 1986, *18* (9), 125–136.

Mensi, R., & Kallel, A. (1990). Utilisation de la margine dans la construction des routes et des pistes agricoles. *Annales de l'equipment*, I, *1*, 61–68 ENIT, Tunisia (in French).

Mercadè, M.E. (1990). *Ph.D. Thesis.* Universidad de Barcelona (in Spanish).

Mercadè, M.E., & Manresa, M.A. (1994). The use of agroinustrial by-products for biosurfactant production. *J. Am. Oil Cem. Soc. (JAOCS)*, *71* (1), 61–64.

Mercadè M.E., Manresa, M.A., Robert, M., Espuny, M. J., Andres, C. de and Guinea, J. (1993). Olive oil mill effluent. New substrate for biosurfactant production. *Bioresource Technol.: Biomass Bioenergy Biowastes Convers. Technol. Biotransform. Prod. Technol.*, *43* (1), 1–6.

Mercer, W.A., Ralls, J.W., & Maagdenberg, H.J. (1971). Reconditioning food processing brines with activated carbon. *Chem. Eng. Progr. Symp Ser.*, *67* (107), 435–438.

Michelakis, N. (1991). Olive processing wastewaters manangement. *Proc. Int. Conf. on Olive Oil Processing Wastewater Treatment Methods.* Hania, Crete, Greece, 1991.

Michelakis, N. (1997). Vegeration water management by evaporation basins. *Proc. Seminar on Olive Growing.* Hania, Crete, Greece, 1997, 97–103.

Michelakis, N., Klapaki, G., & Kasapakis, G. (1999). Olive vegetation water management by evaporation basins. *Acta Horticulturae*, *474*, 757–760.

Millan, B., Lucas, R., Robles, A., García, T., Alvarez de Cienfuegos, G., & Galvez, A. (2000). A study on the microbiotica from olive-mill wastewater (OMW) disposal lagoons, with emphasis on filamentous fungi and their biodegradative potential. *Microbiol. Res.*, *155* (3), 143–147.

Miocic, S., & Milic, I. (2003). *IMPEL Olive Oil Project Report - CMA & NOA.* Number Report 2003/3, Project Manager: Méndez Miguel, Rome, Nov. 2003; number of pages, report: 33 and annexes: 61. European Union Network for the Implementation and Enforcement of Environmental Law (IMPEL) (http://europa.eu.int/comm/environment/impel).

Miranda, M.A., Amat, A.M., & Arques, A. (2001). Abatement of the major contaminants present in olive industry wastewaters by different oxidation methods: Ozone and/or UV radiation versus solar light. *Water Sci. and Technol.*, *44* (5), 325–330.

Miranda, M.A., Galindo, F., Amat, A.M., & Arques, A. (2000). Pyrylium salt-photosensitized degradation of phenolic contaminants derived from cinnamic acid with solar light. Correlation of the observed reactivities with fluorescence quenching. *Appl. Catal. B: Environ.*, *28* (2), 127–133.

Miranda, M.A., Galindo, F., Amat, A.M., & Arques, A. (2001). Pyrylium salt-photosensitized degradation of phenolic contaminants resent in olive oil wastewaters with solar light: Part II. Benzoic acid derivatives. *Appl. Catal. B: Environ.*, *30* (3–4), 437–444.

Mitrakas, M., Papageorgiou, G., Docoslis, A., & Sakellaropoulos, G. (1996). Evaluation of various pretreatment methods for olive oil mill wastewaters. *Eur. Water Pollut. Control.*, *6* (6), 10–16.

Molina Alcaide, E., & Nefzaoui, A. (1995). Recycling of olive oil by products: Possibilities of utilization in animal nutrition. *Proc. Olive Oil Processes and By-Products Recycling*, Granada, Spain, 10–13 Sep. 1995. *Int. Biodeterior. Biodegrad.* 1996, *38* (3–4), 227–235.

Molina Alcaide, E., Yáñez Ruiz, D., Moumen, A., & Martín García, I. (2003). Chemical composition and nitrogen availability for goats and sheep of some olive by-products. *Small Ruminant Res.*, *49* (3), 329–336.

Molinari, R., & Drioli, E. (1988). Processi integrati di ultrafiltrazione e osmosi inversa nel trattamento delle acque reflue da frantoi oleari. *Acqua-Aria, 5*, 579–588 (in Italian).

Moliner, R., Lazaro, M.J., Suelves, I., & Blesa, M.J. (2004). Valorization of selected biomass and wastes by co-pyrolysis with coal. *Int. J. Power Energy Syst.*; *Co-Utilization of Domestic Fuels, 24* (3), 186–193.

Monpezat, G. de, & Denis, J.F. (1999). Fertilization des sols méditteranées avec des issues oléicoles. *Le Nouvel Olivier (OCL), 6* (1), Jan./Feb. 1999, 63–68 (in French).

Montané, D., Salvadó, J., Torras, C., & Farriol, X. (2002). High-temperature dilute-acid hydrolysis of olive stones for furfural production. *Biomass & Bioenergy, 22* (4), 295–304.

Montedoro, G.F., Baldioli, M., & Servili, M. (2001). *Olivo Olio, 4*, 28 (in Italian).

Montedoro, G.F., Bertuccioli, M., & Petruccioli, G. (1975). Effects of enzyme and detaining treatment of olive pastes for single pressure extraction on oil yield, extraction rate and analytical characteristics of oils, fruit waters and waste waters. *Riv. Ital. Sostanze Grasse, 52* (LII) (8), 255–265.

Montedoro, G.F., Petruccioli, G., & Parlati, M.V. (1986). Interventi chimici e fisici sulle acque di vegetazione ed abbattimento parziale del loro tasso inquinamento. *Proc. Round Table on Olive Mill Effluents Disposal.* Spoleto, Italy, 10 Nov. 1986, 29–52 (in Italian).

Montedoro, G.F., Begliomini, A.L., Servili, M., Petruccioli, M., & Federici, F. (1993). Pectinase production from olive vegetation waters and its use in the mechanical olive oil extraction process to increase oil yield and improve quality. *Italian J. Food Sci., 5*, 4, 355–362.

Montemurro, F., Convertini, G., & Ferri, D. (2004). Mill wastewater and olive pomace compost as amendments for rye-grass. *Agronomie 24*, 481–486.

Monteoliva-Sánchez, M., Incerti, C., Ramos Cormenzana, A., Paredes, C., Roig, A., & Cegarra, J. (1996). The study of the aerobic bacterial microbiota and the biotoxicity in various samples of olive mill wastewaters (alpechín) during their composting process. *Proc. Olive Oil Processes and By Products Recycling*; Granada, Spain, 10–13 Sep. 1995. *Int. Biodeterior. Biodegrad.* 1996, *38* (3–4), 211–214.

Montero, M. (1989). *Jornadas sobre innovación tecnológica medio ambiente y desarrollo*, Seville, Spain (in Spanish).

Morelli, A., Rindone B., Andreoni, V., Villa, M., Sorlini, C., & Balice, V. (1990). Fatty acids monitoring in the anaerobic depuration of olive oil mill wastewater. *Biol. Wastes, 32* (4), 253–263.

Moreno, E., Quevedo-Sarmiento, J., & Ramos-Cormenzana, A. (1990). Antibacterial activity of washwater from olive mills. *Encyclopaedie of Environmental Control Technology, 3* (Chapter 26), 731–736 (ed. Cheremisinoff, P.N.), Gulf Publications, USA.

Moreno, E., Pérez J., Ramos-Cormenzana, A., & Martínez, J. (1986). Nutritional analyses of microorganisms isolated from enriched cultures taken from diluted vegetation water and soil: Preliminary results. *Proc. Int. Symp. on Olive By-Products Valorization*, FAO, UNDP (Food and Agriculture Organization of the United Nations), Seville, Spain, 5–7 Mar. 1986.

Moreno, E., Pérez J, Ramos-Cormenzana, A., & Martínez, J. (1987). Antibacterial effect of waste-water from olive oil extraction plants selecting soil bacteria after incubation with diluted waste. *Microbios, 51* (208–209), 169–174.

Moreno, J., González-López, J., Martínez, M.V., Rubia, T. de la, Ramos-Cormenzana, A., & Vela, R.G. (1990). Growth and nitrogenase activity of *Azotobacter vinelandii* in presence of several phenolic acids. *J. Appl. Bacteriol., 69*, 850–855.

Moreno-Castilla, C., Carrasco-Marín, López-Ramón, M.V., & Alvarez-Merino, M.A. (2001). Chemical and physical activation of olive-mill waste water to produce activated carbons. *Carbon, 39* (9), 1415–1420.

Morisot, A. (1979). Utilisation des margines per épandage. *L'Olivier, 19* (1), 8–13 (in French).

Morisot, A., & Tournier, J.P. (1986). Répercussions agronimiques de l' épandage d'effluents et déchets de moulins à huile d' olive. "Agronomic effects of land disposal of waste from olive oil crushers". *Agronomie, 6* (3), 235–241 (in French).

Mosca, L., Marco, C. de, Visioli, F., & Cannella, C. (2000). Enzymatic assay for the determination of olive oil polyphenol content: Assay conditions and validation of the method. *J. Agric. Food Chem., 48*, 297–301.

Mouncif, M., Faid, M., Achkari-Begdouri, A., & Lhadi, R. (1995). A biotechnological valorization and treatment of olive mill waste waters by selected yeast strains. *Grasas y Aceites, 46* (6), 344–348.

Mousa, L., Forster, C.F. (1999). The use of glucose as a growth factor to counteract inhibition in anaerobic digestion. *Process SAF Environ., 77* (B4), 193–198.

Muezzinoglu, A., & Uslu, O. (1986). Alleviation of pollution due to olive oil production. Some practical considerations from Turkey. *Proc. Int. Symp. on Olive By-Products Valorization*, FAO, UNDP (Food and Agriculture Organization of the United Nations), Seville, Spain, 5–7 Mar. 1986, 159–168.

Mulinacci, N., Romani, A., Galardi, C., Pinelelli, P., Giaccherini, C., & Vincieri, F.F. (2001). Polyphenolic content in olive oil waste waters and related olive samples. *J. Agric. Food Chem., 49* (8), 3509–3514.

Murillo, J.M., López, R., Fernández, J.E., & Cabrera, F. (2000). Olive tree response to irrigation with wastewater from the table olive industry. *Irrigation Sci., 19* (4), 175–180.

Nafaâ, A., & Lotfi, M. (2004). Decolourization and removal of phenolic compounds from olive mill wastewater by electrocoagulation. *Chem. Eng. Proc., 43* (10), 1281–1287.

Nefzaoui, A. (1983). Etude de l'utilisation des sous-produits de l'ólivier en alimentation animale en Tunisie. *Animal Production and Health Division*. FAO, Rome (in French).

Nefzaoui, A. (1987). Improving the profitability of olive growing by increasing the value of by-products. *Séminaire CEE-CIHEAM sur l'economie de l'olivier*, Tunisia, 20–22, 1987.

Nefzaoui, A. (1991). Valeur nutritive des ensilages combinés de fientes de volailles et de grignos d'olive. 2. Quantités ingérées, digestibilités, rétentions azotées et transit des particules chez les ovins. *Ann. Zootech., 40*, 133–123.

Nefzaoui, A., Molina, E., Outami, A., & Vanbelle, M. (1985). *Archivos de Zootenica, 33* (127), 219–236 (in French).

Negro Alvarez, M. J., & Solano, M.L. (1996). Laboratory composting assays of the solid residue resulting from the flocculation of oil mill wastewater with different lignocellulosic residues. *Compost Sci. Util., 4* (4), 62–71.

Netti, S., Wlassics, I. (1995). Studio sulle metodologie di smaltimento delle acque di vegetazione. *Riv. Ital. Sostanze Grasse, 72* (LXXII), 119–125 (in Italian).

Nicoletti, G. (1999). Bagazo de oliva. Parte 3: Aspectos tecnicos y economicos de su uso en el cocido de ladrillos. "Olive bagasse. Part 3: Technical and economic aspects of its use in brick baking". *Inf. Tecnol., 10* (2), 29– 40 (in Spanish).

Nogales, R., Thompson, R., Calmet, A., Benítez, E., Gómez, M., & Elvira, C. (1998). Feasibility of microcomposting residues from olive oil production obtained using two-stage centrifugation. *J. Environ. Sci. and Health, Part A: Toxic Hazard. Subst. Environ. Eng., A33* (7), 1491–1506.

Nogales, R., Gallardo-Lara, F., Benítez, E., Soto, J., Hervas, D., & Polo, A. (1997). Metal extractability and availability in soil after heavy application of either nickel or lead in different forms. *Water Air Soil Pollut.*, *94* (1–2), 33–44.

Nogales, R., Melgar, R., Guerrero, A., Lozada, G., Benitéz, E., Thompson, R., & Gómez, M. (1999). Growth and reproduction of *Eisenia andrei* in dry olive cake mixed with other organic wastes. *Pedobiologia*, *43* (6), 744–752.

Nosti Vega, M., Castro Ramos, R. de, & Vázquez Ladron, R. (1982). Composition and nutritive value of some Spanish var. of green olives. V. Influence of the reuse of lyes and omitting the washing in the pickling of green olives. *Grasas y Aceites*, *33* (1), 5–8.

Nosti Vega, M., Vázquez Ladron, R., & Castro Ramos, R. de (1992). Utilization possibilities of by-products from the table olive industry. *Grasas y Aceites*, *33* (3), 135–139.

Notarnicola, L. (1976). Industria Oleicola. Problemi e prospettive della depurazione e della utilizzazione degli effluenti. *Inquinamento*, *5*, 31 (in Italian).

Novak, J.T., & Carlson, D.A. (1970). The kinetics of anaerobic long chain fatty acid degradation. *J. Water Pollut. Control Fed.*, *42*, 1932–1943.

Ntougias, S., & Russell, N.J. (2000). *Bacillus* sp. WW3-SN6, a novel facultatively alkaliphilic bacterium isolated from the wash-waters of edible olives. *Extremophiles*, *4* (4), 201–208.

Ntougias, S., & Russell, N.J. (2001). Alkalobacterium *olivoapovliticus* gen, nov., sp. nov., a new obligately alkaliphilic bacterium isolated from edible-olive wash-waters. *Int. J. Syst. Bacteriol. 51* (3), 1161–1170.

Nunes, J.M., Pereira, S., Albardeiro, A., Silva, C., López-Piñeiro, A., & Pintado, C. (2001). Potentialities of olive mill waste utilisation as organic fertilizer for Mediterranean region soils. R*evista de Ciencias Agrarias*, *24* (3–4), 166–175 (in Portugese).

Obied, H.K., Allen, M.S., Bedgood, D.R., Prenzler, P.D., Robards, K., & Stockmann, R. (2005). Bioactivity and analysis of biophenols recovered from olive mill waste. *J. Agric. Food Chem.*, *53* (4), 823–837.

Oliveira de, J.S. (1974). Effluents from olive oil extracting plants subsidies for the study of the obtention of sealable products with simultaneous elimination of BOD. *12th Int. Congress*, Athens, Greece, 1974.

Oliveira de, J.S., & Raimundo, M.C. (1976). Le traitment des effluents de l'industrie de l'extraction de l'huil d'olive. (*Université Nouvelle de Lisboa*), 31 (in French).

Olori, L. Fulvio, S. de, & Morgia, P. (1990). Acque di vegetazione. Le problematiche ambientali e gli aspetti igienico-sanitari. *Inquinamento*, *1*, 40–46 (in Italian).

Ortega Jurado, A., & Ramos Ayerbe, F. (1978a). Improved solvent extraction of olive foot cake. I. General considerations. *Grasas y Aceites*, *29* (1), 45–49.

Ortega Jurado, A., & Ramos Ayerbe, F. (1978b). Improved solvent extraction of olive foot cakes. II. Separator machines of the pulp-stone fractions. *Grasas y Aceites*, *29* (2), 147–157.

Oukili, O., Chaouch, M., Rafiq, M., Hadji, M., Hamdi F., & Benlemlih, M. (2001). Bleaching of olive mill wastewater by clay in the presence of hydrogen peroxide. *Ann. Chim. Sci. Mat.*, *26* (2), 45–53.

Owen, R.W., Giacosa, A., Hull, W.E., Haubner R., Spiegelhalder, B., & Bartsch, H. (2000). The antioxidant/anticancer potential of phenolic compounds isolated from olive oil. *Eur. J. Cancer*, *36* (10), 1235–1247.

Owen, R.W., Mier, W., Giacosa, A., Hull, W.E., Spiegelhalder, B., & Bartsch, H. (2000). Phenolic compounds and squalene in olive oils: The concentration and antioxidant potential of total phenols, simple phenols, secoiridoids, lignans, and squalene. *Food and Chem. Toxicol.*, *38*, 647–659.

Ozturk, I., Sakar, S., Ubay, G., & Erŏglu, V. (1992). Anaerobic treatment of olive mill effluents. *Proc. 46th Industrial Waste Conf.*, Istanbul, Turkey, 14–16 May 1991, 741–749. Lewis Publishers, Chelsea, Michigan, USA.

Pacifico, A. (1989). Il quadro tecnico-economico del problema. *Agric. et Innov.*, (11), 34–53.

Pacifico, S.M. (1989). Acque di vegetazione. *Agric. et Innov.* Notiziano dell' Enea Renagri, Sep. 11 1989, 33–37.

Pagliai, M., & Vittori Antisari, L. (1993). Influence of waste organic matter on soil micro- and macro-structure. *Bioresource Technol.*, *43*, 205.

Pagliai, M. (1996). Effetti della somministrazione di acque reflue di frantoi oleari sulle caratteristiche fisiche del suolo. *Atti. Sem. Int. su trattamento e riciclaggio in agricoltura dei sottoprodotti dell'industria oleari*, Lecce, Italy, 8–9 Mar. 1996 (in Italian).

Pagliai, M., Pellegrini, S., Vignozzi, N., Rapini, R., Mirabella, A., Piovanelli, C., Gamba, C., Miclaus, N., Castaldini, M., Simone, C. de, Pini, R., Pezzarossa, B., & Sparsoli, E. (2001). Influenza dei reflui oleari sulla qualità sel suolo. *L'Informatore Agrario*, suppl. 1, 50, 13–18 (in Italian).

Paixão, S.M., & Anselmo, A.M. (2002). Effect of olive mill wastewaters on the oxygen consumption by activated sludge microorganisms: An acute toxicity test method. *J. Appl Toxicol.*, *22* (3) 173–176.

Paixão, S.M., Mendonça, E., Picado, A., & Anselmo, A.M. (1999). Acute toxicity evaluation of olive oil mill wastewaters: A comparative study of three aquatic organisms. *Environ. Toxicol.*, *14* (2), 263–269.

Palladius Rutilius Taurus Aemilianus (4th century, A.D.*) The fourteen books of Palladius, Rutilius Taurus Æmilianus, On agriculture*. Translated by, T. Owen, London, printed for, J. White 1807. *De re rustica*. The Middle English translation of Palladius; edited with critical and explanatory notes by Mark Liddell Berlin, E. Ebering, 1896.

Palliotti, A., & Proietti, P. (1992). Ulteriori indagini sull'influenza delle acque reflue di frantoi oleari sull'olivo. *Informatore Agrario*, *39*, 72–76 (in Italian).

Panagou, E.Z., & Katsaboxakis, C.Z. (2006). Effect of different brining treatments on the fermentation of cv. Conservolea green olives processed by the Spanish-method. *Food Microbiol.*, *23* (2), 199–204.

Panelli, G. (2004). Valorizzazione agronomica dei residui. *OlivoeOlio*, 2/2004, 38–42 (in Italian).

Papadelli, M., Roussis, A., Papadopoulou, K., Venieraki, A., Chatzipavlidis, I., Katinakis, P., & Balis, C. (1995). Biochemical and molecular characterization of an *Azotobacter vinelandii* strain with respect to its ability to grow and fix nitrogen in olive mill wastewater; *Proc. Olive Oil Processes and By Products Recycling*; Granada, Spain, 10–13 Sep. 1995; *Int. Biodeterior. Biodegrad.*, 1996, *38* (3–4), 179–181.

Papadimitriou, E.K., & Balis, C. (1996). Comparative study of parameters to evaluate and monitor the rate of a composting process. *Compost. Sci. Util.*, *4* (4) 52–61.

Papadimitriou, E.K., Chatzipavlidis, I., & Balis, C. (1997). Application of composting to olive mill wastewater treatment. *Environ. Technol.*, *18* (1), 101–107.

Papafotiou, M., Phsyhalou, M., Kargas, G., Chatzipavlidis I., & Chronopoulos, J. (2004). Olive-mill wastes compost as growing medium component for the production of poinsettia. *Scientia Horticulturae*, *102* (2), 167–175.

Papaioannou, D. (1988). A method of processing waste gases from the drying of olive press-cake. *Biol. Wastes*, *24* (2), 137–145.

Papamichael-Balatsoura, V.M., & Balatsouras, G.D. (1988). Utilization of modified spent lye as cover brine of Conservolea olives subjected to fermentation as green of Spanish-style. *Grasas y Aceites, 39* (1), 17–21.

Paredes, C. (1998). Compostaje del alpechín. Una solucion agricola para la reduccion de su impacto ambiental. *CEBAS- CSIC.* ISBN 84-00-07721-1 (in Spanish).

Paredes, C., Roig, A., & Bernal, M.P. (2000). Evolution of organic matter and nitrogen during co-composting of olive mill wastewater with solid organic wastes. *Biology and Fertility of Solids, 32* (3), 222–227.

Paredes, C., Bernal, M.P., Cegarra, J., & Roig, A. (2002). Biodegradation of olive mill wastewater sludge by its co composting with agricultural wastes. *Bioresource Technol., 85* (1), 1–8.

Paredes, C., Bernal, M.P., Roig, A., & Cegarra, J. (2001). Effects of olive mill wastewater addition in composting of agroindustrial and urban wastes. *Biodegradation, 12* (4), 225–234.

Paredes, C., Cegarra, J., Bernal, M.P., & Roig, A. (2005). Influence of olive mill wastewater in composting and impact of the compost on a Swiss chard crop and soil properties. *Environ. Int., 31* (2), 305–312.

Paredes, C., Cegarra, J., Roig, A., & Bernal, M.P. (1999a). Composting of a mixture of orange and cotton industrial wastes and the influence of adding olive-mill wastewater. *Organic Recovery & Biological Treatment* (ORBIT 99), Part 1, 147–154, 1999 (eds. Bidlingmaier, W., Bertoldi, M. de, Diaz, L.F., & Papadimitriou, E.K.). Rhombos Verlag, Berlin, Germany. ISBN 3-930894-20-3.

Paredes, C., Cegarra, J., Sánchez-Monedero, M.A., & Galli, E. (1996a). Composting of fresh and pond-stored olive-mill wastewater by the rutgers system. *The Science of Composting, 2,* 1266-1270 (eds. Bertoldi, M. de, Sequi, P., Lemmes B., & Papi, T.), Blackie Academic & Proffesional, Glasgow, UK. ISBN 0-7514-0383-0.

Paredes, C., Bernal, M.P., Roig, A., Cegarra. J., & Sánchez-Monedero, M.A. (1996b). Influence of the bulking agent on the degradation of olive mill wastewater sludge during composting. *Proc. Olive Oil Processes and By-Products Recycling*; Granada, Spain, Sep 10–13, 1995. *Int. Biodeterior. Biodegrad.* 1996, *38* (3–4), 205–210.

Paredes, C., Cegarra, J., Roig, A., Sánchez Monedero, M.A., & Bernal, M.P. (1999b). Characterization of olive mill wastewater (alpechín) and its sludge for agricultural purposes. *Bioresource Technol., 67* (2), 111–115.

Paredes, M.J., Moreno, E., Ramos-Cormenzana, A., & Martínez, J. (1987). Chraracteristics of soil after pollution with waste waters from olive oil extraction plants. *Chemosphere, 16,* 1557–1564.

Paredes, M.J., Monteoliva-Sánchez, M., Moreno, E., Pérez J., Ramos-Cormenzana, A., & Martínez, J. (1986). Effect of wastewaters from olive mill extraction plants on the bacterial population of soil. *Chemosphere, 15,* 659–664.

Pasetti L., Fiorelli, F., & Tomati, U. (1996). *Azotobacter vinelandii* biomass production from olive mill wastewaters for heavy metal recovery. *Proc. Olive Oil Processes and By-Products Recycling,* Granada, Spain, 10–13 Sep. 1995. *Int. Biodeterior. Biodegrad.,* 1996, *38* (3–4), 163–164.

Pérez, J., Ramos-Cormenzana, A., & Martínez, J. (1990). Bacteria degrading phenolic acids isolated on a polymeric phenolic pigment. *J. Appl. Bact., 69,* 38–42.

Pérez, J., Hernández, L.M., Ramos-Cormenzana, A., & Martínez, J. (1987). Characterizacion de fenoles del pigmento del alpechín y transformacion por *Phanerochaete chrysosporium. Grasas y Aceites, 38,* 367–371 (in Spanish).

Pérez, J., Rubia, T. de la, Hamman, O. Ben, & Martínez, J. (1998). *Phanerochaete flavido-alba* laccase induction and modification of manganese peroxidase isoenzyme pattern in decolorized olive oil mill wastewaters *Appl. Environ. Microbiol.*, *64* (7), 2726–2729.

Pérez, J., Rubia, T. de la, Moreno, J., & Martínez, J. (1992). Phenolic content and antibacterial activity of olive oil waste waters. *Environ. Toxicol. Chem.*, *11* (4), 489–495.

Pérez, D.J., & Gallardo-Lara, F. (1987). Effect of the application of wastewater from olive processing on soil nitrogen transformation. *Commun. Soil Sci. Plant Anal.*, *18*, (9), 1031–1039.

Pérez, D.J., & Gallardo-Lara, F. (1989). Sulfur transformation affected by the application of wastewater from olive processing on soil. *Commun. Soil Sci. Plant Anal.*, *20* (1–2), 75–84.

Pérez, D.J., Esteban, E., Gómez, M., & Gallardo-Lara, F. (1986). Effects of wastewater from olive processing on seed germination and early plant growth of different vegetable species. *J. Environ. Sci., Health Part B*, *21* (351), 349–357.

Pérez, D.J., Gallardo-Lara F., & Esteban, E. (1980). Aspetos a considerar en el empleo del alpechín como fertilizante. I. Evaluacion de su efecto fitotoxico inhibidor de la germinacion de semillas. *Cuad. Cienc. Biol.*, *1* (6–7), 59–67 (in Spanish).

Pérez-Torres, J. (1988). Transformacion microbiana de componentes aromaticos del alpechín. *Ph.D. Thesis*. University of Granada (in Spanish).

Perrone, S. (1983). Purification and sewage disposal of water from olive oil extraction. *Ind. Aliment.* (Pinerolo, Italy), *22* (5), 353–356.

Perrone, S. (1989). Lo smaltimento delle acque di vegetazione delle olive. *Inquinamento, 6*, 42–45 (in Italian).

Petarca, L., Vitolo, S., & Bresci, B. (1997). Pyrolysis of concentrated olive mill vegetation waters. *Proc. Int. Conf. on Biomass Gasif. Pyrolysis*, 1997, 374–381. (eds. Kaltschmitt, M., & Bridgwater, A.V.), CPL Press.

Petruccioli, M., Servili, M., Montedoro, G.F., & Federici, F. (1988). Development of a recycle procedure for the utilization of vegetation waters in the olive-oil extraction process. *Biotechnol. Letters*, *10* (1), 55–60.

Petroni, A., Blasevich, M., Salami, M., Papini, N., Montedoro, G.F., & Galli, C. (1995). Inhibition of platelet aggregation and eicosanoid production by phenolic components of olive oil. *Thrombosis Research*, *78*, 151–160.

Petroni, A., Blasevich, M., Salami, M., Papini, N., Montedoro, G.F., & Galli, C. (1997). Inhibition of leukocyte leukotriene B_4 production by an olive-oil-derived phenol identified by mass spectroscopy. *Thrombosis Research*, *87*, 315–322.

Petroni, A., Blasevich, M., Salami, M., Servili, M., Montedoro, G.F., & Galli, C. (1994). A phenolic antioxidant extracted from olive oil inhibits platelet aggregation and arachidonic acid metabolism in *vitro. World Rev. Nut. Diet*, *75*, 169–172.

Piacquadio, P., Stefano, G. de, & Sciancalepore, V. (1998). Quality of virgin oil extracted with the new centrifugation system using a two-phase decanter. *Fet/Lipid*, *100* (10), 472–474.

Picci, G., & Pera, A. (1993). Relazioni su un triennio di ricerche microbiologiche sullo spandimento delle acque di vegetazione dei frantoi oleari su terreno agrario. *Genio Rurale*, *5*, 72–77 (in Italian).

Pinelli, P., Galardi, C., Mulinacci, N., Vincieri, F.F., Tattini, M., & Romani, A. (2000). Quali-quantitiative analysis and antioxidant activity of different polyphenolic extracts from *Olea europea*, L. leaves, *J. Commodity Sci.*, *39* (1).

Pinto, G., Pollio, A., Previtera, L., Stanzione, M., & Temussi, F. (2003). Removal of low molecular weight phenols from olive oil mill wastewater using microalgae. *Biotechnol. Letters*, *25* (19), 1657–1659.

Piperidou, C.I., Chaidou, C.I., Stalikas, C.D., Soulti, K., Pilidis, G., & Balis, C. (2000). Bioremediation of olive oil mill waste water; chemical alterations induced by *Azotobacter vinelandii*. *J. Agric. Food Chem.*, *48* (5), 1941–1948.

Pizzaro-Camacho, D., Soca-Olazabal, N., Linan-Veganzones, M.J. (1999). Dewatering of olive oil mill wastes for combustion of the biomass. *Alimentacion Equipos y Tecnologia*, *18* (4), 143–147.

Pliny the Elder (Gaius Plinius Secundus, A.D. 23–79) "Natural History". Volumes I-X, books 1-37, translated by H. Rackham, Loeb Classical Library, Wm. Heinemann Ltd., London, 1971.

Polcaro, A.M., Mascia, M., Palmas, S., & Vacca, A. (2002). Electrochemical oxidation of p-hydroxybenzoic, & protocatechuic acids at a dimensional stable anode (DSA) in the presence of NaCl. *Annali di Chimica*, *92* (10), 1015–1023.

Pompei, C., & Codovilli, F. (1974). Risultati preliminari sul trattamento di depurazione delle acque di vegetazione delle olive per osmosi inversa. "Preliminary results concerning the depuration treatment of olive vegetation waters by reverse osmosis". *Scienza e Tecnologica degli Alimenti*, *4* (IV) (6), 363–364 (in Italian).

Pomponio, R., Gotti, R., Hudaib, M., & Cavrini, V. (2002). Analysis of phenolic acids by micellar electrokinetic chromatography: application to Echinacea purpurea plant extracts. *Chromatogr. A.*, 945 (1–2), 239–247.

Popov, I.N., & Lewin, G. (1996). Photochemiluminescent detection of antiradical activity. IV: Testing of lipid-soluble antioxidants, *J. Biochemical and Biophysical Methods*, *31*, 1–8.

Potenz, D., Righetti, E., & Volpicella, M. (1980). Effetto inquinante delle acque reflue della lavorazione delle olive su terreno agrario. Nota 3. Caratteristiche e qualita delle sostanze umiche, comportamento di alcune colure erbaces. *Inquinamento*, *22* (2), 65–68 (in Italian).

Potenz, D., Righetti, E., Bellettieri, A., Girardi, F., Antonacci, P., Calianno, L.A., & Pergolesi, G. (1985a). Evoluzione della fitotossicità in un terreno trattato con acque reflue di frantoi oleari. Nota 1. *Inquinamento*, *4*, 49–54 (in Italian).

Potenz, D., Righetti, E., Bellettieri, A., Girardi, F., Antonacci, P., Calianno, L.A., & Pergolesi, G. (1985b). Evoluzione della fitotossicità in un terreno trattato con acque reflue di frantoi oleari. Nota 2. *Inquinamento*, *5*, 49–55 (in Italian).

Potoglou, D., Kouzeli-Katsiri, A., & Haralambopoulos, D. (2003). Solar distillation of olive mill wastewater. *Renewable Energy*, *29* (4), 569–579.

Pozo, C., Martínez-Toledo, M.V., Rodelas, B., & González-López, J. (2002). Effects of culture conditions on the production of polyhydroxyalkanoates by *Azotobacter chroococcum* H23 in media containing a high concentration of alpechin (wastewater from olive oil mills) as primary carbon source. *J. Biotechnol.*, *97* (2), 125–131.

Puebla, M.A., Ruiz, C., Monteoliva-Sánchez, M., et al. (1994). Effect of phenolic-acids from olive oil mill waste-waters on the growth and sporulation of *Bacillus-megaterium ATCC-33085*. *Toxicol. Environ. Chem.* 42 (1–2), 87–92.

Poulios, I., & Kyriacou, G. (2002). Photocatalytic degradation of p-coumaric acid over TiO_2 suspensions. *Environ. Technol.*, *23* (2), 179–187.

Poulios, I., Makri, D., & Prohaska, X. (1999). Photocatalytic treatment of olive mill waste water: Oxidation of protocatechuic acid. *Global Nes: the Int. J.*, *1* (81), 55–62.

Priego-Capote, F., Ruiz-Jiménez, J., & Luque de Castro, M.D. (2004). Fast separation and determination of phenolic compounds by capillary electrophoresis-diode array detection: Application to the characterization of alperujo after ultrasound-assisted extraction. *J. Chromatogr.*, *1045* (1–2), 239–246.

Proietti, P., Cartechini, A., & Tombesi, A. (1988). Influenza delle acque reflue di frantoi oleari su olivi in vaso e in campo. *L'Inormatore Agrario, 45*, 87–91 (in Italian).

Proietti, P., Palliotti, A., Tombesi, A., & Cenci, G. (1995). Chemical and microbiological modification of two different cultivated soils induced by olive oil waste water administration. *Agr. Med., 125*, 160–171.

Psillakis, E., & Kalogerakis, N. (2001). Identification of volatile and semi-volatile compounds in olive-oil mill wastewater by headspace SPME and GC. *7th Inter. Conf. on Environ. Sci. Technol. (CEST2001)*, 453–459, Ermoupolis, Syros Island, Greece, Sep. 2001.

Püppinghaus, K. (1991). Treatment of highly organic-loaded olive oil wastewater for protein recovery. *Gewässerschutz, Wasser, Abwasser, 125* (Industrieabwasser Vermeide Vermindern, Behandeln), 564–567.

Ragazzi, E., & Veronese, G. (1967a). Ricerche sui constituenti idrosolubili delle olive. Nota I: Zuccheri e fenoli. *Ann. Chim., 57*, 1396–1397 (in Italian).

Ragazzi, E., & Veronese, G. (1967b). Ricerche sulle fenolissidasi e sul contenuto in o-difenoli delle olive. *Ann. Chim., 57*, 1456–1492 (in Italian).

Ragazzi, E., & Veronese, G. (1973). Quantitative analysis of phenolic compounds after thin-layer chromatographic separation. *J. Chromatogr., 77*, 369–375.

Ragazzi, E., & Veronese, G. (1982). Indagini sui componenti fenolici degli oli di oliva. *Riv. Ital. Sostanze Grasse, 58* (LVIII), 443–452 (in Italian).

Ragazzi, E., & Veronese, G. (1989). The effect of oxidative coloration on the methanogenic toxicity and anaerobic biodegradability of phenols. *Biol. Wat. 32*, 210–225.

Ragazzi, E., Veronese, G., & Pietogrande, A. (1967). Ricerche sui constituenti idrosolubili delle olive. Nota II: Pigmenti e polisaccharidi. *Ann. Chim., 57*, 1398–1413 (in Italian).

Raimundo, M.C., & Oliveira, de J.S. (1976). Pollution from industrial extraction of olive oil in Portugal. *Special Addendum to NATO-ASI Series, Theory and practice of biological treatment preprints*, Boğaziçi University, Istanbul, Turkey, 13–23 July 1976.

Ralls, J.W., Maagdenberg, H.J., Lemoine, G., & Mercer, W.A. (1971). Alternate storage systems for the production of canned black ripe olives. *J. Food Sci., 36* (3), 408–412.

Ramos Ayerbe, F., & Ortega Jurado, A. (1978). Possible improvement in solvent extraction of olive foot cake. III. Granulating machines for the separated fatty pulp. *Grasas y Aceites, 29* (6), 407–415.

Ramos Ayerbe F., & Ortega Jurado, A. (1980a). Improved solvent extraction of olive foot cakes. IV. Analytical experiments. *Grasas y Aceites, 31* (3), 161–166.

Ramos Ayerbe, F., & Ortega Jurado, A. (1980b). Improved solvent extraction of olive foot cake.V. Economic study and discussion of results. *Grasas y Aceites, 32* (1), 7–12.

Ramos-Cormenzana, A. (1986). Physical, chemical, microbiological and biochemical characteristics of vegetation water. *Proc. Int. Symp. on Olive By-Products Valorization*, FAO, UNDP (Food and Agriculture Organization of the United Nations), 19–40, Seville, Spain, 5–7 Mar. 1986.

Ramos-Cormenzana, A., Juarez Jiménez, B., & García-Pareja, M.P. (1995). Antimicrobial activity of olive mill wastewaters (alpechín) and biotransformed olive oil mill wastewater. *Proc. Olive Oil Processes and By-Products Recycling*; Granada, Spain, 10–13 Sep. 1995; *Int. Biodeterior. Biodegrad.*, 1996, *38* (3–4), 283–290.

Ramos-Cormenzana, A., Monteoliva-Sánchez, M., & López López, M.J. (1995). Bioremediation of alpechin. *Int. Biodet. Biodegr., 35* (1–3), 249–268.

Rampichini, M. (1987). New applications of ultrafiltration: Effluents from synthetic fibers and olive oil production. *Chimica Oggi, 4,* 21.

Ranalli, A. (1989). Il problema dei reflui di frantoio: aspeti tecnici e normative. *L'Informatore Agrario,* 16, 29–52 (in Italian).

Ranalli, A. (1991). L'effluente dei frantoi oleari: proposte per la sua utilizzazione e depurazione con riferimenti alla normativa italiana (I parte). *Olivae, 37,* 30–39 (in Italian).

Ranalli, A. (1992). Ruolo delle microflora selezionata nel processo di trattamento delle acque reflue degli oleifici. *Inquinamento, 5,* 62–70 (in Italian).

Ranalli, A., & Angerosa, F. (1996). Integral centrifuges for olive oil extraction. The qualitative characteristics of products. *(JAOCS) J. Am. Oil Chem. Soc., 73* (4), 417–422.

Ranalli, A., & Martinelli, N. (1995). Integral centrifuges for olive oil extraction at the third millennium threshold: Tranformation yields. *Grasas y Aceites, 46* (4–5), 255–263.

Ranalli, G., Principi, P., Zucchi, M., Da Borso, F., Catalano L., & Sorlini, C. (2000). Pile composting of two-phase centrifuged olive husks: Bioindicators of the process. *Int. Conf. on Microbiology of Composting,* Innsbruck, Austria, 18–20 Oct. 2002. *Microbiol. Compost.,* 165–175.

Ranalli, A., Gomes, T., Delcuratolo D., Contento, S., & Lucera, L. (2003). Improving virgin olive oil quality by means of innovative extracting biotechnologies. *J. Agric. Food Chem., 51* (9), 2597–2602.

Raposo, F., Borja-Padilla, R., Sánchez, E., Martín, M.A., & Martín-Martín, A. (2003). Inhibition kinetics of overall substrate and phenolics removals during the anaerobic digestion of two-phase olive mill effluents (TPOME) in suspended and immobilized cell reactors. *Process Biochem., 39* (4), 425–435.

Raposo, F., Borja-Padilla, R., Sánchez, E., Martín, M.A., & Martín-Martín, A. (2004). Performance and kinetic evaluation of the anaerobic digestion of two-phase olive mill effluents in reactors with suspended and imppobilized biomass. *Water Res., 38* (8), 2017–2026.

Reimers Suarez, G. (1983). Possibilitades de tratamiento del alpechín pour ultrafiltracion y osmosis inversa. *Quimica e Industria,* (C.S.I.C.) de la grasa y sus derivados, Seville, Spain, 273–274 (in Spanish).

Rejano Navarro, L., Brenes Balbuena M., Sánchez, A.H., García-García, P., & Garrido Fernández, A. (1995). Brine recycling: its application in canned anchovy-stuffed olives and olives packed in pouches. *Sciences des Aliments, 15* (6), 541–550.

Ribera, R.G., Monteoliva-Sánchez, M., & Ramos-Cormenzana, A. (2001). Production of polyhydroxyalkanoates by *Pseudomonas putida* KT2442 harboring pSK2665 in wastewater from olive oil mills (alpechín). *Electronic, J. Biotechnol* (http://www.ejb.org), *4* (2), 1–4 (published on line 02/09/2001).

Riccardi, C., Di Basilio, M., Savarese, F., Torrisi, L., & Villarini, M. (2000). Aging-related physico-chemical changes in olive oil mill effluent. *J. Eniron. Sci. Health, Part, A., A35* (3), 349–356.

Richard, D., & Delgado-Nuñez, M. de Lourdes (2003). Kinetics of the degradation by catalytic hydrogenation of tyrosol, a model molecule present in olive oil waste waters. *J. Chem. Techol. Biotechnol., 78* (9), 927–934.

Riffaldi, R., Levi-Minzi, R., Saviozzi, A., & Viti, G. (1997). Carbon mineralization potential of soils amended with sludge from olive processing. *Bull. Environ. Contam. Toxicol., 58* (1), 30–37.

Riffaldi, R., Saviozzi, A., Levi-Minzi, R., & Bertolacci, M. (1992). Efffeto delle acque di vegetazione sulle proprietà di un terreno collinare ad oliveto. *Inquinamento*, *1*, 38–42 (in Italian).

Riffaldi, R., Levi Minzi, R., Saviozzi, A., Vanni, G., & Scagnozzi, A. (1993). Effect of the disposal of sludge from olive processing on some oil characteristics: Laboratory experiments. *Water, Air and Soil Pollut.*, *69* (3–4), 257–264.

Rigoni-Stern, S., Rismondo, R., Szprykowicz, L., & Zilio, G.F. (1988). Anaerobic digestion of vegetation water from olive mills on a fixed biological bed with a biogas production. *Proc. 5th Int. Symp. on Anaerobic Digestion*, Bologna, Italy (eds. Tilche A., & Rozzi, A.), Monduzzi Editore, S.p.A., 561–566. *Ing. Ambientale, 17* (5), 267–271.

Rindone, B., Andreoni, V., Rozzi, A., & Sorlini, C. (1991). Analysis and anaerobic degradation of wool scouring and olive mill wastewaters. *Fresen. J. Anal. Chem.*, *339* (9), 669–672.

Rismondo, R., Sutto, G., Rigoni, S., & Carli, B. de (1978). L'inquinamento da reflui di frantoi di olive nella zona di Massafra. *Monografia Technital-Sezione ecologia applicata* (in Italian).

Rivas, F.J., Beltrán, F.J., Acedo, B., & Gimeno, O. (2000a). Two-step wastewater treatment: Sequential ozonation-aerobic biodegradation. *Ozone-Sci. Eng.*, *22* (6), 617–636.

Rivas, F.J., Beltrán, F.J., Acedo, B., & Gimeno, O. (2001a). Joint treatment of wastewater from table olive processing and urban wastewater. Integrated ozonation -aerobic oxdation. *Chem. Eng. Technol*, 23 (2), 177–181.

Rivas, F.J., Beltrán, F.J., Acedo, B., & Gimeno, O. (2001b). Wet oxidation of wastewater from olive mills. *Chem. Eng. Technol.*, 24 (4), 415–421.

Rivas, F.J., Beltrán, F.J., Frades, J., & Buxeda, P. (2001c). Oxidation of p-hydroxybenzoic acid by Fenton's reagent. *Wat. Res.*, *35* (2), 387–396.

Rivas, F.J., Beltrán, F.J., Gimeno, O., & Frades, J. (2000b). Wet air oxidation of wastewater from table olive-processing industries. *219th Meeting of the American Chemical Society*, San Francisco, California, USA, 26–30 Mar. 2000; abstracts of papers ACS, *219* (1–2), 217.

Rivas, F.J., Beltrán, F.J., Gimeno, O., & Frades, J. (2001d). Treatment of olive oil mill wastewater by Fenton's reagent. *J. Agric. Food Chem.*, *49* (4), 1873–1880.

Rivas, F.J., Beltrán, F.J., Acedo, B., Gimeno, O., & Alvarez, P. (2003a). Treatment of brines by combined Fenton's reagent-aerobic biodegradation: II. Process modeling. *J. Hazard. Mater.*, *96* (2–3), 259–276.

Rivas, F.J., Beltrán, F.J., Gimeno, O., & Alvarez, P. (2001e). Chemical-Biological treatment of table olive manufacturing wastewater. *J. Environ. Eng.*, *127* (7), 611–619.

Rivas, F.J., Beltrán, F.J., Acedo, B., Gimeno, O., & Alvarez, P. (2003b). Optimisation of Fenton's reagent usage as a pre-treatment for fermentation brines. *J. Hazard. Mater.*, *96* (2–3), 277–290.

Rivas, F.J., Beltrán, F.J., Alvarez, P., Frades, J., & Gimeno, O. (2000c). Joint aerobic biodegradation of wastewater from table olive manufacturing industries and urban wastewater. *Bioprocess Eng.*, *23* (3), 283–286.

Robles, A., Lucas, R., Alvarez de Cienfuegos, G., & Galvez, A. (2000b). Biomass production and detoxification of wastewaters from the olive oil industry by strains of *Penicillium* isolated from wastewater disposal ponds. *Bioresource Technol.*, *74* (3), 217–221.

Robles, A., Lucas, R., Alvarez de Cienfuegos, G., & Galvez, A. (2000b). Phenol-oxidase (laccase) activity in strains of the hyphomycete *Chalara paradoxa* isolated from olive mill wastewater disposal ponds. *Enzym. Microb. Technol.*, *26* (7), 484–490.

Rodis, P.S., Karathanos, V.T., & Mantzavinou, A. (2002). Partitioning of olive oil antioxidants between oil and water phases. *J. Agric. Food Chem.*, *50* (3), 596–601.

Rodríguez, M.M., Pérez J., Ramos-Cormenzana, A., & Martínez, J. (1988). Effect of extracts obtained from olive oil mill waste on *Bacillus megaterium* ATCC 33085. *J. Appl. Bacteriol.*, *64* (3), 219–226.

Roig, A., Cayuela, M.L., & Sánchez-Monedero, M.A. (2004). The use of elemental sulphur as organic alternative to control pH during composting of olive mill wastes. *Chemosphere.*, *57* (9), 1099–1105.

Romero Barranco, C., Brenes Balbuena, M., García-García, P., & Garrido Fernández, A. (2001). Management of spent brines or osmotic solutions. *J. Food Eng.*, *49* (2–3), 237–246.

Romero Barranco, C., Brenes Balbuena, M., García-García, P., & Garrido Fernández, A. (2002). Hydroxytyrosol 4-beta-D-glucoside, an important phenolic compound in olive fruits and derived products. *J. Agric. Food Chem.*, *50* (13), 3835–3839.

Rosa, M.F., & Vieira, A.M. (1995). Perspectivas e limitações no tratamento e utilização das águas residuais de lagares de azeite: situação portuguesa. *Boletim de Biotechnologia*, (52), Dec. 1995 (in Portuguese).

Rose, W.W. (1982). Innovation treatment technology. *59th Annual Technical Report of the California Olive Association*, USA.

Rovatti, M., Bisi, M., & Ferraiolo, G. (1992). High added value products from difficult wastes. *Resour. Conserv. Recy.*, *7* (4) 271–283.

Roy, F., Albagnac, G., & Samain, E. (1985). Influence of calcium addition on growth of highly purified syntrophic cultures degrading long chain fatty acids, *Appl. Environ. Microbiol.*, *49* (3), 702–705.

Rozzi, A. (1991). Anaerobic treatment of wastewater from olive oil extraction and the agro-food industry: optimization of anaerobic-aerobic treatment of wastewater from olive oil extraction mixed with municipal sewage. *Ist. Ric. Acque*, *94*, 13.1–13.13.

Rozzi, A., & Malpei, F. (1996). Treatment and disposal of olive mill effluents. *Proc. Olive Oil Processes and By-Products Recycling*, Granada, Spain, 10–13 Sep. 1995. *Int. Biodeterior. Biodegrad.*, 1996, *38* (3–4), 135–144.

Rozzi, A., & Di Pinto, A.C. (1986). Anaerobic treatment of Olive Oil Mill Effluents as Energy Source. *Proc. Int. Symp. on Olive By-Products Valorization*, FAO, UNDP (Food and Agriculture Organization of the United Nations), Seville, Spain, 5–7 Mar. 1986, 182–193.

Rozzi, A., Passino, R., & Limoni, N. (1989a). Anaerobic treatment of olive mill effluents in polyurethane foam bed reactor. *Process Biochem.*, *24* (2), 68–74.

Rozzi, A., Santori, M., & Spinoza, L. (1986). Anaerobic digestion in Italy with special reference to treatment of olive oil mill wastes. *Comm. Eur. Communities, [Rep.] EUR (1986), EUR 9751, Anaerobic Dig. Sewage Sludge Org. Agric. Wastes*, 55–65.

Rozzi, A., Di Pinto, A.C., Limoni N., & Tomei, M.C. (1994). Start-up and operation of anaerobic digesters with automatic bicarbonate control. *Bioresource Technol.*, *48* (3), 215–219.

Rozzi, A., Santori, M., Menegatti, S., & Limoni, N. (1987). Comparison of the anaerobic treatment of olive mill wastewaters in foam-bed and upflow anaerobic sludge blanket reactors. *Ist. Ric. Acque*, *77*, 133–151.

Rozzi, A., Limoni N., Menegatti, S., Boari, G., Liberti, L., & Passino, R. (1988). Influence of sodium and calcium alkalinity on UASB treatment of olive mill effluents, part 1: preliminary results. *Process Biochem.*, *23* (3), 86–90.

Rozzi, A., Boari, G., Liberti, L., Santori, M., Limoni N., Menegatti, S., & Longobardi, C. (1989b). Trattamento combinato anaerobico-aerobico di acque di vegetazione e di scarichi urbani. *Ingegneria Sanitaria, 4,* 44–54 (in Italian).

Rubia, T. de la, González-López, J., Martínez, M.V., & Ramos-Cormenzana, A. (1987). Flavonoids and the biological activity of *Azotobacter vinelandii. Soil Biol. Biochem., 19,* 223–224.

Ryan, D., Robards, K., & Lavee, S. (1998). Determination of phenolic compounds in olives by reverse-phase chromatography and mass spectrometry. *J. Chromatogr.* A, *832,* 87–96.

Ryan, D., Robards, K., & Lavee, S. (1999). Changes in phenolic content of olive during maturation. *Int. J. Food and Technol., 34* (3), 265.

Saglik, S., Ersoy, L., & Imre, S. (2002). Oil recovery from lime-treated wastewater of olive mills. *Eur. J. Lipid Sci. Technol., 104* (4), 212–215.

Saez, L., Pérez, J., & Martínez, J. (1992). Low molecular weight phenolic attenuation during simulated treatment of wastewaters from olive mills in evaporation ponds. *Water Res., 26* (9), 1261–1266.

Sainz H., Benítez, E., Melgar, R., Alvarez, R., Gómez, M., & Nogales, R. (2000). Biotransformación y valorización agrícola de subproductos del olivar -orujos secos y extractados- mediante vermicompostaje, *7* (2), 103–111 (in Spanish).

Saiz-Jiménez, C., Leeuw, J.W. de, & Gómez Alarcón, G. (1986). Sludge from the waste water of the olive processing industry: A potential soil fertilizer? *Adv. in Humic Subst. Res., A Collect. of pap. from the 3rd Int. Meet. of the Int. Humic Subst. Soc.*; Oslo, Norway, 4–8 Aug. 1986. *Sci. Total Environ.,* 1987, *62,* 445–452.

Saiz-Jiménez, C., Gómez Alarcón, G., & Leew, J.W. de (1986). Chemical properties of the polymer isolated in fresh vegetation water and sludge evaporation ponds. *Proc. Int. Symp. on Olive By-Products Valorization,* FAO, UNDP (Food and Agriculture Organization of the United Nations), Seville, Spain, 5–7 Mar. 1986.

Salvemini, F. (1985). Composizione chimica e valutazione biologica di un mangime ottenuto essiccanto termicamente le acque di vegetazione delle olive. *Riv Ital. Sostanze Grasse, 62* (LXII), 559–564 (in Italian).

Sampedro, I., Romero Barranco, C., Ocampo, J.A., Brenes Balbuena, M., & García-Romera, I. (2004). Renoval of monomeric phenols in dry mill olive residue by saprobic fungi. *J. Agric. Food Chem., 52* (14), 4487–4492.

Sampedro, I., Aranda, E., Martín, J., García-Garrido, J.M., García-Romera, I., & Ocampo J.A. (2004). Saprobic fungi decrease plant toxicity caused by olive mill residues. *Appl. Soil Ecology* (a section of Agriculture, Ecosystems and Environment), *26* (2), 149–156.

Sánchez, A.H., García-García, P., Rejano Navarro, L., Brenes Balbuena, M., & Garrido Fernández, A. (1995). The effects of acidification and temperature during washing of Spanish-style green olives on the fermentation process. *J. Sci. Food Agric., 68,* 197–202.

Sánchez-Villasclaras, S., Martínez Sancho, M.E., Espejo Caballero, M.T., & Delgado Pérez, A. (1995). Production of microalgae from olive mill wastewater. *Proc. Olive Oil Processes and By-Products Recycling*; Granada, Spain, 10–13 Sep. 1995. *Int. Biodeterior. Biodegrad.,* 1996, *38* (3–4), 245–247.

Sanjust, E., Pompei, R., Rescigno, A., Rinaldi, A., & Ballero, M. (1991). Olive milling wastewaters as a medium for growth of four *Pleurotus* species. *Appl. Biochem. Biotechnol., 31* (3), 223–235.

Sanjust, E., Pompei, R., Rescigno, A., Rinaldi, A., Scrugli, S., & Ballero, M. (1994). Olive milling wastewater as a substrate for growth of four *Pleurotus* species. *Micologia Ital., 23* (2), 119–121.

Sanna, M., Pelosi, N., & Pietrini, R. (1978). Waste water from factories producing table olives. *Industrie Alimentari, 17* (1), 17–20, 24.

Sansoucy, R. (1984). Utilization de sous-produits de l'olivier en alimentation animale dans le basin Mediterranéen, valorisation de sous-produits de l'olivier. *Reunion du groupe de travail organisee par le projet regional d' amelioration de la production oleicole* (Organisation des Nations Unies pour l'Alimentation et l'Agriculture), Madrid, Spain, 1984, 66 (in French).

Sansoucy, R., Alibes, X., Berge, Ph., Martilotti, F., Nefzaoui, A., & Zoiopoulos, P.E. (1985). Olive by-products for animal feed (Review). *FAO Animal Production and Health Paper, 43*; M-23. ISBN 92-5-101488-4.

Santos-Siles, F.J. (1999). New technologies in table olive processing. *Grasas y Aceites, 50* (2), 131–140.

Saracco, G., Solarino, L., Aigotti, R., Specchia, V., & Maja, M. (2000). Electrochemical oxidation of organic pollutants at low electrolyte concentrations. *Electrochimica Acta, 46* (2–3), 373–380.

Sarika, R., Kalogerakis, N., & Mantzavinos, D. (2005). Treatment of olive mill effluents: Part II. Complete removal of solids by direct flocculation with poly-electrolytes. *Environ. Int., 31* (2), 297–304.

Saviozzi, A., Levi-Minzi, R., & Riffaldi, R. (1990). Cinetica della decomposizione nel terreno del carbonio organico delle acque di vegetazione. *Agrochimica, 34* (XXXIV), 157–164 (in Italian).

Saviozzi, A., Levi-Minzi, R., & Riffaldi, R. (1993). Effetto dello spandimento delle acque reflue del frantoi oleari su alcune propietà del terreno agrario. *Genio Rurale 5*, 68–71 (in Italian).

Saviozzi, A., Levi-Minzi, R., Riffaldi, R., & Cardelli, R. (1997). Caratteristische analitiche delle sanse umide stoccate. *L'Infomratore Agrario, 38*, 44–46 (in Italian).

Saviozzi, A., Levi-Minzi, R., Riffaldi, R., & Lupetti, A. (1991). Effeti dello spandimento di acque di vegetazione sul terreno agrario. *Agrochimica, 35* (XXXV), 135–148 (in Italian).

Saviozzi, A., Levi-Minzi, R., Cardelli, R., Biasci, A., & Riffaldi, R. (2001). Suitability of moist olive pomace as soil amendement. *Water, Air and Soil Pollut., 128* (1–2), 13–22.

Saviozzi, A., Riffaldi R., Levi-Minzi, R., Scagnozzi, A., & Vanni, G. (1993). Decomposition of vegetation-water sludge in soil. *Bioresource Technol., 44*, 223–228.

Sayadi, S., & Ellouz, R. (1992). Decolourization of olive mill wastewater by the white rot fungus *Phanerochaete chrysosporium*: Involvement of the lignin-degrading system. *Appl. Microbiol. Biotechnol., 37*, 813–817.

Sayadi, S., & Ellouz, R. (1993). Screening of white rot fungi for the treatment of olive mill waste-waters. *J. Chem. Technol. Biotechnol., 57* (2), 141–146.

Sayadi, S., & Ellouz, R. (1995). Roles of lignin peroxidase and manganese peroxidase from *Phanerochaete chrysosporium* in the decolorization of olive mill wastewaters. *Appl. Environ. Microbiol., 61* (3), 1098–1103.

Sayadi, S., Allouche, N., & Jaoua, M. (2000) Detrimental effects of high molecular-mass polyphenols on olive mill wastewater biotreatment. *Process Biochem., 35* (7), 725–735.

Scaccini, C., Nardini, M., D' Aquino, M., Gentili, V., Felice, M. de, & Tomasi, G. (1992). Effect of dietary oils on lipid peroxidation and on antioxidant parameters of rat plasma and lipoprotein fractions. *J. Lipid Res., 33*, 627–633.

Schaelicke, D. (1995). Start-up study on anaerobic digestion of olive mill wastewater using a baffled reactor. *Dissertation*. University of Applied Science, Berlin and University of the Aegean, Mytilene, Greece.

Schäfer-Schuchardt, H. (1998). Die Olive. DA Verlag Das Andere GmbH, Nürnberg, 5. Auflage 1998. ISBN 3-922619-26-6 (in German).

Schmidt, A., & Knobloch, M. (2000). Olive oil-mill residues: The demonstration of an innovative system to treat wastewater and to make use of generated bioenergy and solid remainder. *Proc. 1st World Conf. on Biomass for Energy and Industry*, Seville, 5–9 June 2000, 452–454.

Sciancalepore, V., Stefano, G. de, & Piacquadio, P. (2000). Effects of the cold percolation system on the quality of virgin olive oil. *Eur. J. Lipid Sci. and Technol.*, *102* (11), 680–683.

Sciancalepore, V., Colangelo, M., Sorlini, C., & Ranalli, G. (1996). Composting of effluent from a new two-phase centrifuge olive mill. Microbial characterization of the compost. *Toxicol. Environ. Chem.*, *55* (1–4), 145–158.

Sciancalepore, V., Stefano, G. de, Piacquadio, P., & Sciancalepore, R. (1995). Compostaggio in ambiente protetto del residuo della lavorazione delle olive con impianti ad estrazione bifasica. *Ingegneria Ambientale*, *24* (XXIV) (11–12) (in Italian).

Sciancalepore, V., Pizzuto, P., Stefano, G. de, Piacquadio P., & Sciancalepore, R. (1994). Compostaggio della sansa di olive dei nuovi impianti a due fasi. *Rifiuti solidi*, *8* (VIII), 6 (in Italian).

Scioli, C., & Felice, B. de (1993). Impiego di ceppi di lievito nella depurazione dei reflui dell'industria olearia (acque di vegetazione). "Use of yeast strains to depurate effluents from oilve oil mills (vegetable waters)". *Ann. Microbiol. Enzimol.*, *43*, 61–69 (in Italian).

Scioli, C., & Vollaro, L. (1997). The use of *Yarrowia Lipolytica* to reduce pollution in olive mill wastewaters. *Water Res.*, *31* (10), 2520–2524.

Scot, J.P., & Ollis, D.F. (1995). Integration of chemical and biological oxidation processes for water treatment: review and recommendations. *Environ. Prog.*, *14* (2), 88–103.

Senette, C., Melis, P., & Uscidda, D. (1991). Indagini sull'impatto ambientale dei refui oleari a seguito del loro smaltimento nel terreno. *Quaderno di Scienza e Tecnologia* (ed. NIA Ricerche), *2*, 45–51 (in Italian).

Servili, M., & Montedoro, G.F. (1989). Recupero di polifenoli dalle acque di vegetazione delle olive e valutazione del loro potere antiossidante. "Polyphenols extraction from olive vegetation waters and antioxidating capacity analysis". *Ind. Aliment.*, *28*, 14–18, 26 (in Italian).

Servili, M., Baldioli, M., Selvaggini, R., Macchioni, A., & Montedoro, G.F. (1999a). Phenolic compounds of olive fruit: one- and two-dimensional nuclear magnetic resonance characterization of nuzhenide and its distribution in the constitutive parts of fruit. *J. Agric. Food. Chem.*, *47* (1), 12–18.

Servili, M., Baldioli, M., Selvaggini, R., Miniati, E., Macchioni, A., & Montedoro, G.F. (1999b). High performance liquid chromatography evaluation of phenols in olive fruit, virgin oil, vegetation waters and pomace in 1D- and 2D Nuclear Magnetic Resonance characterization. *J. Am. Oil Chem. Soc. (JAOCS)*, *76* (7), 873–882.

Servili, M., Selvaggini, R., Esposto, S., Taticchi, A., Montedoro, G.F., & Morozzi, G. (2004). Health and sensory properties of virgin olive oil hydrophilic phenols: agronomic and technological aspects of production that affect their occurrence in the oil. *J. Chromatogr.*, *1054* (1–2), 113–127.

Servis, D. (1986). The soil as a receiving body of olive mill wastewaters, *M.Sc. Thesis*, Agricultural University of Athens (in Greek).

Setti, L., Maly, S., Iacondini, A., Spinozzi, G., & Pifferi, P.G. (1998). Biological treatment of olive milling waste waters by *Pleurotus ostreatus*. *Ann. Chim.* (Rome), *88* (3–4), 201–210.

Shammas, N.K. (1984). Olive extraction waste treatment in Lebanon. *Effl. Water Treat. J.*, *24* (10), 388–389, 391–392.

Sidal, U., Kolankaya, N., & Kurtonur, C. (2000). Obtaining biosurfactant from olive oil mill effluent (OOME). *Turkish J. Biol.*, *24* (3), 611–625.

Sierra, J., Martí, E., Montserrat, G., Cruãnas, R., & Garau, M.A. (2000). Approvechamiento del alpechín a través del suelo. Estimación del posible impacto sobre las aguas de infiltración. *Edafología*, *7* (2), 91–101 (in Spanish).

Sierra, J., Martí, E., Montserrat, G., Cruãnas, R., & Garau, M.A. (2001). Characterisation and evolution of a soil affected by olive oil mill wastewater disposal. *Sci. Total Environ.*, *279* (1–3), 207–214.

Simone, C. de (1998). Analisi degli effetti genotossici delle acque di vegetazione di frantoi oleari. *Agricoltura Ricerca*, *173*, 75–80 (in Italian).

Simone, C. de, & Marco, A. de (1994). Influence of olive oil wastewaters on genotoxic activity of herbicide maleic hydrazide in Vicia Faba seedlings. *Proc. of 7th Mediterranean Conference on Organic Wastes Recycling in Soils*, Vieste, 22–25 Sep. 1994, Ordine Nazionale dei Biologi (eds. Landi, E. & Dumonet S.).

Singleton, V.L., & Rossi, J.A. Jr. (1965). Colorimetry of total phenolics with phosphomolybdic-phosphotungstic acid reagent, *Am. J. Enol. Vitig.*, *16*, 144–158.

Siniscalco, V., Montedoro, G.F. (1988). *Riv.Ital. delle Sostanze Grasse*, XI, 675 (in Italian).

Siniscalco, V., Montedoro, G. F., Parlati, C., & Petruccioli, G. (1989). Mechanical extraction of olive oil by means of technological aditives note ii. percolation-centrifugation system *Riv. Ital. delle Sostanze Grasse*, *66* (LXVI) (2), 85–90 (in Italian).

Sisto, D. (1989). Fruttificazioni di *Pleurotus eryngii* su sansa di olive. *Micol. Ital.*, *3*, 43–46 (in Italian).

Skarica, B., & Tripodi, B. (1994). A new technology for olive oil manufacture. *Prehrambeno - Tehnoloska i Biotehnoloska Revija 32* (4), 151–155.

Skerratt, G., & Ammar, E. (1999). The application of reedbed treatment technology to the treatment of effluents from olive oil mills. *Final Report. Counrty/Project Number Tunisia 066599003ZH010.*

Slinkard, K., & Singleton, V.L. (1977). Total phenol analysis: Automation and comparison with manual methods. *Am. J. Enol. Vitic.*, *28*, 49–55.

Smith, A.E., & Secox, D.M. (1975). Forerunners of Pesticides in Classical Greece and Rome. *Agric. Food Chem.*, *23* (6), 1050.

Sorlini, C., Andreoni, V., Ferrari, A., & Ranalli, G. (1986). The influence of some phenolic acids present in oil mill waters on microbic groups for the methanogenesis. *Proc. Int. Symp. on Olive By-Products Valorization*, FAO, UNDP (Food and Agriculture Organization of the United Nations), Seville, Spain, 5–7 Mar. 1986, 81–88.

Sorlini, C., Andreoni, V., Daffonchio, D., Bondanti, P.L., Lazzari M., Cardinale, S., Rosso, P., Lupieri, L., Rosa M.F., Pasarinho, P., Vieira, A.M., & Tsezos, M. (1996). Proposal for an integrated process for olive oil mill waste water treatment with energy recovery: Preliminary results. *Proc. 10th Forum for Applied Biotechnology*, 26–27 Sep. 1996, Gent. *Meded. Fac. Landbouwkd. Toegepaste Biol. Wet.* (Univ. Gent), *61* (4A-B), 2069–2076.

Sousa, M. (2003). *IMPEL Olive Oil Project Report - CMA & NOA*. Number Report 2003/3, Project Manager: Méndez Miguel, Rome, Nov. 2003; number of pages, report: 33 and annexes: 61. European Union Network for the Implementation and Enforcement of Environmental Law (IMPEL) (http://europa.eu.int/comm/environment/impel).

Spandre, R., & Dellomonaco, G. (1996). Polyphenols pollution by olive-mill waste waters, *J. Env. Hydrol.*, *4*, 1–13.

Sparapano, L., & Rozzi, A. (1981). Anaerobic fermentation of olive oil mill wastewater. *2nd European Congress of Biotechnology*, Eastbourne, UK, 5–10 Apr. 1981.

Standards, laws regulations (1984). Determination of lipid content of alpechín. *Spanish Standard*. ISBN 0-87055-466-2.

Steegmans, R. (1987). Abwässer aus der Olivenölgewinnung -Anfall, Problematik und Entsorgung. "Sewage from the olive oil extraction -sewage flow, situation and disposal". *Gewässerschutz, Wasser, Abwasser, 95*, 171–186 (in German).

Stefano, G. de, Piacquadio, P., Servili, M., Di Giovacchino, L., & Sciancalepore, V. (1999). Effect of extraction systems on the phenolic composition of virgin olive oils. *Fett/Lipid, 101* (9), 328–332.

Storm, J. (1989). Evaporacion del alpechín. Journadas sobre innovation tecnologica medio ambiente y desarollo. (C.S.I.C.) de la grasa y sus derivados, Seville, Spain (in Spanish).

Suksankraisorn, K., Patumsawad, S., & Fungtammasan, B. (2003). Combustion studies of high moisture content waste in a fluidised bed. *Waste Manage., 23* (5), 433–439.

Sutherland, I.W. (1995). Microbial biopolymers from agricultural products: production and potential. *Proc. Olive Oil Processes and By-Products Recycling*, Granada, Spain, 10–13 Sep. 1995. *Int. Biodeterior. Biodegrad.*, 1996, *38* (3–4), 249–261.

Swain, T. & Hillis, W.E. (1959). The quantitative analysis of phenolic constituents of *Prunus domestica. J. Sci. Food Agric., 10*, 63–68.

Tamburino, V., Zimbone, S.M., & Quatronee, P. (1999). Storage and land application of olive-oil wastewater. *Olivae, 76*, 36–45.

Tejada, M., & González-López, J. (2003). Effects of foliar application of a by-product of the two-step olive oil mill process on maize yield. *Agronomie, 23* (7), 617–623.

Tejada, M., & González-López, J. (2004). Effects of application of a by-product of the two-step olive oil mill process on maize yield. *Agronomy, J., 96* (3), 692–699.

Tejada, M., Ruiz, J.L., Dobao, M., Benítez, C., & González-López, J. (2003). Evolucíon de parámetros fisicos de un suelo tras la adición de distintos tipos de orujos de aceituna. *Actas de Horticultura, 18*, 514–518 (in Spanish)

Tekin, A.R., & Dalgiç, A.C. (2000). Biogas production from olive pomace. *Resour. Conserv. Recy., 30* (4) , 301–313.

Telmini, M., Sanna, M., & Pelosi, N. (1976). *Industrie Alimentari, 15* (11), 123.

Theophrastus (c.372-c.287 B.C.) *De Causis Plantarum*: Volume I, Books (1–2), Jan. 1976 edition, ISBN 0-674-99519-8; Volume II, Books (3–4), Jan. 1990 edition, ISBN 0-674-99523-6; Volume III, Books (5–6), Jan. 1990 edition, ISBN 0-674-99524-4. Edited and translated by Benedict Einarson, G.K.K. Link, Loeb Classical Library, Harvand University Press, 1990.

Tomati, U. (2002). An European regulation about olive mill waste industry. *Int. Conf. Environ. Problems of the Mediterranean Region*, 12–15 Apr. 2002, Near East University, TRNC.

Tomati, U., & Galli, E. (1992). The fertilizing value of waste waters from the olive processing industry. In *Humus et Planta Proc.* 107–126 (Humus, its Structure and Role in Agriculture and Environment, 117–126) (ed. Kubat I.), Elsevier Science, B.V. Amsterdam, The Netherlands. ISBN 0-444-88980-9.

Tomati, U., Madejon, E., & Galli, E. (2000). Evolution of humic acid molecular weight as an index of compost stability. *Compost Sci. Util.*, *8* (2), 108–115.

Tomati, U., Galli, E., Di Lena, G., & Buffone, R. (1991). Induction of laccase in *Pleurotus ostreatus* mycelium grown in olive oil waste waters. *Agrochimica*, *35* (1–3), 275–279.

Tomati, U., Galli, E., Fiorelli, F., & Pasetti, L. (1995a). Fertilizers from composting of olive-mill wastewaters. *Proc. Olive Oil Processes and By-Products Recycling*; Granada, Spain, 10–13 Sep. 1995. *Int. Biodeterior. Biodegrad.*, 1996, *38* (3–4), 155–162.

Tomati, U., Galli, E., Pasetti L., & Volterra, E. (1995b). Bioremediation of olive-mill wastewater by composting. *Waste Manage. Res.*, *13* (6), 509–518.

Tomati, U., Galli, E., Pasetti, L., & Volterra, E. (1996). Olive-mill wastewater bioremediation: Evolution of a composting process and agronomic value of the end product. *The Science of Composting* (eds. Bertoldi, M. de, Sequi, P., Lemmes B., & Papi, T.), Part 1, 637–647, Blackie Academic & Professional, Glasgow, UK.

Tomati, U., Di Lena, G., Galli, E., Grappelli, A., & Buffone, R. (1990). Indoleacetic acid production from olive waste water by *Arthrobacter* spp. *Agrochimica*, *34* (XXXIV) (3), 228–232.

Tomati, U., Madejon, E., Galli, E., Capitani, D., & Segre, A.L. (2001). Structural changes of humic acids during olive mill pomace composting. *Compost Sci. and Util.*, *9* (2), 134–142.

Tomati, U., Belardinelli, M., Galli, E., Iori, V., Capitani, D., Mannina, L., Viel, S., & Segre, A. (2004). NMR characterization of the polysaccharidic fraction from *Lentinula edodes* grown on olive mill waste waters. *Carbohydr. Res.*, *339* (6), 1129–1134.

Topal, H., Atimtay, A., & Durmaz, A. (2003). Olive cake combustion in a circulating fluidized bed. *Fuel*, *82* (9), 1049–1056.

Torres Martín, M., Velasco, E.E., & Zamora, A.M.A. (1980). Aspetos a considerar en el empleo del aplechín como fertilizante. "Studies on the alpechin vegetation waters as fertilizer". *An. Edafol. Agrobiol.*, *39* (7–8), 1379–1384 (in Spanish).

Trichopoulou, A., Katsouyanni, H., Stuver, S., Tzala., Gnardellis, C., Rimm, E., & Trichopoulos, D. (1995). Consumption of olive oil and specific food groups in relation to breast cancer in Greece. *J. Nat. Cancer Inst.*, *87* (2), 110–116.

Tsioulpas, A., Dimou, D., Iconomou, D., & Aggelis, G. (2002). Phenolic removal in olive oil mill wastewater by strains of *Pleurotus* spp. in respect to their phenol oxidase (laccase) activity. *Bioresource Technol.*, *84*, 251–257.

Tsonis, S.P. (1988). Treatment of olive-mill wastewaters. *Ph.D. Thesis*, Department of Civil Engineering, University of Patras, Greece (in Greek), pp. 387.

Tsonis, S.P. (1991). Start up mode of seasonally operating units for the anaerobic digestion of olive oil mill wastewater. *Proc. Int. Conf. on Environ. Pollut. ICEP-1*; Lisbon, Portugal, Apr. 1991, *2*, 733–740 (ed. Nath Bhaskar), Interscience Enterprises Ltd., Switzerland.

Tsonis, S.P. (1997). Olive oil mil wastewater as carbon source in post-anoxic denitrification. *Proc. 2nd IAWQ Int. Conf. on Pretreatment of Industrial Wastewaters*, Athens, Greece, 16–18 Oct. 1996. *Water Sci. Technol.*, *36* (2–3), 53–60.

Tsonis, S.P., & Grigoropoulos, S.G. (1983). Anaerobic treatment of wastewater from olive oil mills. *Advances in modelling, planning, decision and control of energy, power and environmental systems.* (ed. Tzafestas, S.G. & Hamza, M.H.), 282–285, Acta Press, Anheim.

Tsonis, S.P., & Grigoropoulos, S.G. (1988). High-rate anaerobic treatment of olive oil mill wastewater. *Proc. 6th Symp. on Anaerobic Digestion of Wastewater*, Bologne, Italy, 115–124 (eds. Hall, E.R., & Hobson, P.N.), Pergamon, Oxford, UK.

Tsonis, S.P., & Grigoropoulos, S.G. (1993). Anaerobic treatment of olive oil mill wastewater. *Proc. 2nd IAWQ International Symp. on Waste Management Problems in Agro Industries;* Istanbul, Turkey, 23–25 Sep. 1992. *Water Sci. Technol.*, *28*, 35–44.

Tsonis, S.P., Tsola, V.P., & Grigoropoulos, S.G. (1987). Systematic characterization and chemical treatment of olive oil mill wastewater. *Proc. 4th Int. Congress on Environ. Pollut. and its Impact on Life in the Mediterranean Region*, Kavala, Greece, 6–11 Sep. 1987. *Toxicol. Environ. Chem.*, 1989, *20–21*, 437–457.

Tunay, O., Akbatur, N., Orhon, D., & Ozturk, I. (1992). Final treatability of raw and anaerobically treated olive oil wastewater (1992). *Fresen. Environ. Bull.*, *1* (7), 434–438.

Turano, E., Curcio, S., Paola, M.G. de., Calabró, V., & Iorio, G. (2002). An integrated centrifugation-ultrafiltration system in the treatment of olive mill wastewater. *J. Membrane Sci.*, *109*, 519–531.

Ubay, G., & Ozturk, I. (1997). Anaerobic treatment of olive mill effluents. *Proc. 2nd IAWQ Int. Conf. on Pretreatment of Industrial Wastewaters;* Athens, Greece, 16–18 Oct. 1996. *Water Sci. Technol.*, 1997, *36* (2–3), 287–294.

Vaccarino, C., Lo Curto, R., Munao, F., Tripodo, M.M., Patane, R., & Lagana, G. (1986). Processing of olives: How to treat the wastewater. *AES*, *8* (4), 48–51.

Vaccarino, C., Lo Curto, R., Tripodo, M.M., Lagana, G., Patane, R., & Muano, F. (1986). *Vegetation water treatment by aerobic fermentation with fungi.* Proc. Int. Symp. on Olive By-Products Valorization, FAO, UNDP (Food and Agriculture Organization of the United Nations), Seville, Spain, 5–7 Mar. 1986, 23.

Valenzuela, G. (1986). *Proc. Int. Symp. on Olive By-Products Valorization*, FAO, UNDP (Food and Agriculture Organization of the United Nations), Seville, Spain, 5–7 Mar. 1986, 173–177.

Vallini, G., Pera, A., & Morelli, R. (2001). Il compostaggio delle acque di vegetazione dei frantoi oleari. *L'Informatore Agrario*, suppl. 1, 50, 22–26 (in Italian).

Varro Marcus Terentius (c.116–27 B.C.) (and Cato) *De Re Rustica*, translated by, W.D. Hooper, & H.B. Ash, Loeb Classical Library, Wm. Heinemann Ltd., London, 1967.

Vasallo, C. (2003). *IMPEL Olive Oil Project Report - CMA & NOA.* Number Report 2003/3, Project Manager: Méndez Miguel, Rome, Nov. 2003; number of pages, report: 33 and annexes: 61. European Union Network for the Implementation and Enforcement of Environmental Law (IMPEL) (http://europa.eu.int/comm/environment/impel).

Vásquez-Roncero, A., Graciani Constante, E., & Maestro-Durán, R. (1974a). Componentes fenólicos de la aceituna I: Polifenoles del la pulpa. "Phenolic components of olives I: Polyphenols in the pulp" *Grasas y Aceites*, *25*, 269–279 (in Spanish).

Vásquez-Roncero, A., Maestro-Durán, R., & Graciani Constante, E. (1974b). Componentes fenólicos de la aceituna II: Polifenoles del alpechín. "Phenolic components of olives II: Polyphenols in vegetable water". *Grasas y Aceites*, 25 (6), 341–345 (in Spanish).

Vásquez-Roncero, A., Janer del Valle, C., & Janer del Valle, M.L. (1976). Componentes fenólicos de la aceituna. III. Polifenoles del aceite. "Phenolic components of olives III: Polyphenols in olive oil". *Grasas y Aceites*, *27*, 185–191.

Vassilev, N., Vassileva, M., Azcon, R., Fenice, M., Federici, F., & Barea, J.M. (1998). Fertilizing effect of microbially treated olive mill wastewater on *Trifolium* plants. *Bioresource Technol.*, *66* (2), 133–137.

Vegliò, F., Beolchini, F., & Prisciandaro, M. (2003). Sorption of copper by olive mill residues. *Water Res.*, *37* (20), 4895–4903.

Velioğlu, S.G., Curi, K., & Çamillar, S.R. (1987). Laboratory experiments on the physical treatment of olive oil wastewater. *Int. J. Dev. Technol.*, *5* (1), 49–57.

Velioğlu, S.G., Curi, K., & Çamillar, S.R. (1992). Activated sludge treatability of olive oil-bearing wastewater. *Water Res.*, *26* (10), 1415–1420.

Verde Carmona, A., Gutierrez González -Quijano, R., & Flores Lugue, V. (1972). Hexane detection in alpechín (waste liquid from olive oil extraction). *Grasas y Aceites*, *23* (4), 318–321.

Vial, J., Hennion, M.-C., Fernández-Alba, A., & Agüera, A. (2001). Use of porous graphitic carbon coupled with mass detection for the analysis of polar phenolic compounds by liquid chromatography. *J. Chromatogr.*, *937*, 21–29.

Vierhuis, E., Korver, M., Schols, H.A., & Voragen, G.J. (2003). Structural characteristics of pectic polysaccharides from olive fruit (*Olea europaea* cv moraiolo) in relation to processing for oil extraction. *Carbohydrate Polym.*, *51* (2), 135–148.

Vierhuis, E., Servili M., Baldioli, M., Schols, H.A., Voragen, A.G.J., & Montedoro, G.F. (2001). Effect of enzyme treatment during mechanical extraction of olive oil on phenolic compounds and polysaccharides. *J. of Agric. Food Chem.*, *49* (3), 1218–1223.

Vigo, F., & Cagliari, M. (1999). Photocatalytic oxidation applied to olive mill wastewaters treatment. *Riv. Ital. Sostanze Grasse*, *76* (LXXVI) (9), 345–353.

Vigo, F., Avalle, L., & Paz, M. de (1983a). Smaltimento delle acque di vegetazione provenienti da frantoi di olive. Studio del processo di ossidazione electrochimica. "Elimination of vegetation water from olive oil mills. The electrochemical oxidation process". *Riv. Ital. Sostanze Grasse*, *60* (LX), 125–131 (in Italian).

Vigo, F., Giordani, M., & Capannelli, G. (1981). Ultrafitrazione di acque di vegetazione da frantoi di olive. "Ultrafiltration of vegetation waters coming from olive mills". *Riv. Ital. Sostanze Grasse*, *58* (LVIII), 70–73 (in Italian).

Vigo, F., Paz, M. de, & Avalle, L. (1983b). Ultrafitrazione di acque di vegetazione da frantoi di olive. Esperineza gestionale in impianto semi-pilota. "Ultrafiltration of vegetation waters coming from olive mills: Experiments on a semi-pilot plant". *Riv. Ital. Sostanze Grasse*, *60* (LX), 267–272 (in Italian).

Vigo, F., Uliana, C., & Traverso, M. (1990). Acque di vegetazione da frantoi di olive. Trattamenti utilizzabili per una riduzione del carico inquinante. "Treatments useful for reduction of the polution due to vegetation waters coming from olive mills". *Riv. Ital. Sostanze Grasse*, *67* (LXVII), 131–137 (in Italian).

Vinciguerra, V., D'Annibale, A., Delle Monache, G., & Sermanni, G.G. (1993). Degradation and biotransformation of phenolic compounds of olive waters by the white rot basidiomycete *Lentinus edodes*. *Med. Fac. Landbouw* (Unvi. Gent), *58* (4), 1811–1814.

Vinciguerra, V., D'Annibale, A., Delle Monache, G., & Sermanni, G.G. (1995). Correlated effects during the bioconversion of waste olive water by *Lentinus edodes*. *Bioresource Technol.*, *51*, 221–226.

Vinciguerra, V., D'Annibale, A., Gacs Baitz, E., & Delle Monache, G. (1997). Biotransformation of tyrosol by whole cell and cell free preparation of *Lentinus edodes*. *J. Molec. Catal. B: Enzymatic*, *3* (5), 213–220.

Virgil, (Publius Vergilius Maro, 70 B.C.-19 B.C.) *The Georgics* in *Eclogues, Georgics, Aeneid*, translated by, H.R. Fairclough, Loeb Classical Library, Wm. Heinemann Ltd., London, 1974.

Visioli, F., & Galli, C. (1995). Naural antioxidants and prevention of coronary heart disease: the pontential role of olive oil and its minor constituents. *Nut. Metab. Cardiovasc. Dis.*, *5*, 306–314.

Visioli, F., Vincieri, F.F., & Galli, C. (1995a). "Waste-waters" from olive oil production are rich in natural antioxidants. *Experientia, 51* (1), 32–34.

Visioli, F., Bellomo, G., Montedoro, G.F., & Galli, C. (1995b). Low density lipoprotein oxidation is inhibited in *vitro* by olive oil constituents. *Atherosclerosis. 117*, 25–32.

Visioli, F., Romani, A., Mulinacci, N., Zarini, S., Conte, D., Vincieri, F.F., & Galli, G. (1999). Antioxidant and other biological activities of olive mill waste waters. *J. Agric. Food Chem.*, *47* (8), 3397–3401.

Visioli, F., Caruso, D., Plasmati, E., Patelli, R., Mulinacci, N., Romani, A., Galli, G., & Galli, C. (2000). Hydroxytyrosol, as a component of olive mill waste water, is a dose-dependently absorbed and increases the antioxidant capacity of rat plasma. *Free Rad. Res.*, *34* (3), 301–305.

Vitagliano, M., & Pantaleo, V.N. (1975). Una possibile utilizzazione. delle acque di vegetazione delle olive. *Estratto dal Vol. II degli Ati del V simposio Nazionale Sulla Conservazione Della Natura*. Bari, Italy, 22–27 Apr. 1975 (in Italian).

Vitolo, S., Petarca, L., & Bresci, B. (1999). Treatment of olive oil industry wastes. *Bioresource Technol.*, *67* (2), 129–137.

Vlyssides, A.G. (2003). *IMPEL Olive Oil Project Report - CMA & NOA*. Number Report 2003/3, Project Manager: Méndez Miguel, Rome, Nov. 2003; number of pages, report: 33 and annexes: 61. European Union Network for the Implementation and Enforcement of Environmental Law (IMPEL) (http://europa.eu.int/comm/environment/impel).

Vlyssides, A.G., Loizidou, M., & Karlis, P.K. (2004). Integrated strategic approach for reusing olive oil extraction by-products. *J. Clean. Prod.*, *12* (6), 603–611.

Vlyssides, A.G., Loizidou, M., & Zorpas, A. (1999). Characteristics of solid residues from olive oil processing as bulking material for co-composting with industrial wastewaters. *J. Environ. Sci. Health Part A, 34* (3), 737–748.

Vlyssides, A.G., Parlavantza, M., & Balis, C. (1989). Co-composting as a system for handling of liquid wastes from olive oil mills. *Proc. Int. Conf. on Composting*, Athens, Greece, 1989.

Vlyssides, A.G., Bouranis, D.L., Loizidou, M., & Karvouni, G. (1996). Study of a demonstration plant for the co composting of olive oil processing wastewater and solid residue. *Bioresource Technol.*, *56* (2–3), 187–193.

Vlyssides, A.G., Loizidou, M., Gimouhopoulos, K., & Zorpas, A. (1998). Olive oil processing wastes production and their characteristics in relation to olive oil extraction methods. *Fresen. Environ. Bull.*, *7* (5–6), 308–313.

Vlyssides, A.G., Loukakis, H., Israilides, C.J., Barampouti, E.M., & Mai, S. (2003). Detoxification of olive mill wastewater using a Fenton process. *2nd European Bioremediation Conf.*, Chania, Crete, Greece, 30 June–4 July 2003, 531–534, (ed. Kalogerakis, N.).

Wähner, R.S., Mendez, B.A., & Giulietti, A.M. (1988). Olive black water as raw material for butanol production. *Biol. Wastes, 23* (3), 215–220.

Wang, T.S.C., Yang, T.K., & Chuang, T.T. (1967). Soil phenolic acids as plant growth inhibitors. *Soil Sci., 103*, 239–246.

Wang, Y.T. (1992). Effect of chemical oxidation on anaerobic biodegradation of model phenolic compounds. *Wat. Environ. Res., 64* (3), 268–273.

Wing-Hong Chan, & Sai-Cheong Tsao (2003). Preparation and characterization of nanofiltration membranes fabricated from poly(amidesulfonamide) and their application in water-oil separation. *J. Appl. Polym. Sci.*, *87* (11), 1803–1810.

Wlassics, I., & Visentin, W. (1994). Metabolizzazione esaustiva delle acque di vegetazione in vasca biologica tramite l' impiego di H_2O_2 "Exhaustive metabolization of olive oil mill wastewaters in a biological vat preceded by H_2O_2 based pretreatment". *Riv. Ital. Sostanze Grasse*, *71* (LXXI) (1), 21–23 (in Italian).

Wlassics, I., Buzio, F., & Visentin, W. (1992). Perossidasi+ H_2O_2: Un metodo efficace ed ecologico per il trattamento delle acque di vegetazione. *Riv. Ital. Sostanze Grasse*, *69* (LXIX), 141–145 (in Italian).

Wlassics, I., Visentin, W., Cavagnis, E., & Benzoni, G. (1994). Semi pilot scale treatment of olive oil mill waste waters. *Riv. Ital. Sostanze Grasse*, *71* (LXXI) (11), 561–563.

Yaman, I., Sahan, M., Haykiri-Acma, H., Sesen, K., & Kucukbayrak, S. (2000). Production of fuel briquettes from olive refuse and paper mill waste. *Fuel Process Technol.*, *68* (1), 23–31.

Yan, L. (2002). Comment on "Oxidation of p-hydroxybenzoic acid by Fenton's reagent" by Rivas, F.J., Beltrán, F.J., Frades, J., & Buxeda, P. *Water Res.*, *36* (4), 1106.

Yesilada, O., & Fiskin, K. (1996). Degradation of olive mill waste by *Coriolus versicolor*. *Turkish, J. Biol.*, *20* (1), 73–79.

Yesilada, O., & Sam, M. (1998). Toxic effects of biodegraded and detoxified olive oil mill wastewater on the growth of *Pseudomonas aeruginosa*. *Toxicol. Environ. Chem.*, *65* (1–4), 87–94.

Yesilada, O., Fiskin, K., & Yesilada, E. (1995). The use of white rot fungus *Funalia trogii* (Malatya) for the decolorization and phenol removal from olive mill wastewater. *Environ. Technol.*, *16* (1), 95–100.

Yesilada, O., Sik, S., & Sam, M. (1998). Biodegradation of olive oil mill wastewater by *Coriolus versicolor* and *Funalia trogii*: Effects of agitation, initial COD concentration, inoculum size and immobilization. *World J. Microbiol. Biotechnol.*, *14* (1), 37–42.

Yurekli, F., Yesilada, O., Yurekli, M., & Topcuoglu, S.F. (1999). Plant growth hormone production from olive oil mill and alcohol factory wastewaters by white rot fungi. *World J. Microb. Biotechnol.*, *15* (4), 503–505.

Zenjari, B., & Nejmeddine, A. (2001). Impact of spreading olive mill wastewater on soil characteristics: Laboratory experiments. *Agronomie*, *21* (8), 749–755.

Zenjari, B., Hafidi, M., El Hadrami, I., Bailly, J.-R., & Nejmeddine, A. (1999). Traitment aérobie des effluents d'huileries par les micro-organismes du sol. *Agrochimica*, *43* (XLIII), 276–285 (in French).

Zervakis, G., & Balis, C. (1996). Bioremediation of olive oil mill wastes through the production of fungal biomass. *Mushroom Biol. Mushroom Prod.; Proc. 2nd Int. Conf.*, 1996, 311–323 (ed. Royse, D.J.), Publisher: Penn State University, College of Agricultural Sciences, University Park, Pa.

Zervakis, G., Yiatras, P., & Balis, C. (1996). Edible mushrooms from olive oil mill wastes. *Proc. Olive Oil Processes and By-Products Recycling*, Granada, Spain, 10–13 Sep. 1995. *Int. Biodeterior. Biodegrad.*, 1996, *38* (3–4), 237–243.

Zoiopoulos, P.E. (1983). Study on the use of olive by-products in animal feeding in Greece. *Animal Production and Health Divisions*, FAO, Rome, 1983.

Zouari, N. (1998). Decolorization of olive oil mill effluent by physical and chemical treatment prior to anaerobic digestion. *J. Chem. Technol. Biotechnol.*, *73* (3), 297–303.

Zouari, N., & Ellouz, R. (1996a). Microbial consortia for the aerobic degradation of aromatic compounds of olive oil mill effluent. *J. Ind. Microbiol. Biotechnol., 16*, 153–162.

Zouari, N., & Ellouz, R. (1996b). Toxic effect of colored olive compounds on the anaerobic digestion of olive oil mill effluent in UASB like reactors. *J. Chem. Technol. Biotechnol., 66* (4), 414–420.

Zucconi, F., & Bukovac, N.J. (1969). Analisi sul'attivitá biologica delle acque di vegetazione delle olive. *Riv. dell' Ortoflorofruticoltura Italiana, 53*, 443–461 (in Italian).

Zumbo, A., Chiofalo, V., Lanza, M., & Dugo, P. (2001). Effect of vitamin E in lambs fed olive cake: Chemical composition of meat and lipid oxidation. *Proc. of the XIV Congress of ASPA*, Firenze, Italy, 12–15 June 2001, 556–558.

Patents

Patent	Family member	Priority	Inventor	Applicant	Title
CZ9401911 A 17-01-1996	CZ280400 B 17-01-1996 WO9605145 A 22-02-1996 EP722425 A 24-07-1996	CZ19940001911 08-08-1994 WO1995CZ00016 03-08-1995	SiegeL L.; Hrusa E.; Poduska J.	Vodni Stavby Praha	"Process of disposing waste from the production of olive oil"
DE19934834 A 01-02-2001		DE19991034834 24-07-1997	Bondioli P.; Buss D.; Catanho Fernandes J.A.; Müller A.; Simoes P.; Swidersky P.	Müller Extract Company GmBH & Co.	Phytosqualen und Phytosqualan sowie Vorrichtung und Verfahren zu deren Herstellung. "Phytosqualene and phytosqualane as well apparatus and process for their production"
DE19829673 A 05-01-2000	WO0001622 A 13-01-2000 TR200001632T 22-01-2001 CN1287540T T 14-03-2001 EP1102724 A 30-05-2001 GR2001300034T 31- 07-2001 ES2157879T T 01-09-2001 US6391202 B 21-05-2002 AT236092T T 15-04-2003 DE59904865D D 08-05-2003 PT1102724T T 29-08-2003 NZ504846 A 26-09-2003	DE19981029673 03-07-1998	Knobloch M.; Schmidt A.; Koch R.; Peukert V.	Knobloch M.; Schmidt A.; Koch R.; Peukert V.	Verfahren und Anlage zur Behandlung von Abwasser aus der Ölfrüchte- und Getreideverarbeitung. "Method and apparatus for treating wastewaters from olive oil plant and cereal processing"

DE19529404 A 13-02-1997		DE19951029404 10-08-1995	Öcknick C.	Reotec- Abwasswertechnik GmbH	Verfahren zur Abscheidung von Belastungstoffen aus Emulsionen, Suspensionen und Vorrichtung zur Durchfürung des Verfahrens. "Process for removing pollutants from emulsions, suspensions, or dispersions and apparatus for implementing the process"
DE4210413 A 07-10-1993	EP623380 A 09-11-1994 CZ9300899 A 15-12-1994 SK55493 A 08-02-1995	DE19924210413 30-03-1992 CZ19930000899 14-05-1993 EP19930107429 07-05-1993 SK19930000554 01-06-1993	Iniotakis N.; Keutman W.; von der Decken K. B.	Iniotakis N.; Keutman W.; von der Decken K. B.	Membran zur Trennung von Polydispersionen und/oder Emulsionen sowie Verfahren zur Herstellung der Membran. "Membrane for the separation of polydispersions and/or emulsions and process for the manufacture of said membrane"
DE3804573 A 17-08-1989		DE19883804573 13-02-1988	Becherer geb. Becherer E.; Dietz W.	Deutsche Carbone AG	Verfarhen zur Behandlung einer Lösung insbesondere zur Abwasserreinigung. "Process for treating a solution in particular for wastewater purification"
DE2640156 A 16-03-1978		DE19762640156 07-09-1976	Asendorf E.	Asendorf E.	Verfahren zur biologischen Reinigung von hochbelasteten Abwässern aus der Lebensmittelindustrie, insbesondere solchen aus der Gewinnung von Olivenöl. "Process for the biological purification of highly polluted wastewaters from the food industry, especially those from the olive oil production"
EP1378491 A 07-01-2004		FR20020008181 01-07-2002	Pina Michel; Guyot Bernard; Graille hean; Figueroa-Espinoza Maria-Cruz	CENTRE DE COOPERATION INTERNATIONAL EN RECHERCHE AGRONOMIQUE POUR LE DEVELOPPEMENT	Procédé de traitement d'un extrait aqueux d'origine végétale additionné de polymère et poudre d'atomisation obtenue. "Process of treating an aqueous extract of vegetable origin with polymer and powder produced by spraying it"

(*continued*)

Patent	Family member	Priority	Inventor	Applicant	Title
EP1310175 A 14-05-2003	DE10154806 A 22-05-2003	DE20011054806 08-11-2001	Johannisbaouer Wilhelm; Bonakdar Mehdi; Richard-Elsner Christinae	Cognis Deutschland GmbH & Co.	Verfarhen zur Isolierung von Antioxidantien. "Process for the isolation of antioxidants"
EP1216963 A 26-06-2002		IT2000TR00005 13-12-2001	Santori F.; Cicalini A. R.	Isrim S.C.a.r.l.	Process of olive-mill waste water phytodepuration and relative plant
EP1157972 A 28-11-2001	GR2000100177 A 31-01-2002 GR1003914 B 25-06-2002	GR20000100177 26-05-2000	Vlissidis A.; Kyprianou D.	Vlissidis A.; Kyprianou D.	A method of processing oil-plant wastes
EP1097907 A 09-05-2001		EP19990670007 02-11-1999	Costa Guedes Da Silva M. J.; Maggiolly Novais J.; Martins Dias S.	Instituto Superior Técnico; Solvay Interox – Produtos; Peroxidados Lda.	A process for the treatment of liquid effluents by means of clean catalytic oxidation, using hydrogen peroxide, and heterogeneous catalysis
EP811678 A 10-12-1997	WO9747711 A 18-12-1997 AU3173997 A 07-01-1998 EP925340 A 30-06-1999 TR9802550T T 22-02-1999 AU717853 B 06-04-2000 IL124279 A 26-08-2001 US6309652 B 30-10-2001 EP925340 B 16-04-2003 AT237666T T 15-05-2003 DE69721017D D 22-05-2003 PT925340T T	EP19960201590 08-06-1996	Aeschbach R; Bracco U; Rossi P.	SOCIETE DES PRODUITS NESTLE S.A ; NESTEC S.A.	Extraction d'antioxydants. "Extraction of antioxidants "

EP581748 A 02-02-1994	31-07-2003 DE69721017T T 06-11-2003 CA2257814 A 18-12-1997 ES2194199T T 16-11-2003 IT1262967 B 23-07-1996	IT1992RM00586 31-07-1992	Vitti A.	INN. TEC S.r.l.	Machine and method for the production of olive oil without crushing the stones
EP557758 A 01-09-1993	DE4206006 C 16-09-1993 EP557758 B 06-03-1996 DE59301755 G 11-04-1996 ES2084401T T 01-05-1996	DE19924206006 27-02-1992	Düpjohann Josef; Geissen Klemens	Westfalia Separator AG	Vefrahren zur Gewinnung von Olivenöl. "Process for producing olive oil"
EP520239 A 30-12-1992	TR26982 A 12-09-1994 IT1248592 B 19-01-1995 ES2090414T T EP520239 B 14-08-1996 16-10-1996 GR3020821T T 30-11-1996	IT1991MI01796 28-06-1991	Wlassics I.	Ausimont S.p.A.	Detoxification of vegetation liquids
EP471132 A 19-02-1992		EP19900830375 13-08-1990	Ronzoni L.; Benigni G.; La Rovere G.	Societa Meridionale Argille Espanse S.p.A; Omnia Ecologia S.R.L	A thermic lithosynthesis process for disposal of liquid and solid industrial wastes
EP451430 A 16-10-1991	EP451430 B 18-05-1994 ES2055404T T 16-08-1994 IT224926Z Z 30-07-1996	IT19900010504U 09-04-1990	Sapia F.; Sapia P.; Sapia M.; Sapia S.	Sapia F.; Sapia P.; Sapia M.; Sapia S.	Plant to depollute wastewater, particularly, water from olive crushers

(*continued*)

Patent	Family member	Priority	Inventor	Applicant	Title
EP441103 A 14-08-1991	IT1240759 B 17-12-1993	IT19900047611 09-02-1990	Bernardini E.	Societa Ingegneria Bernardini Ernesto S.r.l.	Verfahren zum Abbauen der Verschmutzung in den Verarbeitungsanlagen der von Oliven herrührenden Wässern und diesbezügliche Anlage. "Process and installation for the degradation of the impurities in the wastewaters resulting from olive-treating plants"
EP421223 A 10-04-1991	EP421223 B 14-04-1993 IT1239280 B 19-10-1993 ES2041093T T 01-11-1993	IT19890021919 04-10-1988	Poglio A.	Poglio A.	A process and plant for disposal of organic effluents
EP330626 A 30-08-1989	PT89764 A 04-10-1989	IT19880000607 23-02-1980	Dionigi G.	Dionigi G.	Processus d'élimination, par évaporation, des eaux de décharge des moulins à huile obtenu en employant, dans une installation, du marc d'olives imbibé des susdites eaux. "Process for removing by evaporation discharge waters from oil mills by using olive stones soaked with said waters in a plant"
EP324314 A 19-07-1989		EP19880810013 14-01-1988	Cannnazza S. N.	Cannaza-Cantatore-Luigia	Trägerkörper und Reaktor zur biologischen Behandlung von Flüssigkeitenund Verwendung derselben. "Carrier and reactor for biological treatment of liquids and use thereof"
EP295722 A 21-12-1988	DE3720408 29-12-1988	DE19873720408 19-06-1987	Briccoli B. H.; Briccoli B. S.; Briccoli C.; Hussmann P.	Briccoli B. H.; Briccoli B. S.; Briccoli C.; Hussmann P.	Verfahren und Vorrichtung zur umweltschonenden Beseitigung des bei der Olivenpressung anfallenden Abwassers.

					"Method and apparatus for the pollution free disposal of wastewaters produced when crushing olives"
ES2198215 A 16-01-2004		ES20020001526 01-07-2002	Pinillos Villatoro José Luis; González Gomez Miguel Maria	Pinillos Villatoro José Luis; González Gomez Miguel Maria	Método de neutralización de aceitunas. "Olive neutralization method"
ES2186467 A 01-05-2003	ES2186467 B 16-09-2004	ES20000000516 03-03-2000	Brenes Balbuena Manuel; Castro Gomez-Millan Antonio de		Obtención de sustancias antioxidantes a partir de soluciones del proceso de elaboración de aceitunas de mesa. "Procedure is for obtaining phenolic extract with high concentration of antioxidants and involves ultrafiltration of solutions derived from preparation process of preserved table olives."
ES2180423 A 01-02-2003		ES20010000822 06-04-2001	Molina Alcaide Eduarda	Consejo Superior de Investigaciones Científicas	Pienso compuesto para alimnetación de rumiantes obtenido a partir de orujos secos y extractados procedentes de la extracción del aceite de oliva. "Compound feed for ruminants, based on olive dried stones and extracts from the extraction of olive oil"
ES2150360 A 16-11-2000	ES2150360 B 01-04-2001	ES19980001045 20-05-1980	Olmo Peinado José María	Olmo Peinado José María; Rojas Ruiz Sonsoles; Aigner Josef Konrad	Procedimiento integral para el tratamiento y reciclaje del alperujo. "Integrated process for the treatment and recycling of olive-mill waste from two-phase olive processing (alperujo)"
ES2144359 A 01-06-2000	ES2144359 B 16-10-2001	ES19980000416 26-02-1998	Fiestas Ros de Ursinos J. A.	Fiestas Ros de Ursinos J. A.	Procedimiento biotecnológico para la recuperación del aceite por el orujo húmedo. "Biotechnological method for recovering the oil retained in the moist olive waste"

(*continued*)

Patent	Family member	Priority	Inventor	Applicant	Title
ES2143939 A 16-05-2000	ES2143939 B 16-12-2000	ES19980000413 26-02-1998	Fernández-Bolaños Guzman Juan; Guillén Bejarano Rafael; Odriguez Arcos Rocio; Felizon Becerra Blanca; Heredia Oreno Antonia; Jiménez Araujo Ana	Consejo Superior Investigaciones Científicas	Procedimiento de obtención de manitol a partir se pulpa extractada de aceitunas. "Process for obtaining mannitol from pulp extracted from olives"
ES2139505 A 01-02-2000	ES2162739 A 01-01-2002 ES2162738 A 01-01-2002	ES19970001399 25-06-1997	Ratia Martínez Francisco	Ratia Martínez Francisco	Fertilizante foliar líquido a base de alpechín o jámila. "Liquid foliar fertilizer based on olive-mill wastewater or Jamila"
ES2122927 A 16-12-1998		ES19960000005 03-01-1996	Munoz Coronado Salvador	Investigación y Desarrollo Agro-Industrial, S.L	Procedimiento para la producción de vitamina B_{12} a partir de residuos contaminantes de la industria de la aceituna. "Process for the preparation of vitamin B_{12} from residual contaminants of the oil industry"
ES2116923 A 16-07-1998		ES19960001724 01-08-1996	Vega Cárdenas Enrique	Vega Cárdenas Enrique	Procedimiento de obtención de residuos reciclables derivados de la aceituna. "Process for obtaining recyclable waste products derived from olives"
ES2110912 A 16-02-1998	ES2110912 B 01-10-1998	ES19960000099 17-01-1996	García-Moreno Angel	García-Moreno Angel	Procedimiento integral para la industrialización de alpechines y su depuración en almazaras y centros de depaso de alpeorujos. "Integral process for the industrial utilization of the olive wastewater and its purification in olive-mills and centres for purifying the olive-mill waste from two-phase processing"

ES2108658 A 16-12-1997	WO9747561 A 18-12-1997	ES19960001316	Vila Reyes Joan	Ros Roca S.A,; Bio Specific Systems S.L.; Vila Reyes Joan	Procedimiento biológico de depuración de residuos líquidos de alta carga contaminante y/o alta toxicidad, en especial purines y alpechines . "Biological process for purifying liquid residues with high contaminating content and/or high toxicity, particularly liquid purine and olive-mill wastewater"
ES2103206 A 16-08-1997	ES2103206 B 01-04-1998	ES19950002528	Hidalgo Cicuéndez Arturo	Hidalgo Cicuéndez Arturo	Procedimiento de tratamiento, reciclaje y transformación de alpechín y alpeorujo en fertilizantes orgánicos puros. "Process for the treatment, recycling, and conversion of olive-mill wastewater and olive-mill waste from two-phase processing into pure organic fertilizers"
ES2101651 A 01-07-1997	ES2101651 B 01-03-1998	ES19950001340	Calaf Nolla Domenec	Serveis Tarragonins de Construcción i Arquitectura, S.L	Método de depuración de residuos orgánicos resultantes de la obtención del aceite. "Method for the purification of organic wastes resulting from the production of oil"
ES2092444 A 16-11-1996	ES2092444 B 01-07-1997	ES19950000186	Gómez Castellote Franscisco	Gómez Castellote Franscisco	Procedimiento para obtener energía eléctrica alternativa con la utilización de alpechín. "Process for obtaining alternative electrical energy through the use of olive-mill wastewater"
ES2088340 A 01-08-1996	ES2088340 B 01-05-1998	ES19930000976	Franco Pérez Jesús; Varo Reyes Jose	Franco Perez Jesús; Varo Reyes Jose	Instalación de eliminación de efluentes en las almazaras continúas mediante secado térmico. "Installation for eliminating effluent in continuous olive-mills by means of thermal drying"

(*continued*)

Patent	Family member	Priority	Inventor	Applicant	Title
ES2091722 A 01-11-1996	ES2091722 B 01-06-1997	ES19950000389 28-02-1995	Pedro Fuentes Martos	Pedro Fuentes Martos	Procedimiento para el secado de orujos generados en procesos de obtención de aceite de oliva. "Process for drying the olive waste from two-phase processing"
ES2087032 A 01-07-1996	ES2087827 B 16-07-1996	ES19940002050 29-09-1994 ES19940002391 22-11-1994	Cores Roldan Aldredo; Espejo-Saavedra Santa Eugenia	Ingenieria y Técnica Internacional del Clima Itic, S. L	Procedimiento para el tratamiento descotaminante de los residuos de almazaras e instalación para efectuar dicho tratamiento. "Process for the decontaminating treatment of olive oil mill wastes and installation for effecting such treatment"
ES2084564 A 01-05-1996	EP718397 A 26-06-1996 ES2084564 B 16-11-1996 ES2169985 A 16-07-2002	ES19940001934 13-09-1994	Lara Feria Antonio; Antolin Giraldo Gregorio; Peran González Jose Ramon	Tratamiento Integral de Alpechines Baena, S.L	Procedimiento de depuración y aprovechamiento de residuos líquidos (alpechines) y sólidos (orujos) producidos por una almazara para su aprovechamiento integral. "Process for the purification and utilization of liquid and solid waste products produced by an olive-mill"
ES2076899 A 01-11-1995	ES2076899 B 16-07-1996	ES19940000618 22-03-1994	Rodríguez Prieto Cristian	Fuentes Cardona S. A.	Nuevo método de tratamiento de orujos provenientes de la extracción de aceite de oliva. "New method for the treatment of the solid wastes originating from the extraction of olive oil"
ES2060549 A 16-11-1994	ES2060549 B 01-06-1995	ES19930000945 05-04-1993	Martínez-Nieto Leopoldo; García-Granados López de Hierro Andrés	University of Granada	Procedimiento de obtención de manitol y productos derivados a partir de alpeorujo procedente del proceso de aceituna según el procedimiento de dos fases.

					"Process for obtaining mannitol and derived products from waste from the olive process according to the two-phase process"
ES2056745 A 01-10-1994	ES2056745 B 01-04-1995	ES19930000490 10-03-1993	García-Granados López de Hierro Andrés	University of Granada	Procedimiento de obtención de manitol y productos derivados a partir de las ramas y hojas de olivo y pedúnculos de aceituna.
					"Process for obtaining mannitol and derived products from the branches and leaves of the olive tree and olive-mill wastewater and olive fruit stalks"
ES2051242 A 01-06-1994	ES2051242 B 01-12-1994	ES19920002386 12-11-1992	Fernández del Campo; Cuevas J. A; Soriano Carrillo Jesus	Productos de Bituminosos S.A.	Sistema para el aprovechamiento de alpechines en la estabilización de suelos. "Stabilization of soil using olive-mill wastewater"
ES2051238 A 01-06-1994		ES19920002366 24-11-1992	Calero Barcoj José; Martínez-Nieto Leopoldo; García-Granados López de Hierro Andrés	Ingeniería y Desarrollo Agro Industrial S.A.	Procedimiento de aprovechamiento del alpechín para la obtención de ácidos, fenoles, alcohols y derivados mediante extracción en contracorriente. "Process of using olive-mill wastewater for obtaining acids, phenols, alcohols, and derivatives by means of counter current extraction"
ES2048667 A 16-03-1994	ES2048667 B 16-08-1994	ES19920001727 17-08-1992	Artacho del Pino Antonio	Oleicola El Tejar Ntra. Sra de Araceli, Sdad Coop. Lta.	Procedimiento para la extracción de aceite de orujo de oliva sin utilización de disolventes orgánicos. "Process for extracting oil from olive stones without using organic solvents"
ES2043507 A 16-12-1993	ES2043507 B 16-07-1994	ES19910001437 14-06-1991	Ollero de Castro Pedro	Ollero de Castro Pedro	Equipo de evaporación-concentración del alpechín. "Equipment for evaporating-concentrating olive-mill wastewater (OMWW)"

(continued)

Patent	Family member	Priority	Inventor	Applicant	Title
ES2041220 B 16-05-1994	ES2051244 A 01-06-1994 ES2051245 A 01-06-1994	ES19920000842 21-04-1992	Dupuy de Lome Lozano Enrique; Martínez-Bordiu; Ortega Andre	Cidespa-Centro de Ingeniería Diseño y Electromecánico	Un proceso para la depuración de los efluentes líquidos procedentes de la industria azucarera y de la fabricación de aceite de oliva (alpechines). "Process for purifying the effluents from the sugar and olive-mill industries (olive-mill wastewater)"
ES2037606 A 16-06-1993	ES2037606 B 01-02-1994	ES19910002646 27-11-1991	Arroyo Salas José María; Llamas Marcos Argimiro; Galilea Eguizabal Purificatión	Arroyo Salas José María; Llamas Marcos Argimiro; Galilea Eguizabal Purification	Mezclas alpechín -sustrato orgánico útiles como abonos y su procedimiento de obtención. "Mixtures of olive-mill wastewater-organic substrate used as fertilizers and their process of making"
ES2032162 A 01-01-1993		ES19900002231 21-08-1990	De Lara García Rafael	De Lara García Rafael	Procedimiento complementario para la depuración del alpechín. "Supplementary process for the purification of olive-mill wastewater"
ES2028497 A 01-07-1992		ES19890003273 28-09-1989	Dorsch Serrano Fernando	Dorsch Serrano Fernando	Procedimiento mejorado de tratamiento y depuración de alpechines y aprovechamiento de residuos en fábricas de aceite de oliva. "Improved process for the treatment and purification of olive-mill wastewater and the utilization of wastes in olive oil plants"
ES2024369 A 16-02-1992		ES19910000159 22-10-1991	De Lara García Rafael	De Lara García Rafael	Proceso simultáneo de combustión aplicado a un tratamiento depurativo de alpechín, y equipo necesario. "Simultaneous combustion process applied to a treatment for the purification of olive-mill wastewater and the equipment used"

ES2021191 A 16-10-1991	ES19900000486 19-02-1990	Ballester Diaz Lorenzo; García Vinao Agustin F.; Pérez Amer Jorge Pablo	Fabrica de San Carlos, S.A	Instalación para la depuración integral del alpechín. "Installation for the integral purification of olive-mill wastewater"
ES2019830 A 01-07-1991	ES19900001189 26-04-1990	De Lara García Rafael	De Lara García Rafael	Procedimiento de tratamiento de alpechines para su depuración y obtención de subproductos con aprovechamiento agronómico o industrial. "Procedure for treatment of olive waste water for the purification thereof and obtaining by-products with agronomic or industrial uses"
ES2016471 A 01-11-1990	ES19890002174 21-06-1989	Brenes Balbuena M.; Sánchez Roldán F.; García García P.; Garrido Fernández A.	Consejo Superior Investigaciones Cientificas	Procedimiento para la regeneración de salmueras de fermentación de aceitunas verdes estilo español mediante ultrafiltración-ósmosisc inversa y floculación previa con bentonita. "Process for regenerating Spanish-style green-olive fermentation brines by means of ultrafiltration/osmosis and prior flocculation with bentonite"
ES2016470 A 01-11-1990	ES19890002171 21-06-1989	Brenes Balbuena M.; Sánchez Roldán F.; Garrido Fernández A.	Consejo Superior Investigaciones Cientificas	Procedimiento de regeneración de salmueras de aceitunas y otros productos vegetales para su utilización posterior. "Process for regenerating olive brines and other vegetable products for their subsequent use"
ES2011366 A 01-01-1990	ES19880002356 27-07-1988	Dorsch Serrano Fernando	Dorsch Serrano Fernando	Procedimiento de tratamiento de alpechines en fábricas de aceite de oliva. "Process for the treatment of olive-mill wastewater in olive oil plants"

(*continued*)

Patent	Family member	Priority	Inventor	Applicant	Title
ES2010535 A 16-11-1989		ES19880001257 22-04-1988	Fernández Jorquera Sébastián	Fernández Jorquera Sébastián	Procedimiento físico para la reducción del consumo de agua en las fábricas de extracción del aceite de oliva (almazaras). "Physical process for the reduction of water consumption in olive-mills"
ES2009267 A 16-09-1989		ES19880001291 27-04-1988	Romano Camancho Jose Maria	Romano Camancho Jose Maria	Procedimiento químico-mecánico de limpieza de aguas residuales. "Physico-mechanical process for the cleaning of residual waters"
ES2006904 A 16-05-1989		ES19880001171 15-04-1988	Jiménez Rodríguez José Luis	Jiménez Rodríguez José Luis	Procedimiento de obtención de aceite a partir del orujo . "Process for obtaining oil from olive cake"
ES2004269 A 16-12-1988	GB221772 A 01-11-1989 FR2630730 A 03-11-1989 GB2217729 B 23-10-1991	ES19870000833 25-03-1987	Travería Casanova Tomás	TADEVAL S.A.	Procedimiento para la obtención de escualano. "Process for obtaining squalane"
ES2002555 A 16-08-1988		ES19860003418	Sánchez Moral P.; Catano Vazquez Alejandro	Sánchez Moral P.	Nuevo procedimiento de obtención de abono orgánico. "New process for obtaining organic fertilizer"
ES8708149 A 01-12-1987	IT1214604 B 18-01-1990	IT19850020580 06-05-1985	Pieralisi Gennaro; Fedeii Enzo; Lanzan Armando; Ponzetti Araido	Stazione sperimentale per le industrie degli oli e dei grasse ; Nuova M.A.I.P. – machine agricol industriali	Procedimiento de eliminación de las aguas de vegetación en la extracción de aceite. "Process for the elimination of vegetation waters resulting from olive oil extraction"
ES8706800 A 16-09-1987		ES19860555727 05-06-1986	Casanova Traveria Tomas; Baco	Casanova Traveria Tomas; Baco Mata	Procedimiento para la reutilizacion del alpechín como agua de proceso in

		Mata Jose Enrique; Garzon Martínez Angel Manuel	Jose Enrique; Garzon Martínez Angel Manuel	instalaciones de extraccion continua de aceite de oliva. "Process of reusing olive-mill wastewaters as process water in continuous olive oil extraction"
ES8607039 A 01-05-1986	ES19840533670 13-06-1984	Sempere Eugenio Bellido	Sempere Eugenio Bellido	Nuevo procedimiento e instalación de depuración de alpechín y/o aguas residuales de las industrias de obtención de eceite de oliva. "New process and installation for the purification of olive-mill wastewater and/or residual wates from the olive oil industry"
ES8602102 A 01-03-1986	ES19850539827 25-01-1985	Don Joan Casanelles I Jene	Derivan S.A.	Procedimiento para la obtención de escualeno y escualano a partir de subproductos de la refinación fíica y/o desodorización de aceites vegetale. "Process for preparing squalene and squalane from by-products from the refining of vegetable oils"
ES8505401 A 01-09-1985	ES19830528494 28-12-1983	Gonzalo Valenzuela Ruiz	GonzaloValenzuela Ruiz	Procedimiento ciclico para la recuperatión integral de suproductos de la almazara. "Cyclic process for the recovery of by-products from olive press"
ES8404708 A 01-08-1984	ES19830520734 17-03-1983	Manuel González Longoria	Tecnicas y Sevicios Urbanos S.A.	Procedimiento para la fabrication de un combustible organico solido. "Process for the fabrication of solid organic fuel"
ES8402554 A 01-05-1984	ES19820518595 27-12-1982	Don Francisco M. Vidal Torrents; Don Miguel Montero Puig; Don Jose Lostao Camon	Tecnicas Y Servicios Urbanos S.A.	Proceso para la obtención de un fertilizante de origen organico. "Process for obtaining organic fertilizers "

(continued)

Patent	Family member	Priority	Inventor	Applicant	Title
ES8307286 A 16-10-1983		ES19810502987 11-06-1981	D. Rafael Navarro Gamero; D. Juan Miguel Pulpillo Vilches; José Alba-Mendoza	Consejo Superior Investigaciones Científicas; D. Rafael Navarro Gamero; D. Juan Miguel Pulpillo Vilches	Mejoras introducidas en la patente principal no 497902 por "Procedimiento de depuración por separación-recuperación total de sólidos en suspensión y aceite contenido en alpechines" Improvements introduced to the principal patent no 497902 for "Process for the total purification-recovery of the solids in suspension and olive contained in olive-mill wastewater"
ES820395 A 16-01-1982		ES19800497902 18-12-1980	D. Rafael Navarro Gamero; D. Juan Miguel Pulpillo Vilches; José Alba-Mendoza	Consejo Superior Investigaciones Científicas; D. Rafael Navarro Gamero; D. Juan Miguel Pulpillo Vilches	Procedimiento de depuración por separación-recuperación total de sólidos en suspensión y aceite contenido en alpechines. "Process for the total purification-recovery of the solids in suspension and olive contained in olive-mill wastewater"
ES348517 A 01-09-1969		ES19670348517 22-12-1967	José Garrido Márquez	Patronato de Invesigación Cientifica y Técnica "Juan de la Cierva" del Consejo Superior de Investigaciones Científicas	Procedimiento para la obtención de un abono organico mediante fermentation espontanea o dirigida del alpechín concentrado sobre turbas. "Process for obtaining organic fertilizer by spontaneous or controlled fermentation of concentrated olive-mill wastewater and peat"
FR2838451 A 17-10-2003		FR20020004707 16-04-2002	Bardon Dominique; Perichaud Alain	Sarl Etablissement Bardon	Procédé de dépollution de la margine, utilisation du résidue solide obtenu à titre de colle à bois ou de liant pour la fabrication de panneaux de bois aggloméré. "Process for the depollution of olive-mill wastewater, use of the solid residue

FR2830195 A 04-04-2003	JP2003113069 A WO03028692 A	FR20010012802 03-10-2001	Lintner Karl	SEDERMA SA	obtained as wood glue or for the manufacture of chipboard" Compositions cosmétiques et dermopharmaceutiques pour les peaux a tendance acnéique. "Cosmetic and dermopharmaceutical compositions for skins prone to acne"
FR2825022 A 29-11-2002		FR20010006822 23-05-2001	Stoltz Corinne; García Christine; Schubnel Laurent	Societé d'exploitation de produits pour les industries chimiques – SEPPIC	Composition de polyphenols d'olives. Uutilisation comme actif cosmetique et dietetique. "Composition of olive polyphenols. Use as cosmetics and dietary agents"
FR2724922 A 29-03-1996	WO9609986 A 04-04-1996 AU3570295 A 19-04-1996	FR19940011605 28-09-1994 WO1995FR01246 27-09-1995	Raes Jean Paul; Danda Sylvain; Bonfill Jean; Morales Jose; Pescher Yvette; Castelas Bernard; Rabatel Francois	RHONE– POULENC CHIMIE	Procédé et installation de traitement d'un milieu liquide contentant des déchets organiques. "Method and apparatus for treating a liquid medium containing organic waste"
FR2715590 A 04-08-1995	WO9521136 A 10-08-1995 CA2181939 A 10-08-1995 AU1582695 A 21-08-1995 EP741672 A 13-11-1996 JP9510651T T 28-10-1997 DE69519283D A 22-06-1999 AT197286T T 15-11-2000 DK741672T T 05-03-2001	FR19940001093 01-02-1994 WO1995FR00113 01-02-1995	Raes Jean-Paul; Danda Sylvain; Bonfill Jean; Morales Jose; Pescher Yvette; Castelas Bernard; Rabatel Francois	RHONE– POULENC CHIMIE	Procédé d'épuration d'un milieu contenant des déchets organiques. "Method for purifying an organic waste-containing medium"

(continued)

Patent	Family member	Priority	Inventor	Applicant	Title
FR2688383 A 17-09-1993	GR93100091 A 30-11-1993 ES2043534 A 16-12-1993 ES2043534 B 16-05-1994 PT101109 A 30-06-1994 TR26891 A 22-08-1994 IT1256649 B 12-12-1995	ES19920000542 11-03-1992	Fuentes Martos Pedro	Fuentes Cardona S.A.	Procédé pour l'obtention d'uile d'olive et élimination du liquide résiduel comme sous-produit. "Method for obtaining olive oil and removing the residual liquid as a by-product"
FR2620439 A 17-03-1989		FR19870012694 14-08-1987	Laulan Pierre-Yves; Thelier Yves	Societé Generale pour les techniques nouvelles S. G. N.	Procédé et dispositif de traitement per fermentation méthanique d' eaux résiduaires lipidiques. "Process and device for the treatment of lipid-containing sewage by methane fermentation"
FR2576303 A 25-07-1986		FR19850000799 21-01-1985	Don Joan Casanelles I Jene	Derivan S.A.	Procédé d'obtention de squalène et de squalane a partir de sous-produits du raffinage des huiles végétales. "Process for preparing squalene and squalane from by-products from the refining of vegetable oils"
GR1004115 B 21-10-2003	EP1359125 A 05-11-2003	GR20020100208 30-04-2002	Lazaridi Konstantia Aikaterini; Kyriacou Adamantini Alexandrou; Kotsou M. Maria Georgiou; Tassiopoulou Stavroula	Lazaridi Konstantia Aikaterini; Kyriacou Adamantini Alexandrou; Kotsou Maria Georgiou; Tassiopoulou Stavroula Theodorocu; Pilidis Georgios Alexandrou	Συνδυασμένη βιοτεχνολογική και χημική μέθοδος επεξεργασίας υγρών αποβλήτων από τήν παραγωγή βρώσιμων ελιών. "Combined biotechnological and chemical method of treating liquid wastes derived from factories producing edible olives"

GR1004159 B 20-02-2003	WO3066034 A 14-08-2003 US2003185921 A 02-10-2003	GR20020100072 08-02-2002	Theodorou; Pilidis Georgios Alexandrou Galaris D.; Magiatis P.; Mitakou S.; Panaiteskou L. S.; Skaltsounis A. L.; Foteinos S.	Lavipharm S.A.	Αξιολόγηση της αντιοξειδωτικής και κυτταροπροστατευτικής ικανότητας της υδροξυτυροσόλης και εκχυλισμάτων από ελιές, ελαιόλαδο, απόβλητα ελαιοτριβείων και φύλλα ελαιοδένδρων και οι εφαρμογές τους. "Evaluation of the antioxidant and cell-protecting properties of the hydroxytyrosol and extracts from olives, olive oil, olive-mill waste, and olive tree leaves and applications thereof"
GR1003920 B 27-06-2002		GR20010100256 22-05-2001	Elefsiniotis Georgiou L.	Elefsiniotis Georgiou L.	Καθαρισμός υγρών αποβλήτων επεξεργασίας ελαιοτριβείου με ανάκτηση ελαιολάδου. "Cleaning of liquid effluents from processing in an olive-mill with olive oil recovery"
GR1003611 B 29-06-2001		GR19990100348 08-10-1991	Xenopoulou Katerina	Xenopoulou Katerina	Χρήση του μεσογειακού φύκους *Posidonia* oceanica για την παραγωγή οργανικού κομπόστ ή κομπόστ για τη γεωργία με την μέθοδο της οργανικής κομποστοποίησης, με συγκοποστοποίηση αποβλήτων γεωργικών ή ζωικών ή βιομηχανικών μονάδων. "Use of the mediterranean sea grass *Posidonia oceanica* for the production of organic compost and compost for agriculture with co-composting of organic waste from agricultural, animal, or industrial units"

(*continued*)

Patent	Family member	Priority	Inventor	Applicant	Title
GR1003583 B 22-05-2001		GR20000100059 2-02-2000	Stavrakakis Minoos Emmanouil	Stavrakakis Minoos Emmanouil	"Machine for peeling olive fruit — Olive oil producing method"
GR1003558 B 30-04-2001		GR20000100182 29-05-2000	Venetsianos E. T.	Venetsianos E. T.	Νέο σύστημα κατεργασίας, μεταποίησης και χρήσης λυμμάτων ελαιών-ελαιολάδου για την παραγωγή χημικών προϊόντων ενέργειας και επαναχρησιμοποίησης νερού. "New system for processing, conversion, and use of olive and olive oil effluents for the production of chemical products, energy, and for reusing water"
GR97100075 A 30-10-1998	GR1003486 B 30-11-2001	GR19970100075 26-02-1997	Vlysidis Apostolos	Vlysidis Apostolos	Μέθοδος ωφέλιμης αξιοποίησης υγρών αποβλήτων υψηλού οργανικού φορτίου μετά συγκατεργασίας κομποστοποίησης-χουμοποίησης με στερεά οργανικά απορρίματα και γεωργικά παραπροϊόντα. "Method of useful exploitation of liquid effluents of a high organic load by co-processing composting-topsoil formation with solid or organic waste and agricultural by-products"
GR1002131 B 07-02-1996		GR19950100032 30-01-1995	Georgoudis D .	Georgoudis D.	Olive-derived pulp produced by oil-pressed vegetable liquids
GR1001839 B 20-03-1995			Georgoudis D.	Georgoudis D.	Μέθοδος επεξεργασίας καθαρισμού αποβλήτων ελαιοτριβείου με την διαδικασία φιλτραρίσματος αραίωσης. "Method of purifying olive-mill effluents with the filtration–dilution process"
GR93100432 A 31-07-1995		GR19930100432 03-11-1993	Chatzipavlidis I.; Flouri F.; Balis C.	Chatzipavlidis I.; Flouri F.; Balis C.	"Bio-fertilization of olive-mills liquid wastes"

GR89100788 A 15-03-1991	IT1227676 B 23-04-1991 ES2019015 A 16-05-1991 GR1000829 B 25-01-1993	IT19880022835 02-12-1988	Galvagno Mauro; Penna Gino Della; Robertiello Andrea	Eniricerche S.p.A.	Μέθοδος καθάρσεως των φυτικών υγρών παραγομένων υπο ελαιοπρεσσών. "Method of purifying the vegetation liquors produced by oil presses"
GR88100368 A 08-03-1989	IT1204691 B 10-03-1989 ES2007230 A 01-06-1989	IT19870020811 05-06-1987	Brusadelli Enrico; Canepa Pietro; Rognoni Umberto	Snia Fibre S.p.A.	Διαδικασία για την επεξεργασία των αποβλήτων της βιομηχανίας ελαιοτριβείων. "Process for the treatment of the effluents of the oil-mill industry"
GR88100203 A 31-01-1989		ES2009577 A 01-10-1989 IT1206060 B 14-04-1989 GR1000967 B 16-03-1993	Murenna Fabio; D' Oria Daniele; Chiacchio Raffaele; Volpicelli Gennaro	Risvet-Ricerca Sviluppo e Tecnologia S.r.L. and Millipore S.p.A.	Καθαρισμός αποβλήτων ελαιοτριβείου. "Oil press mill waste purification"
GR871461 A 04-02-1988		GR19870001461 21-09-1987	Vaccarino Carmelo	Sviluppo Nuove Technologie S. N.	"Process for the integral use of olive vegetation liquors and of other agroindustrial waste liquors by mixing with olive husks"
GR870652 A 30-06-1987	IT1190283 B 16-02-1988 ES2005196 A 01-03-1989	IT19860047943 28-04-1986	Grappelli Adriana; Galli Emanuele; Palma Grazio; Tomati Umberto	Consiglio Nazionale Ricerche	Procedimento per la depurazione di reflui vegetali agricoli, in partico lare acque di vegetazione. "Process for the purification of agricultural vegetable waste fluids in particular of vegetation waters"
GR61852 A 30-01-1979		GR19780057633 11-11-1978	Margaritis M.	Margaritis M.	Μηχάνημα αφαίρεσης του ελαιολάδου και των υποπροϊόντων από τον καρπό της ελιάς. "Method and apparatus for oil removal from olive pulp"
HR20010028 A 31-08-2002		HR20010000028 12-01-2001	Wolf Predrag	Opatija Inaeenjering D.O.O.	"Apparatus for the treatment of waste originating from the processing of olives, including the composting of solid waste"

(continued)

Patent	Family member	Priority	Inventor	Applicant	Title
ITRM20040084 A 00-00-2004		IT2004RM000084 00-00-2004		National Research Council; Cilento and Vallo del Daino Antional Park	"Method and apparatus for the treatment of oil mill effluents"
ITTO990151 A 31-05-1999		IT1999TO00151 01-03-1999	Grando Frederico	Grando Frederico	Procedimento di trattamento per acque di vegetazione derivanti da lavorazioni di prodotti alimentari e prodotti fertilizzati ottenuti utilizzando le acque depurate mediante detto trattamento delle di vegetazione. "Process for the treatment of vegetation waters derived from processing of alimentary products and fertilizers obtained by using the wastewater by means of said treatment"
ITRM970098 A 24-08-1998	IT1290945 B 14-12-1998	IT1997RM00098 24-02-1997	Capasso Renato; Colombo Claudio; Scognamiglio Francesco; Violante Antonio	Universita degli studi di Napoli Fede	"Process of absorbing, on a solid matrix, vegetation effluent water from oil crushers"
ITRM950104 A 21-08-1996	IT1278025 B 17-11-1997	IT1995RM00104 21-02-1995	Bologna Mauro; Bologna Claudio	Tecnologie 2000 S.r.l.	Separatore centrifugo orizzontale, in particolare per la produzione di olio, con estrazione accelerata della fase liquida. "Horizontal centrifugal separator for rapid oil extraction from liquid phase"
ITRM950298 A 11-11-1996	IT1276576 B 03-11-1997	IT1995RM00298 10-05-1995	Colantoni Rolando; Bologna Mauro; Bologna Claudio Pa	Tecnologie 2000 S.r.l.	Snocciolatore a lame per la preparazione di pasta di olve e apparato mobile per la produzione non industriale di olio. "Stoner with blade for the preparation of olive paste and mobile equipment for the non-industrial production of oil"

ITBO950012 A 19-07-1996	EP722921 A 24-7-1996 EP722921 B 07-05-2003 DE69627908D D 12-06-2003	IT1995BO00012 19-11-1995	Faccini Giuseppe	Faccini Giuseppe	"Liquid additive for enriching natural and chemical fertilizers"
IT1244520 B 15-07-1994		IT1991MI00201 28-01-1991	Potenz Domenico; Righetti Ettore; Della Monica Mario	Potenz Domenico; Della Monica Mario	Procedimento ed impianto per il trattamnento delle acque di scarico dell' industria dell'olio d'oliva. "Process and plant for the treatment of the wastewaters from the olive oil industry"
IT1231601 B 18-12-1991		IT19890048008 29-05-1989	Bonfitto Emanuele; Giacomo Gabriele Di; Brunetti N icola; Jacoboni Sergio; Re Giovanni Del	E.R.S.A. –Ente Regionale di Sviluppo Agricolo in Abruzzo; E.N.E.A. –Comitato Nazionale per la Ricerca e per lo Sviluppo dell' Energia Nucleare e delle Energia Altrnative; Universita' degli studi de l'Aquila	Procedimento per la depurazione delle acque di vegetazione effluenti dai frantoi oleari. "Process for the purification of olive-mill vegetation water"
IT1214359 B 10-01-1990		IT19860018701 18-02-1986	Vaccarino Carmelo	S. N. C. Sviluppo Nuove Tecnologie	Processo per la depurazione delle acque vegetative dei frantoi di olive e di altri liquidi ad alto carico organico inquinante
IT1211951 B 08-11-1989		IT19870048671 04-12-1987	Ernesto Bernardini	S.I.B.E S.R.L.	Procedimento per il trattamento delle acque di vegetazione provenienti d'alla lavorazione d'elle olive. "Process for the treatment of vegetation water originated from olive processing"

(continued)

Patent	Family member	Priority	Inventor	Applicant	Title
IT1206049 A 05-04-1989		IT19870048071 18-06-1987	Ernesto Bernardini	S.I.B.E S.R.L.	Procedimento per la depurazione delle acque di vegetazione derivanti dalla lavorazione delle olive ed impianto per la sua attuazione "Process and plant for the treatment of vegetation water derived from olive processing"
IT1191528 A 23-03-1988		IT19860064801 15-01-1986	Fortunato Vitorio	Fortunato Vitorio	Trattamento chimico-fisico delle acque discarico dei frantoi oleari e o delle acque di vegetazione delle olive. "Physico-chemical treatment of wastewater obtained during olive oil production"
IT1149119 B 03-12-1986		IT19820049480 12-11-1982	Annesini Maria Cristina; Giona Alessandro Romano; Gironi Fausto ; Pochetti Fausto	S.P.I –Sviluppo Processi Industriali S.R.L	Procedimento per il tratiamento del le acque di scarico degli oleifici per distillazione con recuperodei prodotti volatili. "Process for the treatment of oil-containing wastewater by distillation with recovery of volatile products"
IT1110321 B 23-12-1985		IT19780002102 16-01-1978		Socogin S.R.L.	Impianto per la depurazione delle acque di vegetazione derivate dalla lavorazione delle olive. "Olives processing wastewater treatment"
IT1098424 B 07-09-1985		IT19780027294 04-09-1978	Diefenbach Attilio	Diefenbach Attilio	Procedimento per il trattamento di acque di scarico degli oleifici per la spremitura delle olive, più propriamente chiamate acque di vegetazione
JP2003019192 A 21-01-2003		JP20010207397 A 09-07-2001	Kusumaru Masafumi; Mori Miki; Ikemoto Takeshi	KANEBO LTD	"Deodorant and deodorant for tobacco odor"
JP2000319161 A 21-11-2001		JP19990061373 09-03-1999	Ikemoto Takeshi; Fukubayashi	KANEBO LTD	"Skin cosmetic"

		JP19990312081 02-11-1999	Tomoko; Haratake Akinori; Kaneyama Hiroshi		
NL1005938C C 03-11-1998		NL19971005938 01-05-1997	Bakker Simon Marinus	W. J. Wiendels Beheer B. V.	Werkwijze voor het behandelen van organische producten, zoals bijvoor-beeld olijven, in zoutoplossingen. "Regenerating spent brine used to treat e.g. olives"
PT85790 A 01-10-1987	ES2004816 A 01-02-1989 IT1215079 B 31-01-1990	IT19860018708 24-09-1986	Vaccarino Carmelo	S.N.T. Sviluppo Nuove Tecnologie S.N.C	Processo para a utilização integral do líquido vegetal das azeitonas e de outros refluentes agro-industriais por mistura com o baganho da azeitona.
PT69240 A 01-03-1979	ES477805 A 16-10-1979 GR67273 A 26-06-1981 IT1110457 B 23-12-1985 IT1110458 B 23-12-1985 IT1166611 B 05-05-1987 IT1166612 B 05-05-1987	IT19780020348 16-02-1978 IT19780020349 16-02-1978 T19790019872 02-02-1979 IT19790019873 02-02-1979	Pruna Tudor	Pruna Tudor	Metodo par ottenere la separazione dell' alcol dalla morchia come sottoprodotto, per scopi alimnetari, e per ridurre il potere inquinante delle acque dalla stessa. "Method for alcohol recovery from olive vegetation water"
PT69785 A 01-07-1979	GR69996 A 23-07-1982 IT1162542 B 01-04-1987 IT1109184 B 16-12-1985 IT1109185 B 16-12-1985 IT1165093 B 22-04-1987 ES481766 A 16-01-1980	IT19780024783 21-06-1978 IT19780024782 21-06-1978 IT19790023552 13-06-1979 IT19790023553 13-06-1979	Pruna Tudor	Pruna Tudor	Processi per il recupero e l'utilizzazione dei sottoprodotte discarto dell' industria olivicola ai fini ecologici e della valorizzazione integrale dell' industria olivicola stessa. "Process for the recovery of olive indus-trial waste by-products ensuring better ecological processing"

(continued)

Patent	Family member	Priority	Inventor	Applicant	Title
PT64109 A 22-06-1976	FR2316881 A 04-02-1977	FR19750022280 09-07-1975 PT19750064109 25-07-1975	Vitagliano Michele	Vitagliano Michele	Pâtée pour emplois zootechniques produite à partir des sous-produits des olives et procédé pour sa fabrication. "Paste for feeding animals obtained from the by-products of olives and its process of making"
US2003108651 A 12-06-2003	WO03068171 A 21-08-2003 WO2004005228 11-01-2004	US20020190043 05-07-2002 US20000230535P 01-09-2000 US20010944744 31-08-2001 US20020356847P 13-02-2002	Crea Roberto	Creagri Inc.	Hydroxytyrosol-rich composition from olive vegetation water and methods of use thereof
US5801127 A 01-09-1998		US19970951546 16-10-1997	Duhon J. J. Sr .	Duhon J. J. Sr.	Olive pulp additive in drilling operations
US4370274 B 25-01-1983	US4452744 B 05-06-1984 US4522119 B 11-06-1985	US19800220170 23-12-1980 US19840600134 13-04-1984	Finch H. E. and Trapanese Salvatore P	Finch H. E.; Trapanese Salvatore P.	Olive oil recovery
US3975270 A 17-08-1976		US19750637250 03-12-1975	Roy Teranishi; Donald J. Stern	The United States of America as represented by the Secretary of Agriculture	Process for recovering usable olive-processing liquor from olive-processing waste solution
US3732911 A 15-05-1973		US19710124895 16-03-1971	Edison Lowe; Everett L. Durkee	The United States of America as represented by the Secretary of Agriculture	Process for reconditioning spent olive-processing brines
WO2005003037 A 13-01-2005	GR2003100295 A 28-03-2005	GR2003100295 08-07-2003	Castanas E.; Andricopoulos N.; Mposkou G.;	EMERGO (CYPRUS) LTD; Castanas E.;	A method for the treatment of olive-mill wastewaters

WO2005002364 A 13-01-2005		WO2003ES00333 03-07-2003	Pinillos Villatoro José Luis; González Gomez Miguel Maria	Andricopoulos N.; Mposkou G.; Vercauteren J. Pinillos Villatoro José Luis; González Gomez Miguel Maria	Método de neutralización de aceitunas. "Olive neutralization method"
WO2004110171 A 23-12-2004		GB20030014294 19-06-2003	Zumbe A lbert	NatraceuticaL S. A.; Zumbe Albert	Olive powder
WO2004064978 A 05-08-2004		AU20030900226 21-01-2003	Lobban Sarah Elisabeth Chenery; Lobban Mark Richard	Lobban Sarah Elisabeth Chenery; Lobban Mark Richard	A filter system
WO2004009206 A 29-02-2004	US2004176647 A 09-09-2004 ES2199069 A 01-02-2004	ES20020001671 01-02-2004	Mercedes Ballesteros Perdices; Maria Jose Negro Alvarez ; Paloma Manzanares Secades; Ignacio Ballesteros Perdices; Jose Miguel Oliva Domínguez	Centro de Investigaciones Energeticas	Process to exract phenolic compounds from a residual plant material using a hydrothermal treatment
WO03080006 A 02-10-2003	DE10213019 A 02-10-2003	DE20021013019 22-03-2002	Bruchwald-Werner Sybille; Griesbach Ute	Cognis Deutschland GMBH & CO. KG.; Bruchwald-Werner Sybille; Griesbach Ute	Verwendung von Extrakten des Olivenbaumes als Antischuppenmittel. "Use of extracts from olive trees as antidandruff agents"
WO03079794 A 02-10-2003	DE10213031 A 02-10-2003	DE20021013031 22-03-2002	Bruchwald-Werner Sybille; Griesbach Ute	Cognis Deutschland GMBH & CO. KG.; Bruchwald-Werner Sybille; Griesbach Ute	Verwendung von Extrakten des Olivenbaumes in Wasch- Spüll- und Reinigungs-mitteln. "Use of olive tree extracts in detergents, rinsing agents, and cleaning agents"

(continued)

Patent	Family member	Priority	Inventor	Applicant	Title
WO3066034 A 14-08-2003	US2003185921 A 02-10-2003 GR2002100071 A 15-10-2003 GR1004402 B 19-12-2003	GR2002100071 08-02-2002 GR2002100072 08-02-2002	Aligiannis N.; Galaris D.; Magiatis P.; Mitakou S.; Panaiteskou L. S.; Skaltsounis A. L.; Foteinos S.	Lavipharm S.A.	Compounds and compositions derived from olives and methods of uses thereof
WO03000601 A 03-01-2003	ES2182704 A 01-03-2003	ES20010001504 25-06-2001	Toro Gálvez José	Toro Gálvez José	Sistema de depuración de aguas residuales procedentes del procesado de la aceituna mediante aireación-neutralización-filtracion en carbón-ozonización. "System for purifying wastewater originating from olive processing by means of aeration , neutralization, active carbon filtration, and ozonation"
WO02064537 A 22-08-2002	ES2177457 A 01-12-2002 ES2172429 A 16-09-2002 EP1369407 A 10-12-2003	ES20010000346 15-02-2001 ES20000002422 15-02-2001	Fernández-Bolaños Guzmán Juan; Guillén Bejarano Rafael; Rodríguez Arcos Rocío; Rodríguez Gutiérrez Guillermo; Heredia Moreno Antonia; Jiménez Araujo Ana	VONSEJO SUPERIOR DE INVESTIGACIONES-CIENTIFICAS	Prodedimiento de obtención de hidroxitirosol purificado a partir de productosy subproductos derivatos del olivo. "Method for obtaining purified hydroxytyrosol from products and by-products derived from the olive tree"
WO0218310 A 07-03-2002	AU8858001 A 07-03-2002 US2002058078 A 16-05-2002 US6416808 B 09-07-2002 US2002198415 A 26-12-2002	US20000230535P 01-09-2000	Crea Roberto	Creagri Inc.	Method of obtaining a hydroxytyrosol-rich composition from vegetation water

WO0212159 A 14-02-2002	CA2420893 A 07-03-2003 US200380651 A 12-06-2003 AU4468301 A 18-02-2002 CA2419041 A 10-02-2003 EP1310478 A 14-05-2003	JP20000240347 08-08-2002 WO2001JP02788 30-03-2001	Kuno Noriyasu; Shinohara Gou	Nisshin Oil Mills Ltd.	Process for producing oleanolic acid and/ or maslinic acid
WO0145514 A 28-06-2001	AU2267401 A 03-07-2001 US2002004077 A 10-01-2002 US6358542 B 19-03-2002 US6361803 B 26-03-2002	US19990467439 20-12-1999 WO2000US34096 15-12-2000 US20000656949 07-09-2000	Rabovskiy Alexandre; Cuomo John	Usana Inc.	Antioxidant compositions extracted from olives and olive by-products
WO0004794 A 03-02-2000	AU5121699 A 14-02-2000 US6165475 A 26-12-2000 US6197308 B 06-03-2001 EP1098573 A 16-05-2001 AU746712 B 02-05-2002	US19980093818P 23-07-1998 US19990359150 22-07-1999 WO1999US16549 22-07-1999 US20000491680 26-01-2000	Crea Roberto; Caglioti Luciano	Creagri Inc.	Water-soluble extract from olives
WO9935097 A 15-07-1999	AU1678899 A 26-07-1999 GR98100001 A 30-09-1999 GR1003258 B 12-11-1999 EP968137 A 05-01-2000	GR19980100001 02-01-1998 WO1999GR00001 04-01-1999	Siskos Dimitrios	Siskos Dimitrios	Wastewater treatment plant for olive oil processing effluents comprising a rotating biological contactor with the addition of linear or circular motion

(*continued*)

Patent	Family member	Priority	Inventor	Applicant	Title
WO9804679 A 05-02-1998	ES2112215 A 16-03-1998 AU3623997 A 20-02-1998 EP857779 A 12-08-1998 ES2112215 A 01-10-1998	ES19960001666 1 26-07-1996 WO1997ES00189 24-07-1997	Ferrer Munoz Estrella; Gibello Prieto Alicia; Martín Fernández Margarita; Sanz Perucha Jesus ; Blanco Alvarez Jesus	Universidad Complutense de Madrid	Procedimiento para la biodegradacion de compuestos aromáticos y sintesis de pigmentos y colorantes, alcaloides y polimeros utilizanto la cepa recombinante *Escherichia* coli . P-260. "Process for the biodegradation of aromatic compounds and synthesis of pigments and colorants, alkaloids, and polymers, with the use of the recombinant strain *Escherichia coli.* P-260"
WO9804331 A 05-02-1998	CA223269 A 05-02-1998 ES211148 A 01-03-1998 EP894517 A 03-02-1999 JP11513042T T 09-11-1999 US6037492 A 14-03-2000 IL123795 A 19-03-2001 AT226470T T 15-11-2002 DE69716593D D 28-11-2002 PT894517T T 31-03-2003	ES19960001652 25-07-1996 WO1997ES00190 24-07-1997	García-Granados López De Hierro Andres	Universidad de Granada	Procedimiento de aprovechamiento industriel de los ácidos oleanolico y maslinico contenidos en los subproductos de la molturación de la aceintura. "Process for the industrial recovery of oleanolic and maslinic acids in the olive milling by-products"

WO9728089 A 07-08-1997		WO1996GR00002 31-01-1996	Georgoudis Dianellos	Georgoudis Dianellos	Method of extraction of olive paste from vegetable water and its use as a foodstuff
WO921120 A 09-07-1992		NO19900005473 9-12-1990 NO19910002460 24-06-1991	Knudsen Carl-Henrik Larsen Stein-Thore	Ticon VVS AS	Process and plant for purification of agricultural waste material
WO8904355 A 18-05-1989	EP409835 A 30-01-1991; JP3502202T T 23-05-1991; US5084141 A 28-01-1992; EP409835 B 07-04-1993; DE3880155G G 13-05-1993; US5330623 A 19-07-1994	GB19870026397 11-11-1987; US19920826126 27-01-1992	Holland K. M.	Holland K. M.	Destructive distillation of organic material especially rubber — involving preheating and pyrolysis in microwave discharge zone

Glossary

AAS: Atomic absorption spectroscopy uses the absorption of light to measure the concentration of gas-phase atoms. Since samples are usually liquids or solids, the analyte atoms or ions must be vaporized in a flame or graphite furnace. The atoms absorb ultraviolet or visible light and make transitions to higher electronic energy levels. The analyte concentration is determined from the amount of absorption. Applying the Beer-Lambert directly in AAS is difficult due to variations in the atomization efficiency from the sample matrix, and non-uniformity of concentration and path length of analyte atoms (in graphite furnace AAS). Measurements are usually determined from a working curve after calibrating the instrument with standards of known concentration. The light source is usually a hollow-cathode lamp of the element that is being measured. Lasers are also used in research instruments. The disadvantage of these narrow-band light sources is that only one element is measurable at a time.

absorption: Process of transferring molecules of gas, liquid, or a dissolved substance to the surface of a solid where it is bound by chemical or physical forces.

acid detergent lignin (ADL): Lignin in the residue determined following extraction with acid detergent; the most commonly used method in animal science and agronomy.

acid detergent fiber (ADF): Insoluble residue following extraction of herbage with acid detergent (van Soest). The ADF fraction is a measure of cellulose but not hemicellulose (cell wall constituent minus hemicellulose); compounds that are only partially utilized by the animal. ADF is more reliable than NDF for estimating ration digestibilty. Grains and feeds that contain high levels of starch or fat will have low ADF values. Forages have high ADF values, especially those that are mature or contain few leaves relative to stems. The ADF value is expressed as a percentage.

acidity: The capacity of an aqueous solution to neutralize a base.

activated carbon: A highly adsorbent form of carbon used to remove dissolved organic matter from water and wastewater, or to remove odors and toxic substances from gaseous emissions.

activated sludge: Term referring to the brownish flocculent culture of organisms developed in an aeration tank under controlled conditions. Also, sludge floc produced in raw or settled wastewater by the growth of zoological bacteria and other organisms in the presence of dissolved oxygen.

activated sludge process: A biological waste-water treatment process where a mixture of wastewater and biologically enriched sludge is mixed and aerated to facilitate aerobic decomposition by microbes.

ADF: Acid detergent fiber.

ADL: Acid detergent lignin.

adsorption: Process of transferring a substance from a liquid to the active sites on the surface of a solid substance (adsorbent) where it is bound by chemical or physical forces. As the active sites are occupied by adsorbed molecules, the adsorbent becomes progressively exhausted. Desorption of the adsorbent is required to regenerate the active sites.

advanced oxidation process (AOPs): A process using a combination of disinfectants, such as ozone and hydrogen peroxide, to oxidize toxic organic compounds to non-toxic form.

aeration: Incorporation of air into a liquid or solid material by exposure (passive), mixing, agitation, chemical means, or direct injection with the aim of transferring oxygen to the material.

aerobic: Condition characterized by the presence of free oxygen.

aerobic digestion: Sludge stabilization process involving direct oxidation of biodegradable matter and oxidation of microbial cellular material.

alkalinity: The capacity of water to neutralize acids, a property imparted by the water's content of carbonates, bicarbonates, hydroxides, and occasionally borates, silicates, and phosphates.

almazaras: (Spanish) Common name given in Spain to the facilities where the olives are processed to obtain oil through physical separation techniques; olive mills.

alpechín: (Spanish) OMWW.

alpechín-2: (Spanish) Liquid fractions from secondary 2POMW (alperujo) treatments (second decanting, repaso, etc.); margine-2 (French); Jamila-2 (Italian).

alperujo: (Spanish) Term used in Spain to define the olive waste produced during the two-phase extraction of olive oil; alperujo; orujo de dos fases; orujo humende; 2POMW.

AMBERLITE®: Commercial name of a polymeric adsorbent material (Rohm and Haas GmbH). The AMBERLITE® resin is a macroreticulated cross-linked copolymer having a plurality of microscopic channels resulting from the liquid expulsion of a precipitating agent during polymerization of a monomer mixture under suspension conditions. These resins are typically styrenic, acrylic, or phenolic-based. Among the AMBERLITE® resins the polystyrene-based resins are preferred and especially the grades, AMBERLITE® XAD-2, AMBERLITE® XAD-4, AMBERLITE® XAD-7, and AMBERLITE® XAD-16.

amurca: Latin term used to describe the watery bitter-tasting liquid residue obtained when the oil is drained from compressed olives; olive lees; amorgi (αμόργη) in ancient Greek.

amino acids: Carboxylic acids that contain an amine function.

anaerobic: Condition characterized by the absence of free oxygen.

anaerobic digestion: Sludge stabilization process where the organic material in biological sludges are converted to methane and carbon dioxide in an airtight reactor.

anion: A negatively charged ion that migrates to the anode when an electrical potential is applied to a solution.

anionic polymer: A polyelectrolyte with a net negative electrical charge.

anode: The positive electrode where current leaves an electrolytic solution.

anoxic: A biological environment that is deficient in molecular oxygen, but may contain chemically bound oxygen, such as nitrates and nitrites.

anoxic process: A denitrification process by which nitrate–nitrogen is converted to nitrogen gas.

aqueous olive effluent: OMWW.

aquifer: A subsurface geological formation containing a large quantity of water.

artificial wetlands: Constructed wetlands.

atherosclerosis: Type of arteriosclerosis. It comes from the Greek words athero (meaning gruel or paste) and sclerosis (hardness). It involves deposits of fatty substances, cholesterol, cellular waste products, calcium, and fibrin (a clotting material in the blood) in the inner lining of an artery. The build-up that results is called plaque.

atomic absorption spectroscopy: See AAS.

Bacillus: An aerobic, rod-shaped, gram-positive, spore-producing bacterium, often occurring in chainlike formations.

bacteria: Microorganisms without a cell nucleus. They are structured as rod-shaped, sphere-shaped, or spiral-shaped. They decompose and stabilize organic matter in wastewater. Bacteria can be classified in several ways: By type of cell wall (Gram-positive or Gram-negative), or by oxygen requirements (aerobic bacteria need oxygen to grow; anaerobic bacteria do not). Some typical bacterial species are listed below:

Aerobic	Anaerobic
• *Pseudomonas* sp.	• *Desulfovibrio* sp.
• *Citrobacter* sp.	• *Clostridium* sp.
• *Klebsiella* sp.	
• *Proteus* sp.	
• *Escherichia* sp.	

***Bactrocera oleae* (Gmelin) (Insecta: Diptera: Tephritidae):** The olive fruit fly (formely known as *Dacus oleae*) is a serious pest of olives in most of the countries around the Mediterranean sea. The larvae are monophagous and feed exclusively on olive fruits. Adults feed on nectar, honey dew, and other opportunistic sources of liquid or semi-liquid food. The damage caused by tunneling of larvae in the fruit results in about 30% loss of the olive crop in Mediterranean countries, and especially in Greece and Italy where large commercial production occurs.

bagasse: Term used to describe the fibrous material remaining after the extraction of the juice from the sugar cane; olive press-cake.

Bardenpho process: A biological nutrient removal process. This is the most efficient process for removing high levels of nitrogen and phosphorous while producing the least amount of sludge. The Bardenpho process consists of an initial anaerobic contact zone followed by four alternating stages of anoxic and aerobic conditions. In the anaerobic zone, all of the raw wastewater is mixed with the return sludge. The anaerobic condition in the initial contact zone is necessary to effect phosphorous removal. The first anoxic zone follows the anaerobic zone. Nitrates and nitrites (NO_x) are supplied to the anoxic zone by recycling nitrified mixed liquor from the following aerobic zone. The organic material in the raw wastewater is used as a carbon source by the denitrifying bacteria in the denitrifaction zone. The first aerobic (oxic) zone is followed by a second anoxic zone where any remaining nitrites in the mixed liquor are reduced by the endogenous respiration of the activated sludge. The final stage is aerobic where the mixed liquor is reaerated before reaching the final clarifier. The dissolved oxygen of the wastewater effluent is increased to prevent further denitrification in the clarifier and to prevent the release of phosphates to the liquid in the clarifier.

basidiomycete: Name of fungal group that all bear spores on basidia (club-shaped sexual spores producing cells that characterize the basidiomycetes).

batch process: A non-continuous treatment process in which a discrete quantity or batch of liquid is treated or produced at one time.

batch reactor: A reactor where the contents are completely mixed and flow is neither entering, nor leaving the reactor vessel.

bentonite: Colloidal clay-like mineral that can be used as a coagulant aid in water treatment systems. Also, sometimes used as the earth component or soil amendment for construction of a pond or landfill liner because of its low permeability.

bicarbonate: A chemical compound containing an HCO_3 group.

biochemical oxygen demand (BOD): The quantity of oxygen utilized in the biochemical oxidation of organic matter under standard laboratory procedure in five days at $20°$ in terms of milligrams per liter (mg/l).

biocide: A chemical used to inhibit or control the population of troublesome microbes.

biodegradable: Term used to describe organic matter that can undergo biological decomposition.

biofilm: A complex layer of active microbes associated with or attached to a solid surface, e.g. stones or synthetic plastic media.

biogas: Gases produced by the anaerobic decomposition of organic matter.

biological filter: A bed of sand, stone, or other media through which wastewater flows that depends on biological action for its effectiveness.

biomass: The mass of biological material contained in a system.

bioreactor: Vessel or tank in which whole cells or cell-free enzymes transform raw materials into biochemical products and/or less undesirable by-products.

bioremediation: Process by which living organisms act to degrade hazardous organic contaminants or transform hazardous inorganic contaminants to environmentally safe levels in soils, subsurface materials, water, sludges, and residues.

biosurfactant: A surface-active agent produced by microorganisms.

blanching: The process of pretreating vegetables before drying by deactivating the enzymes that cause ripening and eventual decay.

BOD$_5$: Five-day carbonaceous or nitrification-inhibited BOD; see also "biochemical oxygen demand".

bract: A modified leaf, often highly colored and sometimes mistaken for a petal. Examples of house plants with showy bracts are Poinsettia, Aphelandra, and Bougainvillea.

brine: Water saturated with, or containing a high concentration of salts, usually in excess of $36 \, g/l$.

buffer: A substance that stabilizes the pH value of solutions.

CaCO$_3$: See calcium carbonate.

cake: Dewatered sludge with a solids concentration sufficient to allow handling as a solid material.

calcareous soil: A soil containing enough calcium carbonate, or related minerals, so that it effervesces (bubbles off CO_2) when treated with acid (e.g. HCl). Usually formed from shells or chemical precipitation, these soils tend to be coastal in occurrence.

calcium carbonate: A white, chalky substance which is the principle hardness and scale-causing compound in water. Chemical formula is $CaCO_3$; see also limestone.

calcium hypochlorite: A chlorine compound frequently used as a water or wastewater disinfectant. Chemical formula is $Ca(OCl)$.

Carbon/Nitrogen ratio: The proportion of carbon to nitrogen which affects how quickly microorganisms work. The ideal "C/N" ratio is in the range of 25/1 to 35/1.

cathode: The negative electrode where the current leaves an electrolytic solution.

cation: A positively charged ion that migrates to the cathode when an electrical potential is applied to a solution.

cationic polymer: A polyelectrolyte with a net positive electrical charge.

centrifuge: A dewatering device relying on centrifugal force to separate particles of varying density such as water and solids.

chemical oxygen demand (COD): A measurement of biodegradable and non-biodegradable (refractory) organic matter, widely used as a means of measuring the pollutional strength of domestic and industrial wastewaters.

chitin: A highly insoluble aminopolysaccharide occurring widely in the external skeleton of many insects and crustaceans. A polymer of N-acetyl-2-amino-2-deoxy-D-glucose units joined by β-1,4'-links. The structure is therefore, that of cellulose in which the hydroxyl groups on carbon-2 are replaced by — $NHCOCH_3$. The usual crystalline form α-chitin, has a unit cell similar to that of cellulose. The polymer may be deacetylated to chitosan.

chitosan: Deacetylated chitin, prepared by treatment of chitin with hot concentrated alkali. Partial hydrolysis of chitosan to oligosaccharides (chitobiose, etc.) established the structure of chitin.

chlorophyta: Division of the kingdom Protista consisting of the photosynthetic organisms commonly known as green algae. The organisms are largely aquatic or marine. The various species can be unicellular, multicellular, coenocytic (having more than one nucleus in a cell), or colonial.

C/N ratio: See Carbon/Nitrogen ratio.

coagulant: Flocculant.

coagulation: Flocculation.

coalesce: The merging of two droplets to form a single, larger droplet.

COD: Chemical oxygen demand.

colloid: Dispersion of distinguishable particles in the size range of 0.01 to $10\,\mu m$ in a medium that may be regarded as a structureless continuum.

compost: The end product of composting.

composting: Stabilization process relying on the aerobic decomposition of organic matter in sludge by bacteria and fungi.

co-composting: Composting of a mixture of two or more wastes, e.g. manure and OMWW.

constructed wetlands: Man-made structures designed for wastewater treatment and typically have a relatively impermeable bottom and a layer(s) of soil, muck, gravel, or other media to support the roots of aquatic plant species. Two types of constructed wetlands are currently used for wastewater treatment: free-water surface (FWS) and subsurface flow systems (SFS); artificial wetlands.

coryneform bacteria: Term used to describe aerobically growing, asporogenous, non-partially-acid-fast, irregularly shaped gram-positive rods; group of bacteria that are morphologically similar to the organisms of the genus *Corynebacterium,* which includes many animal and plant pathogens, such as the causative agent of diphtheria; corynebacteria; called also *coryneform group*.

CP: crude protein.

crude olive cake: Residue which remains after the first pressing of the olives through traditional and continuous machines. There is still a small amount of oil in his cake. If not going on for further processing, this cake is often used for heating, for animal feed supplement, or returned to the olive grove as mulch.

crude protein (CP): Determined by measuring the total nitrogen contained in a feed. The value is termed "crude" protein because not all nitrogen present in a feed is in the form of true or available protein. In most feeds and forages, nitrogen is 16% of the weight of proteins. Crude protein is calculated as 6.25 (100/16) times the %N in a sample and expressed as a percentage.

Dacus oleae (Gmelin): See *Bactrocera oleae*.

DAF: Dissolved air flotation.

Dalton (Da): Unit of mass equal to the unified atomic mass (atomic mass constant) [IUPAC Compendium]. After John Dalton (1766–1844), British chemist and physicist. Frequently used in biochemistry to express molecular mass, although the name and the symbol [Da] have not been approved by Comité international des poids et mesures (CIPM) or International Organization for Standardization (ISO); kDa.

denitrification: Biological process in which nitrates are converted to nitrogen.

denitrifying bacteria: Bacteria that convert nitrate to nitrogen (N_2) or nitrous oxide (N_2O) gas. Denitrifiers are anaerobic, meaning they are active where oxygen is absent, such as in saturated soils or inside soil aggregates.

digester: A tank or vessel used for sludge digestion; (bio)reactor.

digestible protein (DP): The crude protein in feeds is not completely digested by animals. The amount of crude protein that is digested and made available for use by an animal is termed "digestible protein," calculated from crude protein using standards determined in feeding trials and expressed as a percentage. The use of digestible protein is advisable when forages are the total diet. Feeding standards for beef cattle may utilize DP values.

digestion: Biological oxidation of organic matter in sludge resulting in stabilization; biodegradation.

dissolved air flotation (DAF): Clarification of flocculated material by contact with minute bubbles causing the air/floc mass to be buoyed to the surface, leaving behind clarified water. Use of a gas other than air is referred to as dissolved gas flotation or DGF.

dissolved organic carbon (DOC): Fraction of TOC that is dissolved in a water sample.

distillate: A liquid product condensed from vapor during distillation.

distillation: Process of boiling a liquid solution, followed by condensation of the vapor, for the purpose of separating the solute from the solution.

DM: Dry matter.

dormancy: The state of temporary cessation of growth and slowing down of other activities in whole plants, usually during the winter.

dry matter (DM): Expressed a percent of the sample as received. Dry matter contains the nutrients that are important in feeding programs. The value is determined by oven-drying to a constant weight at 60°C.

DUOLITE®: Commercial name of an adsorbent polystyrene-based resin, which has been investigated for OMWW decolorization (Duolite Company). The Duolite XAD

761^{TM} grade is used industrially for the adsorption of mono- and poly-aromatic compounds. It removes color, protein, iron complexes, tannins, hydroxymethyl furfural, and other ingredients responsible for off-flavors.

dysentery: A disease of the gastrointestinal tract usually resulting from poor sanitary conditions and transmitted by contaminated food or water.

EC$_{50}$: The median effective concentration (ppm or ppb) of the toxicant in the environment (usually water) that produces a designated effect in 50 percent of the test organisms exposed.

ecology: The relationship of living things to one another and their environment.

ecosystem: The total community of living organisms, together with their physical and chemical environment.

effluent: Partially or completely treated water or wastewater flowing out of a basin or treatment plant.

eklima: (Greek) "Εκλυμα"; Effluent.

electrical conductivity (EC$_{25}$): The most common measurement of salinity, in soil and water. It is directly related to the sum of the cations (or anions), as determined chemically and is closely correlated, in general, with the total salt concentration. Electrical conductivity is a rapid and reasonably precise determination, and values are always expressed at a standard temperature of 25°C to enable comparison of readings taken under varying climatic conditions. The unit of electrical conductivity is dS/m (or mS/cm).

electrolysis: Passage of electric current through an electrolyte resulting in chemical changes caused by migration of positive ions towards the cathode, and negative ions to the anode.

electrolyte: A substance that dissociates into two or more ions when it dissolves in water.

emulsion: A heterogeneous mixture of two or more mutually insoluble liquids that would normally stratify according to their specific gravities.

epuvalisation: Term used to describe a biologic wastewater treatment technique which uses plants. Based on the Nutrient Film Technique (NFT), this technique has the advantage, not only to purify, but also to produce plants. The name comes from the contraction of two French words: "epuration" and "valorization".

eutrophication: Nutrient enrichment of water, causing excessive growth of aquatic plants and eventual deoxygenation of the water body.

evaporation: Process in which water is converted to a vapor that can be condensed.

evaporation pond: A natural or artificial pond used to convert solar energy to heat to accomplish evaporation.

evaporator: A device used to heat water to create a phase change from the liquid to the vapor phase.

exhausted olive cake: The residue that is left after the above crude olive cake has any remaining oil extracted from it by using solvents such as hexane. This cake is also often used for heating, for animal feed supplement or returned to the olive grove as mulch; deoiled or exhasuted orujillo.

fats: Triglyceride esters of fatty acids that are usually solid at room temperature.

fermentation: Biological process where a microorganism, such as yeast, is grown on a substrate (e.g. glucose) and, in so doing, produces compounds or materials (e.g. alcohol) which can then be harvested.

ferric chloride: An iron salt commonly used as a coagulant. Chemical formula is $FeCl_3$.

ferric sulfate: An iron salt commonly used as a coagulant. Chemical formula is Fe_2SO_4.

ferrous sulfate: An iron salt commonly used as a coagulant. Chemical formula is $FeSO_4$.

fertirrigation: Discharge of a treated liquid effluent containing nitrogen and phosphorous compounds onto agricultural land as fertilizer.

filter: A device utilizing a granular material, woven cloth, or other medium to remove suspended solids from water, wastewater, or air.

filter cake: Layer of solids that is retained on the surface or upstream side of a filter.

filter press: A dewatering device where water is forced from the sludge under high pressure.

filtrate: Liquid remaining after removal of solids through filtration.

flesh: See pulp.

float: The concentrated solids at the surface of a dissolved air flotation unit.

floc: Small, gelatinous masses formed in water by adding a coagulant, or in wastewater through biological activity.

flocculant: An organic polyelectrolyte, used alone or with metal salts, to enhance floc formation and increase the strength of the floc structure.

flocculation: Agent-induced aggregation of particles suspended in liquid media into larger particles. Essentially, it can be described as the destabilization process of a stable colloidal dispersion by the addition of a chemical known to effect the destabilization. The terms flocculation and coagulation are both used in connection with formation of aggregates, frequently interchangeably and sometimes with distinctions that vary among professional disciplines. Although no distinction is made in this review, the more common types of distinction appearing in literature are enumerated: (1) based on mechanisms for destabilization of a suspension and/or type of aggregate formed, coagulation implying formation of compact aggregates and flocculation implying formation of lose or open network aggregates; (2) based on chemical agents used, coagulation for inorganic materials, and flocculation for organic polymers; (3) based on engineering process step, coagulation representing conditioning the particles with the chemical agent, and flocculation representing the mechanical particle transport step (collisions between conditioned particles) leading to aggregates; and (4) based on another engineering usage, coagulation representing the overall aggregation process and flocculation again representing the particles transport step.

flocculator: A device used to enhance the formation of floc through gentle stirring or mixing.

flotation: A treatment process where gas bubbles are introduced into water and attach to solid particles creating bubble-solid agglomerates that float to the surface.

flux: The flow rate of material passing through the membrane per area. It has a unit of volume per unit area per unit time. Flux is typically reported in gallons per square foot of membrane per day (gfd) or liters per square meter of membrane per hour (LMH).

foot cake: Olive cake or 2POMW.

Fourier-transform infrared spectrometer: See FTIR.

FTIR: Fourier-transform infrared spectrometer; see Infrared spectroscopy.

fungi: Small, multicellular non-photosynthetic organisms that feed on organic matter. There are two types of fungi: molds and yeasts. Some examples of species of each type are shown below.

Molds
- *Aspergillus* sp.
- *Cephalsporium* sp.
- *Fusarium* sp.
- *Trichoderma* sp.
- *Trichosporon* sp.

Yeasts
- *Candida* sp.
- *Saccharomyces* sp.

fungicide: A substance used to kill, or inhibit the growth of, fungi or molds.

furfural: (furfuraldehyde) A colorless, transparent, oily liquid with the characteristic odor of bitter almonds. Furfural is formed from the acid hydrolysis or heating of polysaccharides which contain pentose and hexose fragments, and has been detected in a broad range of fruits and fruit juices, wines, whiskeys, coffee, olives, and tea; used industrially as a solvent and as a raw material for synthetic resin.

GPC: Gel Permeation Chromatography.

Gram-negative: See bacteria.

Gram-positive: See bacteria.

granular activated carbon (GAC): A granular form of activated carbon used in filter beds or contactor vessels to absorb organic compounds.

grignons: (French) Olive cake.

heavy metals: Metallic elements with high molecular weights, generally toxic in low concentrations to plant and animal life. Such metals are often residual in the environment and exhibit biological accumulation. Examples include mercury, chromium, cadmium, arsenic, and lead.

herbaceous plant: A non-woody plant of which the stem perishes at the end of the growing season, while the roots remain permanent and send forth a new stem in the following season. It is chiefly applied to perennials, although botanically it also applies to annuals and biennials.

herbicide: A chemical used to control or kill weeds.

heterogeneous catalysis: Type of catalysis where the catalyst and the reactants are present in separate phases.

High Performance Liquid Chromatography: See HPLC.

homogeneous catalysis: Type of catalysis where the catalyst and the reactants are present in the same phase.

HPLC: High Performance Liquid Chromatography. A separation technique that uses small particle size, narrow bore columns, and high inlet pressures to achieve separation in short periods of time (several minutes to an hour). HPLC is used to separate and quantitate compounds present in complex solutions such as OMWW.

HRT: Hydraulic retention time.

humic acid: Organic acids that are by-products of decomposing organic matter that colors water.

humus: Dark or black decomposing organic matter in soil.

husk: See stone.

husks: See olive cake.

hydrate: A compound formed by the union of water with another substance.

hydrated lime: The calcium hydroxide product that results from mixing quicklime with water. Chemical formula is $Ca(OH)_2$.

hydraulic retention time: The resident time for wastewater in the various unit operations of wastewater at a given rate of flow.

hydrogen peroxide: An oxidizing agent used for odor control and disinfection. Chemical formula is H_2O_2.

hydroxide ion: A negatively charged ion consisting of a hydrogen atom and an oxygen atom. Chemical formula is OH^-.

hypochlorite: Chlorine anion commonly used as an alternative to chlorine gas for disinfection. Chemical formula is OCl_3^-.

incineration: Process of reducing the volume of a solid by burning of organic matter.

infrared spectroscopy (IR): IR spectroscopy is measurement of the wavelength and intensity of the absorption of mid-infrared light by a sample. Mid-infrared light (2.5–50 μm, 4000–200 cm^{-1}) energetic enough to excite molecular vibrations to higher energy levels. The wavelengths of IR absorption bands are characteristic of specific types of chemical bonds and IR spectroscopy finds its greatest utility for identification of organic and organometallic molecules. Modern IR instruments more commonly use Fourier-transform techniques with a Michelson interferometer and they are called Fourier-transform infrared spectrometers (FTIR).

inoculum: Material used to initiate a microbial culture.

insecticide: A pesticide compound specifically used to kill or prevent the growth of insects.

ion: An electrically charged atom, molecule, or radical.

ion exchange: (1) A chemical process involving reversible interchange of ions between a liquid and a solid, but no radical change in structure of the solid. (2) A chemical process in which ions from two different molecules are exchanged. (3) The reversible transfer or sorption of ions from a liquid to a solid phase by replacement with other ions from the solid to the liquid.

ion-exchange chromatography: Any chromatographic technique that separates ionic substances through the use of an insoluble ion exchanger as the stationary phase. When the ionic solution is passed over the solid phase, the ions on it exchange with those in the solution. The two main types are cation and anion exchangers.

iron salt: An iron-based coagulant used in water and wastewater treatment.

irrigation: Artificial applications of water to meet the requirements of growing plants or grass that are not met by rainfall alone.

isomer: One of two or more molecules that have the same chemical formula but different atomic arrangements.

jamila: (Spanish/Italian); OMWW.

jamila-2 (Italian) alpechín-2; margine-2 (French).

katsigaros: (Greek) "κατσίγαρος"; OMWW.

kernel: Seed.

Kjedahl nitrogen: The sum of the organic plus ammonia nitrogen in a water sample.

kDa: KiloDalton; see Dalton.

laccase: (E.C.1. 10.3.2 para-diphenol: oxygen oxidoreductase) A multi-copper oxidase able to catalyze the one-electron oxidation of a wide array of substrates, such as phenols, aromatic amines, benzenethiols, hydroxy-indoles, and phenothiazinic compounds, with simultaneous reduction of oxygen to water.

Lactase: An enzyme which catalyzes the hydrolysis of lactose to D-galactose and D-glucose

LC: Liquid chromatography used to separate analytes in solution including metal ions and organic compounds. The mobile phase is a solvent and the stationary phase is a liquid on a solid support, a solid, or an ion-exchange resin.

LC50: The median lethal concentration; the concentration that kills 50% of the test organisms, expressed as milligrams (mg) or cubic centimeters (cc, if liquid) per animal. It is also the concentration expressed as parts per million (ppm) or parts per billion (ppb) in the environment (usually water) that kills 50% of the test organisms exposed.

leaching: The process or an instance of separating the soluble components from some material by percolation. The process or an instance of removing nutritive or harmful elements from soil by percolation.

lignans: Group of phenolic hormone-like compounds characterized by the coupling of two C_6C_3 units (propylbenzene).

lignin: The most abundant natural aromatic organic polymer found in all vascular plants. Lignin together with cellulose and hemicellulose are the major cell wall components of the fibers of all wood and grass species. Lignin is composed of coniferyl, *p*-coumaryl, and sinapyl alcohols in varying ratios in different plant species.

limestone: Sedimentary rock consisting primarily of calcium carbonate.

lipids: A loosely defined term for substances of biological origin that are soluble in non-polar solvents. They consist of saponifiable lipids, such as glycerides (fats and oils) and phospholipids, as well as non-saponifiable lipids, principally steroids.

liquid chromatography: See LC.

loam: A term used for soil of medium texture, often easily worked, that contains more or less equal parts of sand, silt, and clay, and is usually rich in humus. If the proportion of one ingredient is high, the term mab be qualified as silt-loam, clay-loam, or sandy-loam.

lyophilization: The process of freeze-drying, which is the removal of liquid from heat-sensitive materials. The material is frozen, placed under a high vacuum, and maintained at a low temperature. The pressure generated by the vacuum causes the ice to turn from a solid to a gaseous form without passing through a liquid state.

lysimeter: container of soil to measure the water movement, gains, or losses through that block of soil, usually undisturbed or *in situ*.

macerate: To chop or tear.

mannitol: White crystalline, sweetish, water-soluble, carbohydrate alcohol, $HOCH_2$ $(CHOH)_4CH_2OH$, occurring in three optically differently forms.

marc: (French) pomace.

margine: (French) OMWW.

margine-2: (French) alpechín-2; Jamila-2.

maslinic acid: 2-α-,3-β-dixydroxy-28-carboxy-oleanene; also known as crazegolic acid, it has been found in a dozen of plants, inter alia the *Olea europaea*. It has antihistammic and anti-inflammatory activity.

membrane: A thin barrier that permits passage of particles of a certain size or of particular physical or chemical properties.

mesocarp: See pulp.

methane: A colorless, odorless combustible gas that is the principle by-product of anaerobic decomposition of organic matter in wastewater. Chemical formula is CH_4.

microfiltration (MF): A low pressure (100–400 kPa, 15–60 psi) membrane filtration process which removes suspended solids and colloids generally larger than 0.1 μm diameter.

(micro)algae: marine and freshwater plant-like organisms (including most seaweeds) that are single-celled, colonial, or multi-celled, with chlorophyll but no true roots, stems, or leaves and with no flowers or seeds; see also chlorophyta.

microbe: Short version of the word micro-organism.

microorganism: Living organisms that can only be seen under a microscope. There are 5 basic groups of microbes:

a. bacteria

b. fungi
   b1   yeasts
   b2   molds
c. viruses
d. protozoa
e. (micro)algae.

mineralization: The breakdown of organic materials into inorganic materials brought about by microorganisms; ammonification.

mixed liquor: Mixture of microbial solids and wastewater present in aeration tanks of activated sludge plants.

molds: Filamentous fungus. Molds can form protective slime coatings and can develop large numbers of spores that serve to spread infestation.

mulch: Any material such as straw, sawdust, leaves, plastic film, or loose soil that is spread on the surface of the soil to protect the soil and the plant roots from the effects of raindrops, soil crusting, freezing, and evaporation.

municipal waste: The combined solid and liquid waste from residential, commercial, and industrial sources.

mycorrhizae: Soil fungi that live in beneficial association with plant roots.

nanofiltration: A specialty membrane filtration process, which rejects solutes larger than approximately one nanometer (10 Å) in size.

NDF: Neutral detergent fiber.

neutral detergent fiber (NDF): The NDF fraction measures the cell wall constituents, including hemicellulose, cellulose, lignin, and silica. NDF is a good predictor of ration consumption.

neutralization: The restoration of the hydrogen (H^+) or hydroxyl (OH^-) ion balance in solution so that the ionic charge of each are equal. The chemical process that produces a solution that is neither acidic, nor alkaline.

nitrate: Form of nitrogen commonly found in the soil and used by plants for building amino acids, DNA, and proteins. It is commonly produced by the chemical modification of nitrite by specialized bacteria. A stable, oxidized form of nitrogen having the formula NO_3^-.

nitrification: Biological process in which ammonium is converted first to nitrite and then to nitrate.

nitrifying bacteria: Bacteria that change ammonium (NH_4^+) to nitrite (NO_2^-) then to nitrate (NO_3^-) — a preferred form of nitrogen for grasses and most row crops. Nitrate is leached more easily from the soil, so some farmers use nitrification inhibitors to reduce the activity of one type of nitrifying bacteria. Nitrifying bacteria are suppressed in forest soils, so that most of the nitrogen remains as ammonium.

nitrite: Form of nitrogen commonly found in the soil. It is commonly produced by the chemical modification of ammonium by specialized bacteria. This form is toxic to plants and animals at high concentrations. Chemical formula for nitrite is NO_2^-.

nitrosation: An intermediate stage in nitrification during which ammonium salts are biologically oxidized to nitrites.

oil skimmer: A device used to remove oil from water's surface.

oil-foot: 2POMW.

oils & grease: Common term used to include fats, oils, waxes, and related constituents found in wastewater.

oleanolic acid: 3-β-hydroxy-28-carboxyoleanene; a triterpenic acid found in almost a hundred plants, inter alia the *Olea europaea*. It has a number of proven biological activities (abortifacient, anticariogenic, anti-fertility, antihepatotoxic, anti-inflammatory, antisarcomic, cancer-preventive, cardiotonic, diuretic, hepatoprotective and uterotonic).

oleuropein: The main phenolic compound in the olive fruit, a heterosidic ester of elenolic acid and 3,4-dihydroxyphenyl ethanol. The empirical formula of oleuropein($C_{25}H_{32}O_{13}$) makes it a member of the iridoid group, a uniquely structured chemical class that contains a carbohydrate component appearing as D-glucose.

olive cake: The solid phase left after oil separation comprising ground pulp or flesh and stones. It contains of around 3% olive oil by mass and has a moisture content around 40–50%; cake; pomace; grignons; pirina; husks; marc; orujo.

olive lees: See amurca.

olive-mill wastewater (OMWW): The mixture of the own water of the olives (vegetation water) together with the water used in the different stages of oil elaboration (washing and processing); alpechín; margine, jamila.

OMWW: Olive-mill wastewater.

orujo: (Spanish) Olive cake.

orujo de dos fases: (Spanish) See 2POMW.

orujo humedo: (Spanish) See 2POMW.

orujillo: (Spanish) Exhausted or deoiled 2POMW (alperujo); exhausted or deoiled olive cake (orujo).

osmosis: Movement of water from a dilute solution to a more concentrated solution through a permeable membrane separating the two solutions.

osmotic pressure: Excess pressure that must be applied to a concentrated solution to produce equilibrium and prevent the movement of a more dilute solution, through a semi-permeable membrane, into the more concentrated solution.

oxidant: A chemical substance, such as chlorine or ozone, that is capable of promoting oxidation.

oxidation: (1) A chemical reaction in which an element or ion loses electrons. (2) The biological or chemical conversion of organic matter into simpler, more stable forms.

oxidation-reduction potential (ORP): The potential required to transfer electrons from an oxidant to a reductant that indicates the relative strength potential of an oxidation-reduction reaction.

ozonation: Process of using ozone in water or wastewater treatment for oxidation, disinfection, or odor control.

ozone: An unstable, blue gas with pungent odor. It is a powerful oxidizing agent with disinfection properties similar to chlorine, also used in odor control and sludge processing. Chemical formula is O_3. The gas is made by passing oxygen through a silent electric discharge: $3O_2(g) \rightleftarrows 2O_3(g)$.

PAHs: Polycyclic aromatic hydrocarbons are natural products of the incomplete combustion of carbon compounds.

parenteral: Taken into the body in a manner other than through the digestion canal.

partly destoned olive cake: Produced if some of the crushed olive seeds are removed from the paste after processing. This cake is also often used for heating, for animal feed supplement, or returned to the olive grove as mulch.

pathogen: Highly infectious, disease producing microbes commonly found in sanitary wastewater.

pathogenic: Capable of causing diseases.

pectin: A polysaccharide composed of galacturonic acid subunits, partially esterified with methyl alcohol, and capable of forming a gcl. Pectin is used as a gelling agent, an emulsifier, and stabilizer. Plant tissues contain protopectins cementing the cell walls together. As fruit ripens, protopectin breaks down to pectin, and finally to pectic acid under the influence of enzymes. Thus, over-ripe fruit loses its firmness and becomes soft as the adhesive between the cells breaks down.

percolation test: Test used to determine the water absorbing capacity of soil where the drop in water level in a test hole is measured over a fixed time period.

permeate: Liquid that passes through a membrane.

peroxidase: An enzyme, which catalyzes the transfer of oxygen from the hydrogen peroxide to a suitable substrate and, thus brings about oxidation of the substrate.

pH: The reciprocal of the logarithm of the hydrogen ion concentration in gram moles per liter. On the 0 to 14 pH scale, a value of 7 at 25°C (77°F) represents a neutral condition. Decreasing values indicate increasing hydrogen ion concentration (acidity), and increasing values indicate decreasing hydrogen ion concentration (basicity).

phenols: Organic pollutant, also known as carbolic acid, occurring in industrial wastes from petroleum processing and coal coking operations.

physico-chemical treatment: Treatment processes that are non-biological in nature.

phytodepuration: Phytoremediation.

phytoremediation: Remediation performed by plants; the removal of pollutants from soil or water using plants that either absorb or degrade the pollutants; phytodepuration.

pirina: (Greek/Turkish) Olive cake.

pit: Stone; endocarp.

pollutant: A substance, organism or energy form present in amounts that impair or threaten an ecosystem to the extent that its current or future uses are precluded.

pollution: The presence of a pollutant in the environment.

polyelectrolytes: Complex polymeric compounds typically composed of synthetic macromolecules that form charged species (ions) in solution. Insoluble polyelectrolytes are used as ion exchange resins.

pomace: (Crude) Olive cake.

pomace olive oil: Term given generically to oils obtained through solvent extraction (benzene or hexane) of the pomace remaining from the olive oil obtaining procedures. It is considered an inferior grade and is used for soap making or industrial purposes.

2POMW: Semi-solid waste produced during the two-phase extraction of olive oil having a moisture content in the range (55–70)%; two-phase olive mill waste; alperujo; orujo de dos fases; orujo humende; foot cake.

powdered activated carbon (PAC): A powered form of activated carbon fed as slurry to water to absorb organics, particularly taste and odor-causing constituents.

precipitate: A solid that separates from a solution.

precipitation: The phenomenon that occurs when a substance held in solution passes out of solution into a solid form.

preliminary treatment: Treatment steps including comminution, screening, grit removal, preaeration and/or flow equalization, which prepare wastewater influent for further treatment.

proteins: Naturally occurring polypeptides that contain more than 50 amino acids units — most proteins are polymers of 100 to 300 amino acids.

pulp: Residual paste, which is produced if the whole olive stones are removed from the paste prior to processing. This residual paste has very high water content and is difficult to store or dispose of; flesh; mesocarp.

pv.: Abbreviation for pathovar.

pyrolysis: The thermal decomposition of biomass at high temperatures (greater than 200°C or 400°F) in the absence of air. The end product of pyrolysis is a mixture of solids (char), liquids (oxygenated oils), and gases (methane, carbon monoxide, and carbon dioxide) with proportions determined by operating temperature, pressure, oxygen content, and other conditions; destructive distillation, carbonization.

quicklime: A calcium oxide material produced by calcining limestone to liberate carbon dioxide, also called calcined lime or pebble lime, commonly used for pH adjustment. Its chemical formula is CaO.

raw OMWW: Olive-mill wastewater before it receives any treatment.

raw pomace olive oil: Oil coming directly from extraction; this is the name given to the oil before refining.

redox potential: Oxidation-reduction potential.

reduction: A chemical reaction where an element or compound gains electrons causing a decrease in valence.

reed bed: A large area of marsh plants used to treat wastewater.

refined pomace olive oil: Raw pomace olive oil that has been refined.

recalcitrant: Resistant to microbial attack.

rejection: The ability of an reverse osmosis (RO) nanofiltration (NF) membrane/system to hinder solutes from passing through the membrane. Mathematically, it is the quantity of solutes in the feed water subtracted from the quantity of solutes passing through a semipermeable membrane, which is then divided by the quantity of solutes in the feed water, typically expressed as a percentage.

residence time: The period of time that a volume of liquid remains in a tank or system.

reverse osmosis: (RO) A method of separating water from dissolved salts by passing feed water through a semi-permeable membrane at a pressure greater than the osmotic pressure caused by the dissolved salts.

reactor: A tank where a wastewater stream is mixed with bacterial sludge and biochemical reactions occur.

16S rRNA: A large polynucleotide (about 1500 bases) which functions as a part of the small subunit of the ribosome of prokaryotes and from whose sequence evolutionary information can be obtained; the eukaryotic counterpart is 18S rRNA.

salmonella: Aerobic bacteria that are pathogenic in humans and chiefly associated with food poisoning.

secoiridoids: Phenolic compounds characterized by the presence of either elenolic acid or elenolic acid derivatives in their molecular structure. Oleuropein, demethyloleuropein, ligstroside, and nüzhenide are the most abundant secoiridoids glucoside in the olive fruit.

sedimentation: Removal of settleable suspended solids from water or wastewater by gravity in a quiescent basin or clarifier. It is typically accomplished by reducing the velocity of the liquid below the point at which it can transport the suspended material. It can be variously classified as discrete, flocculant, hindered, and zone sedimentation. It may be enhanced by flocculation; settling.

seed: Softer, inner part of the nut; kernel.

sepiolite: A mineral, hydrous magnesium silicate ($H_4Mg_2Si_3O_{10}$), occurring in white, clay-like masses, used for ornamental carvings, for pipe bowls, etc; meerschaum.

sessile: A microorganism attached to solid surfaces (opposite of planktonic).

settling: See sedimentation.

silica: A mineral composed of silicon and oxygen.

sludge: (1) Accumulated solids separated from liquids during the treatment process that have not undergone a stabilization process. (2) Removed material resulting from chemical treatment, flocculation, sedimentation, flotation, or biological oxidation of water or wastewater. (3) Any solid material containing large amounts of entrained water collected during water or wastewater treatment.

slurry: A suspension of a relatively insoluble chemical in water, usually having a suspended solids concentration of 5000 mg/l or more.

solid-state fermentation (SSF): Fermentation processes on solid matrices in the absence of free water.

sp.: Abbreviation for "species" (singular). It refers to a particular species in a genus even though the identity of the species is unknown. In this context "degradation of the phenolic compounds in OMWW by *Phanerochaete* sp.," actually means the degradation of the phenolic compounds in OMWW by a particular species of *Phanerochaete* (whose full name is unknown).

spp.: Abbreviation for "species" (plural). It is used for a group of species that belong to a particular genus; e.g. "detoxification of OMWW by *Aspergillus* spp.," is the same as saying "detoxification of OMWW by *Aspergillus* species," and is actually indicating the purification of OMWW by any species in the genus.

sporulation: The formation of spores by bacteria. Division of fungi into many small spores.

squalane: ($C_{30}H_{62}$) Hydrogenated squalene. It is mainly used in the formulation of cosmetics and as a carrier of lipid soluble drugs.

squalene: ($C_{30}H_{50}$) A symmetrical 30-carbon polyprenyl compound containing six prenyl (also known as isoprenoid or isoprene) units. It is a naturally occurring compound, primarily known for its key role as an intermediate in cholesterol synthesis. It received its name because of its occurrence in shark liver oil (*Squalus* ssp.), which contains large quantities and is considered the richest source of squalene.

stone: Nut, hard part of the olive. It can be used for heating, building materials or for activated charcoal; pit; husk; endocarp.

surfactant: A surface-active agent such as a detergent which, when mixed with water, generally increases its cleaning ability, solubility, and penetration, while reducing its surface tension.

suspended solids: (SS) Solids captured by filtration through a glass wool mat or $0.45\,\mu m$ filter membrane.

tannins: Colored compounds that form when plant matter degrades in water.

TKN: Total Kjeldahl nitrogen.

TOC: Total organic carbon.

tocopherol: Group of compounds of interest for the food industry owing to their antioxidant activity. The tocopherols, well known components of vitamin E, have been detected in olive by-products. The most widely available isomer is α-tocopherol, which is fat-soluble.

TS: Total solids.

total Kjedahl nitrogen (TKN): The sum of the organic plus ammonia nitrogen in a water sample which is determined by digesting and distilling the sample, then measuring the ammonia concentration in the distillate.

total dissolved solids (TDS): The weight per unit volume of all volatile and non-volatile solids dissolved in a water or wastewater after a sample has been filtered to remove colloidal and suspended solids.

total solids (TS): The sum of dissolved and suspended solids in a water or wastewater. Matter remaining as residue upon evaporation at 103 to 105°C. It comes from the vegetation water and the soft tissues of the olive fruits.

two-phase olive-mill waste: See 2POMW.

toxic: Capable of causing an adverse effect on biological tissue following physical contact or absorption.

toxicity: The property of being poisonous, or causing an adverse effect on a living organism.

UF: Ultrafiltration.

ultrafiltration (UF): A low pressure (200–700 kPa, 20–100 psi) membrane filtration process, which separates solutes in the 20 to 1000 Å (up to 0.1 μm) size.

ultraviolet light (UV): Light rays beyond the violet region in the visible spectrum, invisible to the human eye.

UV: Ultraviolet light.

vegetation water: Water, which originates mainly from the soft tissues (flesh) of the olive fruit.

verbascoside: A caffeyl glucoside reported to be an ortho-diphenolic compound generally accompanying the oleuropein in a number of olive cultivars ($C_{29}H_{36}O_{15}$).

vermiculite: Any group of platy minerals, hydrous silicates of aluminum, magnesium, and iron; vermiculites should be classed as montmorillonoids; they expand markedly on being heated and used in the expanded state for heat insulation.

Vitamin E: Generic term that refers to all entities (eight found so far) that exhibit biological activity of the isomer tocopherol.

waste activated sludge (WAS): Excess activated sludge that is discharged from an activated sludge treatment process.

wastewater: Liquid or waterborne wastes polluted or fouled from households, commercial, or industrial operations, along with any surface water or storm water.

wet air oxidation (WAO): Process where sludge and compressed air are pumped into a pressurized reactor and heated to oxidize the volatile solids without vaporizing the liquid.

wetlands: Areas of marsh, fen, peatland, or water, whether natural or constructed (artificial), permanent or temporary, with water that is static or flowing, fresh, brackish or salt. In ecological context, wetlands are intermediate between terrestrial and aquatic ecosystems. Wetland treatment systems are a form of phytoremediation that is, they use living plants to solve a variety of water pollution problems.

yeast: single-celled fungus.

Index

List of Authors

List of Patents

US4452744 269, 434
US4522119 269, 434
US5084141 439
US5330623 439
US5801127 291, 434
US5914040 425
US6037492 438
US6165475 437
US6197308 437
US6309652 412
US6358542 437
US6361803 437
US6391202 410
US6416808 436

WO0001622 410
WO0004794 112, 269, 270, 437
WO0145514 263–265, 437
WO02064537 265, 436
WO0212159 270, 437
WO0218310 102, 269, 270, 436

WO03000601 230, 318, 436
WO03028692 425
WO03079794 267, 435
WO03080006 267, 435
WO2004064978 120, 121, 435
WO2004110171 254, 255, 435
WO2005002364 303, 435
WO2005003037 112, 264, 434
WO3066034 427, 436
WO8904355 134, 439
WO9211206 43, 109, 142, 143,
   150, 151, 154
WO9521136 425
WO9605145 410
WO9609986 425
WO9728089 111, 439
WO9747561 417
WO9747711 412
WO9804331 270, 438
WO9804679 290, 438
WO9935097 188, 437